MW00760470

RADIATION PHYSICS RESEARCH PROGRESS

RADIATION PHYSICS RESEARCH PROGRESS

AIDAN N. CAMILLERI
EDITOR

Nova Science Publishers, Inc.
New York

For permission to use material from this book please contact us:
Telephone 631-231-7269; Fax 631-231-8175
Web Site: http://www.novapublishers.com

Library of Congress Cataloging-in-Publication Data

Radiation physics research progress / Aidan N. Camilleri (editor).
p. cm.
ISBN-13: 978-1-60021-988-7 (hardcover)
ISBN-10: 1-60021-988-8 (hardcover)
1. Radiation--Research. I. Camilleri, Aidan N.
QC475.5.R33 2007
539.2--dc22 2007047181

Published by Nova Science Publishers, Inc. ✦ New York

CONTENTS

PREFACE

This new book focuses on the latest research from around the globe on: Fundamental processes in radiation physics; radiation sources and detectors and applications of radiation physics in fundamental research.

Chapter 1 - The review contains brief description of history, basic concepts, principal achievements and recent developments of radiation methods for oil refining. Special attention is paid to expansion of radiation methods to upgrading and deep processing of a greater variety of heavy oil feedstock and production of various designed products together with the dramatic reduction of the process temperature.

Chapter 2 - Since the discovery of the positron, low to intermediate energy (< 1 keV) positron impact ionization studies have provided insight into how antimatter and matter interact. Plus, by comparing positron and electron impact data, insights and deeper understanding of matter-matter interaction dynamics and mechanisms are achieved. Until recently however, the vast majority of experimental studies for positron impact have been limited to measurements of integral cross sections; only a handful of experiments which provide differential ionization information have been performed. The primary reason for this is associated with the low-intensity positron beams that are available. This results in extremely weak differential signal rates and precludes using standard experimental methods and techniques. This paper gives an overview of differential studies that have been performed to date with emphasis on the methods used and examples of data acquired. This is followed by a discussion of ongoing work at the University of Missouri-Rolla where for the past several years the authors have been developing techniques which have enabled us to obtain highly differential ionization information using subfemtoamp positron beam intensities. Our basic method involves crossing a positron beam with a simple gas jet. A specially designed spectrometer equipped with a position sensitive detector is used to detect post-collision positrons for a range of scattering angles and energy losses. Ionization events are identified by using a weak electric field to extract target ions from the interaction region. Time of flight techniques determine their charge state, i.e., the degree of target ionization. A second position sensitive channelplate detects electrons ejected into a wide range of angles parallel and perpendicular to the beam direction. Coincidences between these post collision particles have allowed us to study single and multiple ionization as a function of energy loss and scattering angle, to obtain fully kinematic information about single ionization, and to measure the differential electron emission resulting from double and triple ionization by positron impact. Our method has also provided the first look at "ultra inelastic" collisions where the incoming particle transfers all of its available energy to a target electron. Experimental approaches and modifications leading to our present setup plus various examples of data the authors have obtained are provided. Where appropriate, comparisons with electron and photon impact data are made to illustrate differences and help interpret the positron data. Finally, limitations of our present methods plus possible future studies and needs are discussed.

Chapter 3 - This chapter is devoted to the theory of radiative ionization of target atoms in energetic collisions with highly stripped projectiles. It is set into context with the simultaneously occurring processes of nonradiative electron capture to continuum and radiative capture to bound projectile eigenstates. Among other processes linked by inverse kinematics, particular emphasis is laid on the relation between radiative ionization and electron-nucleus bremsstrahlung. Specific features of the electron and photon spectra and their angular distributions as well as the photon linear polarization are reviewed. In addition, new results are presented and are compared with experiments using relativistic uranium beams. The validity of the theoretical model is also inferred from a comparison with accurate partial-wave calculations for bremsstrahlung and radiative electron capture.

Chapter 4 - This chapter describes the production of thin films of Nd^{3+}-doped fluorides of high optical quality, to be used as devices in non linear optics, via the pulsed laser deposition technique. The innovative aspect of this research is the use of the same monocrystalline undoped fluoride as the substrate for the deposition. The technique is briefly recalled and the experimental apparatus is described in some detail. The authors describe how two different kinds of Nd^{3+}-doped fluoride films, namely $Nd^{3+}:YF_3$ and $Nd^{3+}:LiYF_4$, can be obtained on a pure $LiYF_4$ substrate from the ablation of a $Nd^{3+}:LiYF_4$ bulk crystal by changing some ablation/deposition parameters such as the substrate temperature and the presence or absence of a buffer gas in the ablation/deposition chamber. The onset of a film on the substrate is checked by interferometric measurements. The optical characterization of the films includes both polarized laser induced fluorescence spectroscopy analysis, which testifies of the kind of fluoride film produced, and the $^4F_{3/2}$ Nd^{3+} manifold lifetime measurement, which is related to the Nd^{3+} ions concentration. Both the laser induced fluorescence spectra and the lifetime measurements are compared with the corresponding ones obtained in the bulk crystal. Some data on the films thickness and on their morphology are also presented together with some checks on the ablation plume constituents and on their expansion dynamics.

Chapter 5 - The effect of radiation induced chemistry as a potential alternative to conventional chemistry for air and water pollution control is studied in this paper. The precipitation of sulphur dioxide (SO_2) and nitrogen oxides (NO_X) by irradiation with accelerated electron beams and/or 2.45 GHz microwave was studied in the presence of ammonia and water as reagents. The same technology was studied for the preparation of polymers used for the treatment of wastewaters from the food industry. An original experimental unit easily adaptable for both separate and simultaneous irradiation with accelerated electron beams and microwaves was built for this research. It was shown that the application of synergetic methods such as combined accelerated electron beam (EB) and microwave (MW) results to a significant reduction of electrical energy consumption for the cleaning of flue gases as well as to the obtaining of polymers with improved properties such as higher molecular weight and good water solubility.

Chapter 6 - The generalized mathematical model of the electron-beam volatile organic compounds (VOCs) treatment in industrial flue gases is proposed. The model gives theoretical description of physical and chemical processes induced by the electron-beam irradiation of gaseous mixture and resulting in VOCs decomposition: the formation of active species (ions, atoms and radicals) by the action of accelerated electrons on gas macro-components; the gas-phase oxidation of main admixtures, SO_2 and NO, in reactions with active species yielding sulfuric and nitric acids; the formation of the aerosol droplets upon the binary volume condensation of the sulfuric acid and water vapors; the heterogeneous oxidation of SO_2 in the aerosol droplets; VOCs decomposition and their formation from precursors under influence of active species. The most toxic representatives of VOCs such as polycyclic aromatic hydrocarbons, their chlorinated and nitrated derivatives, polychlorinated dibenzo-*p*-dioxins and dibenzofurans, and polychlorinated biphenyls are included in consideration. Results of calculations are compared with available experimental data.

Chapter 7 - Irradiation is an excellent method to ensure the microbial safety, but causes color changes, lipid oxidation and off-odor production in irradiated meat. Irradiation of ground beef under aerobic conditions oxidized myoglobin and drastically reduced color a*-values. Under vacuum or non-oxygen conditions, however, irradiation did not influence the redness of ground beef. Also, the red color of ground beef was maintained even after the irradiated beef was exposed to aerobic conditions. Vacuum-packaged irradiated ground beef had lower met-myoglobin content and lower oxidation-reduction potential than the aerobically packaged ones. Irradiating ground beef under vacuum-packaging conditions was also advantageous in preventing lipid oxidation and aldehydes production. Vacuum-packaged irradiated beef, however, produced high levels of sulfur volatiles during irradiation and maintained their levels during storage, which resulted in the production of characteristic irradiation off-odor. Double-packaging (V3/A3: vacuum-packaging during irradiation and the first 3 days of storage and then aerobic-packaging for the remaining 3 days) was an effective alternative in maintaining original beef color (red), and minimizing lipid oxidation and irradiation off-odor. The levels of off-odor volatiles in double-packaged irradiated ground beef were comparable to that of aerobically packaged ones, and the degree of lipid oxidation and color changes were close to those of vacuum-packaged ones. Ascorbic acid at 200 ppm level was not effective in preventing color changes and lipid oxidation in irradiated ground beef under aerobic conditions, but was helpful in minimizing quality changes in double-packaged irradiated ground beef. This suggested that preventing oxygen contact from meat during irradiation and early storage period (V3/A3 double-packaging) and double-packaging+ascorbic acid combination are excellent strategies to prevent quality changes in irradiated ground beef.

Chapter 8 - The processes of arising of characteristic and bremsstrahlung x-ray, conditioned by flux of photo- , Auger and Compton electrons in a irradiated sample are considered. The theoretical dependencies are obtained which allow carrying out calculations of intensity of arising radiation.

It is shown that in long-wave range of x-ray fluorescent spectrum at x-ray tube primary radiation these processes can be insignificant.

The contribution of ionizing action of photo- , Auger electrons of different shells of atoms of containing matrix is estimated in forming of x-ray fluorescence. It has appeared that for fluorine, oxygen and nitrogen this contribution is essential, and for carbon, boron and beryllium it becomes to be dominant.

The role of photo- and Auger electrons increases at the decreasing concentration of elements with low atomic number and is defined by composition of containing matrix. As rival process of cascade transition is considered which transform energy of primary photons to long-wave L-radiation of matrix elements. This radiation is able to effectively excite atoms of elements with low atomic number. For Fe, Zn, Ni matrixes the contribution of processes of cascade transitions to excitation of x-ray fluorescence of these elements was estimated. The conditions are studied allowing to obtain maximum intensity of analytical signal. The results of theoretical calculations are confirmed by comparison to experimental values of relative intensities of carbon measured for different carbonaceous bonds.

The bremsstrahlung of photo- , Auger and Compton electrons arising at its interaction with nuclei of atoms of matrix elements is studied. The realized calculations are allowed to define dependence of separate components of this radiation from composition of irradiated matrix. It is shown that bremsstrahlung of Compton electrons becomes to be significant only at high energies of primary photons. At usual voltages of x-ray tubes (up to 60 kV) this component can be neglected. The comparison with intensity of scattered polychromatic radiation of x-ray tube is allowed to define wave-length range where bremsstrahlung of considered electrons becomes to be significant. The bremsstrahlung of these electrons is considered in two approximations. The first approximation supposes that electrons brake in point of primary photon absorption. The second approximation takes into account that track-length of considered electrons can appear to be comparable with depth of penetration of long-wave bremsstrahlung. The bremsstrahlung of photo- , Auger and Compton electrons can appear to be dominant in forming of x-ray background of analytical lines of elements with low atomic number. The background genetically concerned with origination of x-ray fluorescence therefore it is not eliminated.

Chapter 9 - Scintillation crystals such NaI:Tl, BGO, CsI, BaF_2, CeF_3, PbF_2 and PWO are widely used in high-energy and nuclear physics for photon detection: electromagnetic calorimeters (EMC) made of scintillation detectors achieve the best energy resolution for photons and electrons. These scintillation detectors have also been considered among the detectors for applications in nuclear medicine, particularly in computerised tomography scanners (CTS), because of their density and large light yield. New scintillation crystals such GSO and LSO are also considered as imaging detectors for single proton emission computed topmography (SPECT) and positron emisson tomographs (PET).

One of the most important characteristics of scintillation crystals is their radiation hardness, i.e. their ability to retain scintillation efficiency and uniformity of light output over the crystal volume after exposure to ionizing radiation. In this paper the authors report on experimental study of radiation hardness of these crystals at accumulated doses of low-energy gamma-rays up to 10^5 Gy . The radiation sensitivity has been examnined by the measurement of the change in optical transmission spectra before and after irradiations. The resuls are discussed in the framework of the light output degradation and appropriate induced absorption coefficiens. The recovery from radiation damage has been treated as a multistep process.

Chapter 10 - Different chemical compounds have been studied aiming to optimize dosimetric systems in radiation processes. The absorbed dose measurement that induces beneficial changes in the irradiated material characteristics is the most efficient method to monitoring the radiation process quality. The chemical dosimetry is a dosimetric method very useful in high doses measurements. Chemical changes radiation induced in liquid, solid or gaseous can be measured and quantified by means of espectrophotomety, termoluminescence or electronic paramagnetic resonance techniques and then the absorbed dose can be determined. A variety of chemical dosimeters is available however, same time these materials are not adequate for monitoring all types of products and irradiation conditions. The necessity of developing high quality dosimeters whose cost is viable to the irradiation installations to assure the quality of their processes, have stimulated the researchers to study new compounds that can be used in the irradiations routine control. When the used dosimeter and the studied species have approximately the same density and atomic composition the dosimetry is facilitated; the chemical dosimeters offer these conditions. Aiming to improve the dosimetry processes used in Brazil the High Doses Laboratory - HDL of the Instituto de Pesquisas Energéticas e Nucleares – IPEN-CEN/SP developed and studied four different chemical dosimeters: bromocresol green solutions, potassium nitrate (KNO_3) pellets pure and mixed with manganese dioxide (MnO_2), Fricke gel dosimeter (FXG) and dyed polimethylmethacrylate radiochromic films. The optical absorption (OA) measures were chosen to study the dosimetric properties and evaluate the advantages and disadvantages of each dosimeter. The main dosimetric properties studied were: incident radiation energy and angle dependence, ambient conditions influence and dose rate response; stability before and after irradiation; reproducibility and response accuracy. The preparation method, the cost of raw materials and use easiness were evaluated. The obtained results shown that the addition of MnO_2 in the KNO_3 pellets extend the dose range of KNO_3 dosimeter from between 1 and 150 kGy to 150 and 600 kGy. PMMA films produced using dye Macrolex® red 5B can be applied to measure dose range between 5 and 100 kGy, bromocresol green solution doses from 1 to 15 kGy and Fricke gel dosimeter (FXG) presents useful dose a range between 5 and 50 Gy. These four dosimeters are of easy obtaining, low cost and simple use and can be applied to ^{60}Co monitoring in industrial irradiation processes routine dosimetry.

Chapter 11 - In this chapter PTFE acidic cationic exchange fiber have been prepared via the radiation grafting of acrylic acid, maleic acid or the sulfonation of the grafted polystyrene via radiation. It shown the special properties for the radiation grafted PTFE fiber for example, the acidity and superacidity, the pH titration curves and the excellent adsorption, desorption and separation properties for some metal ions.

The NIPA grafted chitosan also have been prepared via irradiation. The grafted products shown good thermo and pH activity, it has the potential use in the drug releasing and the other areas.

Chapter 12 - This work is devoted to the theoretical and experimental study of the feasibility of fast 14MeV neutron production from thermal neutrons by using an in-core cylindrical converter. Prediction of the characteristics of this alternate 14MeV neutron source was performed by means of analytical calculations and by our Monte Carlo based code. The neutron flux density and energy dependent spectrum have been measured, at the irradiation position, by foil activation detectors in conjunction with the iterative unfolding method adopted in the code SANDII. Measurements show that the efficiency of the hollow cylindrical ^6LiD converter, under study, is 1.310^{-4} when considering neutrons of energy above 12MeV and 1.510^{-4} while the energy range is extended to 10MeV. The validity of our Monte Carlo based code was proved by comparison with the available data in the literature. In fact, our code reproduces correctly the efficiency and the neutron energy spectrum due to an ^6LiD converter, in plane geometry, designed to be used in thermal column of reactor. Good agreement was obtained between our Monte Carlo calculation and the converter efficiency the authors have measured. However, slight discrepancy was observed in spectra. An application of the in-core converter to averaged cross sections of some (n,γ) captures and threshold reactions was performed.

In: Radiation Physics Research Progress
Editor: Aidan N. Camilleri, pp. 1-9
ISBN: 978-1-60021-988-7

Expert Commentary A

ALTERATIONS IN HERBS SUBMITTED TO RADIATION DECONTAMINATION

T.J.A. Pinto[1,]*, J.M.F. Silva[1], I.S. Kikuchi[1] and M.R. Nemtanu[2]

[1] Department of Pharmacy, Faculty of Pharmaceutical Sciences,
University of São Paulo, São Paulo, Brazil

[2] National Institute for Lasers, Plasma and Radiation Physics,
Bucharest-Măgurele, Romania

Introduction

Herbal medicines are the oldest known forms of health care. The World Health Organization (WHO) has estimated that traditional medicines, as phytotherapeutics (herbal medicinal or phytomedicines), have extensive applications in some other aspects of primary health care. Nevertheless herbs (plants or plant parts) can present high contamination by microorganisms, insects, organic matter, what may occur during harvesting, handling, transportation and storage. The microbiological contamination of herbs can change some sensorial characteristics (color, flavor, aroma, texture) and constitute a serious problem to the production of therapeutics preparations. Although microbiological quality requirements concerning final products may be achieved by means of different methods of decontamination, some conventional preservation techniques present disadvantages such as generation of toxic residues (for example, ethylene oxide). In addition, the contents and characteristics of biological active substances such as essential oils, flavonoids, glycosides, anthocyans, poliphenoloacids, triterpene, saponins and oleanosides might be modified significantly after decontamination treatment. It is also known that some preservation techniques applied to herbs cause physicochemical changes to these. Consequently, it is necessary to analyze and certify herbal products before and/or after processing to ensure the quality of the products for the consumers. Radiation treatment has been recognized as a suitable decontaminating method, but it has received little attention in its application to

* E-mail address: tjapinto@usp.br; Tel: 55-11-30913674; Fax: 55-11-30318986; Academic address: Department of Pharmacy – Faculty of Pharmaceutical Sciences, University of São Paulo, Av. Prof. Lineu Prestes, 580 – Bloco 13A, 05508-900 São Paulo, SP, Brazil (Corresponding author)

medicinal plants, especially fresh herbs. Gamma-rays irradiation is a fast and effective decontamination process against various microorganisms and does not promote radioactivity release from raw materials and final products after their treatments. Sterilization of packed materials by ionizing irradiation can induce physicochemical changes in polymeric materials (crystallinity percentage, molecular weight and density) especially when applying high doses of radiation. The amplitude of radiation-induced changes depends on the polymer type, the additives added to compose the polymers, and the products to be radiosterilized. This type of sterilization has proved to be very effective. However, its application must be rigorous and cautious to prevent decomposition or change of physicochemical, pharmacological or therapeutic properties.

A Brief History

In the history of medicines, herbal substances have played an important role in the treatment of diseases. Despite the fact that only recently in some countries natural products have been considered one of main sources of therapeutic products [Fang and Wu, 1998], oriental and indigenous cultures have for a long time been developing therapies based on herbal substances, applied until today [Kultur, 2007; Teklehaymanot et al., 2007]. The World Health Organization estimates that about 65-80% of the world's population living in developing countries depends essentially on plants for primary health care [Calixto, 2000]. In addition, many expressive effects of herbs have been verified and studied to treat several diseases [Kroll and Cordes, 2006]. However, more research work should be developed, using classical and advanced techniques, with a view to identify important active principles, how they act and improve their therapeutic properties.

Herbal Decontamination

Although herbal medicines may be prepared by different means, decoction and infusion are mostly used, in spite of the increase of the growth of industrialized forms – pills, extracts, tinctures and others. However, important aspects of herbal products should be given special attention, such as the change of their microbiological, physicochemical, pharmacological, and therapeutic properties [Owczarczyka et al., 2000; Byun et al., 1999], what can be a consequence of inadequate irrigation systems or water processing, or unsuitable harvesting, handling, packaging, decontamination, and storage conditions [Li et al., 2007; Rodriguez et al., 2006; Johnston et al., 2005; Tanko et al., 2005]. Microbiological contamination in botanical raw materials can change sensorial characteristics as flavor, aroma, texture and color [Beaulieu et al., 1999], and it is necessary to determine which microbial population is contaminating those products to define the bioburden and establish protocols to their decontamination (ANSI, 1994).

The increase in the consumption of herbal products enhances the responsibility of manufacturers and regulatory agencies [Stickel and Schuppan, 2007; Soriani et al., 2005] which should control their microbiological quality and reduce their bioburden, employing safe decontamination methods.

Although vegetable products are generally called phytotherapeutics, only those submitted to clinical assays and approved by regulatory agencies can be considered medicines. The use of herbs to treat diseases has been known by several cultures for a long time and the pharmaceutical industry has put together popular knowledge and technology and invested large sums to produce medicines. These present the same safety of those traditionally manufactured and obtained by chemical synthesis [Febrafarma, 2004]. Since ethylene oxide was no more allowed as a sterilization process in herbs, the legal use of the irradiation process is increasing in Brazil [Koseki et al., 2002].

Raw plant materials normally carry a great number of bacteria and fungi from the soil. The current practices of harvesting, handling, storage and production may cause additional contamination and microbial growth. The microbial contamination of products of vegetable origin makes them inadequate for food, pharmaceutical and cosmetic applications, offering potential hazards to consumers. Thus the evaluation of the hygienic quality of medicinal plants, as well as the use of decontamination methods are important steps towards the consumer safety and therapeutical efficiency.

Sterilization by steam and by ethylene oxide are two classical and efficient methods, but the inherent drawbacks have induced limitations in their application as herbal decontamination methods [Fang and Wu, 1998]. Since 1978, the Food and Drug Administration (FDA) have restricted continuous exposition of certain drugs and medical devices to ethylene oxide due to the generation of toxic compounds, such as ethylene chlorohydrin (2-chloroethanol), ethylene glycol, and ethylene residues. Besides, the potential risk of mutagenicity in patients and the high temperatures of processes can promote a loss of effective components of herbs if these are volatile [Wu et al., 1995].

The Radiation method, as sterilization technology, was created in the years 1950-1960 for the sterilization of medical products and devices [Dorpema, 1990], and gamma radiation is becoming increasingly popular as a method of herb decontamination [Crawford, 1996] mainly regarding the conversion from ethylene oxide processes to ionizing radiation.

Why irradiate herbs? Important features in radiation technology make it advantageous if compared to other methods, such as fast activity, short processing times, immediate product release, low temperature, quantitative and qualitative alterations of bioburden, what make this method applicable to different areas [Soriani et al., 2005]. In spite of these characteristics, some drawbacks may be able to compromise the quality of final products.

Decontamination Dose Selection

The most important element in radiation processes is the dose to obtain efficient decontamination and safe products. The minimum and maximum exposure doses for various commodities must be defined by rigorous processes of validation in order to assure specifics aspects in plants, facilities, sterilization processes and products, for example, types and quantity of herbs. However, the application of high doses of radiation does not eliminate some limitations as viruses, which can not be affect, and the survival of spore forming microorganisms (*Clostridium botulinum)* [Crawford, 1996].

Many research papers have evaluated the capacity of sterilization methods to promote microbiological reduction, and in minor rate, endotoxins [Dutkiewiczet al., 2001]. However, the concentration of free radicals species generated during irradiation [Signoretti et al., 1998] represents the major factor of damage, especially the contents of biological active substances such as essential oils, flavonoids, glycosides, anthocyans, poliphenoloacids, triterpene, saponins, and oleanosides. Herbs in aqueous solution are highly likely to lose these constituents by radiolysis, if compared to dry herbs [Fang and Wu, 1998], besides the humidity to promote the growth of fungi in contaminated product [Tournas et al., 2006]. To avoid this damage, rigorous quantitative and qualitative validation must be accomplished, as well as the determination of adequate microbial kill and irradiation dose ratio.

Ražem and Katušin-Ražem (2002) evaluated the dose requirements for microbial decontamination of botanical materials by irradiation and verified that the resistance to irradiation of contaminating microflora increases ongoing from flowers and leaves to fruits and seeds, to dry and liquid extracts. The authors reported on microbial contamination levels typical of botanicals harvested in moderate climate. The contamination and the pertaining resistivity to irradiation are inversely related, and the dose required to reduce initial contamination to tolerable level amounts to between 4 and 30 kGy under a typical scenario, and between 8 and 40 kGy under the worst-case one.

Many researchers have investigated the effects of gamma irradiation on medicinal herbs and reported that doses at 10 kGy [Owczarczyk et al., 2000; Byun et al., 1999], at 17.8 kGy [Soriani et al., 2005], at 30 kGy [Koseki et al., 2002] did not change pharmacological activities.

Minea et al. (2004) presented the results regarding the electron beam irradiation of fresh *Salvia officinalis* and *Calendula officinalis*. Herbs were decontaminated without any important alteration in active principles up to 1 kGy (permissible level according to Romanian Pharmacopoeia) but they lost their fresh aspect earlier than non-irradiated ones.

The radiation treatment of vegetable drugs is accepted by Brazilian legislation and with a previous evaluation of possible alteration in the raw material [Brasil, 2001, 2004]. Therefore, studies concerning the stability of the active principles after the process are required [Soriani et al., 2005].

Soriani et al. (2005) evaluated the effects of different radiation doses on the microbial burden and chemical constituents of ginkgo (*Ginkgo biloba* L.) and guarana (*Paullinia cupana* H.B.K.) and the results show that gamma irradiation can be considered effective to improve the microbial quality of ginkgo and guarana. Moreover after irradiation up to 17.8 kGy the content of the main active principles was not modified.

Al-Bachir et al. (2004) detected alterations in mineral oil, glycyrrhezinic acid and maltose concentrations in solution produced from irradiated licorice root products but no significant differences in sensory evaluation.

Food irradiation is recognized as a means of reducing food losses due to microbial contamination and insect damage but qualitative and quantitative tests for dose estimation and to establish the duration and nature of post irradiation treatment are needed to support good radiation practice where this is permitted [Sanderson et al., 1989]. Laboratory tests, which can identify irradiated samples, are needed to support restrictions in countries prohibiting this technique, and to enforce labeling rules in those countries allowing it. Among the methods used for detections of irradiated foods, thermoluminescence is one of those based on the production of free radicals in the irradiated products [Delincée, 1993]. Termoluminescence emissions were mainly originating from the mineral contaminates adhering to various food materials and not from the organic matrix [Sanderson et al., 1989].

Degradation of Biomolecules and Toxins

The validation of analytical methodologies – Thin Layer Chromatography (TLC), Gaseous Chromatography (GC), High Performance Liquid Chromatography (HPLC) [Fang and Wu, 1998], Thermoluminescence (TL) [Mamoon et al., 1994; Bögl, 1990], Electron Paramagnetic Resonance (EPR) [Raffi et al., 2000], Electrophoresis (EP), Mass Spectrometry (MS) [Nazzaro et al., 2007], and Electron Parametric Resonance (EPR) [Brasoveanu et al., 2004] – of biomolecule rates has been developed to promote the detection of irradiated products, regarding the increase of commercialization of products inappropriately labeled (radiation doses, radiation source, raw materials and storage conditions) [Beaulieu et al., 1999; ANSI, 1994; Delincée, 1993]. Irradiation processes have demonstrated potent inhibitory effects on fungi toxins [Zeinab et al., 2001]. Besides this, compounds as proteins, nucleic acids, lipids, carbohydrates, and vitamins can have their rates in foods after these processes [Crawford, 1996]. Thus, it is fundamental to study and determine conditions to avoid a decrease of nutritional quality.

The application of appropriate radiation doses on herbs has achieved satisfactory results, providing the maintenance of active principle rates and its effectiveness in therapies [Brasoveanu et al., 2005; Minea et al., 2004; Koseki et al., 2002; Owczarczyka et al., 2000; Kim et al., 2000].

Packaging Materials, Radiation Stabilizers and Additives

The sterilization process by irradiation can affect packaging materials, modifying their physicochemical properties and, depending on the type of polymers used, the process can promote radiation absorption, production of volatile residues [Radziejewska-Kubzdela et al., 2007; Demertzis et al., 1999] or a new class of materials [Pentimalli et al., 2000]. Similar facts can be related concerning the decontamination of products containing high levels of microorganisms. When radiation doses up 20 kGy may be required to enhance safety for the consumers [Farkas, 1984], alterations in packaging materials were found as polyethylene, polypropylene, poly (ethylene terephthalate), polyamide, polystyrene, and poly (vinyl chloride) [Demertzis et al., 1999]. It is known that hydroxyl radicals may be produced during the irradiation process and cause damages to users. Fang and Wu (1998) reported the application of a scavenger substance to protect products against irradiation damages and introduced some carbohydrates before treatment of traditional Chinese medicines. Then, the development of a specific scavenger substance and radioprotectans would promote the increase of radiation processes (gamma and electron beam radiation) and guarantee safe consumption of medicines.

Radiation sterilization technology has provided good applications in fields where products based on herbal substances are developed. However, more studies are required for standardization and modernization of Good Agricultural Practices (GAPs) and Good Manufacturing Practices (GMPs) so that these are efficiently prepared to assume all the steps of production (harvesting, handling, decontamination, transportation and storage), and adequate manufacturing conditions.

Perspectives

The Brazilian Pharmaceutical industry has more than an US$ 8 billion market and has been the 8th largest in the world since 2006. About 200 enterprises in Brazil have produced phytotherapeutic medicines involving around US$ 550 millions and it is estimated that this market can increase to US$ 1 billion until 2010. Brazil has been known to have about 50,000 different plants and this biodiversity confers it the greatest potential in this field. In the Amazon, for example, we have 20% of all freshwater of the world, 35% of superior plants and 220,000 living organisms classified [Febrafarma, 2004].

An increasing interest for medicinal plants has provided investments in research in this field by pharmaceutical industries. As a consequence, the number of patents has gradually increased in Brazil although all potential has not yet been explored [Febrafarma, 2004].

Several authors have reported the treatment of medicinal herbs with ionizing radiation but another elegant alternative sterilization technique is the use of plasma (ionized gas). The plasma promotes an efficient inactivation of the microorganisms, minimizes the damage to the materials and presents very little danger for personnel and the environment. Pure oxygen reactive ion etching type of plasmas were applied to inactivate a biologic indicator, the *Bacillus stearothermophilus*, to confirm the efficiency of this process [Moreira et al., 2004]. Artichoke (*Cynara scolymus* L.) and ginkgo (*Ginkgo biloba* L.) were treated in plasma system using a gas mixture of oxygen and hydrogen peroxide and the results showed a reduction in the microorganism number such as 3 and 4 logarithmic cycles, suggesting the treatment as a suitable method to decontaminate them [Kalkaslief-Souza et al., 2007] but more studies are required to understand better this mechanism.

References

[1] Fang, X., Wu, J. (1998). Feasibility of Sterilizing Traditional Chinese Medicines by Gamma-Irradiation. *Radiat. Phys. Chem.* **52**, 53-58.

[2] Kultur, S. (2007). Medicinal plants used in Kırklareli Province (Turkey). *J. Ethnopharmacol.* **111**, 341-364.

[3] Teklehaymanot, T., Giday, M., Medhin, G., Mekonnen, Y. (2007). Knowledge and use of medicinal plants by people around Debre Libanos monastery in Ethiopia. *J. Ethnopharmacol.* **111**, 271–283.

[4] Calixto, J.B. (2000). Efficacy, safety, quality control, marketing and regulatory guidelines for herbal medicines (phytotherapeutic agents). *Braz. J. Med. Biol. Res.* **33**(2), 179-189.

[5] Kroll, U., Cordes, C. (2006). Pharmaceutical prerequisites for a multi-target therapy. *Phytomedicine.* **13**, 12–19.

[6] Owczarczyka, H.B., Migdal, W., Kedzia, B. (2000). The pharmacological activity of medical herbs after microbiological decontamination by irradiation. *Radiat. Phys. Chem.*. **57**, 331-335.

[7] Byun, M.-W., Yook, H.-S., Kim, K-S., Chung, C-K. (1999). Effects of gamma irradiation on physiological effectiveness of Korean medicinal herbs. *Radiat. Phys Chem.*. **54**, 291-300.

[8] Li, S-L., Yan, R., Tam, Y-K., Lin, G. (2007). Post-Harvest alteration of the main chemical ingredients in *Ligusticum chuanxiong* HORT. (Rhizoma Chuanxiong). *Chem. Pharm. Bull.* **55**(1), 140-144.
[9] Rodriguez, O., Castell-Perez, M.E., Ekpanyaskun, N., Moreira, R.G., Castillo, A. (2006). Surrogates for validation of electron beam irradiation of foods. *Int. J. Food Microbiol.* **110**, 117-122.
[10] Johnston, L.M., Jaykus, L-A., Moll, D., Martinez, M.C., ANCISO, J., Mora, B., Moe, C.L. (2006). Field study of the microbiological quality of fresh produce. *Int. J. Food Microbiol.* **112**, 83-95.
[11] Tanko, H., Carrier, D.J., Duan, L., Clausen, E. (2005). Pre- and post-harvest processing of medicinal plants. *Plant. Genet. Resour.* **3**(2), 304-313.
[12] Beaulieu, M., D'Aprano, M.B.G.; Lacroix, M. (1999). Dose Rate Effect of γ-Irradiation on Phenolic Compounds, Polyphenol Oxidase, and Browning of Mushrooms (*Agaricus bisporus*). *J. Agric. Food Chem.* **47**, 2537-2543.
[13] American National Standard Institute (1994). ANSI/AAMI/ISO 11137-1994, *Sterilization of health care products–Requirements for validation and routine control–Radiation sterilization.*
[14] Stickel, F., Schuppan, D. (2007). Herbal medicine in the treatment of liver diseases. *Digest. Liver Dis.* **39**, 293-304.
[15] Soriani, R.R., Satomi, L.C., Pinto, T.J.A. (2005). Effects of ionizing radiation in ginkgo and guarana. *Radiat. Phys. Chem.* **73**, 239-242.
[16] FEBRAFARMA – Federação Brasileira da Indústria Farmacêutica. (2004). A indústria farmacêutica no Brasil. Brazil. p.11.
[17] Koseki, P.M., Villavicencio, A.L.C.H., Brito, M. S., Nahme, L.C., Sebastiao, K. I., Rela, P.R., Almeida-Muradian, L.B., Mancini-Filho, J., Freitas, P.C.D. (2002). Effects of irradiation in medicinal and eatable herbs. *Radiat. Phys. Chem..* **63**, 681-684.
[18] Food and Drug Administration. Department of Health, Education, and Welfare. Ethylene oxide, ethylene chlorohydrin, and ethylene glycol - *Proposed Maximum Residue Limits and Maximum Levels of Exposure* (1978). p.12.
[19] Wu, J., Xujia, Z., Rongyao, Y., Yongke, H. (1995). Radiolysis of herbs. *Radiat. Phys. Chem..* **46**(2), 275-279.
[20] Dorpema, J.W. (1990). Review and state of the art on radiation sterilization of medical devices. *Radiat. Phys. Chem.* **35**, 357-360.
[21] Crawford, L.M., Ruff, E.H. (1996). A review of the safety of cold pasteurization through irradiation. *Food Control.* **7**(2), 87-97.
[22] Dutkiewicz, J., Krysinska-Traczyk, E., Skorska, C., Sitkowska, J., Prazmo, Z., Golec, M. (2001). Exposure to airborne microorganisms and endotoxin in herb processing plants. *Ann. Agric. Environ. Med.* **8**, 201-211.
[23] Signoretti, E.C., Valvo, L., Santucci, M, Onori, S., Fattibene, P., Vincieri, F.F., Mulinacci, N. (1998). Ionizing radiation induced effects on medicinal vegetable products. Cascara bark. *Radiat. Phys. Chem..* **53**, 525-531.
[24] Tournas, V.H., Katsoudas, E., Miracco, E.J. (2006). Moulds, yeasts and aerobic plate counts in ginseng supplements. *Int. J. Food Microbiol.* **108**, 178-181.
[25] Räzem, D, Katusin-Raäzem, B. (2002). Dose requirements for microbial decontamination of botanical materials by irradiation. *Radiat. Phys. Chem..* **63**, 697-701.

[26] Byun, M.W., Yook, H.S., Kim, K.S., Chung, C.K. (1999). Effects of gamma irradiation on physiological effectiveness of Korean medicinal herbs. *Radiat.Phys.Chem.* **54**, 291-300.

[27] Minea, R., Nemtanu, M.R., Brasoveanu, M., Oproiu, C. (2004). Accelerators use for irradiation of fresh medicinal herbs. *Proceedings of EPAC 2004*, Lucerne, Switzerland, 2371-2373.

[28] Brasil. Ministério da Saúde. Agência Nacional de Vigilância Sanitária. Diretoria Colegiada. Resolução n. 21 de 26 de janeiro de 2001. Lex Coletânea de Legislação e Jurisprudência, São Paulo: LEX editora S/A, v. 65, p. 870-874, january 2001.

[29] Brasil. Ministério da Saúde. Agência Nacional de Vigilância Sanitária. Diretoria Colegiada. Resolução n. 48 de 16 de março de 2004. Diário Oficial, Brasília, Seção 1, v.53, p.39-41, 18th march 2004.

[30] Al-Bachir, M., Al-Adawi, M.A., Al-Kaid, A. (2004). Effect of gamma irradiation on microbiological, chemical and sensory characteristics of licorice root product. *Radiat. Phys. Chem.* **69**, 333-338.

[31] Sanderson, C.C.W., Slater, C., Cairns, K.J. (1989). Thermoluminescence of foods: origins and implications for detecting irradiation. *Radiat. Phys. Chem.* **34**, 915-924.

[32] Delincée, H. (1993). Control of irradiated food: recent developments in analytical detection methods. *Radiat. Phys. Chem.* **42**, 351-357.

[33] Bögl, B.W. (1990). Methods for identification of irradiated food. *Radiat. Phys. Chem.* **35**, 301-310.

[34] Mamoon, A., Abdul-Fattah, A.A., Abulfaraj, W.H. (1994). Thermoluminescence of irradiated herbs and spices. *Radiat. Phys. Chem..* **44**, 203-206.

[35] Raffi, J., Yordanov, N.D., Chabane, S., Douifi, L., Gancheva. V., Ivanova, S. (2000). Identification of irradiation treatment of aromatic herbs, spices and fruits by electron paramagnetic resonance and thermoluminescence. *Spectrochim. Acta [A]*. **56**, 409- 416.

[36] Nazzaro, F., Fratianni, F., Picariello, G., Coppola, R., Reale, A., Di Luccia, A. (2007). Evaluation of gamma rays influence on some biochemical and microbiological aspects in black truffles. *Food Chem.* **103**, 344-354.

[37] Brasoveanu, M., Grecu, M.N., Nemtanu, M.R. (2006). Electron paramagnetic resonance study on irradiated green coffee. *Rom. Journ. Phys.* **51**(1-2), 147–150.

[38] Zeinab, E.M.E.B., Hala, A.F., Mohie, E.D.Z. E.F., Seham, Y.M.E.T. (2001). Inhibitory effect of gamma radiation and Nigella sativa seeds oil on growth, spore germination and toxin production of fungi. *Radiat. Phys. Chem..* **60**, 181-189.

[39] Kim, M.J., Yook, H.S, Byun, M.W. (2000). Effects of gamma irradiation on microbial contamination and extraction yields of Korean medicinal herbs. *Radiat. Phys. Chem..* **57**, 55-58.

[40] Brasoveanu, M., Nemtanu, M.R., Minea, R., Grecu, M.N., Radulescu, E.M. (2005). Electron beam irradiation for biological decontamination of Spirulina platensis. *Nucl. Instrum. Methods Phys. Res. B.* **240**, 87-90.

[41] Radziejewska-Kubzdela, E., Czapski, J., Czaczyk, K. (2007). The effect of packaging conditions on the quality of minimally processed celeriac flakes. *Food Control.* **18**, 1191-1197.

[42] Demertzis P.G., Franz, R., Welle, F. (1999). The Effects of g-Irradiation on Compositional Changes in Plastic Packaging Films *Packag. Technol. Sci.* **12**, 119-130.

[43] Pentimalli, M., Capitani, D., Ferrando, A., Ferri, D., Ragni, P., Segre, A.L. (2000). Gamma irradiation of food packaging materials: an NMR study. *Polymer.* **41**, 2871-2881.

[44] J. Farkas. (1984). Radiation Decontamination of Dry Food Ingredients and Processing Aids. *J. Food Eng.* **3**, 245-264.

[45] Moreira, A.J., Mansano, R.D., Pinto, T.J.A., Ruas, R., Zambon, L.S., Silva, M.V., Verdonck, P.B. (2004). Sterilization by oxygen plasma. *Appl. Surf. Sci.* **235**, 151-155.

[46] Kalkaslief-Souza, S.B., Mansano, R.D., Moreira, A.J., Nemtanu, M.R., Pinto, T.J.A. (2007). *Microbial decontamination study of some medicinal plants by plasma treatment.* I. International Medicinal and Aromatic Plants Conference on Culinary Herbs. Abstract. Antalya, Turkey.

In: Radiation Physics Research Progress
Editor: Aidan N. Camilleri, pp. 11-15
ISBN: 978-1-60021-988-7

Expert Commentary B

RADIATION IN CORK TREATMENTS

***C. Pereira*[*] *and L. Gil*[†]**
INETI, Estrada do Paço do Lumiar 22, 1649-038 Lisboa, Portugal

Abstract

The problem of odours and tastes in wine is complex. Some of them are considered negative by the consumers. The main responsible is the 2,4,6-trichloroanisole (TCA) which has a very low detection threshold. There are a few preventive and curative processes in order to avoid TCA presence. Some of them are based on radiation methods, namely electron beam radiation and gamma ray irradiation. Furthermore, like several other food products, cork stoppers should also be sterilized, and this treatment is well performed using radiation techniques. Radiation levels are different for each purpose. TCA may be produced on cork through the conversion of some chlorine compounds in TCA by microorganisms as fungi. The sterilization by irradiation can be a preventive method but further contamination, *e.g.* by environment existing TCA, must be treated with a curative process which eliminates TCA. This can be achieved by irradiation techniques, using specific conditions. As these techniques can be used with cork stoppers already completely packed, avoiding further contamination, these are excellent for this purpose. Some treatments are also foreseen for contaminated wine.

Introduction

The quality of wine, from an organoleptic point of view, depends a lot on its aromas which are directly related with the presence of volatile organic compounds. The mouldy/musty smell in wine is today one of the concerns in the wine production activity and, therefore, of related activities, mostly upstream, as is the case of the cork industry. The problem of aromas and flavours is complex. These flavours and/or aromas are described in several ways, e.g. "moist paper/card", "chemical products", "mouldy" etc (Gil, 2002; Gil, 2006). Organized campaigns have even been responsible for the decline of certain markets, an increase of rejections and even an anti-cork behaviour. It is a

[*] E-mail address: carlos.pereira@ineti.pt; Tel. 217165141; Fax 217166939; (Corresponding author)
[†] E-mail address: luis.gil@ineti.pt;

economically a important problem, which is related to the improvement of wine quality, more purified and masking less the contaminants, and also to a greater knowledge and exigency of consumers. Although 2,4,6-TCA is the main cause of problematic flavours and odours in wine, it is not toxic or dangerous to humans in the concentrations normally found in wine (ppt), but it usually depreciates the product.

In 1982, Buser *et al.* (1982) identified 2,4,6–TCA as the main component responsible for causing the mouldy odour, detected at concentrations of up to 10 ppt. The 2,4,6-TCA formation process is closely related to the presence of chlorinated compounds in cork, which can have different origins: free atmospheric chlorine (e.g. maritime fog), chlorinated products of the agri-forestal treatments, chlorine from the "traditional" cork stopper washing operation, chlorophenols of wood treatments (wooden frames, pallets etc) and chlorine from cleaning (hypochlorite and bleach).

There is a great variety of processes for the 2,4,6 – TCA elimination/reduction, which can include: chemical, physico-chemical and biological processes (Pereira *et al.*(A)., 2006). Some focus on eliminating the causes through the elimination of the microorganisms present and/or the presence of chlorinated agents, while other act directly on the levels of TCA present in cork.

Cork Treatments Using Radiation

Gamma radiation has been used since the 80's in cork stopper sterilization. Due to the fact that some of the existing micro-organisms may give rise to metabolites which may be the origin of chlorinated compounds such as 2,4,6–TCA, its reduction/elimination may contribute for the decreasing of "cork taint", but they do not "act" if 2,4,6 – TCA is already present.

Following 2,4,6 – TCA identification, in 1982, as being the main cause of" cork taint" Zehnder et al. (1984) published a study on the gamma irradiation of cork stoppers, aiming at the prevention of the conversion of 2,4,6 – trichlorophenol in 2,4,6 – trichloroanisole through biometilation by micro-organisms. One of the biggest disadvantages of this technique is that it does not remove 2,4,6 – TCA already found in cork. However, as the microbiological contamination is eliminated the probability of 2,4,6 – TCA formation decreases, and so this is an indirect method for 2,4,6 – TCA reduction.

Some years later Botelho *et al.* (1988) conducted a preliminary study on the utilization of gamma radiation for cork stoppers sterilization. The study aimed at determining several microorganisms resistance to gamma radiation, in cork, as well as the sterilization level that can be obtained using this method. Mould was the main contaminant found in the studied samples. With doses of 15 kGy it is possible to assure a sterilization level equivalent to the probability of finding one non sterilized cork stopper in a group of ten thousand.

Some studies have been carried out on 2,4,6–TCA and cork behaviour under the influence of radiation - in this case an electron beam. Careri et al. (2001) examined the behaviour of 2,4,6-TCA solutions under the influence of electrons beams of varying doses, in the presence of cork. The results obtained showed that, under the effect of an electron beam, the 2,4,6–TCA is degraded with doses of 25-50 kGy. The degradation products are mainly mono and dichloroanisoles. The high 2,4,6 – TCA degradation level, the low quantity of by-products and the absence of toxicity of these products makes it possible to conclude that this method is able to reduce the quantity of 2,4,6-TCA in alcoholic solutions of this compound, without toxicity problems. However, nothing was said about the application to cork stoppers, where others substances may interfere, and to the application of other radiation doses.

Mazzoleni et al. (2000) ascertained that cork irradiation at 10 kGy using electron beam controlled several fungi strains. The irradiation of the same material using doses of 1000 kGy reduces the contents of cafeic, coumaric and ferulic (phenolics) and an increase of the saturated hydrocarbons, also reporting a decrease in chloroanisoles and related compounds. The same author (Mazzoleni, 2001) carried out tests for the elimination and he obtained a reduction of the 2,4,6–TCA level in wine lower than 3 mg/l, using doses greater than 100 kGy. This researcher did not carried out any study using gamma radiation.

It is well know that the irradiation technologies using gamma radiation or electron beam are different. The former is more penetrating and suitable for fairly large packed materials such as cork stoppers packed in plastic bags inside cardboard boxes, allowing all the material inside to be treated. In places were gamma radiation facilities already exist it will not necessary to build new electron beam facilities (Janeiro J. *et al.*, 2004).

A recent work (Pereira C. *et al.*(B), 2006) has demonstrate that it is possible to eliminate or transform 2,4,6 – trichloroanisole (2,4,6 – TCA) in cork stoppers, using gamma radiation with an doses that leads to the degradation of the 2,4,6–TCA molecule transforming it in molecular residues which don't have the same organoleptic/odour characteristics. This patented process when applied to packed cork stoppers ready to use, inside their sealed packages, avoiding later contamination, assures the elimination/reduction of the problem, which is an enormous advantage over prior art processes for eliminating TCA. Studies were carried out with cork stoppers naturally contaminated and using radiation doses ranging from 15 kGy to 400 kGy (Pereira C.(A) *et al.*, 2005). The results showed elimination/reduction rates of 2,4,6–TCA higher than 90%, at a concentration lower than its detection limit. For radiation doses similar or higher than 100 kGy the reduction percentages of 2,4,6-TCA were always greater than 57%, with a maximum at 100 kGy. For radiation doses lower than 100 kGy this percentage was much lower as is the case of the 1,4% reduction for a dose of 15,4 kGy. The sterilization of cork stoppers and other materials is usually carried out with doses of radiation ranging from 15-20 kGy, i.e., this value is not sufficient for 2,4,6-TCA "elimination"(Pereira C.(B) *et al.*, 2005). The best result was obtained for a radiation dose of 100 kGy, corresponding to a reduction percentage of 2,4,6-TCA of 97,9%. The values obtained after treatment, considering that the lower detection limit of 2,4,6-TCA is of 1,4 ng/l (Gil L., 2002), are of about 2 ng/l, which are very near that limit value. At a trade level it is usual to accept batches having 3 ng/l. It should also be noted that, for the experimental conditions used in all performed tests, no 2,3,4,6-tetrachloroanisole was detected. Dichloroanisoles were detected in some experiments and their content usually increases in cork stoppers after treatments (gamma radiation) indicating that these may be some of the by-products of 2,4,6-TCA degradation. Nevertheless the by-products do not have the same organoleptic taste/odour and their detection limit is much higher. Pentachloroanisole was only detected as traces. This patented process when applied to packed cork stoppers ready to use, inside their sealed packages, avoiding later contamination, assures the elimination/reduction of the problem, which is an enormous advantage over prior art processes for eliminating TCA.

These studies allow the conclusions that gamma radiation provided a remarkable reduction on 2,4,6-TCA levels in cork stoppers. The percentages of elimination depend on the radiation dose and the optimal point seems to be near 100 kGy. The capacity of this technology to reduce/eliminate 2,4,6-TCA from cork and subsequently "cork taint" in wine was demonstrated (Gil L. *et al.*, 2005).

Related with these aspects some studies, not yet published, were also carried out using gamma irradiation in order to lower or eliminate TCA level in contaminated wine. The radiation doses were 10 KGy, 5 KGy and 2,5 KGy. The first results showed that over 5,0 KGy TCA was not detected but the wine tasted very old and oxided. Under this level organoleptic results were better.

Due to the increasing level of exigency of consumers and therefore of cork stoppers users (bottlers) and their quality demands, and having in mind that cork treatments with radiation are mainly related with sterilization and reduction of taint problems, it would be expected that this technology should be increasingly used for this purpose.

References

[1] Botelho, M. L.; Almeida-Vara, E.; Tenreiro, R.; Andrade, M. E., 1988, "*Searching for a strategy to Gamma-Sterilize Portuguese Cork Stoppers – Preliminary Studies on Bioburden, Radioresistance and Sterility assurance Level*", *Radiation Physics and Chemistry*, **31**(4-6), 775-781).

[2] Buser, H.; Zanier, C.; Tanner, H.; 1982, "Identification of 2,4,6-Trichoroanisole as a potent compound causing cork taint in wine", *Journal of Agricultural and Food Chemistry*, **30** (2), 359-362.

[3] Careri M., Mazzoleni V., Musci M., Molteni R., 2001, "Study of Electron Beam Irradiation Effects on 2,4,6-Trichloroanisole as a Contaminant of Cork by Gas Chromatography – Mass Spectrometry", *Chromatographia*, **53** (9-10), 553-557).

[4] Gil, L., 2002. A rolha de cortiça e a sua relação com o vinho. Ed. APAFNA – Agrupamento de Produtores Agrícolas e Florestais do Norte Alentejo, Portalegre.

[5] Gil L., Pereira C. and Silva P., 2005, "New Cork Products patented by INETI", Suberwood 2005 – New challenges for the integration of cork oak forests and products, Universidade de Huelva – Oral Presentation.

[6] Gil, L. 2006. A cortiça e o vinho, Ed. CEDINTED/INETI, Lisboa.

[7] Janeiro J., Matos P., Pereira C., 2004, "Tratamento de materiais por radiação gama", Jornadas de Ciência e Tecnologia de Materiais/Materiais de Origem

Florestal/Cortiça “Realidades Tecnológicas e Potencialidades, INETInovação – Oral Presentation.

[8] Mazzoleni, V. ; Molteni, R.; Furni, M.D. ; Musci, M., 2000, “*Effect of accelerated electron beam irradiation on cork used for stopper production*”, *Ind. Bevande*, **29** (167), 247-257.

[9] Mazzoleni, V. ; Molteni, R.; Furni, M.D. ; Musci, M., 2001, “Reduction of cork taint in wine and other beverages by using electron beam irradiation”, Patent DE 10022535 A1, 4 pp.

[10] Pereira C.(A) and Gil L., 2005, “Redução do teor em 2,4,6-tricloro-anisole em rolhas de cortiça utilizando radiação gama”, CONGRESSO INTERNACIONAL “SOBRAIS, FÁBRICAS E COMERCIANTES. PASSADO, PRESENTE E FUTURO DA ACTIVIDADE CORTICEIRA” Palafrugell (Girona) – Oral Presentation.

[11] Pereira C.(B) and Janeiro J., 2005, “Clean Cork Stoppers Treated with Gamma Radiation”, International Conference – Off-Flavors: Musty Taint in Wine and Other Beverages: Origin, Detection & Control, Tarragona – Oral Presentation.

[12] Pereira, C.(A); Gil, L.; 2006, “O problema do odor a mofo nas rolhas de cortiça e processos para a sua redução / eliminação”, Silva Lusitana, Vol. 14, Nº1, 2006, p. 101-111.

[13] Pereira, C.(B); Gil, L.; Carriço, L., 2006, “Reduction of the 2,4,6-triclhloroanisole content in cork stoppers using gamma radiation”, *Radiation Physics and Chemistry*, Vol. 76, Nº4, p. 729-732.

[14] Zehnder, H.J.; Buser, H.R.; Tanner, H., 1984, “*Cork Taint Formation in Wine and Its Prevention by an Irradiation Treatment of the Corks*”, Deutsche Lebensmittel - Rundscha

In: Radiation Physics Research Progress
Editor: Aidan N. Camilleri, pp. 17-103

ISBN: 978-1-60021-988-7

Chapter 1

NEW TRENDS IN THE RADIATION PROCESSING OF PETROLEUM

Yuriy A. Zaikin and Raissa F. Zaikina
PetroBeam, Inc., USA

Abstract

The review contains brief description of history, basic concepts, principal achievements and recent developments of radiation methods for oil refining. Special attention is paid to expansion of radiation methods to upgrading and deep processing of a greater variety of heavy oil feedstock and production of various designed products together with the dramatic reduction of the process temperature.

Introduction

During the past decade, the use of gamma ray and electron beam technologies for the refining and upgrading of oil have been intensely developed. It has been reported that at elevated temperatures (350-450^{0}C), irradiated petroleum can undergo thermally-activated chain-breaking reactions. These kinds of propagating reactions could be the basis for a commercially viable high-rate oil processing technology. In the 1990s, studies of Radiation Thermal Cracking (RTC) were focused on heavy petroleum feedstock consisting of complex hydrocarbon constituents that are particularly difficult to process by conventional methods. These studies showed that RTC could efficiently processing heavy oil feedstock such as high-viscous crude oil, fuel oil, tar, bitumen, heavy residua of oil primary distillation, and wastes of oil extraction. Subsequently, further work has demonstrated that radiation technologies also offer enormous potential for environment protection against pollution by oil contaminants, especially for desulphurization and the regeneration and refining of used oil products.

Compared to conventional thermocatalytic methods for oil cracking, the RTC process is characterized by higher production rates, higher degrees of conversion, greater flexibility in feedstock and target products, much lower energy consumption, and lower capital and operational costs. RTC has successfully reached pilot-plant and small-scale industrial production rates in test facilities. Beyond the early work on RTC, work continues to development technologies based on the synergistic effects of radiation and ozone containing air formed as a result of air ionization under the electron beam. This type of oil processing is not only an energy-saving process but uses, otherwise lost, by-products of a radiation facility's operation.

Still more efficient is the PETROBEAM™ (USA) technology for high-rate radiation processing of heavy oil. This technology is based on the low-temperature radiation-induced generation of self-sustaining chain-scission reactions in oil feedstock. This process offers the highest energy savings and the lowest capital and operational expenses compared with any existing methods for oil refining. Unlike conventional technologies, the low-temperature upgrading and cracking technology removes the cost and safety issues of alternative processes based on heat, pressure, or catalysts to "crack" petroleum feedstock.

Today, radiation methods for oil processing cover a wide range of different approaches intended to solution of the most acute problems of oil industry. The principal achievements in this field of applied radiation science are briefly described in this review.

Early Studies of Hydrocarbon Radiolysis

Since the early 1950s, researchers have studied the effects of various kind of radiation on hydrocarbons, in general, and oil/petroleum products, specifically. This included studies of kerosene, aviation fuel and crude oils from West Texas, Salt Lake (Rangely), Los Angeles Basin Crude, and other fields around the world.

A review of the US Patent literature from the mid-1940s [1,2] through to 1963 indicates that a good number of irradiation process patents were issued covering various uses of this technology in the manufacture of chemicals and in the petrochemical industry. Around 190 US patents in this timeframe were found to be related to irradiation polymerization, graft polymerization, chemical processes and refining or petroleum processes. Over one-fourth of these (around 56 patents) were issued to Esso Research and Engineering (now Exxon) alone.

Other petroleum or petrochemical companies were also active during this period: Standard Oil of Indiana (Amoco), Shell Oil, Phillips Petroleum, Gulf Oil, Cities Service, Union Oil (names which are now blurred through mergers and acquisitions).

The process interests of the chemical companies and of the petroleum and petrochemical companies did not evolve into viable businesses. The petrochemical companies were producing products on a continuous basis with their various catalysts and thermal reactors. At that time, the reliability of industrial electron beams was not as acclaimed as today, so that the risk of using EB processing in a 24 hour, 7 day a week operation was considerably high. Much research work was conducted using low dose-rate Cobalt-60 irradiators and in batch type operations. Generally, none of these studies showed any possibility of a commercially viable process based on any form of ionizing radiation.

In these early studies, the concept of radiolysis covered any structural or chemical changes that occurred in hydrocarbons under the action of ionizing irradiation. However, further studies have shown that mechanisms of radiation-chemical reactions in hydrocarbons undergo considerable changes at heightened temperatures and high dose rates of irradiation.

In early 1960s, the phenomenon of radiation-thermal cracking (RTC) of hydrocarbons was discovered and described by L.S. Polak, A.V. Topchiev, K.P. Lavrovsky and other researchers [3, 4] who revealed the basic mechanisms of radiation-induced chemical conversion in hydrocarbons and regularities of the chain reactions of radiation-thermal cracking (RTC) in particular. It was the first evidence of highly efficient self-sustaining chain reactions of hydrocarbon decomposition initiated by combined radiation and thermal action. The discovery was practically important because only chain reactions can provide the high rates of deep oil processing necessary in industrial conditions.

Since then, different aspects of ionizing radiation application to deep radiation processing of oil feedstock are the subject of research and technological investigations. Up to the beginning of 1990s these studies [5-7], with a few exceptions [8, 9], were limited by consideration of radiation-induced decomposition of model hydrocarbons and light oil fractions.

Radiation-Thermal Cracking of Hydrocarbons

Desirable conditions of any cracking process assume rejection of using high pressure and temperature that considerably raise production costs and make lower its safety.

The thermal cracking of normal alkanes usually takes place at high temperatures (for example, 500-600^{0}C for heptane). According to the generally accepted theory [3], this process is accomplished in two steps: (1) initiation of the reaction by radicals produced by dissociation of a molecule of the starting compound; and (2) propagation of the chain. The latter consists of dissociation of the large initial-free radicals into an olefin and shorter radicals and the reaction of the latter with starting molecules.

The chain propagation process results in formation of a reaction product molecule and of a new large radical. Because of its low activity, this radical cannot interact with another starting molecule, i.e. it cannot propagate the chain. However, it can dissociate into an olefin molecule and a shorter, and correspondingly, more active radical which will be able to propagate the chain. The initiation step requires the activation energy of about 250 kJ/mole, i.e. the reaction can proceed at an appreciable rate only at the temperatures of 500-600^{0}C. The chain propagation step, controlled by dissociation of the radical, requires the activation energy of about 80 kJ/kg, i.e. it requires a much lower temperature.

Therefore, there are two main conditions and two stages necessary for chain cracking reaction:

(1) Formation and maintenance of relatively low concentration of chain carriers (light radicals, such as $H^{\bullet}, CH_3^{\bullet}, C_2H_5^{\bullet}$) necessary for cracking initiation;

(2) Formation and maintenance of sufficient concentrations of excited molecules necessary for chain propagation caused by interaction of radicals with excited molecules and their disintegration.

In the case of thermal cracking (TC), both stages of the process are thermally activated. In the case of radiation-thermal cracking (RTC), the first stage is radiation-initiated: chain carriers are created by irradiation. The absorbed electron energy, necessary for formation of radical concentration sufficient for cracking initiation is only about 0.4 kJ / mole. The propagation stage of RTC is still thermally activated.

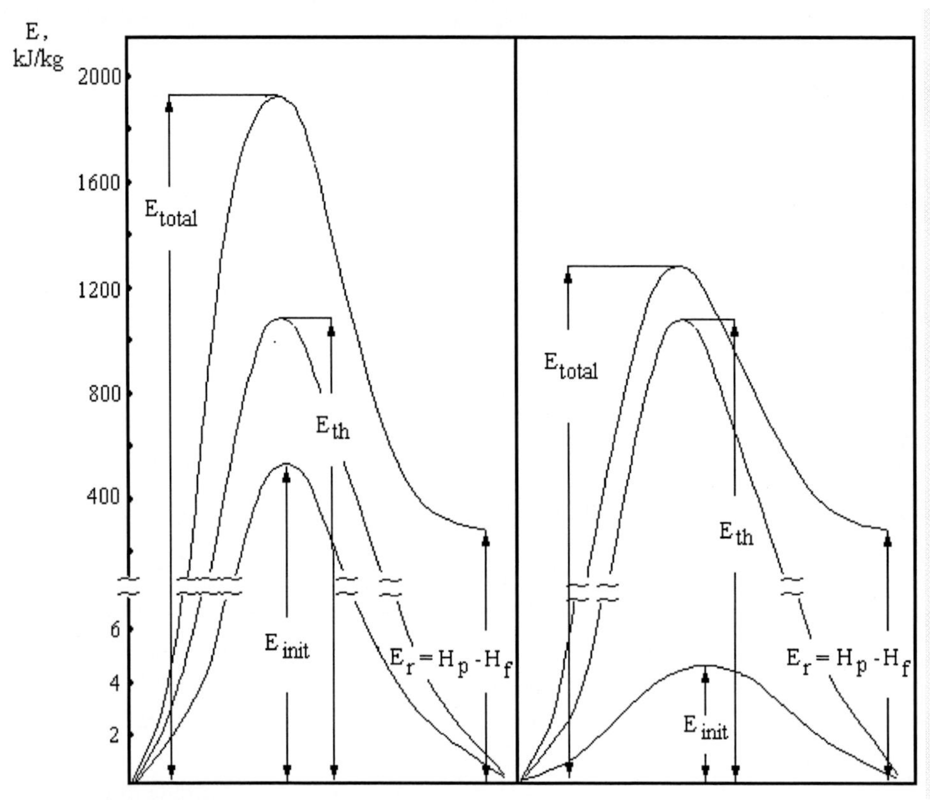

p^{20} = 0,95 g/cм^3, M = 390 g/mol
E_{total} is total consumed energy;
E_t is energy consumed for feedstock heating up to reaction temperature (400°C);
E_{init} is energy consumed for cracking initiation;
$E_r = H_p - H_f$ is increase in the product energy after processing

Figure 1. Heat balance diagram for conventional thermal cracking of fuel oil at 650°C and radiation-thermal cracking at 400°C [10].

The radical mechanism of RTC assumes that radiation-induced chain initiation does not depend on temperature and the rate of radical generation depends only on irradiation dose rate. Conventional sources of ionizing radiation, such as electron accelerators and isotope sources provide generation of radicals in concentrations sufficient for initiation of chain reactions in hydrocarbons.

Application of RTC releases the most energy consuming stage of cracking associated with initiation of chain carriers, thereby it lowers cracking temperature by 200-250°C compared with the thermal process (Fig.1).

The total energy expended in the radiation-thermal process is significantly less (at least by ~ 40%) than that required by the standard heating thermal process alone because initiation energy E for the radiation-induced cracking is much lower than that for thermal cracking.

Fractional contents of the overall products of RTC and TC are compared in Table 1.

Table 1. Hydrocarbon fractions of refined products from fuel oil [10]

Boiling_temperature (^{0}C)	Feed fractions mass %	TC mass %	RTC mass %
<200	none	10	15
200-300	2	15	27
300-360	8	15	18
360-450	38	25	20
>450	52	30	10
gases	none	5	10

In the case of RTC, contents of valuable commodity products are essentially higher. Therefore, RTC application provides additional energy savings for production of the designed products, proportionally to the ratio of their yields for the two kinds of processing. Subject to this factor, energy consumption of RTC for production of gasoline (boiling temperature below 200^{0}C) and diesel fuel (boiling temperature in the range 200-360^{0}C) is at least by 60% lower than that for the thermal process.

There are five main types of radical reactions which considerably contribute to RTC of liquid hydrocarbons [11]:

$$
\begin{array}{ll}
R + R \longrightarrow H & \text{- recombination} \\
R + R \longrightarrow OL + RH & \text{- disproportionation} \\
R \longrightarrow R1 + OL & \text{- dissociation} \\
R + RH \longrightarrow R2 + RH & \text{- molecule break} \\
R + OL \longrightarrow \text{polymerization} &
\end{array}
\quad (1)
$$

where *R* is a radical, *OL* is an olefin, *RH* is a hydrocarbon molecule.

Radiation-thermal cracking is characteristic for the high reaction rate. The initial rate of RTC controlled by the radical mechanism can be written as

$$W = K_1 [R] \exp\left(-\frac{E_p}{kT}\right) + K_2 \exp\left(-\frac{E_p + E_i}{kT}\right), \quad (2)$$

where K_1, K_2 are the rate constants of the combined radiation-thermal and the mere thermal process: $K_1, K_2 \approx 4 \,.\, 10\text{-}10\ cm3/molecules \cdot s\ (\sim 10^{12}\ s^{-1})$; E_i and E_p are activation energies for cracking initiation and propagation, respectively: $E_i \approx 250\, kJ/mole$; $E_p \approx 80\, kJ/mole$.

Dynamically equilibrium concentration of chain carriers (radicals), *[R]*, generated by ionizing irradiation at the dose rate *P*, is defined by the second order recombination kinetics:

$$G_R P = K_r [R]^2 \quad (3)$$

where K_r is the rate constant of radical recombination; G_R is radiation-chemical yield (G-value) of radicals.

The relative contributions of radiation-thermal and merely thermal reactions depend on the rate ratio of thermal, W_T, and radiation, W_R, generation of chain carriers:

$$W_T = K_1 \exp(-E_p / RT),$$
$$W_R = G_R P. \quad (4)$$

The ratio W_R/W_T decreases as the temperature goes up and increases at higher dose rates (Table 2).

Table 2. Ratios of the total rates of the radiation and thermal processes for heptane [3]

Temperature, ^{0}C	Ratios of the rates of radiation and thermal cracking initiation	Ratios of the total rates of radiation and thermal processes
550	820	28.6
600	41	6.5
650	3.7	2.2
700	0.41	1.2
800	0,008	1.00

Combination of the high rate of chain propagation and the high ratio of the total rates of radiation and thermal processes is provided in the temperature range of 350-550^{0}C for the most part of petroleum feedstock.

Dependence of the product G-values on temperature and irradiation dose rate is given by the equation:

$$G, molecules\ /100\,eV = \frac{100\,e\,N_A}{PM} W =$$
$$= \frac{100\,e\,N_A}{PM}\left[\left(\frac{G_R\,P}{K_r}\right)^{1/2} K_1 \exp\left(-\frac{E_p}{kT}\right) + K_2 \exp\left(-\frac{E_i + E_p}{kT}\right)\right], \quad (5)$$

where e is electron charge, M is product molecular mass.

In the temperature range of the prevailing radiation-thermal process, characteristic G-values of RTC liquid products lie in the range of 1000-20000 molecules per 100 eV of absorbed radiation energy.

Different mechanisms of cracking initiation and propagation in thermal and radiation-thermal processes result not only in the different reaction rates but also in different hydrocarbon contents of the cracking products.

For example, olefins are the natural products that limit the yields of light liquid fractions for any cracking process, including thermal cracking (TC), thermocatalytic cracking (TCC), and radiation-thermal cracking (RTC). However, very high olefin concentrations in liquid fractions are usually undesirable, except special cases.

In different types of initiated cracking, olefin concentrations in products are very different. The highest olefin content is observed after the conventional thermal cracking. The olefin concentration in the TCC liquid product is noticeably lower. Due to the effect of catalyst, the processes of hydrogen redistribution and skeleton isomerization in hydrocarbons are much more pronounced in the case of TCC, and the probability of the olefin decomposition is much higher. RTC results in still lower olefin concentrations compared with TCC. In this case, the role of catalyst is played by ionizing radiation.

The essential RTC unit is an electron accelerator used as a source of radiation for basic initiation of different reactions in hydrocarbons. Isotope gamma sources can also be used for oil radiation processing, however, they usually do not provide high production rates of the industrial process.

Linear electron accelerators for technological purposes are presently available on the world market and are used in different scientific and industrial fields of (e.g. sterilization of medical items, food conservation, sewage water purification, polymer production, etc).

Due to mostly reliable protection of electron accelerators no injurious after-effects of electron beams technological application on maintenance-staff can be observed (many types of modern accelerators have internal protection). Application of electron beams of energy up to 10 MeV excludes possibility of any residual radioactivity that makes both processing and products quite safe.

Radiation-Thermal Processing of Heavy Oil Feedstock

Systematic experiments on radiation processing of heavy oil feedstock were started in the beginning of 1990s [12-28]. Studies of radiation-chemical transformations in complex hydrocarbon mixtures and technological approaches to radiation-thermal processing have shown that radiation technologies are most advantageous and promising for large-scale industrial application when they are applied to processing of high-viscous or high-paraffin crude oils, bitumen, wastes of oil extraction, used oil products, and heavy residua of oil primary processing. Processing of such types of oil feedstock by traditional methods is not economical or comes across considerable technological difficulties. In this connection, physical and chemical aspects of radiation processing of heavy hydrocarbon feedstock are of special technological interest. The large-scale application of high-efficient and environmentally friendly radiation methods for heavy oil processing could be a radical solution of many acute problems of oil extraction, transportation and refining [29-32].

In contrast to light crude oil, the specificity of self-sustaining radiation–chemical transformations at heightened temperatures in heavy hydrocarbon feedstock of complex chemical composition displays itself in strong synergetic effects that accompany radiation-thermal cracking. Such effects are provoked by redistribution of radiation energy between the original components of a complex hydrocarbon mixture and a great number of intermediate reaction products. These phenomena facilitate side reactions of non-destructive character that considerably affect the rate of feedstock conversion and the contents of final products [18,19, 24].

The two most important synergetic effects, revealed and investigated in experiments with the problem sorts of high-paraffin and high-viscous oil, are the phenomena of radiation-induced isomerization and radiation-induced polymerization in conditions of RTC. Statement of conditions and evaluation of the intensity of these effects are important for obtaining high yields and quality control of the commodity oil products produced as a result of radiation processing of heavy oil feedstock.

Radiation-Induced Isomerization in the Process of Radiation-Thermal Cracking of High-Viscous Oil

Isomerization by-effects in conditions of induced cracking were noted in the works on radiolysis and photolysis [3, 37] of light oil fractions. The phenomenon of strong radiation-enhanced isomerization in the process of RTC was first observed in the experiments on radiation-thermal processing of high-viscous oil from Karazhanbas field, Kazakhstan [19, 25]. Together with the considerable paraffin component, this sort of oil is characteristic for high concentrations of resins and heavy aromatic compounds.

Karazhanbas oil pertains to the sort of crude from some of Kazakhstan fields that are hardly extracted, transported and processed due to its high viscosity (99.7 mm^2/s), density (0.93-0.95 g/cm^2), considerable contents of sulfur (about 2 mass %) and vanadium (100-120 mcg/g). This sort of oil has a low point of solidification (-18°C) as a result of high contents of pitch-aromatic components combined with a small amount of solid paraffins (1.4-1.5 mass %).

Concentration of fractions with the boiling temperature up to 200°C is 1.3- 6.5 mass %, the content of fractions boiling in the temperature range of 200-350°C is 15.1-20.2 mass %, and that in the range of 350-490°C is 22.6-25.7 mass %. According to chromatography data, the hydrocarbon content of the gasoline fraction with boiling temperature up to 100°C is represented by isoparaffin (29.5 mass %), naphtene (30.2 mass %), and aromatic hydrocarbons (15,0 mass%); concentration of n-paraffins (mainly HC with C_8-C_9 chain length) is 5.9 mass %.

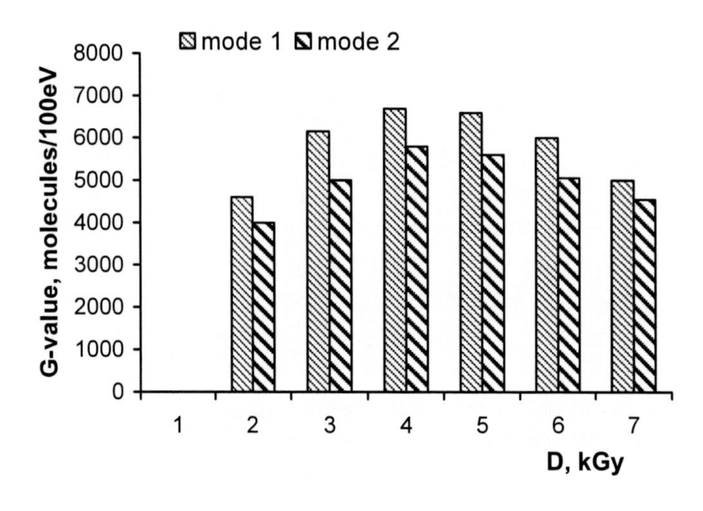

Figure 2. Typical dose dependence of G-values for gas oil fractions with boiling temperature up to 350°C obtained by RTC [19]:
Mode 1 - T = 425°C, P = 1,5 kGy/s;
Mode 2 - T=375°C, P = 1,25 kGy/s

The main products obtained by radiation-thermal processing of Karazhanbas oil [19] were the liquid gas oil fraction with boiling temperature from 60 to 350°C, the heavy coking rest (pitch, asphaltenes, solid coke particles), and the gaseous fraction (4-7 mass % hydrogen, 35-40 mass % methane, 18-21 mass % ethane, 10-12 mass % butane, 10-12 mass % ethylene, 8-12 mass % propylene and other gases).

Oil samples were irradiated by electrons with the energy of 2 MeV from a linear electron accelerator ELU-4. The time-averaged dose rate of electron irradiation was varied in the range 0.5-1.5 kGy/s.

Fig.2 shows the typical dose dependencies of G-values for gas oil fractions with boiling temperature up to 350°C obtained as a result of RTC of Karazhanbas oil in two different modes [19]. Mode 1 provides conditions for intense molecular destruction, while mode 2 is more favorable for isomerization reactions.

The maxima of G-values corresponding to the yields of liquid fractions of 80 and 52.5 mass % in modes 1 and 2, respectively, were observed at the same dose values of about 6 kGy. The further dose rise makes G-values of the liquid RTC products lower due to effect of polymerization reactions. Intramolecular isomerization stabilizes alkyl radicals, increases activation energy necessary for their disintegration, and, therefore, also contributes to the reduction of the rate of radiation-induced chemical conversion. As irradiation dose grows, alterations in structure and length distribution of paraffin chains cause changes in the share of radicals stabilizing in a unit of time, and this affects the rates and the yields of products in radiation-induced reactions.

Table 3. Hydrocarbon contents (mass %) of gas oil fraction with boiling temperature from 70 to 350°C obtained by RTC of Karazhanbas crude oil [19]

Dose kGy	Mode	Concentration of the gas oil fraction ($70^{\circ}C<T_{boil}<350^{\circ}C$) in the total product	Hydrocarbons numerator – mass % in respect to the analyzed fraction denominator – mass % in respect to the total product			
			C_6-C_{10}	C_{11}-C_{13}	C_{14}-C_{18}	C_{19}-C_{21}
0	-	26.8	17.9 / 4.8	19.8 / 5.3	24.6 / 6.6	37.7 / 10.1
2	1	65.0	13.2 / 8.6	37.1 / 24.1	43.4 / 28.2	6.3 / 4.1
	2	45.0	15.0 / 6.7	42.7 / 19.2	34.6 / 15.6	7.7 / 3.5
6	1	80.0	40.2 / 32.2	33.6 / 26.9	23.7 / 19.0	2.5 / 2.0
	2	52.5	28.5 / 15.0	45.7 / 24.0	24.6 / 12.9	1.2 / 1.0

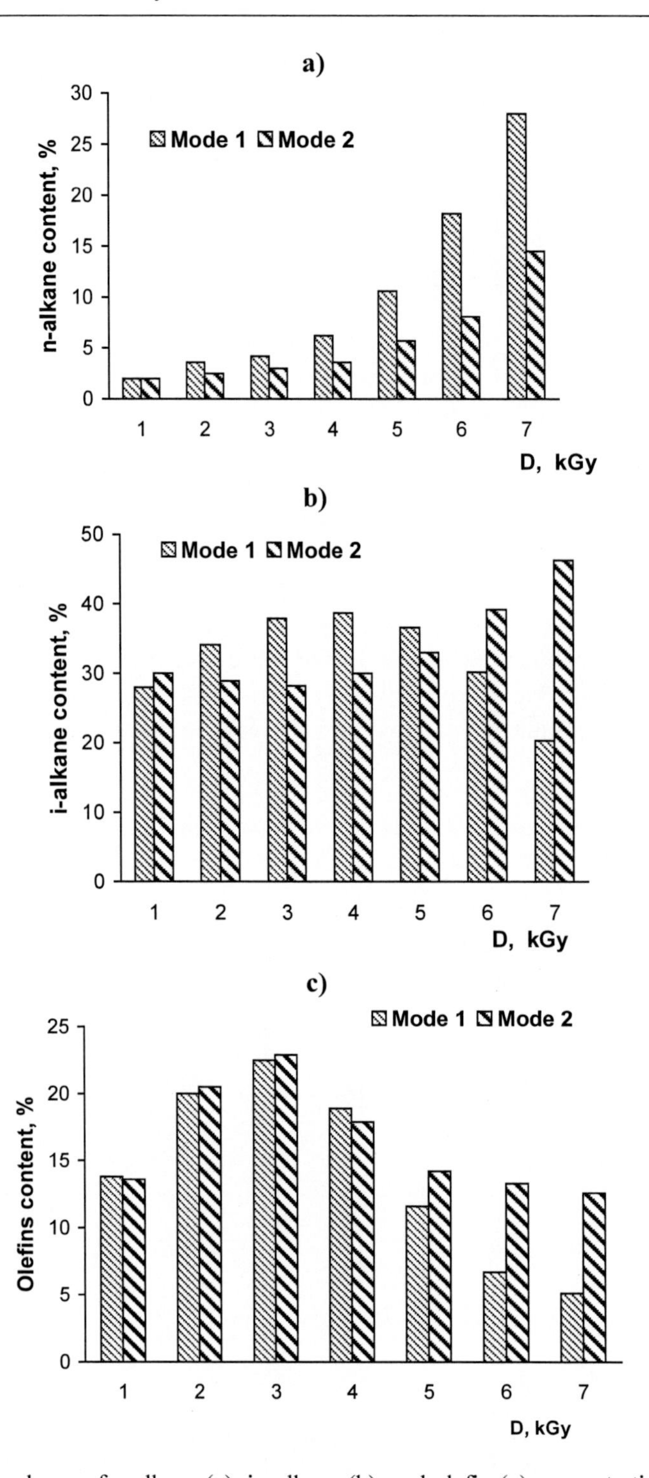

Figure 3. Dose dependence of n-alkane (a), isoalkane (b), and olefin (c) concentrations in the gasoline fractions with start of boiling at 100°C obtained by RTC of Karazhanbas oil in two irradiation modes [19].

Contents of liquid fractions obtained in the two RTC modes were determined from GLC data (Table 3). The maximum yield of gasoline fractions $C_6 - C_{10}$ (boiling temperature up to

180°C) in mode 1 characterized by the heightened rate of heavy fraction disintegration was 32.2 mass % (in respect to the feed mass), while in mode 2 it attained only 15 mass %. Chromatography data show that the gasoline yield from the condensate of light fractions in mode 2 is higher than that in mode 1 for the doze of 2 kGy. However, the lower yield of liquid fractions in mode 2 makes the resulting yield of the gasoline fraction in respect to the feed mass lower than that in mode 1 (Table 3). The gasoline fractions produced in the two modes essentially differ in their hydrocarbon contents due to different mechanisms of radiation-induced reactions.

The yields of different hydrocarbon groups (n-alkanes, isoalkanes, and olefins) are shown in Fig.3 versus irradiation dose for gasoline fractions (start of boiling at 100°C) produced by RTC of Karazhanbas oil in the two modes.

Fig.3 shows that the amount of n-alkanes in the gasoline fraction grows rapidly in mode 1, while the yields of isoalkanes and olefins reach maximum and then decrease as the dose rises. Such behavior of the yield curves testifies to the high degree of hydrocarbon disintegration and the prevailing role of polymerization reactions in limitation of unsaturated hydrocarbon accumulation.

The unusually high yields of iso-paraffins in the RTC modes characterized by relatively low values of temperature and dose rate were attributed to the effects of energy transfer from paraffin to aromatic component of the hydrocarbon mixture.

Decrease in the reaction temperature and the doze rate in mode 2 displays the "protective" role of heavy aromatic compounds that take the excess radiation energy away from big alkyl radicals. It leads to increase in the lifetime of alkyl radicals which becomes sufficient to transform primary alkyl radicals to more thermodynamically stable secondary (tertiary) radicals. Fig.3 demonstrates that concentration of n-alkanes in the gasoline fraction produced in mode 2 increases slower with irradiation dose compared with mode 1 that is accompanied by the cymbate increase in isoalkane yields.

Taking into account rather considerable decrease in RTC product yields in mode 2 compared with mode 1 (nearly proportional to decrease in the dose rate), it was assumed that the fraction of alkyl radicals stabilizing due to intramolecular isomerization grows as the dose increases and contributes to some limitation of the rate of destructive conversion, together with polymerization.

These results can be compared with the data on isomerisation processes observed during oil thermal processing without irradiation. Studies of cracking mechanisms and kinetics in hydrocarbons [3, 5] have shown that isomerization of primary radicals plays an important role in thermal disintegration of high-molecular alkanes (C_6 and higher). Activation energy for isomerization can be much lower than the energy necessary for radical disintegration that would influence the equilibrium between primary and secondary isomer forms.

Decrease in temperature provokes intensification of isomerization processes. It is evaluated [38] that equilibrium concentration of secondary butyl radicals is about 9 times higher than the concentration of primary radicals in the case of conventional thermal cracking (900 K) while in the case of initiated cracking (700 K) this factor is estimated as 16. According to calculations [39, 40], concentrations of tertiary isobutyl radicals exceed concentrations of primary isobutyl radicals by the factor of 16 in the case of thermal cracking and by the factor of 35 in the case of initiated cracking.

The absolute rates of hydrocarbon isomerization during thermal cracking are very low and isomer concentrations in a hydrocarbon mixture usually do not exceed 1-2% after processing. To reach higher isomer concentrations catalytic cracking is usually applied. Three temperature intervals characteristic for catalytic isomerization were registered [41]. The lower temperature range (up to 600°C is characterized by predominant isomerization of the components originally present in the mixture. At higher temperatures, increase in the rates of carbon cation reactions leading to isomerization of the appearing decomposition products was observed; isoparaffin concentrations increased to 26-28 mass %. Further temperature rise (above 800°C) caused drastic fall of isoparaffin hydrocarbon yields.

Isomerization rates and isomer yields after radiation-induced cracking are much higher than those in the case of standard thermal cracking. It was stated [3] that the isobutane yield after radiation- thermal cracking of the light gasoline fraction with boiling temperature up to 140°C was 0.09 mass % for the cracking temperature of 350°C. At the same time no measurable amounts of thermal cracking products were detected for reaction temperatures below 500°C. The isobutane yield observed after radiation-induced cracking at 500°C was 1.4 mass %, i.e. 200 times higher than that in the case of thermal cracking.

Radiation-induced cracking opens unique opportunities for intensification of intramolecular isomerisation due to combination of high rates of radical generation and rather low rates of thermal processes, all of them regulated in wide ranges.

Isomerization intensity observed in mode 2 is defined not only by RTC conditions but also by specific hydrocarbon contents of the oil feedstock characterized by heightened concentrations of pitch-aromatic components. The latter are the compounds of high radiation resistance that absorb a considerable part of radiation energy, and thus can lower the rate of radiation-induced chemical conversion and provoke isomerization processes.

This conclusion can be confirmed by the results of RTC of the light oil of Kok-Zhide field, Kazakhstan, in similar irradiation conditions [19]. This feedstock is characterized by higher paraffin contents (51.5 mass % in the gasoline fraction) and lower concentrations of naphtenes and arenes in light fractions (32.2 mass % in the gasoline fraction compared with 45.2 mass % for Karazhanbas oil). The yields of n-alkanes, isoalkanes and olefins obtained in the result of Kok-Zhide oil processing in conditions of mode 2 are more characteristic for the mode 1 less favorable for isomerization. So, reactions of polymerization and isomerization can be considered as the main factors limiting the rates of radiation-induced chemical conversion and essentially affecting the yields and the hydrocarbon contents of RTC products. The relative contributions of these processes depend on RTC conditions and feed contents.

Thus, high isomerization rate in RTC of hydrocarbon compositions with high concentrations of heavy aromatics was explained by availability of the lower temperature and dose rate limits for noticeable RTC reactions. Transfer of the excess radiation energy to an alkyl radical in the excited paraffin composition assists its disintegration and impedes intromolecular isomerization. On the contrary, the addition of heavy aromatics allows combination of rather high dose rates and temperatures with favorable conditions for isomerization. Aromatic compounds, known for high radiation resistance, can absorb the excess energy of a considerable part of radiation-generated radicals. In this case, many alkyl radicals can have enough time to stabilize their electron structure and to form isomers before their disintegration or recombination.

This interpretation was confirmed with the experiments on bitumen which are characterized by higher concentrations of heavy aromatics and, therefore, still more pronounced effects of enhanced isomerization [27].

Experimental data have shown that the most part of iso-alkanes is concentrated in the gasoline fraction of the liquid product of radiation processing. Maximal iso-alkane yields were observed at the lowest values of dose rate and temperature sufficient for noticeable chain reaction. At the given dose and dose rate, they increase as density of the feedstock becomes higher.

Maximal yields of isomers in gasoline fraction obtained by RTC of hydrocarbon mixtures with very high concentrations of aromatic compounds, such as bitumen and highly viscous oil, can be approximately estimated using the following empiric equation [19, 25]:

$$Y^{par}_{i-par} = \frac{1040}{PT} \frac{[(1+190/T)\tilde{\rho} - \rho_0]}{\rho_0}, \tag{6}$$

where Y^{par}_{i-par} is isoparaffin concentration in the paraffin part of the gasoline fraction; P is dose rate, kGy; T – absolute temperature, K; $\tilde{\rho}$ - feedstock density at 20^0C, kg/m^3, and ρ_0 - density of the gasoline fraction (in calculations below ρ_0 was taken equal to 780 kg/m^3).

Table 4 shows a good agreement of experimental and calculated concentrations of iso-alkanes in the RTC products after processing different types of feedstock.

Formula (6) was based by the following non-rigorous consideration in paper [25] where energy transfer from light paraffin to heavy aromatic molecules was considered in the hypothetical two-component hydrocarbon mixture. It was assumed that the paraffin component can be partially or completely formed by radiation-induced detachment of alkyl substituents from the aromatic structure. It was also supposed that an alkyl radical can form an isomer in the only case when the light paraffin fraction transmits a part of its excess energy to aromatic molecules.

The isomer yield from the feedstock, Y_i^{feed}, can be written in the form:

$$Y_i^{feed} = \int_{\rho_{min}}^{\rho_{hf}} \frac{P_2(\rho_{hf}) - P_1(\rho)}{P_1(\rho)} f(\rho, T, P)\, C_{hf}\, C(\rho)\, d\rho, \tag{7}$$

where $f(\rho)$ is the probability density of intramolecular isomerization of an alkyl radical in the absence of energy transfer from for the fraction characterized by “partial density” (concentration) ρ ($\rho = M/V$, where M is molecular mass of the fraction and V is molecular volume); $P_1(\rho)$ is probability of excitation energy absorption by a molecule of the light fraction characterized by “partial density” ρ ; $P_2(\rho_{hf})$ is probability of excitation energy absorption by heavy aromatic fraction; C_{hf} and $C(\rho)$ are atomic concentrations of the “heavy” and the “light” fractions, respectively.

For the sake of simplicity it was supposed that $f(\rho)$ is the δ -function:

$$f(\rho, T, P) = \frac{\alpha}{PT} \delta(\rho - \rho_0), \tag{8}$$

where α is constant; ρ_0 is of the fraction where isomers are concentrated, in the case of RTC product of bitumen it is gasoline and partially kerosene fraction.

Table 4. Isoparaffin concentration in the paraffin part of gasoline fraction of synthetic oil produced by radiation processing of bitumen and high-viscous crude oil [25]

Feedstock	Reaction tempera-ture T, K	Dose rate, P, kGy	Feedstock density $\tilde{\rho}$, kg/m^3	Isoalkane concentration in gasoline fraction Y^{gas}_{i-par}, %	Calculated isoalkane concentration in paraffin part of gasoline $(Y^{par}_{i-par})_{calc}$, %	Experimental isoalkane concentration in paraffin part of gasoline $(Y^{par}_{i-par})_{exp}$, %
Fuel oil	673	1.1	939	46.7	76.4	75
Heavy Crude Oil (Karazhanbas, Kazakhstan)	648	1.25	942	35.5	72.1	67-73
Athabasca bitumen - 1	683	1.3	970	38.2	68.9	65-68
Bitumen (Mortuk, Kazakhstan)	683	1.5	976	33.4	60.3	59.6
Bitumen (Shilikty, Kazakhstan)	693	1.5	998	36.1	63	61-65
Athabasca bitumen - 2	693	1.5	1020	40.2	66.7	66-68

Substitution of $f(\rho)$ to expression (2) gives:

$$Y_i^{feed} = \frac{\alpha}{PT} \frac{P_2(\rho_{hf}) - P_1(\rho_0)}{P_1(\rho_0)} C_{hf}\, C(\rho_0). \qquad (9)$$

In suggestion that isomerization proceeds only in the fraction with the average density ρ_0 that transfers its excess energy to the remaining part of material with density ρ_{hf},

C_{hf}=1-C(ρ_0), and

$$Y_i^{feed} = \frac{\alpha}{PT} \frac{P_2(\rho_{hf}) - P_1(\rho_0)}{P_1(\rho_0)} [1 - C(\rho_0)] C(\rho_0). \qquad (10)$$

In supposition that probability of energy absorption by the each of the two components of the hydrocarbon mixture is proportional to their "partial densities":

$$P(\rho) \sim \rho, \tag{11}$$

equation (10) can be rewritten as follows:

$$Y_i^{feed} = \frac{\alpha}{PT}\frac{\rho_{hf} - \rho_0}{\rho_0} C(\rho_0)[1 - C(\rho_0)]. \tag{12}$$

Proportionality of the probability of excitation energy transfer to the substance density corresponds to the quantum mechanical solution of the problem when static dipole-dipole interaction is considered in the case of the ideal resonance of two oscillators [3]. We shall note that linear losses of radiation energy that can be considered as a measure of substance ability to accumulate excitation energy are also proportional to the substance density.

Introduction of the average feedstock density

$$\tilde{\rho} = \rho_0 C(\rho_0) + \rho_{hf}[1 - C(\rho_0)] \tag{13}$$

allows reduction of equation (10) to the simple expression:

$$Y_{i-par}^{par} = \frac{Y_i^{feed}}{C(\rho_0)} \approx \frac{\alpha}{PT}\frac{\tilde{\rho} - \rho_0}{\rho_0}. \tag{14}$$

Thus, an expression derived is of the same form as equation (1). Formula (6) shows that degree of paraffin isomerization during radiation-thermal cracking is determined by transfer of excess excitation energy to the more dense and radiation resistant medium. The factor *(1+190/T)* can be considered as a correction for different thermal expansion of the "heavy" and the "light" fractions; 1040/PT =1 for the characteristic mode of radiation (P=1.5 kGy, T=420^{0}C) favorable for paraffin isomerisation in the gasoline fraction of bitumen.

If linear energy losses for excitation energy transfer are proportional to substance density, then volume energy losses should be proportional to $\rho^{3/2}$. Basing on analysis of experimental data on hydrocarbon radiolysis inhibition by different chemical additions, it was stated [3] that inhibitor capacity for energy absorption is proportional to $\rho^{3/2}$; in this case expression $\frac{\tilde{\rho} - \rho_0}{\rho_0}$ in formula (14) should be replaced by $\left(\frac{\tilde{\rho} - \rho_0}{\rho_0}\right)^{3/2}$. However, the available experimental data on isomer yields can be satisfactorily described by the linear correlation (6).

Table 5. Fractional, hydrocarbon contents and octane numbers of gas-condensate gasoline after bremsstrahlung X-ray irradiation [44]

Sample	Processing conditions	Hydrocarbon contents					Octane number
		n-alkanes	iso-alkanes	arenes	naphtenes	olefins	
GCG	feedstock	25.7	29.3	12	32.2	0.8	54
GCG	X-ray D=6-8 kGy	24.6	30.5	10.3	29.1	5.5	63
GCG + 6 mass% bitumen	X-ray D=6-8 kGy	24.5	38.6	11.8	22.4	2.7	67
GGG + 15 mass% bitumen	X-ray D=6-8 kGy	25.1	39.2	12.3	21.1	2.3	68

GCG – gas-condensate gasoline

The favorable conditions for isomerization are lowered dose rates of ionizing irradiation and lowered temperatures. Therefore, the effect of radiation-enhanced isomerization in presence of heavy aromatics should be considerable in the case of low dose rate gamma-irradiation of aromatic-rich hydrocarbon mixtures at lowered temperatures.

For experimental verification of this conclusion, special experiments were conducted on radiation processing of low-octane gas-condensate gasoline at room temperature [44]. Gasoline was irradiated with the bremsstrahlung X-rays from the 2-MeV electron beam. The experiments have shown that radiation-enhanced isomerization became especially pronounced when heavy aromatic hydrocarbons were added to relatively lighter feedstock and this mixture was irradiated with moderate X-ray doses. In these experiments, heavy residua of bitumen radiation processing were used as additional agents for initiation of isomerization.

Table 5 [44] summarizes the observed changes in fractional and hydrocarbon contents of gasoline extracted from gas condensate and demonstrates the effect of aromatics addition on radiation-induced isomerization.

Experimental data of Table 5 and Fig.4 show that effect of paraffin isomerization becomes apparent only in presence of heavy aromatic compounds and only as a result of mixture radiation processing. Mixing gas-condensate gasoline with 15 mass% residue of bitumen vacuum distillation and subsequent X-ray irradiation at room temperature provides increase in iso-alkanes concentration by 33.8% and increase in gasoline octane number from 54 to 67 without any chemical additions.

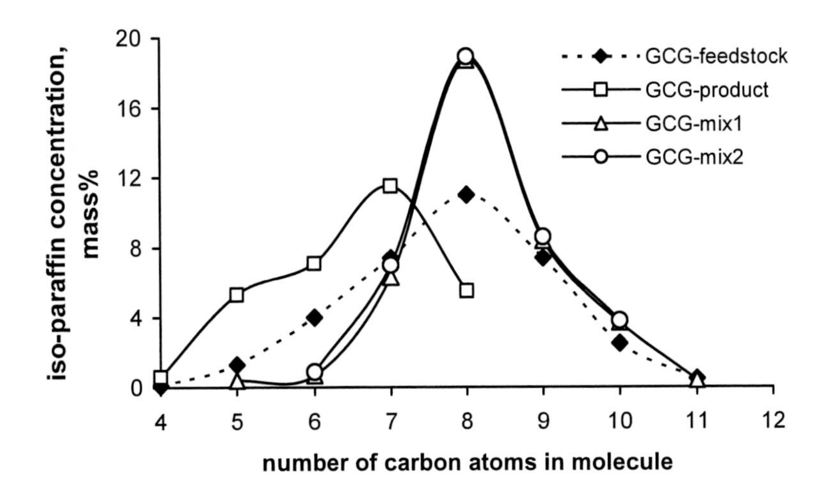

Figure 4. Molecular mass distribution of iso-paraffin concentrations in mixtures of gas-condensate gasoline [44]; Designations are the same as in Table 2.

Radiation-Induced Polymerization in Conditions of Radiation-Thermal Cracking of High-Paraffin Oil

This effect, most clearly pronounced in oils with high contents of high-molecular paraffins, was observed in experiments on processing of crude oil from Kumkol field, Kazakhstan, with accelerated electron beams and described in papers [18, 24, 25]. This type of oil is characterized by high contents of C_{15} - C_{22} hydrocarbons of paraffin series with solidification temperature 10-44°C with their total concentration of 60-77%. The contents of polycyclic aromatic hydrocarbons and asphalt-pitch substances are very low.

High temperatures of crude oil solidification are due to high concentrations of heavy paraffins. The contents of heavy aromatic compounds in fractions 400-450^{0}C and 450-500^{0}C are 5-7%, while the total concentration of paraffin and naphthene hydrocarbons is 80-90%. Considerable amount of paraffins with high melting point affects the rheological properties of oil.

Oil samples were irradiated [24] by 2 MeV electrons from an electron accelerator ELU-4, using current densities from 1 to 3 $\mu A/cm^2$. Different conditions of radiation-thermal processing were provided by variation of main processing parameters: temperature in the range 340-450^{0}C, irradiation dose in the range 1-4 kGy, and dose rate in the range 1-4 kGy/s.

The results of these experiments were unexpected. Kumkol oil is highly paraffin, i.e. it contains hydrocarbons mostly subjected to decomposition under irradiation. High concentration of high-molecular paraffins in this feedstock testifies to great hydrogen reserves that principally assumes high yields of light fractions. However, experiments have shown that radiation-thermal processing of mixtures of light and heavy paraffins at moderate temperatures and dose rates does not result in the expected high yields of light fractions.

The main products were the liquid gas oil fraction with boiling temperature from 60 to 350°C (70-80 mass %), the gaseous fraction (8-10 mass %) and the heavy paraffin residue

(15-20 mass %) with the strong tendency to polymerization. Product yields varied depending on conditions of radiation-thermal cracking.

According to the GLC data the gasoline fraction separated from the feedstock contained 38.0 mass % normal alkanes, 27.0 % isoalkanes, 29.0% naphthenes, 3.5 % aromatics, and 1.8 % unsaturated hydrocarbons. After radiation-thermal processing of Kumkol oil in different conditions, 30-35 mass % n-alkanes, 28-33% isoparaffins, 25-30% naphthenes, 5-7% aromatics, and 2.5-4.2% olefins were identified in the hydrocarbon composition of the gasoline fraction (C_4-C_{10}) from the RTC product. Cyclic olefins and dienes were also present but their concentrations did not exceed 1%.

Dependence of the yields of gasoline and kerosene fractions on the electron dose is shown in Fig. 5. It is obvious that radiation-thermal processing of the mixture of heavy and light paraffins in these conditions does not raise distillation yields of motor fuels compared with those from the feedstock (D=0). Gasoline originally available in such feedstock is partially converted into gases and partially absorbed by the excited heavy paraffin substance. As a result, yields of light fractions are much lower than those observed after RTC of highly viscous Karazhanbas oil under the same processing conditions (Figs.2, 3).

a)

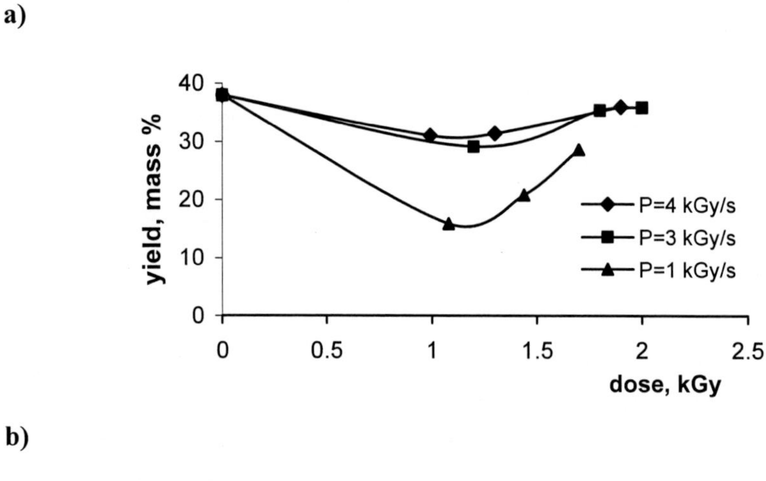

b)

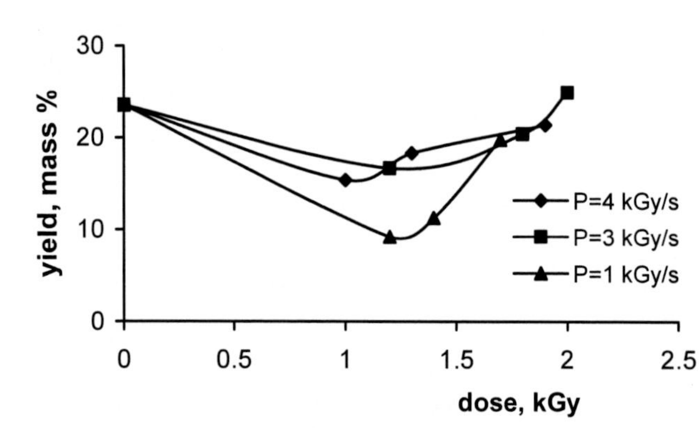

Figure 5. Dose dependence of the yields of gasoline (a) and diesel (b) fractions after RTC of Kumkol crude oil [24]. T=400^0 C.

a)

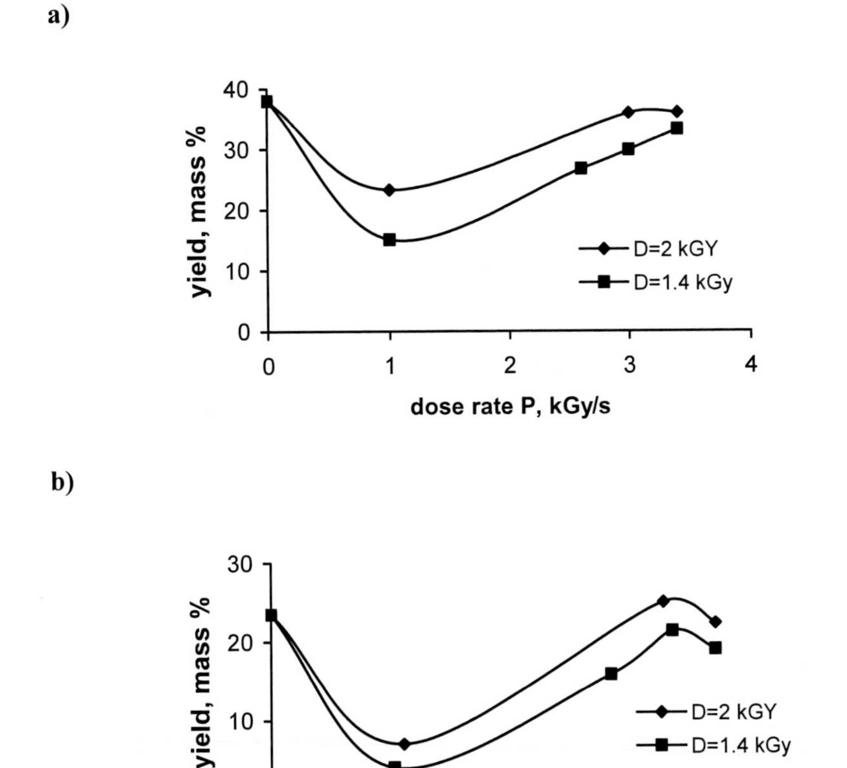

b)

Figure 6. Dependence of gasoline (a) and diesel (b) yields after RTC of Kumkol crude oil on dose rate [24]; T=400^0 C.

Fig. 6 shows dependence of gasoline and diesel yields on dose rate. It demonstrates that increase in the dose rate up to 2 kGy/s intensifies destruction processes in the gasoline fraction and leads to a decrease in the yields of light fractions. When the dose rate is higher than 2 kGy/s, the light fractions increase in their yields due to heavy paraffin destruction.

Radiation-chemical yields of n-alkanes in the gasoline fraction decrease with increasing dose (Fig.7). However, there is a noticeable trend to an increase in the relative contribution of heavier paraffins (n-pentane and n-hexane for low doses, and n-nonane and n-decane for heightened doses). Fig.8 shows the higher rates of the yields of heavy alkanes with the increasing dose rate.

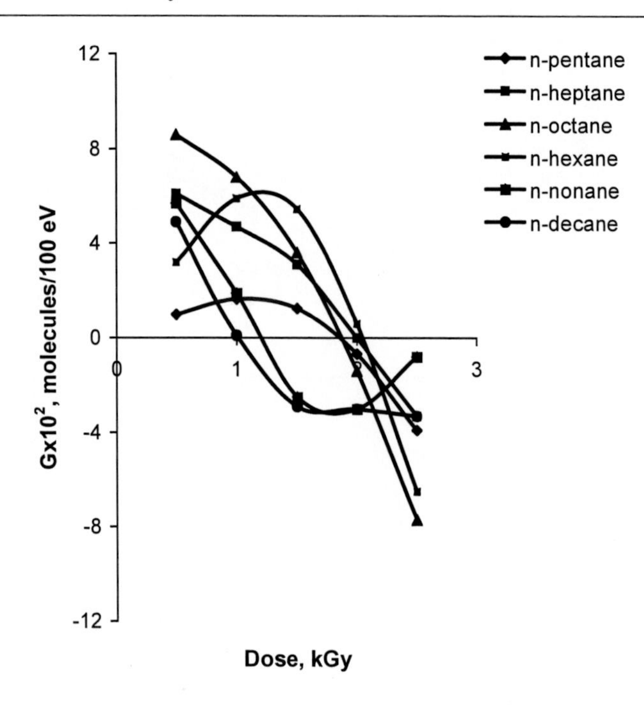

Figure 7. Dose dependence of n-alcane (C7-C10) G-values after RTC of Kumkol crude oil [24]; P=3 kGy/s; T=400^0 C.

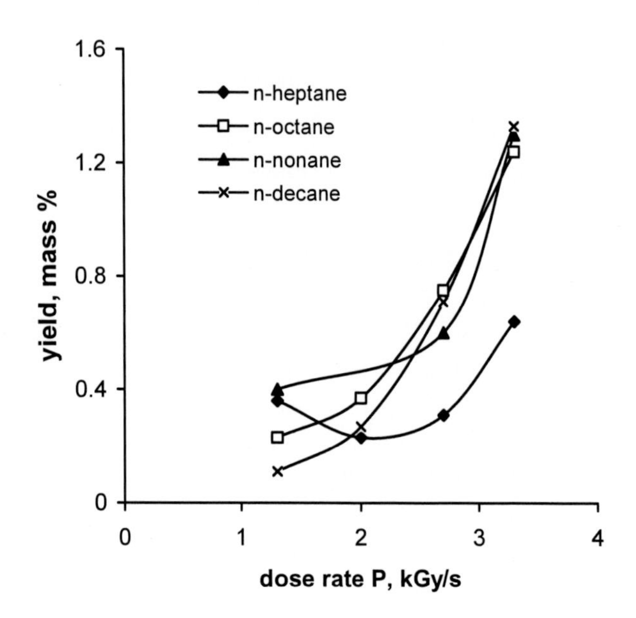

Figure 8. Dependence of n-alkane yields on dose rate after RTC of Kumkol crude oil [24]; D=1.5 kGy; T=400^0 C.

Generally, increase of light fractions molecular weight with the rise of electron dose rate is a characteristic phenomenon for RTC of the paraffin Kumkol oil. Fig.9 shows that increase in the dose rate leads to increase in concentrations of fractions with high boiling temperature in gasoline.

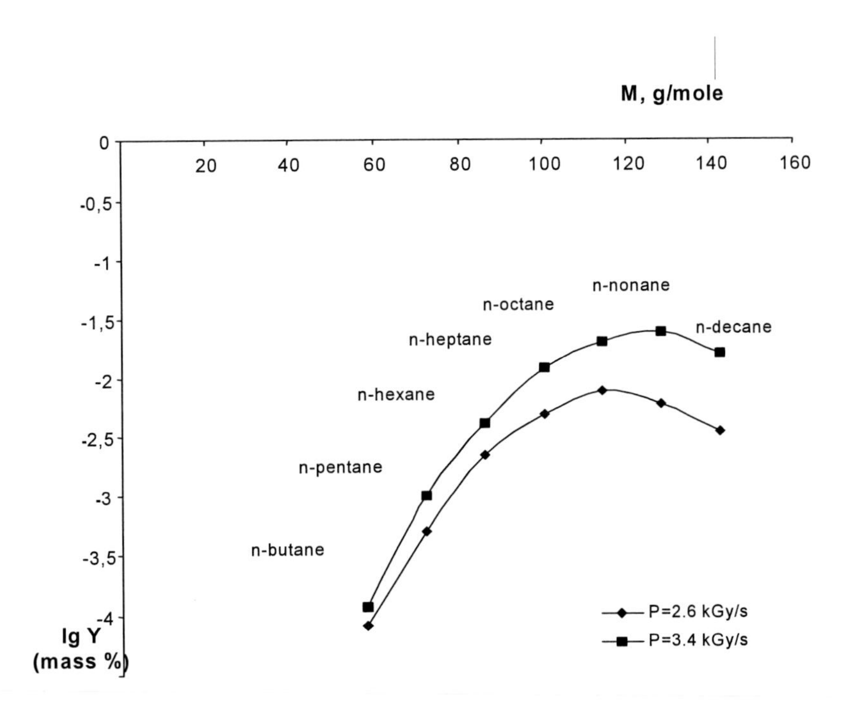

Figure 9. Mass distribution of n-alkane yields (Y) in gasoline produced by RTC of Kumkol crude oil [24]; D=2 kGy; T=400^0 C.

The effect of the carbon chain length on the yields of n-alkane radiolysis products was studied in the work [31]. It was shown that increase in the number of C-C bonds in a molecule (from six to sixteen) leads to excitation energy redistribution over the greater amount of bonds that diminishes efficiency of C-C bonds breaks. This causes increase in the yields of parent radicals and reduces the yields of fragmentary radicals.

The increase in the molecular weight of the gasoline fraction observed in [24] could be explained by increased destruction of the paraffins in the middle of molecules as the dose rate increased. It raises the probability of alkyl radical recombination with subsequent formation of paraffin molecules lighter than the molecule destroyed but heavier with respect to the gasoline fraction.

Compared with the hydrocarbon contents of gasoline fractions produced as a result of RTC of high-viscous Karazhanbas oil (see previous section), the RTC gasoline of Kumkol oil contained lower concentrations of isoparaffins. The study of Karazhanbas oil has shown that irradiation modes characterized by low values of temperature and dose rate provided high level of paraffin isomerization. It was reasonable to suppose that share of alkyl radicals stabilizing per unit of time due to intramolecular isomerisation was higher in these conditions. It caused a steady increase in the isoalkane concentration in the gasoline fraction with increasing dose.

Fig. 10 shows that in similar irradiation conditions (P=1kGy/s) isomerization observed during RTC of Kumkol oil is much less pronounced. The maximum increase in isoalkane concentration is about 7% compared with 15% in the case of Karazhanbas oil. When the dose is higher than 1.5 kGy, isoalkane concentration decreases with increasing dose, while its monotonous growth is observed for Karazhanbas oil in similar conditions.

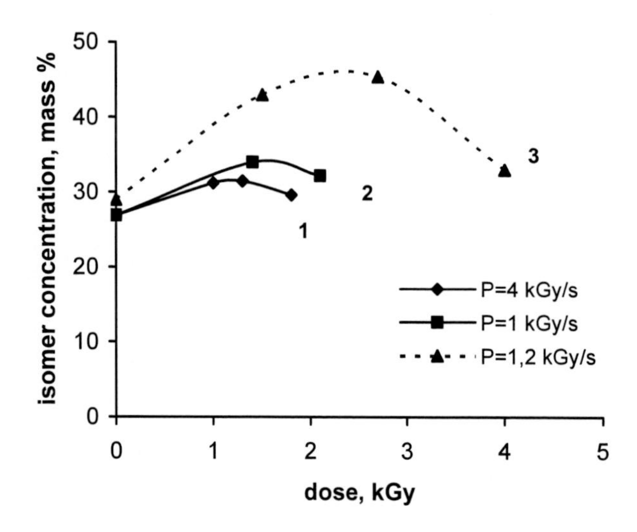

Figure 10. Dose dependence of isomer contents in the paraffin part of RTC products of Kumkol crude oil (1,2) and Karazhanbas crude oil (3) [24]; T=400^{0} C.

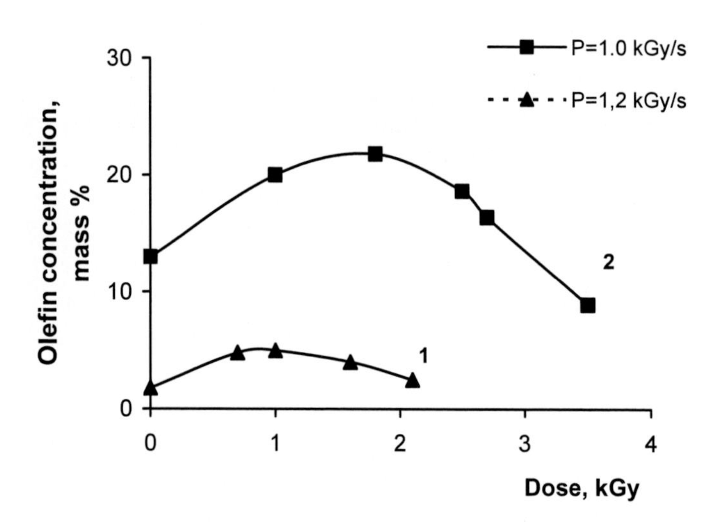

Figure 11. Olefin concentrations in gasoline fraction after RTC of Kumkol crude oil (1) and Karazhanbas crude oil (2); T=400^{0} C [24].

These observations show the synergetic nature of the isomerization processes in complex hydrocarbon mixtures when isomerization rate depends not only on conditions of radiation-thermal processing but also on the specific hydrocarbon contents of the feedstock.

Higher concentrations of pitch-aromatic components in the case of Karazhanbas oil are responsible for the higher yield of isomerization. In the case of Kumkol oil with very low concentrations of aromatic hydrocarbons, this yield of isomerization cannot be achieved even at low dose rates.

Important characteristics of RTC in the paraffinic oil can also be drawn from analysis of the yields of unsaturated hydrocarbons.

Dose dependence of olefin concentrations in the gasoline fractions separated from liquid products after RTC of Kumkol oil is shown in Fig.11 in comparison with those in Karazhanbas oil. It shows that unsaturated hydrocarbons that appear during radiation processing are practically entirely concentrated in the gasoline fractions, i.e. they are mainly low-molecular compounds.

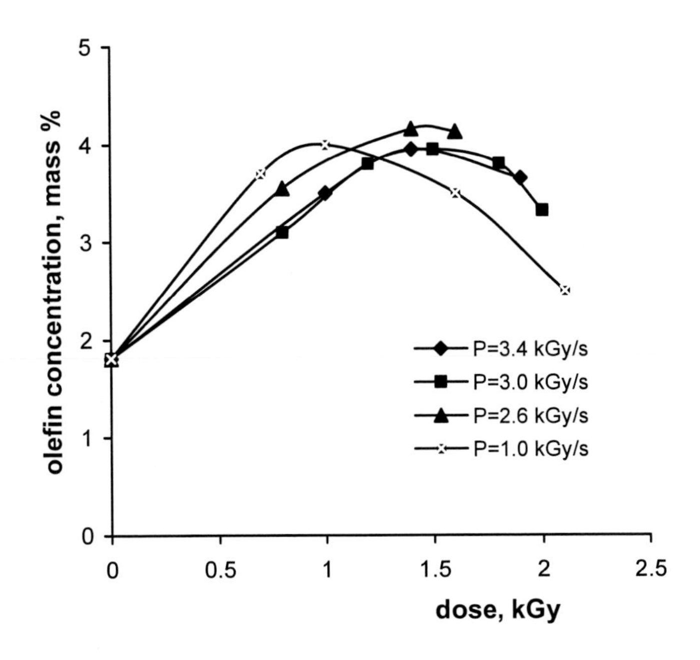

Figure 13. Olefin concentrations in gasoline fraction after RTC of Kumkol crude oil [24]; T=400^0 C.

Fig.12 shows that the maximum olefin concentration in the gasoline fraction after RTC of Kumkol oil does not exceed 4-5%, it follows from Fig.12 that it is 3-5 times lower than that in Karazhanbas oil. These values are very low for RTC gasoline fractions. Another characteristic feature of the results shown in Fig.13 is the low dose of electron irradiation corresponding to the olefin maximum (0.8-1.5 kGy).

Such composition of cracking products and its dose dependence is characteristic neither for RTC of relatively light paraffins [3-5] nor for RTC of high-viscous oil with high concentrations of aromatic compounds [19, 25]. Note that similar RTC conditions in the case of Karazhanbas oil provide gasoline containing 20-25% olefins with the dose of maximum olefin yields being about 2.5 kGy.

Radiation - chemical conversion of the model paraffin hydrocarbons $C_{16}H_{34}$ and $C_{17}H_{36}$ in different aggregate states were studied in works [41, 7, 42]. According to the data by Földiak [41], the rate of $C_{16}H_{34}$ decomposition is practically constant up to about 523 K.

Further increase in temperature causes a sharp increase of the reaction rate as a result of its transformation to a chain process (RTC).

These studies have shown that the main products of radiolysis of these hydrocarbons in the liquid phase are dimers forming mainly as a result of secondary radical recombination. Our studies of high-paraffin Kumkol oil show that the increase in molecular mass of the paraffin feedstock gives a considerable rise to polymerization during and after RTC.

The observed regularities of olefin accumulation in the gasoline fraction of high-paraffin oil can be qualitatively interpreted in frames of the simplest model of RTC kinetics [24], where this process is reduced to disintegration of paraffin molecules with the given average molecular mass that leads to olefin accumulation limited by reactions of polymerization. Paraffin disintegration and olefin polymerization are initiated by radiation generation of radicals, their quasi - stationary concentration being calculated in the approach of the square-law chain break and assumed to be independent on irradiation dose in this rough model.

In accordance with this simplified standard mechanism:

$$\frac{d[Par]}{dt} = -k[R][Par], \tag{15}$$

$$\frac{d[Ol]}{dt} = k[R][Par] - k_p[R][Ol],$$

where $[Par]$ and $[Ol]$ - are concentrations of paraffins and olefins, respectively, and $[R]$ is the concentration of carbon centered radicals.

The rate constants of olefin formation k and polymerization k_p are defined by the usual exponential relations:

$$k = k_0 \exp\left(-\frac{E}{k_B T}\right);$$

$$k_p = k_{op} \exp\left(\frac{E_p}{k_B T}\right),$$

where E, E_p are activation energies of the processes; k_0, k_{0p} are pre-exponential factors, and k_B is the Boltzman constant.

Radical concentration $[R]$ can be found from the equation

$$[R] = \left(\frac{G\dot{D}}{k_r}\right)^{\frac{1}{2}} \tag{16}$$

where $\dot{D}$ is the dose rate, *G* is the radiation-chemical yield for radicals, k_r is the radical recombination rate constant.

Solution of the system of equations (15) with the initial condition

$$[Par]_{t=0} = [M]$$

leads to the following equation for olefin concentration:

$$[Ol] = k[R][M] * \\ * \frac{\exp(-kt) - \exp(-k_p t)}{k_p - k}, \tag{17}$$

The maximum olefin concentration

$$\frac{[OL]_{max}}{[M]} = \frac{1}{\frac{k_p}{k} - 1} \left[\exp\left(\frac{1}{1 - \frac{k_p}{k}} \right) \ln\left(\frac{k_p}{k} \right) - \\ - \exp\left(\frac{\frac{k_p}{k}}{1 - \frac{k_p}{k}} \right) \ln\left(\frac{k_p}{k} \right) \right] \tag{18}$$

should be observed at the time t_0 corresponding to irradiation dose D_0:

$$D_0 = \dot{D} t_0 = \frac{\dot{D}}{k[R]} \varphi\left(\frac{k_p}{k} \right) = \\ = \frac{\dot{D}^{\frac{1}{2}}}{\left(\frac{G\dot{D}}{k_r} \right)^{\frac{1}{2}} k_0 \exp\left(-\frac{E}{k_B T} \right)} \varphi\left(\frac{k_p}{k} \right), \tag{19}$$

where

$$\varphi\left(\frac{k_p}{k}\right) = \frac{1}{\frac{k_p}{k} - 1} \ln\left(\frac{k_p}{k}\right). \tag{20}$$

The plot of the dependence (19) of olefin concentration in the cracking product on the ratio of constants of paraffin scission and olefin polymerization is shown in Fig.14.

Fig. 14 shows that the maximum olefin concentration in cracking products increases non-linearly as the ratio $\frac{k_p}{k}$ decreases. In the simple model under consideration this effect does not depend on the dose rate. According to formula (19) the dose corresponding to maximum olefin yield is proportional to $P^{1/2}$ and essentially dependent on the $\frac{k_p}{k}$.

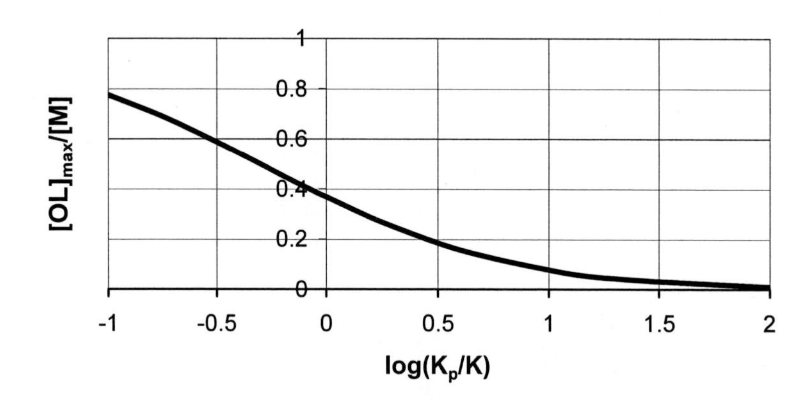

Figure 14. Dependence of maximum olefin concentration on ratio *Kp/K* [24].

These regularities indicate to considerable increase in the polymerization constant during RTC of high-paraffin Kumkol oil compared with Karazhanbas oil with high contents of aromatic compounds. Qualitatively, equations (18-20) describe the observed dose dependence of olefin yields.

For the rough quantitative estimation, the following values for reaction constants characteristic for hydrocarbons were used [49, 50]:

$$E = 35 \text{ kJ/mol};$$

$$k_p = k_r = 3*10^{-19}\,\text{m}^3\text{molecule}^{-1}\text{s}^{-1};$$

$$G = 3.1 \text{ radicals}*\text{J}^{-1}.$$

The average molecular mass of the feedstock and its density were asuumed as $M = 0.35$ kg/mol and $\rho = 800$ kg / m^3, respectively.

For the given above values of constants, equation (19) can be written in the form:

$$D_0(Gy) = \dot{D}t_0 = 8.3 * 10^5 \dot{D}\varphi\left(\frac{k_p}{k}\right). \tag{21}$$

where the $\dot{D}$ dimension is Gy/s.

In the case of Karazhanbas oil, the maximum olefin concentration of 20% corresponds to the value $\frac{k_p}{k} \cong 3$ (Fig.14); in the case of Kumkol oil the maximum olefin concentration of 4-5% corresponds to the value $\frac{k_p}{k} \cong 20$, that testifies to the increase in the polymerization rate constant by 6-7 times.

Substituting these values of $\frac{k_p}{k}$ to formula (22) we obtain the values of 0.6 kGy for Karazhanbas oil and 2.0 kGy for Kumkol oil for the doses corresponding to maximum olefin yields. These estimations approximately agree with the observed alterations in characteristics of dose dependence of olefin concentrations in different types of oil feedstock.

Correlation of polymerization rate in the RTC process with the average length of the paraffin molecule in the original mixture was considered in paper [18]. The values of k_p/k were calculated using experimental data on maximal olefin concentrations in gasoline fractions after RTC of crude oils and oil products with different average molecular masses of the paraffin component. The value of k_p/k can be easily determined from the graph in Fig.14 [24].

The results of calculations presented in Fig.2 quantitatively characterize the decrease in the olefin polymerization constant with the densification of paraffin mixtures subjected to RTC (Table 6).

If the ratio k_p/k is known, formula (20) can be used for calculation of the irradiation dose D_0, characteristic for the maximum olefin yield.

Calculated and experimental values of the irradiation dose, D_0, corresponding to the maximum olefin yields are give in Table 1 for various types of oil. Comparison of the data allows estimation of applicability of the simplified model used and, therefore, reliability of the correlation shown in Fig.15.

Table 6. Calculations of the dose corresponding to the maximum olefin yields for petroleum feedstock of different original composition

Type of feedstock	Average number of carbon atoms in the paraffin chain	Maximum olefin concentration [OL], mass %	Dose corresponding to the maximum olefin yield, D kGy	
			Experimental value	Calculated value
Crude oil (Kumkol, Kazakhstan)	16	4,2	2	1,6
Crude oil (Mangyshlak, Kazakhstan)	18	2,9	1,0	1,2
Crude oil (Southen Tatarstan)	10	7,3	2,2	1,8
Crude oil (Karazhanbas, Kazakhstan)	8	6,5	2,0	2,1
Crude oil (Kok-Zhide, Kazakhstan)	7	8,4	2,5	1,8
Crude oil (Bugulma, Tatarstan)	6	9,9	2,2	2,0
Fuel oil	5	11,5	2,5	3,0

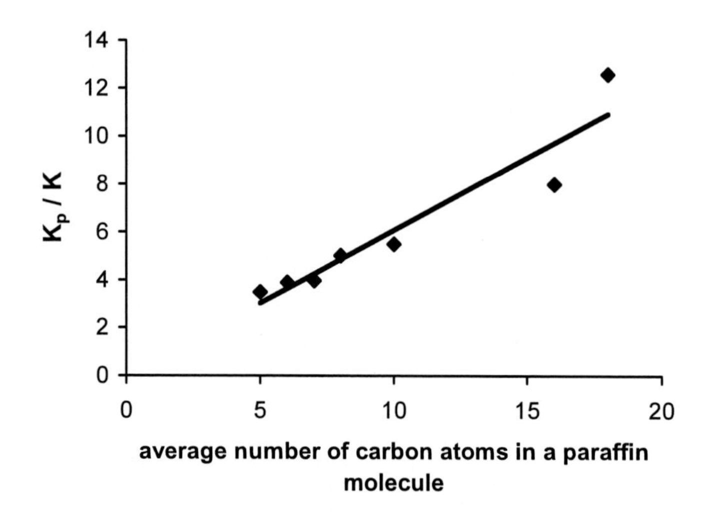

Figure 15. Dependence of the ratio of the rates of olefin polymerization and formation during RTC on the length of a paraffin chain [18].

It shows that for the most part of the samples of crude oil and oil products calculated and experimental values of the dose for the maximum olefin yield are in a satisfactory agreement.

Intense polymerization of RTC products was also observed in our earlier experiments with such wastes of oil extraction as asphalt-pitch-paraffin sediments (APPS) [17]. This feedstock contains high concentrations of heavy paraffins and, hence, great reserves of hydrogen. Therefore, it has high potential yields of light fractions. However, reactions of polymerization and chemical adsorption limited the maximum yields of light RTC products to 40%.

A specific feature of APPS radiation-induced destruction is formation of a big amount of the reactive paraffinic residue. The products of APPS destruction are unstable and maintain the tendency to polymerize if kept in contact for several days. In the absence of stirring a distinct interface is formed between light fractions and heavy paraffin residue. As a result of interaction with the polymerizing residue the light liquid fractions were completely absorbed and the heavy residue got denser and solidified after several days of exposure at room temperature. When separated the light fraction was stable, but the residue continued increasing in molecular weight.

Thus low olefin contents in RTC products and the high yield of their polymerization are the characteristic features of RTC of highly paraffinic oils. The effects observed can be explained taking into account the known features of heavy paraffin radiolysis. The G-value of radical generation in n-alkanes is almost independent of molecular mass, being usually about $5*10^{17}$ radicals$*J^{-1}$. Nevertheless, distribution of radiolysis products as a function of molecular mass differs significantly for different hydrocarbons [45].

In particular, as we pass from n-butane to n-hephtane, the yields of methyl and ethyl radicals drop down by nearly an order of magnitude and the yield of radicals containing more than six carbon atoms makes more than 80% of the heavy and light alkyl radicals that appear during n-heptane radiolysis.

According to the radical-chain cracking theory [3, 42] the excited n-alkane molecules disintegrate as a result of C-C or C-H bond breaks. In the latter case, there is a high probability of formation of the secondary carbon valence while a hydrogen atom of excess energy participates in the fast abstraction reactions forming another radical of this type.

According to Kossiakoff-Rice theory [46], it is probable that the big alkyl radical formed in RTC, for example, a primary radical $C_{10}H_{21}$ can convert to equilibrium state through an intermediate (for example, six member) secondary alkyl radical before it disintegrates:

CH₂ / H₂C, CH₂ / H₂C•, CH / H, R → CH₂ / H₂C, CH₂ / H₃C, •CH / R →

$$\longrightarrow CH_3CH_2{=}CH_2 + RHC^{\bullet}$$

At 300-500°C, there are the well-documented chain reactions of long-chain alkanes. Of course, the degradation reactions compete with the reverse reactions of chain addition of alkyl radicals to the olefins. Also the secondary radicals, in hypothesis of radical-radical recombination, yield branched alkanes [3-7, 43, 46].

Thus, formation of high concentrations of relatively long living alkyl radicals that cannot directly be the RTC chain carriers should cause the observed intensification of polymerization and isomerization in heavy paraffin fractions. These reactions result in limitation of the yields of light RTC products, low olefin concentrations in light fractions and increase in their molecular weights as the dose rate grows.

A considerable increase in polymerization rate for heavy paraffins in RTC conditions is analogous to the well-known phenomenon of acceleration of radiation-induced polymerization as the polymer molecular mass grows due to decrease in mobility of polymer radicals [47].

Generally, the following characteristic features of RTC should be noticed in oil with high contents of heavy paraffins:

- the low level of isomerization in light RTC fractions;
- the high polymerization rate and low olefin contents in RTC products;
- relatively low yields of light fractions at low irradiation dose rates;
- increase in the molecular weight of the gasoline fraction as the irradiation dose rate grows.

These observations are attributed to the behavior of heavy alkyl radicals that initiate polymerization and isomerization in heavy paraffin fractions.

Radiation-induced polymerization always limits the yields of light products of RTC of heavy hydrocarbon feedstock. In high-paraffin oils, the effect is so strong that efficient oil upgrading by means of RTC becomes difficult. The problem of paraffin oil radiation processing was resolved in PetroBeam™ technology where the optimal processing conditions at lowered temperatures take into account the structural state of the paraffin feedstock.

Dependence of the described above synergetic effects on hydrocarbon contents of the original feedstock and irradiation conditions was used in the new technological approaches for control of yields and quality of the target products of heavy oil radiation processing.

Radiation-Thermal Processing of High-Viscous Oil

Application of the technology based on RTC to heavy fuel oil (residue of crude oil primary distillation) is illustrated in Fig.16. RTC proceeds at nearly atmospheric pressure and temperatures 150-200°C less than those for conventional thermal cracking.

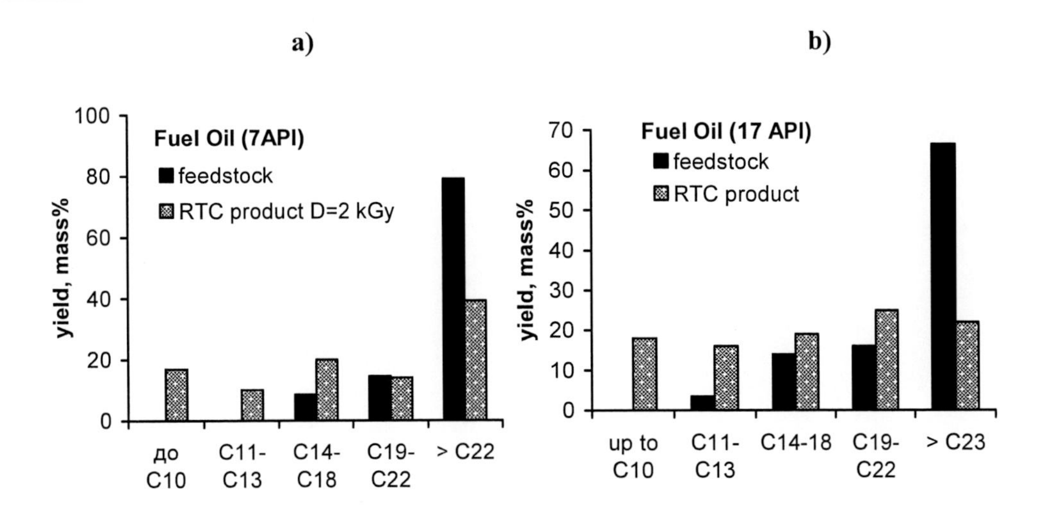

Figure 16. Fractional contents of different sorts of fuel oil and products of its radiation-thermal processing [36].

X-axis shows the average number of carbon atoms in a molecule

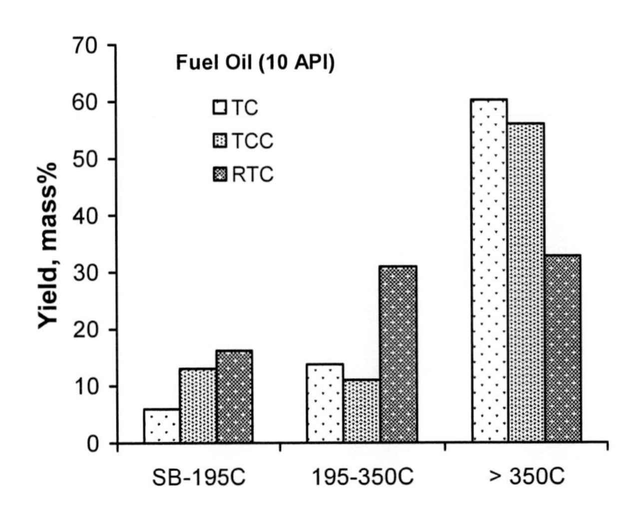

TC – thermal cracking in lab conditions
TCC – thermocatalytic cracking in the same conditions (iron oxide systems) [33]
Figure 17. Comparison of fractional contents of liquid product obtained by different methods of thermal processing [36]

The optimum conditions of radiation processing afford production up to 80 mass% motor fuel including up to 20% gasoline and 60% diesel fractions.

A by-product of radiation-induced cracking is heavy coking residue (up to 10%) and the gas mixture (up to 10%), containing ethylene and other unsaturated compounds valuable for chemical industry.

Fig.17 shows the higher efficiency of RTC in comparison with the one-through catalytic processing in similar conditions.

Fractional contents of the liquid product obtained by RTC of heavy crude oil from Karazhanbas field, Kazakhstan, and mass distribution of paraffin molecules in its gasoline fractions are shown in Fig.18.

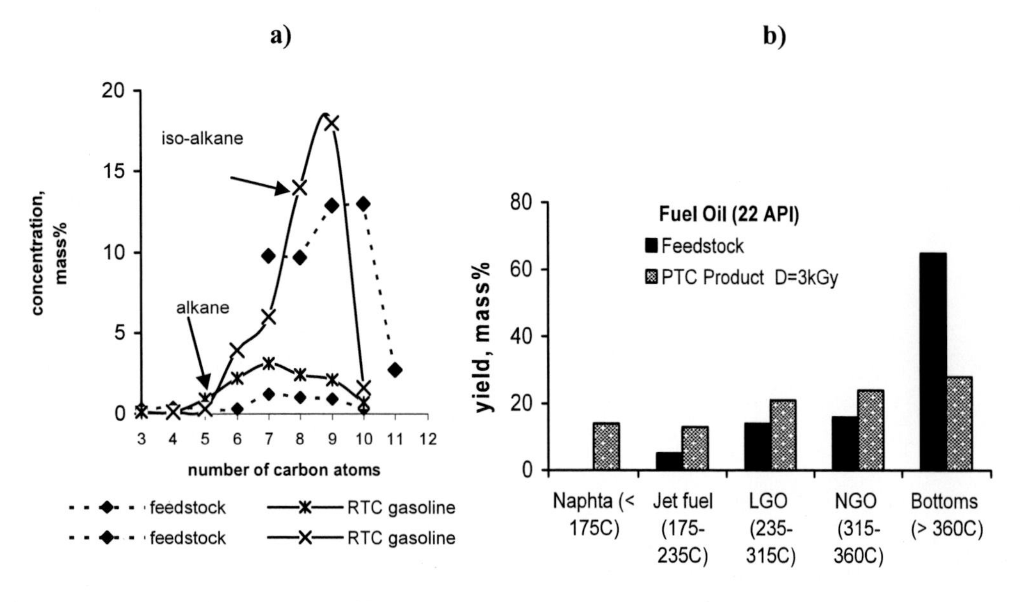

Figure 18. Mass distribution of paraffin molecules in the gasoline fraction after RTC of heavy oil [36].

Table 7. Characteristics of gasoline fractions obtained by RTC from Karazhanbas oil and from fuel oil compared with standard characteristics of gasoline after catalytic reforming

Characteristics	Ordinary Gasoline USA ON 97	European standard ON 96	RTC gasoline	
			Karazhanbas Oil ON 98	Fuel Oil ON 80
Benzol contents, vol. %	2,0	5,0	<0,5	<1,0
Contents of aromatics, vol.%	32,0	NA	7,5	15,8
Olefin contents, vol.%	9,5	NA	17,2	12
Oxygen contents, vol.%	NA	NA	< 2,0	<1,0
Sulfur contents, ppm	339	500	440	430

Due to the effect of radiation-induced isomerization, the gasoline fraction of the product is remarkable for high concentrations of iso-paraffins and high octane numbers. Table 7 and Fig.19 demonstrate high quality of RTC products obtained by the one-run processing of crude oil without application of any catalysts or special additives.

The first two types of the commercial gasoline represented in Fig.19 were obtained after a series of special operations and application of special aromatic additives for the increase in the octane number. On the contrast, gasoline fractions of the liquid product of radiation processing were not subjected to any additional treatment and no additives were used. Gasoline fractions obtained by radiation processing are stable and meet the requirements for commercial gasoline; olefin concentrations are moderate for a cracking process. In the most of the used processing modes, they were lower than those in the straight-run gasoline separated from the feedstock.

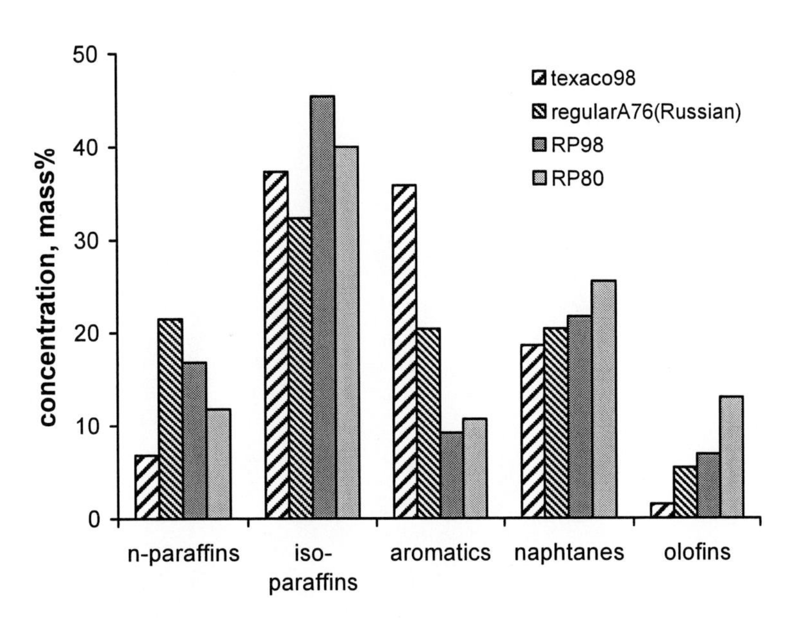

Texaco-98 – commercial Texaco gasoline (octane number -98)

RegularA76 – regular gasoline (Russian production; octane number - 76)

RTC 98 – gasoline separated from the liquid product of one-stage radiation processing of crude oil (octane number - 98)

RTC 80 – gasoline separated from the liquid product of one-stage radiation processing of crude oil in the other irradiation conditions (octane number - 80)

Figure 19. Comparison of hydrocarbon contents of commercial gasoline and gasoline obtained from high-viscous oil (Karazhanbas crude oil) by radiation processing (RP)

Bitumen Radiation Processing

Approaches developed and tested on heavy crude oil and oil residua were applied to processing of still heavier and more viscous feedstock, such as bitumen from different fields of Kazakhstan, Russia and Canada [36, 47].

According to conventional classification, density of heavy crude oil is about 0.95 g/cm^2 while bitumen density is nearly 1.00 g/cm^2. Oil residua of atmospheric or vacuum fractionation fall into the same class of heavy petroleum feedstock. Traditional methods of deep oil processing are mainly purposed for processing of oil feedstock with density up to 0.90 g/cm3. A general characteristic of bitumen is an enormous molecular mass of its components that can be higher than 500 g/ mole, for example, in the case of high concentrations of asphaltenes.

Oil bitumen rocks are important alternative sources of hydrocarbons wide-spread all over the world. Depending on bedding depth they contain from 5 to 45% bitumen organic part that can be used as valuable feedstock for production of engine fuels, lubricants, coke, etc.

However, economic bitumen utilization come across technological difficulties and high cost of their processing.

Bitumen radiation processing was directed to production of synthetic oil with properties that answer demands for oil commodities and ecological requirements.

The modes of radiation-thermal cracking remarkable for high yields of light fractions are illustrated in Figs. 20, 21 on the example of Athabaska bitumen.

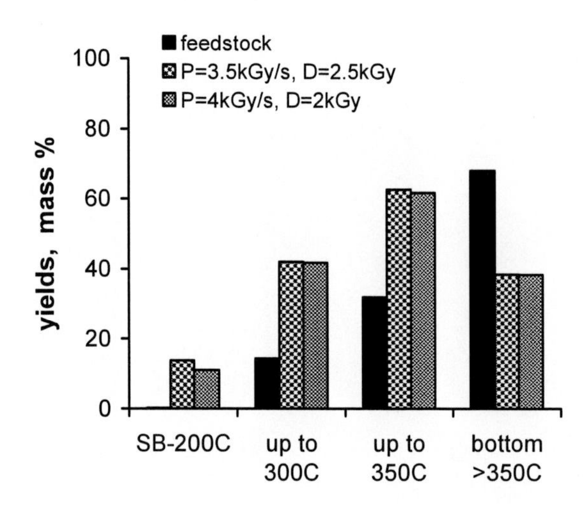

Figure 20. Fractional contents of synthetic oil produced by RTC of Athabaska bitumen at 400^{0}C [36].

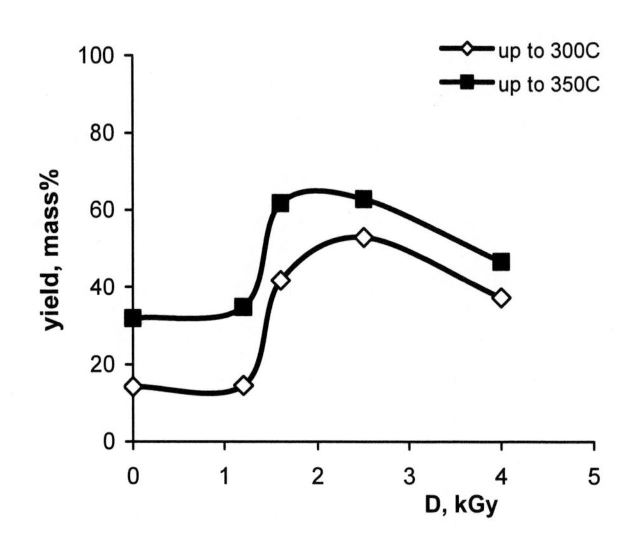

P=3.5- 4.0 kGy/s

Figure 21. Dose dependence of fraction yields after RTC of Athabaska bitumen at 400^{0}C [36].

Hydrocarbon molecular mass distributions in bitumen from Shilikty field, Western Kazakhstan, (density ~ 1.02 g/cm3) is shown in Fig. 22 before and after radiation-thermal processing.

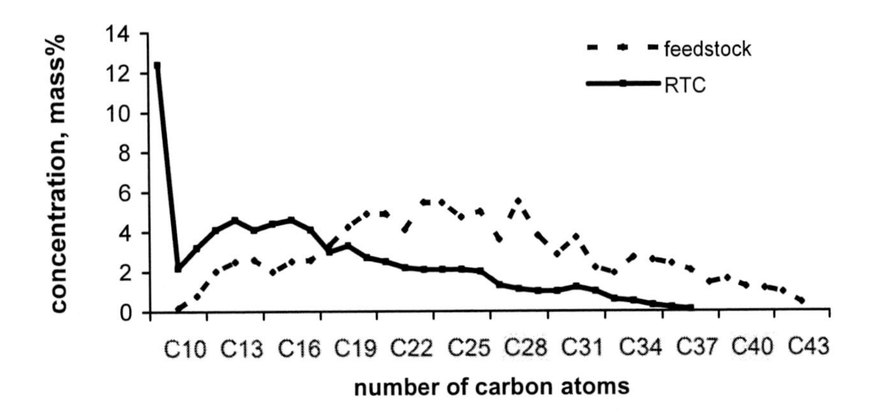

(P=1.5 kGy, D=3.5 kGy, T=410°C)

Figure 22. Fraction distribution in respect to the number of carbon atoms in Shilikty bitumen before and after radiation-thermal processing [27].

Fig.22 shows that the one-stage bitumen radiation processing provides almost twice reduction in the average molecular mass of the liquid fraction that allows production of synthetic oil with the high contents of the gasoline-diesel fraction.

Comparison of light fraction yields obtained by application of different methods for deep bitumen processing are shown in Fig.23 on the example of bitumen from Mortuk field (Western Kazakhstan)

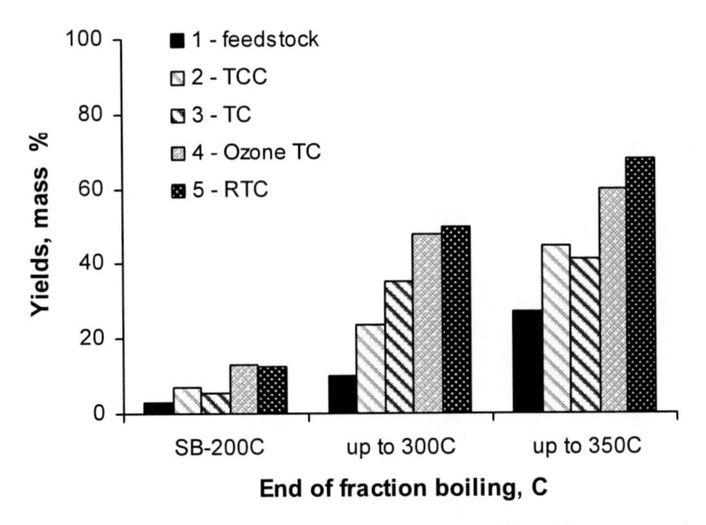

1 -Original feedstock (natural bitumen). Fractions were thermally distillated by a conventional method [34];

2- TCC – thermocatalytic cracking using facility [35] for bitumen processing in the temperature range of T = 400-500°C; clay and sand particles in the rock served as catalysts.

3 – TC – thermal cracking [34];

4 - Ozone TC – preliminary bitumen ozonization by introduction of 34g of O_3 per 1 kg of feedstock and subsequent thermal processing at 350°C [34];

5 - RTC – radiation-thermal cracking [27].

Figure 23. Radiation-thermal cracking of Mortuk bitumen (Kazakhstan).

Fig.23 shows that RTC yields of light fractions are higher than those obtained by other methods, including thermocatalytic cracking and such effective, though rather expensive technology as combination of ozonolysis with subsequent thermal processing at 350°C.

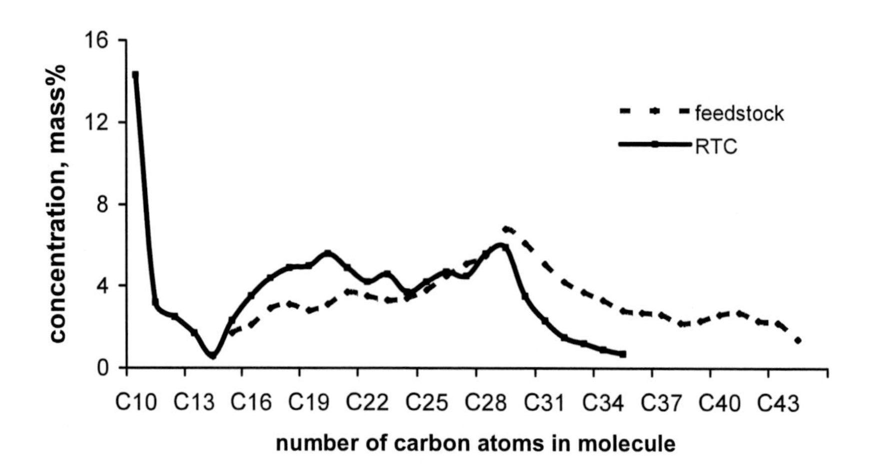

D = 2.5 kGy, T = 370°C

Figure 24. Fraction distribution in respect to the number of carbon atoms in Tyubkaragan bitumen before and after radiation-thermal processing [27].

Fractional contents of synthetic oil produced from Tubkaragan bitumen, Kazakhstan, are shown in Fig.24 in respect to the number of carbon atoms in a molecule. Fig.25 shows the fraction yields of RTC and thermocatalytic cracking in respect to fraction boiling temperatures. Conversion of the heavy residue with the start of boiling at 450°C is twice higher in the case of bitumen radiation processing.

Typically, yields of synthetic oil produced by RTC of different types of bitumen reached 82-86 mass %; yields of the coking residue (feedstock for coke production) were up to 10 mass %; yields of gases (H_2, CH_4, C_2H_4, etc.) were 4-8 mass%. The average density of the synthetic oil was 0.86 g/cm^3.

Fractional contents of synthetic oil obtained by means of gas-liquid chromatography are shown in Table 8. Data on hydrocarbon contents of gasoline fractions from synthetic oil obtained by thermocatalytic (TCC) and radiation-thermal cracking (RTC) of Mortuk and Shilikty bitumen are compared in the Table 9.

The table shows that the content of gasoline produced by RTC essentially differs from that obtained by themocatalytic method.

Concentration of iso-paraffins after TCC is almost twice lower than that of n-paraffins due to the highly pronounced effect of radiation-induced isomerization. In the case of RTC, iso-paraffin concentration is almost twice higher compared with n-paraffin concentration in RTC gasoline and iso-paraffin concentration in the TCC product. Octane numbers are correspondingly higher in the case of RTC gasoline.

Another characteristic feature of the gasoline fraction obtained by RTC of bitumen is a higher concentration of paraffin hydrocarbons both of the linear and the branched structure and, therefore, lower concentrations of aromatics and unsaturated hydrocarbons compared with gasoline produced by thermocatalytic method. Since alkyl substituants in aromatic rings are the main source of aliphatic compounds, predominance of paraffins in hydrocarbon contents of light RTC products is the evidence of more efficient substituent detachment during RTC.

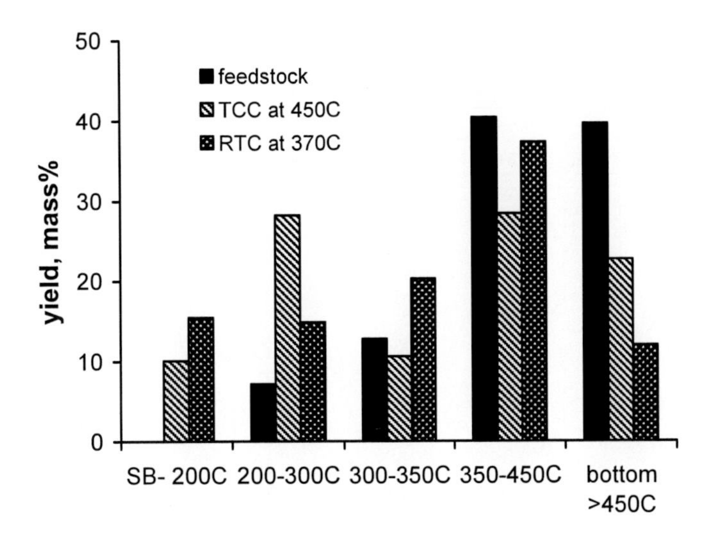

Figure 25. Fractional contents of synthetic oil produced by thermocatalyitic technology (TCC) and by means of radiation processing (RTC) of bitumen from Tyubkaragan field [27].

Table 8. Fractional contents of synthetic oil [36]

Fraction yield, mass % (fraction were separated on the basis of their boiling temperature, ^{0}C)	Field					
	Mortuk (ρ_{20} = 978 kg/m^3)		TyubKaragan (ρ_{20} = 982 kg/m^3)		Shilikty (ρ_{20} = 1,02 kg/m^3)	
	feedstock	RTC	feedstock	RTC	feedstock	RTC
Start of boiling – 195^0 C	2,6	10,6	2,8	14,0	2,0	16,4
195 - 235^{0}C	7,6	12,4	5,8	14,8	3,2	11,8
235 - 300^{0}C	7,9	27,0	8,0	22,4	6,8	19,6
300 – 350^{0}C	10,8	18,0	10,6	14,8	10,0	16,7
350 - 450^{0}C	15,4	20,0	16,4	16,0	14,8	16,8
> 450^0 C	54,6	12,0	58,8	18,0	63,2	18,7

Hydrocarbon contents of the aromatic part of gasoline fraction generally correspond with the original molecular structure of bitumen [48]. The latter is characterized by a number of polycyclic and benzol aromatic hydrocarbons $C_{21} - C_{42}$ with the maximum of molecular mass distribution in the region of C_{31}. There are many cycloalkyl structures that can be either substituents of other hydrocarbons, e.g. cycloalkylbenzols, or bicyclic naphtene structures with cycloalkyl groups divided by long chains. Together with alkyl substituents, many compounds have cycloalkyl and phenylalkyl substituents.

Table 9. Hydrocarbon content of gasoline fraction produced by bitumen processing [36]

Field, type of processing	Hydrocarbon contents, mass %					octane number
	n-paraffins	iso-paraffins	aromatic hydrocarbons	naphtene hydrocarbons	Unsaturated hydrocarbons	
Mortuk, TCC*	29,0	17,0	22,6	13,4	19,0	68
Mortuk, RTC	17,9	33,4	21,8	13,8	10,4	75
Shilikty, TCC*	28,2	17,3	22,5	12,7	19,3	66
Shilikty, RTC	20,6	36,1	15,5	13,0	14,8	72
TyubKaragan, TCC*	30.1	16,4	22	13,5	18,0	68
TubKaragan, RTC	20,0	40,9	11,6	19,1	8,4	76

Most of the components observed in the aromatic part of gasoline fraction after RTC, such as isooctane (2,2,4-trimethylpetane), cycloolefins, etc., can be formed as a result of destruction of substiuent substructure during radiation-thermal processing. However, higher absolute yields of aromatics compared with thermocatalytic process do not exclude appearance of monoaromatic hydrocarbons as a result of direct radiation-induced destruction of polyaromatic compounds.

The general bitumen characteristic is the deficiency of hydrogen that limits yields of light fractions after any type of processing. In bitumen radiation processing, water was used as an additional source of hydrogen [27]. Application of water addition allows considerable increase in the yields of synthetic oil, even up to the level higher than original feedstock mass [6].

Fractional contents of synthetic oil produced from different types of bitumen with water addition are shown in Fig. 26. Addition of water does not considerably change fractional contents of synthetic oil obtained by bitumen radiation-thermal processing. There is a slight increase in the yield of liquid fraction boiling out below 350^0C. The effect is most noticeable for the least heavy of the bitumen processed – Mortuk bitumen ($\rho^{20}{}_4 = 976$ kg/m^3).

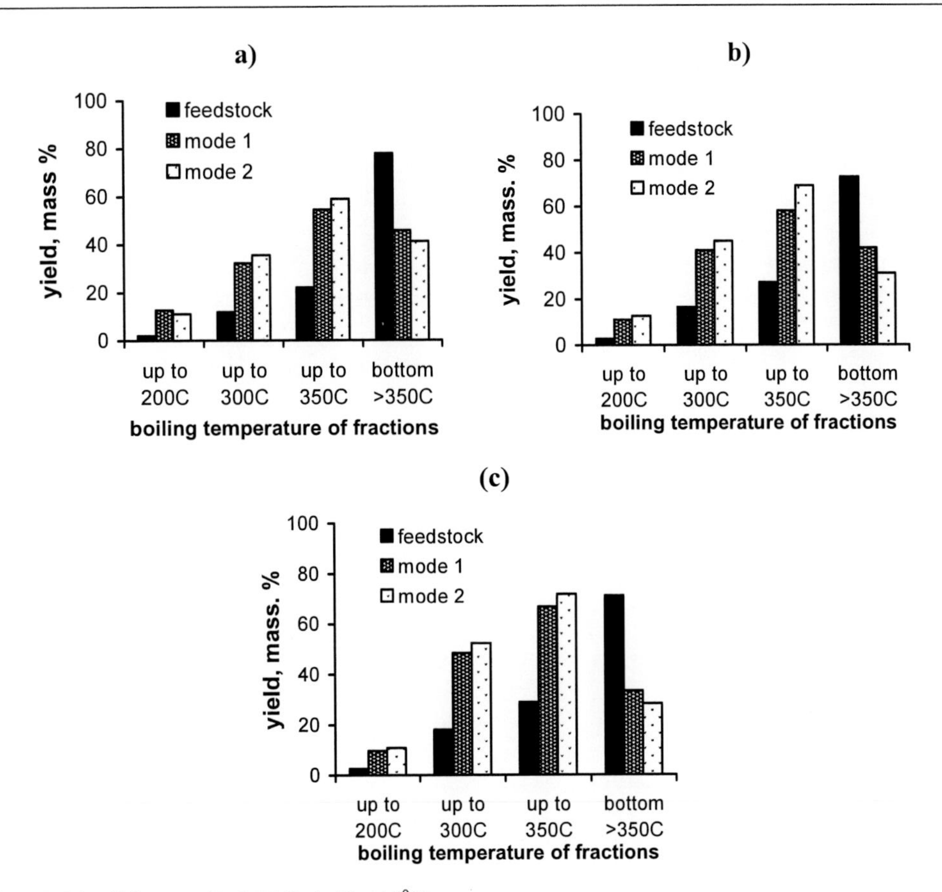

Mode 1: RTC of bitumen (P=1.5 kGy/s, T=410^{0}C);
Mode 2: RTC of bitumen mixed with 5 mass % of water
Fraction boiling out at 200^{0}C - gasoline; fraction boiling out at 300^{0}C - gasoline, kerosene, and light diesel fuel; fraction boiling out at 350^{0}C - gasoline, kerosene, and diesel fuel
Shilikty bitumen ($\rho^{20}{}_4$ = 1020 kg/m^3); b) TyubKaragan bitumen ($\rho^{20}{}_4$ = 982 kg/m^3);
c) Mortuk bitumen ($\rho^{20}{}_4$ = 978 kg/m^3)

Figure 26. Fractional contents of synthetic oil produced by radiation-thermal processing of bitumen from Kazakhstan fields [36].

Considerable increase of the total product mass due to water addition and observed changes in its chemical composition are the evidence of the intense radiation-induced reactions with water participation.

Apparently, most chemically active particles in these reactions are hydrogen atoms H and radicals OH and HO_2 that can initiate a great number of reactions, e.g.

$$RH + H \rightarrow R + H_2,$$
$$HO_2 + RH \rightarrow ROOH + R,$$
$$HCOOH + H \rightarrow H_2 + COOH,$$
$$HCOOH + OH = H_2O + COOH, \text{ etc.}$$

Bitumen feedstock usually contains considerable amount of water (up to 30 mass %) that must be taken into account in the mass and elemental balances of bitumen processing as well as in the forecast of the yields of light fractions.

Typical material and elemental balances of bitumen radiation-thermal processing are shown in Tables 10, 11.

Table 10. Yields of synthetic oil in material balances of products obtained by processing of bitumen from Tyubkaragan field by means of TCC and RTC [36]

Product	Yields, mass%	
	TCC at 450^{0}C	RTC at 370^{0}C
Synthetic Oil	82	92
Coke	6	2
Gases	12	6

Table 11. Typical elemental balance of the overall product produced by radiation-thermal cracking of the organic part of the bitumen rock [36]

Products and their yields, mass%	H	C	S	N+O	C/H
Feedstock-	84.480	9.000	1.700	4.820	9.400
Bitumen + 6% H_2O	79.410	9.120	1.600	9.800	8.700
70% - light gas oil	57.290	7.130	0.635	4.945	8.000
25% - heavy gas oil	20.880	1.780	0.850	1.490	11.700
5% gases	1.240	0.210	0.115	3.435	5.900

As an approach to evaluation of synthetic oil commercial value and development of methods for utilization of heavy coking residue (if it is available in the target product) distributions of vanadium-containing compounds in products of bitumen radiation-thermal processing were determined. Properties of the coking residue after radiation-thermal processing were studied in different irradiation modes [36].

Data on concentrations of vanadium compounds in products of bitumen processing obtained using different experimental methods were compared and analyzed. These experiments were intended to find out how conditions of radiation-thermal cracking affect chemical conversion of vanadium-containing molecules, including such compounds of high commercial value as vanadium porthyrins.

The remarkable feature of the motor fuels (gasoline and diesel fuel) produced by means of radiation-thermal cracking is the absence of metals in these products.

It was shown that concentration of the 4-valent vanadium in the coking residue after RTC varies in the range 800-1200 g/ton, while its content in the original feedstock was 160g/ton. It was found that concentration of the 4-valent and the 5-valent vanadium in feedstock and RTC products are different. These results testify to intense radiation–induced vanadium oxidation and can be compared with the available data on vanadium accumulation in coke during tar thermo-destruction [49]. In this work, oil residua were subjected to thermal destruction at temperatures higher than 400^{0}C in presence of particles (carriers) of different nature. Asphalt-pitch and metal-containing compounds precipitated on the surface of the carriers, transformed into coke and in such a way were removed from the products forming. Together with thermopolycondensation of asphaltenes and pitches, this process was accompanied by formation of up to 70% distillate fractions.

To achieve higher vanadium extraction, solid particles of iron ore concentrate were used as carriers in the coking process. Oxidation of coked carriers was studied at 575^{0}C [49]. Oxidation time was 30-90 minutes; rate of air supply was 2-3 ml/s per 1 g of carrier. In these experiments, maximum of observed vanadium concentration in coke particles was 17 x 10^{-2} mass%, i.e. 1700 g / ton, that is about 10 times higher than that in the original feedstock.

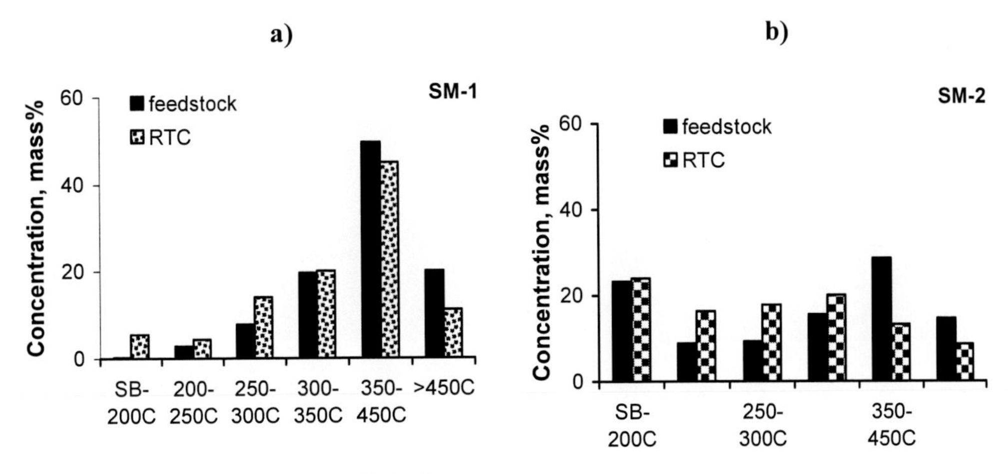

SM-1 – 65 mass% bitumen, 35 mass% fuel oil;
SM-2 - 40 mass% bitumen, 60 mass% fuel oil.

Figure 28. Fractional contents of the products after radiation-thermal processing of the mixtures of bitumen and high-paraffin fuel oil [36].

Thus, without application of any additional technique, radiation bitumen processing provides practically the same efficiency of vanadium transfer to coke as specially developed methods using oxidation of different additives. This result shows that intense bitumen oxidation that accompany RTC causes not only sulfur transformation and its predominant concentration in the heavy residue of processing but also metal transfer to the coking residue.

Interesting results were obtained in the experimental studies on irradiation of bitumen mixed with highly paraffinic residua of paraffinic oil [36]. The total yield of motor fuels (fractions boiling out at the temperatures below 360^{0}C) has made 40-60 mass%.

Radiation facility for heavy oil and bitumen radiation processing with the production rate of 200 kg of feedstock per hour [36] is shown in Fig. 29. The long-time tests have shown its stable and reliable work and high efficiency in processing of heavy oil and the organic part of bitumen.

Generally, radiation-thermal processing of bitumen allows obtaining very high yields of synthetic oil together with the reliable control of its fractional and hydrocarbon contents more efficient compared with conventional methods of bitumen refining.

Figure 29. Experimental facility for radiation-thermal oil processing [36].

Regeneration and Refining of Used Oil Products

Environment contamination by ecologically dangerous components of used oil products and, first of all, by lubricants is the problem of a global scale. Annually about 6 millions tons of oil products are introduced into biosphere all over the world, including more than 50% used lubricants. Their toxic components spread in the atmosphere, water, and soil, fall into food circuits and meals. Used oil products are not regenerated in sufficiently large scales because of the complicacy and low economic efficiency of existing technologies.

Application of radiation methods for used oil products cleaning and reprocessing is one of the most promising approaches to solution of this acute environmental problem [14, 21, 26, 32].

Unlike other methods for heavy feedstock radiation processing, the used oil products do not require deep conversion. Therefore, the most efficient for regeneration of used oil products are conditions combining low-degree hydrocarbon destruction with deep olefin polymerization. In this case, the role of the polymerization process can be explained by comparison with a well-known radiation technology for purification of sewage water in chemical industry. The conditions required for regeneration of used oil products are provided by radiation processing at the temperatures some lower than those necessary for the start of the intense RTC reaction.

The technology was tested for different types of used oil products, such as mixtures of used diesel fuels and detergents from railway stations (MFD 1 and MFD 2); mixtures of used oil products (diesel and heavy gas oil fractions from precipitation tanks) (MFD 3); mixtures of used motor fuels (ML 1); mixtures of used motor oils collected from cars (ML 3, ML 4) and trucks (ML 4*) and mixtures of used transformer oils (ML 2*).

Table 12. Yields of regenerated and refined products after radiation processing [53]

Product of radiation-thermal processing	Yields in mass % for different types of used oil products as feedstock	
	MFD	ML
Refined product (liquid fractions)	94-96	92- 94
Coking residue	3-4	4-5
Gases and other losses	1-2	2-3

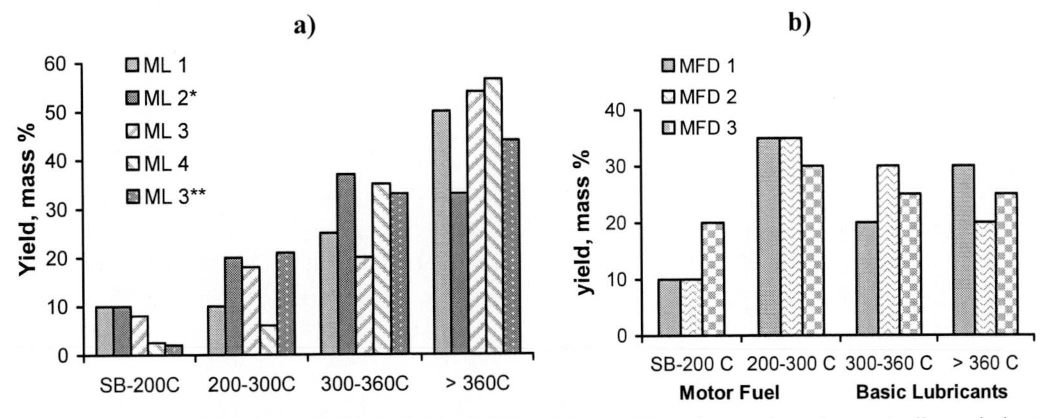

(a) Mixtures of used lubricants (ML), including ML*- mixtures of transformer from the waste disposal plant; ML**- mixtures of engine fuels used in trucks;
(b) mixtures of used diesel fuel and detergents (MFD)

Figure 30. Fractional contents of upgraded mixtures after radiation-thermal processing of used oil products [53].

The processing rate was varied in a wide range up to the values characteristic for the industrial scale (from 1 to 200 kg / hour). The temperature of radiation processing was in the range 340-380^{0}C, the dose rate was 1-2 kGy/s, and the dose varied from 6 to 10 kGy

Yields of basic products of radiation processing, i.e. purified motor fuels (fraction “start of boiling – 360^{0}C”) and basic lubricants (the residue of processing with boiling temperature higher than 360^{0}C) are given in Table 11.

Characteristics of the “lubricant” fraction with boiling temperature higher than 360^{0}C produced by radiation processing of different types of used oil product mixtures are given in Table 13. The product characterized in Table 13 satisfies standard requirements for basic lubricants. Commodity lubricants can be produced by special additions to basic lubricants.

Table 13. Characteristics of basic lubricants after radiation processing (RP) [53]

Characteristics	Types of used oil products			
	MFD	after RP	ML	after RP
Density at 20 C, g/cm3	0.8895	0.8764	0.928	0.782
Viscosity, cSt at 50^{0}C at 100^{0}C	 4.2 1.8	 28.2 6.4	 20.0 37.1	 NA 7.6
Index of viscosity	NA	90	NA	92
Pour Point, ^{0}C	NA	-15	NA	-17
Flash Point, ^{0}C	50	206	140	209
Sulfur, wt.%	NA	0.3	NA	0.2
Acid number	NA	0.015	NA	0.02
Solid particles	0.1%	-	0.14%	
Water	0.6 %-	-	> 10%-	-

Fractional contents of upgraded products obtained after processing of different types of used oil products and their mixtures are shown in Fig.30.

The necessary conditions for a large-scale industrial application of radiation technology for regeneration and refining of used oil products is the availability of the continuous feed sources (a great car fleet, the developed net of railways and other industrial objects), and well organized collection of used oil products.

If decentralized regeneration of lower volumes of used oil products (less than 20000 tons a year) is desirable, application of less expensive gamma-isotope sources (Co-60, Cs-137) could be expedient.

Radiation Processing of Heavy Paraffin Oil Wastes

An important ecological problem of oil industry in oil-exploring countries is the environment pollution by oil extraction wastes and residua of oil refining, the scales of contamination being tremendous. In particular, a number of pills of so-called "lake oil" formed as a result of failures and damages in oil extraction and transportation are spread over Kazakhstan territory. Since efficient methods for oil lake processing are not available, these ponds are annually overfilled with additional oil wastes.

Additional source of environment pollution are asphalt-pitch-paraffin sediments (APPS). Annually up to 150-200 thousands tons of APPS are accumulated all over the world without any utilization. Partially APPS are buried in swamps, mucks, holes, barns, and collectors and partially burned. Anyway it causes contamination by oil wastes and contravenes environmental equilibrium [9].

The technology, that is applied today for paraffin sediments processing, includes the stage of feed preparation (removal of water, light oil fractions and solid impurities), refining by means of traditional oil-chemistry methods using clays and sulfuric acid, propane desulfurization, crystallization in acetone-benzol, etc. In this technology, the product of APPS processing is ceresine with boiling temperature up to 80^{0}C.

The choice of technology for APPS processing is defined by APPS composition and properties that are very different for different fields. Generally, the existing methods for APPS processing are characterized by the low degree of feed conversion and low economic efficiency; they do not solve the environmental problems associated with APPS accumulation.

A promising new approach to oil wastes processing is the application of radiation technologies [10] characterized by flexibility in respect to the type of oil wastes processed. To make radiation technology advantageous, specific features of radiation-induced reactions in this type of feedstock are to be taken into account. This feedstock contains extremely high concentrations of pitches, asphaltenes, heavy paraffins; to a great extent, it is contaminated by different impurities.

Such type of heavy oil feedstock as lake oil is also characteristic for its considerable tend to coking during radiation-thermal processing that should be taken into account in the facility design, in the choice of irradiation conditions and in the rate of the feed flow through the pipe-lines of the given cross-section.

The samples of lake oil and APPS were irradiated in flow and stationary conditions by 2-4 MeV electrons.

Table 14. Physical and chemical properties of the original lake oil and products of its radiation-thermal processing

Sample	Density at 20^0C, g/cm^3	Viscosity, cSt		Flash temperature T, ^{0}C	Solidification temperature T, ^{0}C
		at 50^0C	at 95^0C		
Feedstock	0,85	27,0	-	27	±31
Product 1 Oil fraction (SB-360^0C)	0,83	3,68	1,89	70	-
Product 2 Paraffinic Oil fraction (> 360^0C)	0,86	6,86	3,0	118	-

Table 15. Elemental contents of lake oil and products of its radiation-thermal processing

Sample	Concentration, mass %				
	C	H	N+O	S	C/H
Feedstock	85,90	13,18	0,46	0,45	6,51
Product 1 Oil fraction (SB-360^0C)	85,07	14,02	0,35	0,56	6,07
Product 2 Paraffinic Oil fraction (> 360^0C)	85,33	13,81	0,53	0,33	6,17
Pitchy coking residue	86,43	12,19	0,68	0,70	7,09

Table 16. Fractional contents of product 1 (D= 3 kGy, P=5 kGy/s, T= 380 ^{O}C)

Fractional contents of product	Fraction yield, mass %
Up to C 10 (start of boiling -180 ^{0}C)	7,60
C 10 -C 13 (180-240 ^{0}C)	34,50
C 13 -C 18 (240-320 ^{0}C)	40,11
C 18 -C 22 (320-360 ^{0}C)	17,80

As a result of radiation-thermal processing of lake oil, the following products can be obtained: 40 mass % liquid fractions with boiling temperature up to 360°C); 22% lube fraction with boiling temperature higher than 360°C and solidification temperature T > 23^{0}C; 25.8% pitchy coking residue; and 12.2% gases. Physico-chemical properties of the original lake oil and RTC products are given in Table 15.

The maximum yield of light fractions after lake oil irradiation was 60-65% for irradiation temperatures of 360-390^{0}C, the dose rate of 1 kGy/s and doses of 3-5 kGy (1 Gy = 1 J/s). The fractional contents of products separated from RTC products (D=3kGy; T=380^{0}C) are shown in Table 16.

The process kinetics is complicated with the concomitant polymerization and other intermediate reactions. However, it is predominantly defined by the kinetics of the basic chain RTC reactions that are described in many papers and monographs in details [10, 11].

The most noticeable characteristic feature of APPS radiation-thermal cracking is availability of the intensive competing processes of polymerization of hydrocarbon destruction products. Polymerization limited the maximum yield of light fractions that did not exceed 40%. Unlike the final products of lake oil processing, the products of APPS destruction are unstable and tend to polymerization after radiation processing. As a result of interaction with the reactive polymerizing residue, liquid fractions were entirely absorbed during exposure at the room temperature and the heavy residue got dense and solid. Therefore, light fractions were separated immediately after radiation processing.

The residue of APPS radiation processing can be reprocessed for deeper conversion in the more severe radiation conditions or used as a basic lubricant after its chemical stabilization.

The described above technological approaches based on radiation-thermal cracking of hydrocarbons are remarkable for such unique advantages as

- high efficiency of energy transfer to the feedstock processed;
- ability of radiation to create high concentrations of reactive particles for initiation of various chemical reactions at any temperatures;
- relative easiness of control of radiation-chemical reactions, and
- minimum energy consumption.

Technology of radiation-thermal cracking of oil feedstock allows economically, technologically and environmentally overcoming many acute problems of oil industry. However, chain propagation in radiation-thermal cracking is still thermally activated; therefore, RTC should be conducted at heightened temperatures (350-420^{0}C in the case of heavy oil feedstock). Therefore, further research was directed to achieve radical decrease in the temperature of oil radiation processing.

Complex Radiation-Thermal Treatment and Radiation Ozonolysis of Petroleum Feedstock

A new promising approach to processing of high-viscous and high-sulfuric oil is application of synergetic action of the two types of initiated cracking: radiation-thermal cracking and cracking initiated by ozonolysis [42, 50-53]. This approach is based on the ability of ozonides and sulphoxides to initiate radical chain reactions that can be effectively used for intensified thermal destruction of heavy, high-sulfuric crude oil, bitumen and different types of oil wastes. An additional advantage of such combination is a possibility to utilize ionized ozone-containing air as a by–product of a radiation facility operation, instead of expensive ozone.

It is known that temperature of the ozone-initiated cracking is lower than that of the conventional thermal cracking (TC) by 150-200 K and close to the temperature of radiation-thermal cracking (RTC) [3, 34]. It means that principally the two types of initiated cracking can be combined in a one-stage process.

In a series of experiments [42], oil samples were irradiated by 2 MeV electrons and bremsstrahlung X-rays from a linear electron accelerator ELU-4.

Two basic processing modes were used:

- -preliminary feedstock bubbling by ionized air at the room temperature in the field of X-ray radiation and subsequent radiation-thermal cracking (electron irradiation) at a heightened temperature;
- -simultaneous feedstock bubbling by ozone-containing air and X-ray (or electron) irradiation at the temperature of 20-70^{0}C.

The experiments have shown that effective radiation methods for oil refining have a potential for considerable further improvement if combined with such promising processing as radiation ozonolysis. This combination preserves and amplifies advantages of both types of initiated cracking and makes oil processing more efficient and economic.

Low-Temperature Radiation Processing

The synergetic effect of ionizing radiation and ozonolysis on oil feedstock conversion near room temperature to a very great extent depends both on processing conditions and, especially, on the original contents of the feedstock processed [42, 50-53].

In the temperature range of 20-40^{0}C strong synergetic effects accompanied by high radiation-chemical yields of light fractions characteristic for chain cracking reactions were observed for furnace oil. The fractional contents of two sorts of furnace oil used in the experiments are shown in Table 17. They are characterized by

(a) high concentration (~30 mass %) of heavy aromatic compounds (availability of the fraction boiling out at temperatures higher than 250^{0}C), and

(b) availability of the gasoline fraction boiling out at the temperature below 200^0 C (not less than 10%).

Table 17. Fractional contents of furnace oil [42]

Boiling temperature, ^{0}C	Fraction concentration, mass%	
	Furnace oil from Southern Tatarstan	Furnace oil from Northern Kazakhstan
SB -200	17.6	23.1
200-300	32.6	58.3
300-350	20.0	17.1
> 350	29.8	1.5

Irradiation of this type of feedstock by 2 MeV electrons or by bremsstrahlung X-rays in the condition of continuous bubbling by ionized ozone-containing air lead to considerable increase in the yields of light fractions and profound alterations in composition of the heavy residue.

Changes in fractional contents of furnace oil after different types of processing are shown in Fig.31. The original hydrocarbon composition of the feedstock allowed observation of synergetic effects due to the combined action of radiolysis and radiation ozonolysis.

Fig.31 shows that X-ray irradiation of furnace oil accompanied by continuous feedstock bubbling by ionized air results in the increase of gasoline concentration by 28.4 mass %. The result of combined action of the two factors (X-ray irradiation and continuous ozonolysis) is higher than the sum of the effects of their separate action by 1.7 times, i.e. the synergetic effect makes 170%.

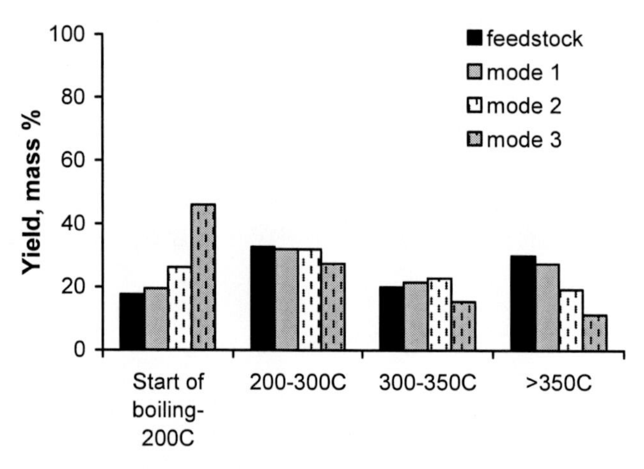

Mode 1: bubbling by ozone-containing ionized air in the accelerator room for 30 minutes at the rate of 20mg/s per 1 kg of feedstock

Mode 2: X-ray irradiation of feedstock preliminarily bubbled with ozonized air during 2 hours; P=0.275 Gy/s; D=2 kGy

Mode 3: X-ray irradiation of the sample processed in mode 2 combined with continuous bubbling by ionized air during 2 hours; P=0.275 Gy/s; D=2 kGy

Figure 31. Fractional contents of Tatarstan furnace oil before and after its processing studied by different methods [42].

In this case, the value of radiation-chemical yield (RCY) of gasoline is 13800 molecules/100eV, that is characteristic for chain cracking reactions. It was suggested that reactions responsible for hydrocarbon cracking initiation are radiation-induced chain reactions of aromatic compounds oxidation accompanied by the attachment of C-C β-bonds to substituents and ozone-initiated radical chain oxidation of saturated hydrocarbons and molecular fragments with subsequent oxidation of peroxides through the possible canals [34]; under the action of ionizing radiation these reactions can proceed with rather high rates near the room temperature in conditions of continuous ozone saturation of the feedstock.

It is interesting to compare the results of low-temperature radiation processing of the two sorts of furnace oil characterized in Table 17. Furnace oil produced in Southern Tatarstan is a product of primary crude oil distillation with higher concentrations of sulfur and heavy aromatic compounds compared to furnace oil produced at the Kazakhstan refinery.

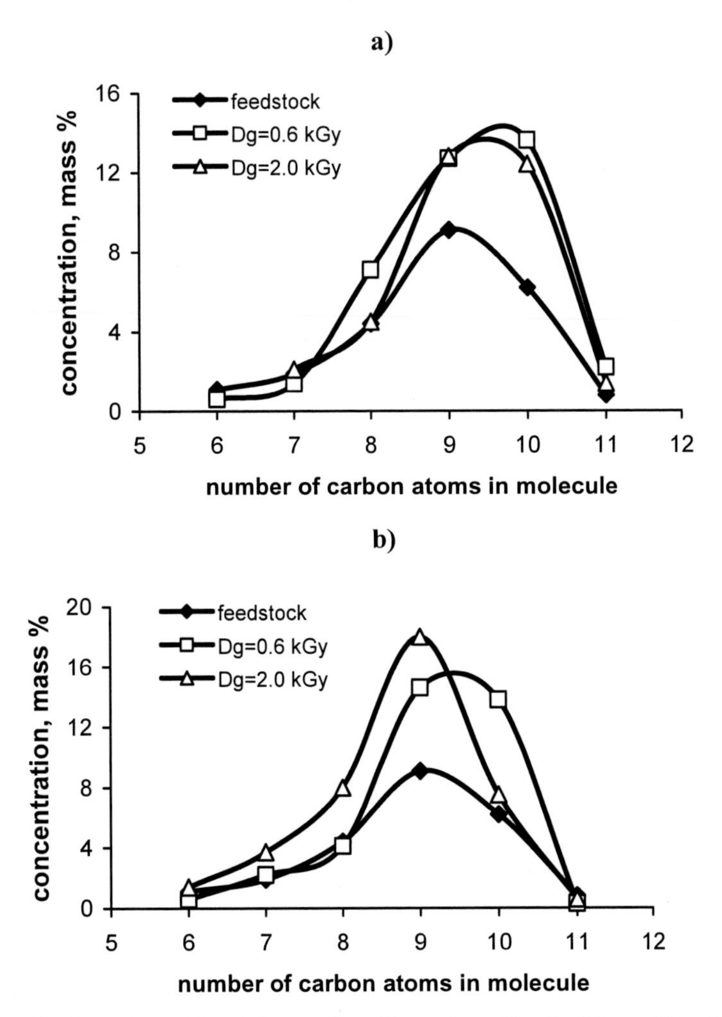

a) rate of bubbling V_1=20 mg/s per 1 kg of feedstock for X-ray doses D_1= 0.6 kGy and D_2= 2.0 kGy;
b) rate of bubbling V_2=40 mg/s per 1 kg of feedstock for the same doses

Figure 32. Molecular mass distribution of isoparaffin concentration in Kazakhsrtan furnace oil after bremsstrahlung X-ray irradiation combined with radiation ozonolysis [42]

Application of gamma-irradiation and radiation ozonolysis to Tatarstan furnace oil (Fig.31) not only increases gasoline concentration but significantly upgrades its quality due to the changes of gasoline hydrocarbon contents. On the example of Kazakhstan furnace fuel subjected to combined X-ray-irradiation and radiation ozonolysis, Fig.32 shows that the rate of feedstock bubbling by ionized air is an essential factor that affects radiation-induced isomerization.

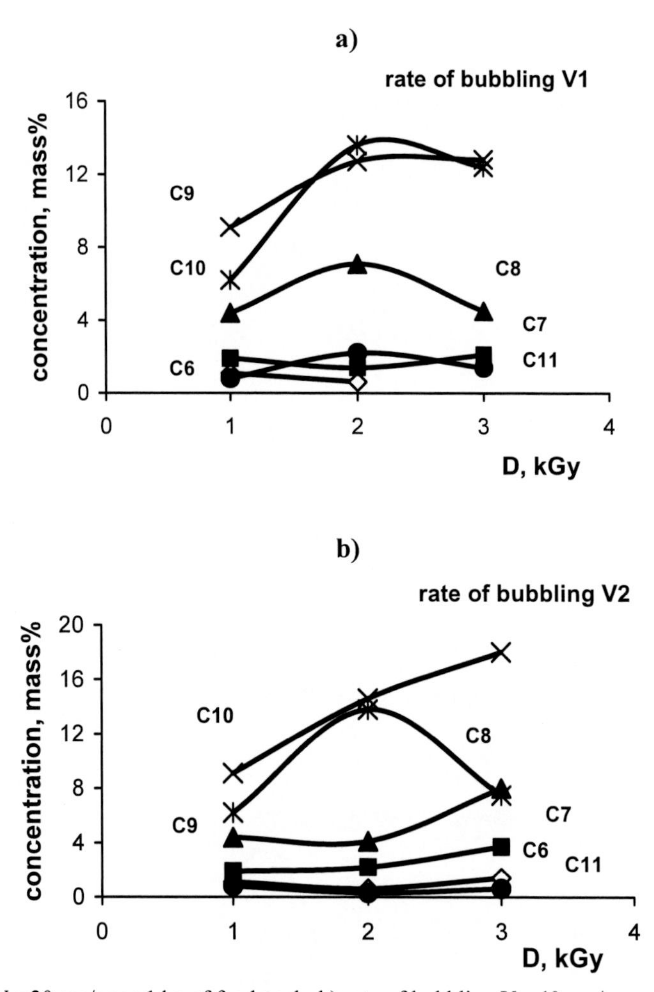

a) rate of bubbling V_1=20 mg/s per 1 kg of feedstock; b) rate of bubbling V_2=40 mg/s per 1 kg of feedstock

Figure 33. Dose dependence of isoparaffin concentrations in Kazakhstan furnace oil after X-ray irradiation combined with radiation ozonolysis [42]

At the lower rate of air bubbling effect of paraffin isomerization comes to saturation at the irradiation dose of 0.6 kGy. In the case of twice higher rate of bubbling isomer concentrations are some greater at the same dose of gamma-irradiation and continue to grow as the dose increases. It indicates that radiation-induced chain oxidation reactions destabilize paraffin structures and facilitate isomerization processes.

Fig.33 shows that concentrations of paraffins C8-C9 reach maximum at the X-ray dose of 2 kGy. Apparently, such alkanes of the medium molecular mass can most easily change their electron structure and have a pronounced tendency to isomerization. Increase in the rate of air bubbling intensifies this process. Feedstock bubbling by non-ionized air has no considerable effect both on product yields and quality and on RTC conditions.

Similar dependencies are shown in Fig.34 for Tatarstan furnace oil, a product with the higher concentration of heavy aromatic compounds. Comparison of Figs. 33 and 34 shows that for the same conditions of radiation processing concentrations of isomers in Tatarstan furnace oil are higher, while the dose corresponding to their maximum concentration (about 2kGy) is nearly the same than that for Kazakhstan furnace oil (about 2 kGy).

Evidently, dealkylation (detachment of alkyl substituemts from aromatic rings) is the very substantial mechanism that affects formation of light fractions. Moreover, alkyl substituents detached from aromatic rings are the main source of potential isomers. Therefore, the higher is concentration of heavy aromatics, the higher is maximum isomer concentration.

Fig.35 shows that the yield of the gasoline fraction is more than 7 mass % higher in the heavier Tatarstan furnace oil. It confirms our conclusion that due to the easier dealkylation the conversion degree is higher after radiation processing of mixtures with higher concentration of heavy aromatics. The higher yield of the gasoline fraction correlates with higher concentration of isomers after radiation processing of the heavier product.

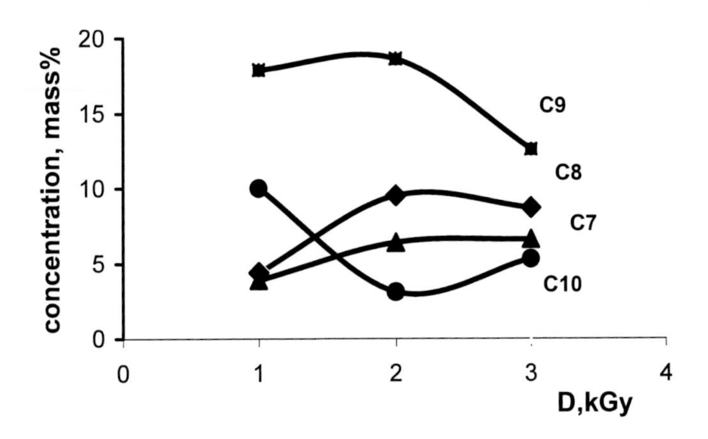

Figure 34. Dose dependence of isoalkane concentrations in Tatarstan furnace oil after gamma-radiation combined with radiation ozonolysis [42].

Figs.36 a,b show that a strong dose dependence is characteristic for concentration of aromatic compounds C9 just as it was observed for paraffins. There are reasons to suggest that this molecular mass corresponds to the fragments of branched aromatic structures that can be most easily detached under combined action of X-rays and radiation ozonolysis. The concentration maximum observed at the dose of about 0.6 kGy can be related to the loss of these substituents by aromatic structures. Decrease in C9 aromatics at higher doses can be associated with their slow decomposition of under irradiation and their elimination in the liquid phase.

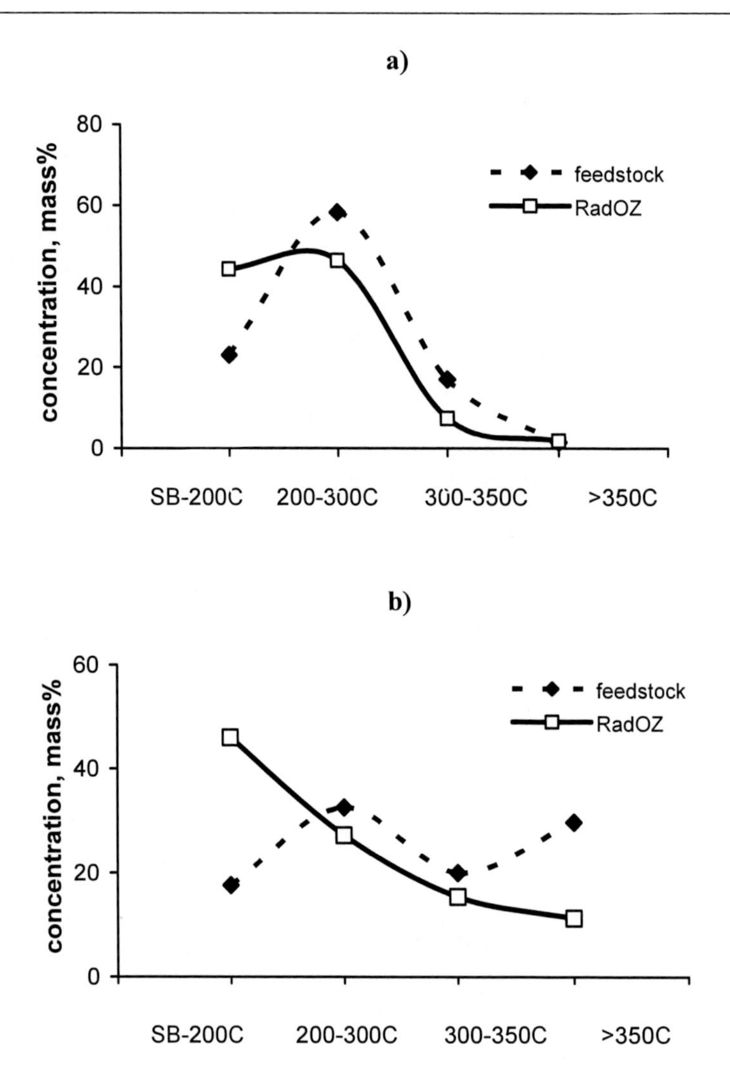

RadOZ - combined X-ray irradiation and radiation ozonolysis

Figure 35. Fractional contents of Kazakhstan (a) and Tatarstan (b) furnace oil after radiation processing [42]

Fif.36b demonstrates that twice increased rate of air bubbling does not lead to considerable changes in dose dependence of the concentrations of aromatic compounds. There are only slight differences in Figs.15 (a) and (b): some higher concentration of C7 and some lower concentration of C9 aromatics. Comparison with Figs. 32-34 shows that increase in the time of radiation processing and increase in the bubbling rate affect rather isomerization processes than contents of the aromatic part of the product.

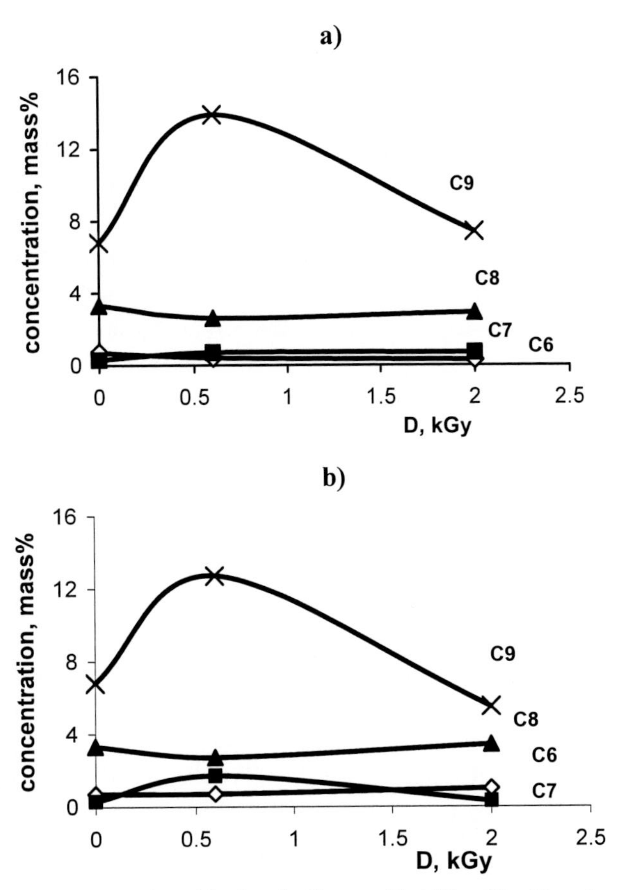

a) rate of bubbling V_1=20 mg/s per 1 kg of feedstock; b) rate of bubbling V_2=40 mg/s per 1 kg of feedstock

Figure 36. Dose dependence of aromatic compounds concentrations in Kazakhstan furnace oil after gamma-radiation combined with radiation ozonolysis [42]

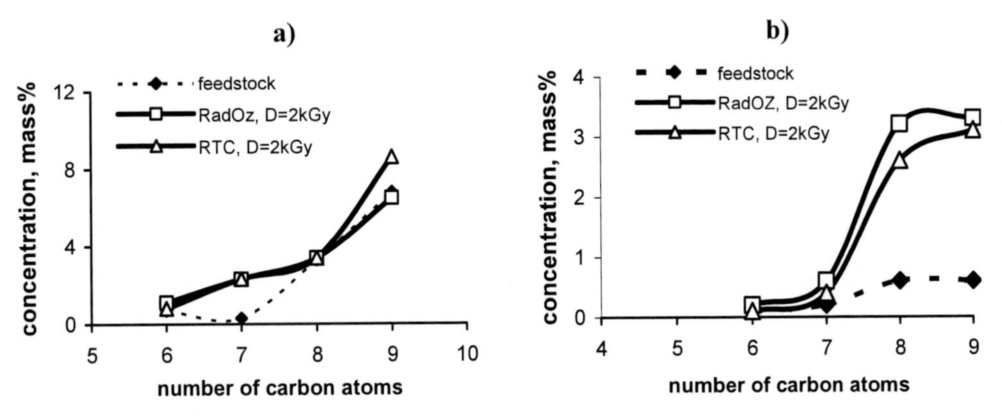

RadOZ - rate of bubbling V=20 mg/s per 1 kg of feedstock for gamma-irradiation dose [42]
D= 2.0 kGy;
RTC – radiation-thermal cracking at 400°C, electron dose D= 2.0 kGy

Figure 37. Molecular mass distribution of aromatic compounds concentration in Kazakhstan (a) and Tatarstan (b) furnace oil after X-ray irradiation combined with radiation ozonolysis.

Fig.37a shows that combined X-ray irradiation and radiation ozonolysis of Kazakhstan furnace oil up to the dose of 2 kGy practically do not change mass distribution of aromatic compounds in the gasoline fraction. However, processing of Tatarstan furnace oil with the higher concentration of heavy aromatics in similar conditions essentially raises concentration of C7-C9 aromatic compounds, even more considerably than high-temperature electron-initiated radiation-thermal cracking. Therefore, in the heavier feedstock relatively light alkyl substituents in aromatic rings can be easier torn away from heavy aromatic structures. This conclusion is important for interpretation of the mechanism of radiation-chemical conversion of hydrocarbon mixtures under combined action of ionizing irradiation and radiation ozonolysis.

Table 18 shows that as a result of radiation processing, enrichment of gasoline fraction with light aromatic compounds and isoalkanes raises gasoline octane number by 15 units.

Similar synergetic effect was observed in such wide-spread and cheap feedstock as mixtures of low-quality diesel and furnace fuel. Processing of these types of feedstock in optimal conditions provides considerable changes in their hydrocarbon contents and improves their quality.

Table 18. Hydrocarbon content of gasoline aromatics and gasoline octane numbers in the product of low-temperature radiation processing of Tatarstan high-sulfuric fuel oil [42]

Aromatic hydrocarbon	feedstock (ON - 60)	Bubbling by ionized air (ON – 64)	Bubbling by ionized air + gamma-irradiation (ON –75)
Benzene	-	0.03	0.19
Toluene	0.23	0.92	1.20
Ethylbenzene	-	-	1.18
n-xylene	-	1.24	0.80
n-xylene	0.86	-	1.42
o -xylene	-	0.11	0.41
1-methyl-2-ethylbenzene		-	1.03
1- methyl -3- ethylbenzene	-	0.94	0.66
1- methyl -4- ethylbenzene	0.3	-	0.67
1,3,5 trimethylbenzene	-	-	0.92
1,2,4 trimethylbenzene	-	-	0.40

Fig.38 shows that increase in the rate of air bubbling leads to the rise in the yields of gasoline and kerosene fractions and, therefore, improves hydrocarbon contents of the feedstock processed.

The results of radiation processing for the mixture of low-quality diesel and furnace fuel are illustrated in Fig.39. The first three dots in the left of the graph show concentration of gasoline (boiling out below 200^0C), kerosene (boiling temperature 200-230^0 C) and light diesel fraction (boiling temperature 230-280^0 C) in the original mixture. The second set of dots relates to measurements in the mixture subjected to bubbling by ionize ozone-containing air without irradiation (the X-ray background was very low). Other measurements were made for the samples subjected to different doses of bremsstrahlung X-rays.

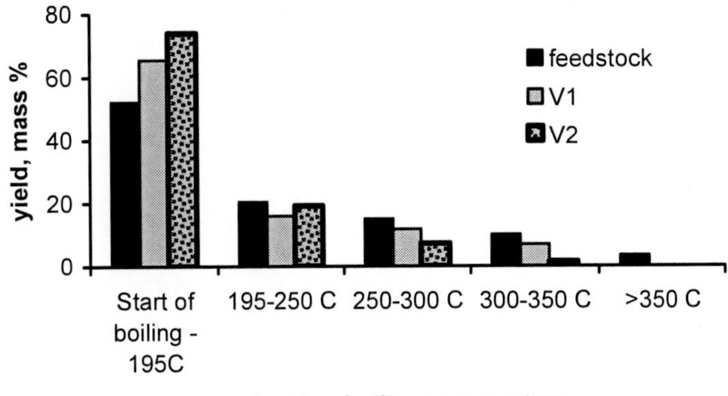

rate of bubbling V_1=20 mg/s per 1 kg of feedstock; V_2=40 mg/s per 1 kg of feedstock

Figure 38. Dependence of distillate fraction yields on the rate of bubbling of diesel and furnace fuel mixture by ionized air in the field of X-ray radiation [42].

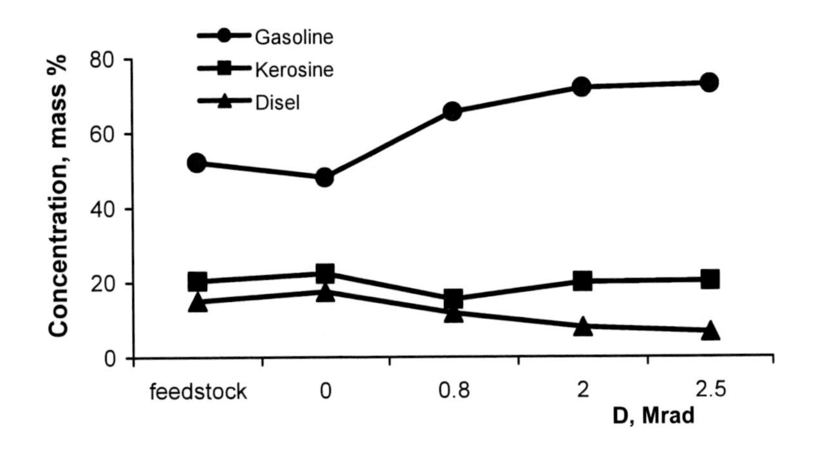

Figure 39. Dose dependence of concentration of light motor fuels in the mixture of low-quality diesel and furnace oil after combined action of X-rays and radiation ozonolysis [42]

Fig. 40 shows that air bubbling without irradiation does not stimulate favorable changes in the feedstock hydrocarbon contents. Due to combination of air bubbling and X-ray irradiation, the processed mixture becomes essentially lighter, the most considerable are changes in concentration of the gasoline fraction. These changes in the hydrocarbon contents of the mixture come to saturation at the dose about 2 MGy.

Quality of the gasoline fraction obtained to a great extent depends on concentration of isomers that can be essentially increased and controlled in this type of processing. Fig.40 gives comparison of isoparaffin molecular mass distributions in the gasoline fraction of the original feedstock, in the product of X-ray irradiation combined with ionized air bubbling, and in the liquid product of radiation-thermal cracking.

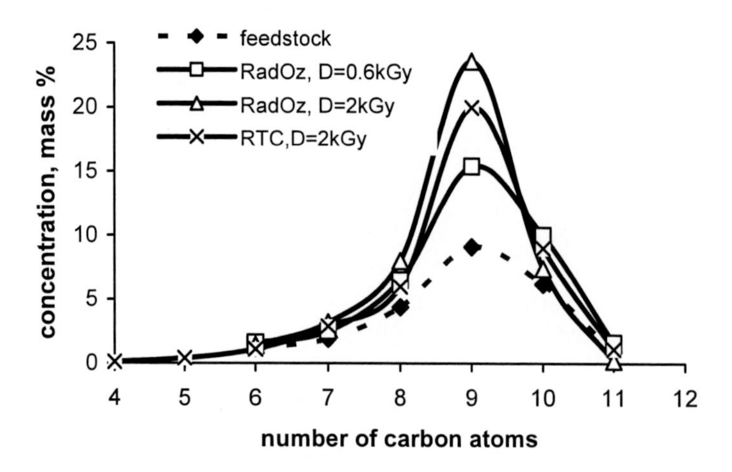

Figure 40. Isoparaffin concentration in the gasoline fraction of the mixed low-quality diesel and furnace fuel before and after its radiation processing [42].

It is obvious that conditions of radiation ozonolysis combined with X-ray irradiation are most favorable for isomerization processes in this type of oil feedstock and, therefore, for the improvement of gasoline quality.

Table 19. Hydrocarbon contents of gasoline produced by radiation processing of mixed low-quality diesel and furnace oil

Type of processing	D, kGy	Hydrocarbon contents, mass%					Octane number
		n-alkanes	iso-alkanes	aromatics	naphtenes	oleffins	
Feedstock	0	48.5	23.7	12.8	13.0	2.0	56
RadOz	0.6	30.7	37.6	17.6	12.8	1.3	72
RadOz	2.0	26.5	43.8	15.1	12.9	1.7	74
RTC at 370^{0}C	2.0	26.6	40.6	11.5	18.4	2.9	70
Bubbling with ionized air and RTC at 370^{0}C	2.0	26.5	39.3	10.3	21.9	2.0	76

RadOZ - combined X-ray irradiation and radiation ozonolysis

Table 19 shows that radiation processing provokes considerable changes in gasoline hydrocarbon contents together with the corresponding increase in gasoline octane numbers.

Hydrocarbon contents of gasoline fractions produced as a result of combined action of radiation ozonolysis and Xray irradiation on low-quality diesel fuels are shown in Figs. 41-43. Concentrations of n-alkanes - linear paraffin molecules (Fig.9), iso-alkanes - branched paraffin molecules (Fig.41) and aromatic hydrocarbons – cyclic hydrocarbons with benzene rings (Fig.42) are plotted as a function of the number of hydrocarbon atoms in a molecule. Concentration ratios for these types of hydrocarbons substantially define the gasoline quality. Generally, Figs. 40-43 demonstrate considerable changes in hydrocarbon contents of the gasoline fraction.

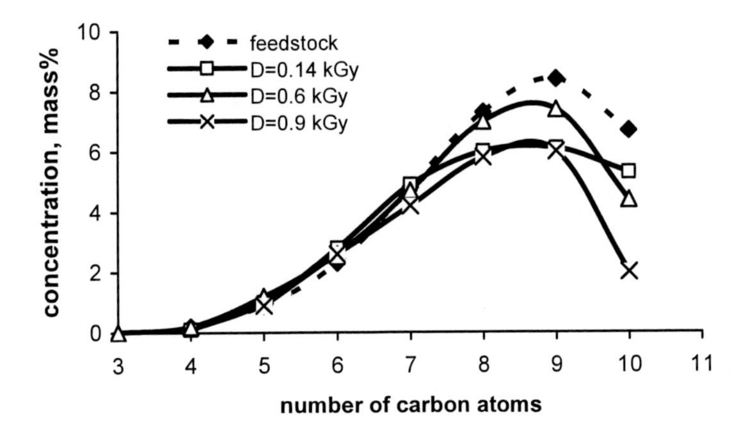

Figure 41. Molecular mass distribution of n-paraffin concentration after radiation processing of low-grade diesel fractions.

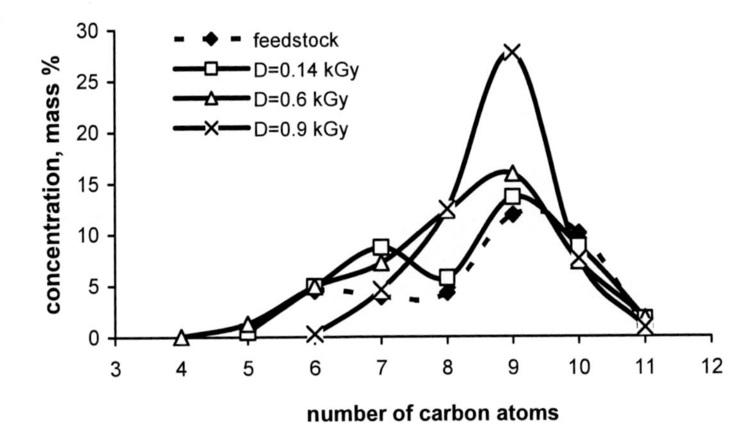

Figure 42. Molecular mass distribution of iso-paraffin concentration after radiation processing of low-grade diesel fractions [42].

Figs.41 and 42 show that iso-alkanes are concentrated in the gasoline fraction of the hydrocarbon mixture with the maximum at 9 C-atoms. Radiation processing does not shift this maximum but makes it considerably higher. Together with the rise in the total iso-alkane concentration, some increase of their average molecular weight is observed that positively affects the gasoline properties. As evident from Fig.43, radiation-induced isomerization is accompanied by decomposition of aromatic structures that raises the concentration maximum of aromatic hydrocarbons and shifts it to lower molecular mass values. These changes also contribute to the increase in the gasoline octane numbers.

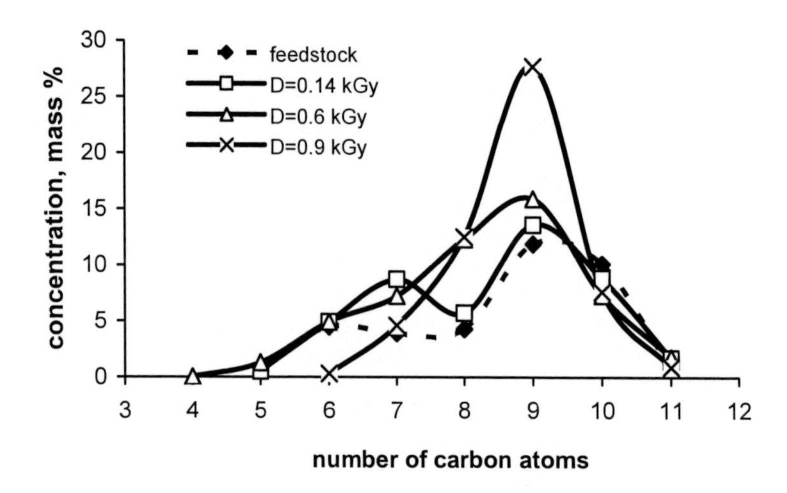

Figure 43. Molecular mass distribution of aromatic hydrocarbons concentration after radiation processing of low-grade diesel fractions [42].

Application of Radiation Ozonolysis for High-Temperature Radiation-Thermal Petroleum Processing

Advantages of oil preliminary bubbling by ionized air before radiation-thermal cracking can be summarized as follows:

- radiation ozonolysis leads to decrease in the irradiation dose and temperature necessary for the maximum yield of liquid RTC products:
- it improves hydrocarbon contents of the liquid RTC product (higher gasoline yields), provokes desulfurization and upgrades gasoline quality.

Some examples of the favorable action of preliminary radiation ozonolysis on contents and quality of RTC products [42, 53] are given below.

Fig.44(a) shows molecular mass distributions of paraffin hydrocarbons in the liquid product of RTC of highly viscous Tatarstan oil for different doses of 2 MeV electron irradiation.

The same dependences in Fig.44(b) are given for the case of preliminary air bubbling of the feedstock in the field of low X-ray background and subsequent RTC in similar conditions.

Comparison shows that molecular mass distributions in Fig.44b are more uniform for all irradiations doses and have maxima shifted to lower values of molecular mass. Therefore, application of preliminary bubbling by ionized air provides higher degree of feedstock conversion and more stable hydrocarbon contents of the gasoline fraction at lower irradiation doses.

Similar changes in contents, quality parameters and optimal irradiation conditions after feedstock bubbling by ionized air were observed for all the types of hydrocarbon mixtures studied, including the heaviest feedstock – bitumen (Table 20).

a)

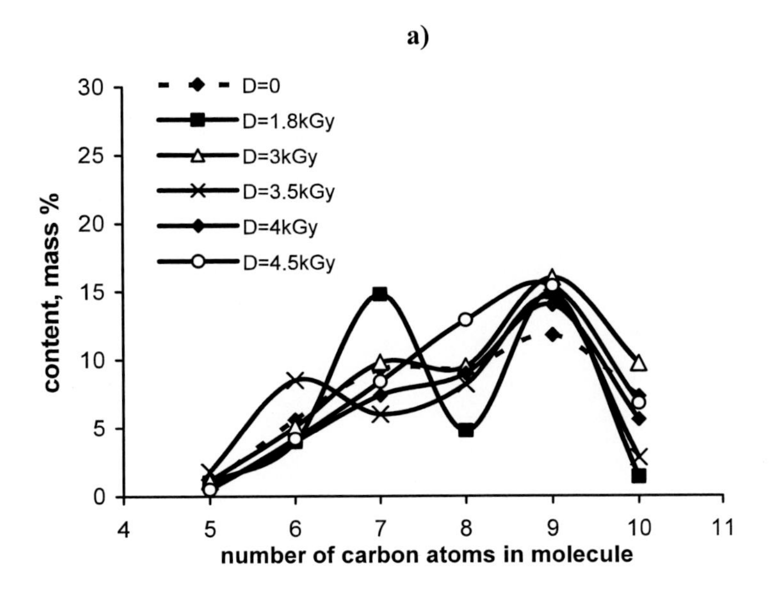

b)

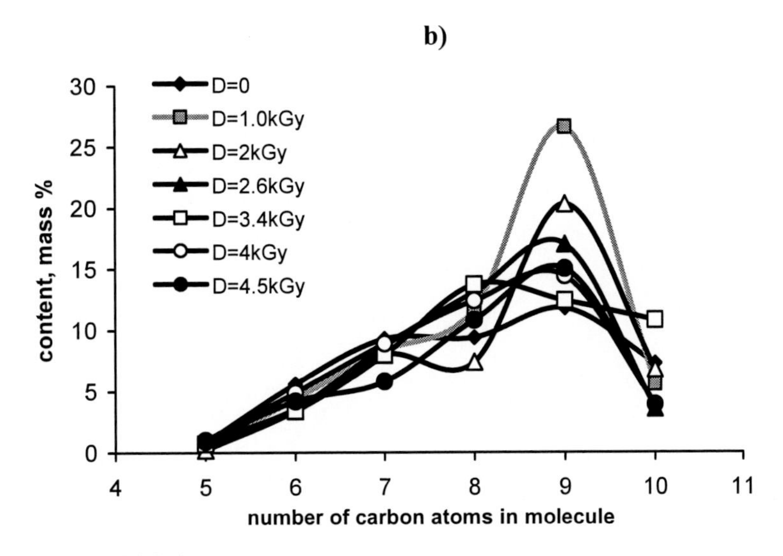

Electron dose rate P=0.5 kGy/s

a) RTC at 370-400^{0}C;

b) preliminary bubbling by ionized air for 30 minutes at the rate of 20 mg/s kg at 20^{0}C and RTC

Figure 44. Molecular mass distribution of paraffin concentrations after radiation-thermal cracking in different conditions

Data of Table 19 show that application of preliminary bubbling by ionized air raises the yield of the liquid RTC product not higher than by 5%. However, it noticeably lowers RTC temperature and the necessary electron dose. The essential effect of radiation ozonolysis is considerably higher product desulfurization. Increase in the temperature of preliminary bubbling affects rather RTC conditions than yields and quality of the product.

Table 20. Bitumen high temperature radiation processing using preliminary radiation ozonolysis (RadOz)

Dose rate P=1.3 kGy/s

Processing conditions	Yield of liquid product, mass %	Desulfurization, %	Processing temperature, ^0C	Dose, kGy
RTC	84	27	410	3.5
RadOz at 20^0C + RTC	88	53	400	3.2
RadOz at 70^0C + RTC	89	52	370	2.8
Bubbling by non-ionized air +RTC	85	28	410	3.4

Mechanism of Combined Action of Low-temperature Ionizing Irradiation and Radiation Ozonolysis

The pronounced effect of the simultaneous action of radiation ozonolysis and ionizing irradiation can be observed only in hydrocarbon mixtures of certain chemical composition. Experiments using the heaviest types of oil feedstock with the highest concentrations of heavy branched aromatics (bitumen and heavy residua of their vacuum distillation) help separation of the roles of ionizing radiation and low-temperature radiation ozonolysis of hydrocarbon mixtures.

The results of bitumen low-temperature radiation processing (LTRP) are represented in Fig. 45 where fraction yields after bitumen bubbling by ionized air are shown. The samples were kept in a container protected against X-ray radiation so that integral irradiation dose did not exceed 0.2 kGy; preliminary heating at 60-80^0C was necessary for lower bitumen viscosity.

Fig.45 demonstrates that on the contrast to lighter oil feedstock, radiation ozonolysis of bitumen without additional action of X-ray radiation provides considerable yields of light fractions.

Subsequent electron irradiation of ozonized samples at 400^0C practically does not change bitumen hydrocarbon contents; increase in temperature of preliminary air bubbling does not affect RTC yields, too. It shows that the “work” on bitumen molecules destruction that requires radiation processing at 400^0C can be done by bitumen ionized air bubbling at nearly room temperature.

Analysis of fractional and hydrocarbon contents of LTRP products have shown that light fractions of the product are always enriched with light aromatic compounds, their concentrations increasing as molecular mass of the original aromatic fractions becomes higher. Therefore, in the heavier feedstock, relatively light alkyl substituents in aromatic rings can be easier torn away from heavy aromatic structures.

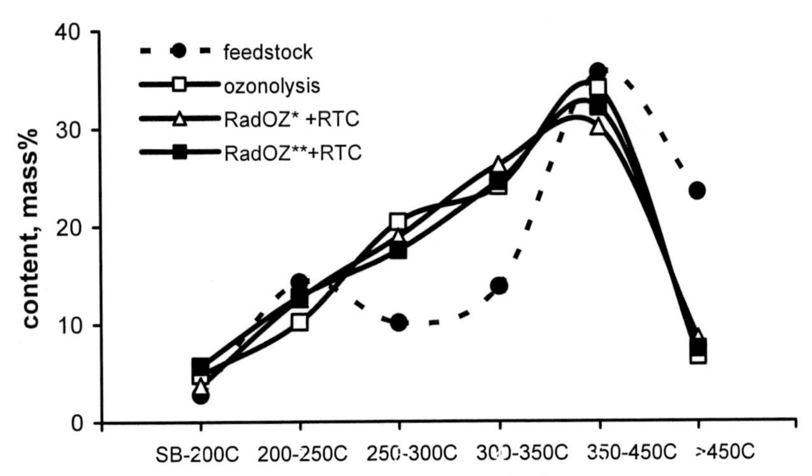

Ozonolysis –bubbling by ionized air, gamma-irradiation dose D=0.2 kGy
RadOZ* - bubbling by ionized air at 50^0C and RTC at 400^0C
RadOZ** - bubbling by ionized air at 100^0C and RTC at 400^0C

Figure 45. Fraction yields after low-temperature and high-temperature bitumen radiation processing

Comparing this observation with the listed above regularities in LTRP of hydrocarbon mixtures, we come to conclusion that the main role of oil mixture saturation by ionized air is destabilization of heavy aromatic structures that facilitates detachment of alkyl substituents. Simultaneous gamma or electron irradiation provides high rates of destructive chain reactions in which gasoline fraction can be presumably considered as a source of light alkyl radicals that appear as a result of gasoline radyolisis. Evidently, radiation-induced cracking of ozonized oil mixtures with destabilized aromatic structure requires much lower concentrations of light radicals compared with high-temperature radiation-thermal cracking. Therefore, activation energy for chain continuation for these types of reactions is much lower than that characteristic for RTC; it strongly depends on the average molecular mass of aromatic fraction and tends to zero in highly branched heavy aromatic structures.

Low-temperature radiation processing based on combination of X-ray or electron irradiation and radiation ozonolysis can be applied practically to any type of oil feedstock mixed with suitable cheap oil products, such as low-grade crude oil, bitumen, wastes of oil extraction and low-quality products of oil primary processing. Optimal concentrations of additives and optimal processing conditions for the specific type of hydrocarbon mixtures can be stated in laboratory conditions.

Oil Demercaptanization and Desulfurization

Because of the trend in supplying heavier and more sulfuric raw material combined with hardening environmental requirements, the problem of clearing oil distillates and oil residue from sulfuric compounds is of great importance. Today it is solved by using complex technique that assumes application of catalysts complicated with hydrogen expense and processing at high temperatures and pressures.

The existing technologies for oil desulfurization can be principally brought to two basic approaches.

In the first approach, sulfuric compounds are reduced to H_2S which is then removed from the processed product and reduced to elemental sulfur (Klaus process). Both of these stages are technologically complicated, comprise multi-step catalyst reactions, and cause environmental problems associated with utilization of processing wastes (acids, alkalies, etc.). In this approach, simple technological processes cannot provide a high desulfurization level.

The other approach presumes oxidation of the sulfuric species in oil with subsequent extraction of the oxidized high-molecular sulfuric compounds. The most prominent problem is to achieve the high level of sulfur oxidation since the second stage of oxidized species extraction is well advanced and does not come across considerable technological difficulties. The higher is oxidation degree of the sulfuric compounds, the more perfect is the appropriate technology.

Both of the two approaches or their combination are used in radiation methods for oil desulfurization depending on sulfur concentration in the original feedstock [15, 16, 20, 26].

Radiation processing can be efficiently applied for demercaptanization and desulfurization of crude oil and petroleum products [20]. The technology is especially simple when initial sulfur concentration does not exceed 2.0 mass%. The proposed method of desulfurization [20] includes two stages. The first stage is radiation processing, and the second one is the standard procedure for extraction of deeply oxidized sulfuric compounds. At the first stage, no desulfurization of the overall product is observed. It results in the strong oxidation of mercaptans and other light sulfuric species to sulphones, sulfur oxides and acids that does away with their chemical aggressiveness and releases their removal. Besides, it causes sulfur redistribution in the overall product leading to the partial desulfurization of its light fractions.

Contact of the product of radiation-thermal processing with reactive ozone-containing air at the output of the radiation-chemical reactor allows combination of radiation-thermal cracking with intense oxidation of sulfur compounds. The process is accompanied by sulfur transfer to high-molecular compounds, such as sulphoxides and sulphones.

The high level of sulfur oxidation is due to the double activation of oxidation processes both by radiation activation of the feedstock and by activation of the atmospheric air.

One of the channels for mercaptan oxidation and transformation can be illustrated by the following scheme:

$$R^{\cdot} + R_1SH \longrightarrow R_1S^{\cdot} + RH$$

$$R_1S^{\cdot} + R_1S^{\cdot}(R_2^{\cdot}) \longrightarrow R_1SSR_1\ (R_1S\,R_2)$$

$$R_1S\,R_2 + O_3 \longrightarrow R_1R_2S^{+}\text{-O-O-O}^{-} \dashrightarrow R_1\overset{\overset{\displaystyle O}{||}}{S}R_2 + O_2$$

sulfoxide

$$R_1\overset{\overset{O}{||}}{S}R_2 + O_3 \longrightarrow R_1R_2\overset{\overset{O}{||}}{S^{+}}\text{-O-O-O}^{-} \longrightarrow R_1\underset{\underset{O}{||}}{\overset{\overset{O}{||}}{S}}R_2 + O_2$$

sulphone

$$R_1S^{\cdot} + H \longrightarrow R_1SH$$

Accumulation of hydrogen hampers mercaptan conversion but the controlled access of ozone-containing air into the reactor allows regulation of the mercaptan conversion.

One of the experimental tests is illustrated in Figs. 46 and 47. High-viscous crude oil was irradiated by 2 MeV electrons in the two modes:

Mode 1: (P=6 kGy/s, D=30 kGy) implies "severe" conditions for deep radiation destruction processing and results in the high yields of motor fuels.

Mode 2: (P=2 kGy/s, D=70 kGy) is "milder", it causes lesser changes in hydrocarbon contents and appears to be more favorable for conversion of sulfuric compounds.

Sulfur concentration in the feedstock was 2 mass%

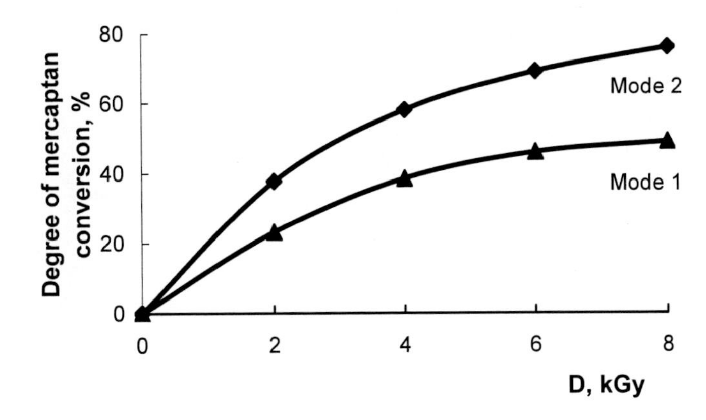

Figure 46. Degree of mercaptan conversion in high-viscous crude oil versus irradiation dose for different irradiation conditions [20].

The degree of 80% mercaptan conversion was reached in Mode 2 and more than 90% of total sulfur concentrated in the heavy liquid fraction with boiling temperature higher than 350^{0}C. Similar mercaptan conversion was observed after radiation processing of lake oil, a waste of crude oil extraction with nearly the same original sulfur concentration.

Deep conversion of the oil feedstock in the described modes is accompanied by considerable desulfurization of light fractions as a result of total sulfur redistribution in the overall processed product.

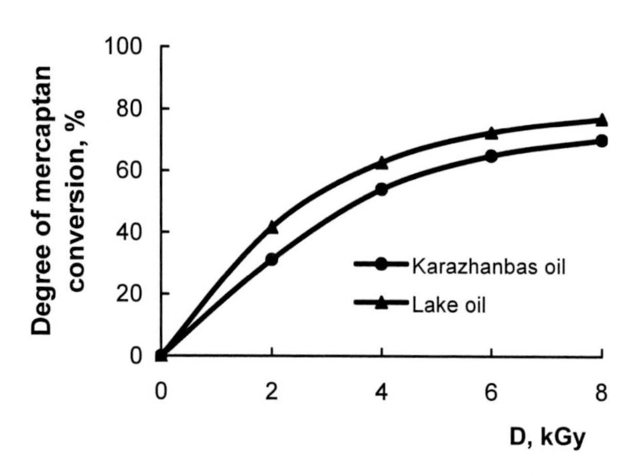

Figure 47. Degree of mercaptan conversion in mode 2 versus irradiation dose for different types of feedstock [20]

The sulfur content in the gasoline fraction of fuel oil decreased by 40 times compared ith its concentration in the original feedstock and made 430 ppm without additional extraction of sulfuric compounds.

After radiation-thermal cracking of highly viscous crude oil from Karazhanbas field, Kazakhstan, sulfur concentration in gasoline did not exceed 440 ppm that is close to European standards for gasoline.

If deep oil destruction is undesirable and the purpose of oil processing is desulfurization, appearance of by-products (gases, cocking rest) could be considered as product yield losses that should be reduced to minimum.

In this case, other conditions of radiation processing can be applied for deep oxidation of sulfuric compounds.

When the irradiation dose rate or the processing temperature are below the ascertained threshold values, the rate of propagation of the cracking chain reaction becomes comparable with the recombination rate of hydrocarbon radicals generated by radiation.

In these conditions, the intensity of hydrocarbon decomposition drops down sharply while the intensity of radiation-induced oxidation remains high. These conditions are optimal for oil desulfurization with the minimum product yield losses (1-5%).

As a result of radiation-induced conversion sulfur concentrates in heavy oil fractions in the forms of high-molecular oxidized compounds (sulphones, sulphoxides, sulphonic acids) that can be relatively easily extracted by well-known methods using appropriate solvents (water solutions of sulfuric acid, acetone, ethylene glykol, monoethyle alcohol).

The extracted high-molecular sulfuric compounds are valuable commodity products. They are applied in mining and metallurgy as extraction agents in the processes of extraction and separation of radioactive and rare metals (uranium, thorium, zirconium, hafnium, niobium, tantalum, rare-earth elements, tellurium, rhenium, gold, palladium, etc.)

It was predicted that the wide application of oil sulphoxides will have a great impact to ore enrichment with rare and noble metals. Oil sulphoxides appear to be much more efficient than tributylphosphate and traditional extraction agents. The oil sulphoxides are used in hydrometallurgy being the best extraction agents due to high efficiency and selectivity in extraction of gold, palladium and silver. In oil refining, they can be applied as selective solvents for paraffin hydrocarbon purification from unsaturated and organic sulfuric compounds.

Radiation processing of high-sulfuric oil with sulfur contents of 3 mass% and higher becomes more difficult when feedstock contains high concentrations of dissolved hydrogen sulphide together with mercaptans.

One-stage high-temperature radiation processing of such feedstock can provoke oxidation of hydrogen sulphide and its transformation into mercaptans. Some increase in sulfur concentration in the light RTC products can result from disintegration of sulfur-containing hetero-aromatic cycles of pitches and asphaltenes.

One of the most essential problems of bitumen, heavy crude oil, and oil wastes processing is reduction of harmful sulfuric compounds and total sulfur contents in the refined products. Combination of radiation-thermal treatment with other types of processing is a promising way to reach this goal. Due to synergetic action of ionizing irradiation and reactive ionized air, high initial concentration of sulfuric compounds in heavy oil feedstock and bitumen distillate fractions can be reduced by several times. At the same time cracking temperature can be considerably lowered that improves technological conditions of oil processing.

Preliminary bubbling of high-sulfuric oil with ionized air produced as a result of accelerator operation was used in the work [26]. At the next stage, oil samples were irradiated in a flow by 2 MeV electrons from a linear electron accelerator.

This type of processing was studied using the samples of highly sulfuric crude oil from the fields of Southern Tatarstan and products of its primary distillation: light oil distillate (gasoline-diesel fraction C_4–C_{22}); furnace oil (broad gas oil fraction fraction C_4–C_{30}); fuel oil (heavy residue C_{20}–C_{42} of oil primary distillation). The samples of crude oil and oil products considerably differed by their density and concentrations of sulfur and heavy aromatics (Table 21).

Table 21. High-sulfuric feedstock characterization [42]

Type of sample	**Fractional contents, mass %**		**Con-centration of gasoline fraction, mass%**	**Density ρ, g/ cm^3**	**Sulfur concentration**	
	Number of carbon atoms	**Boiling temperature, ^{0}C**			**Total sulfur, mass %**	**Mercaptan sulfur, mass %**
Feedstock	Up to C_{40}	below 550	12	0,923	3,90	NA
Fraction boiling out below 300^{0}C	$C_4 - C_{22}$	34 - 350	23	0,785	1,59	0,045
	$C_4 - C_{30}$	38 - 450	17,6	0,904	3,71	0,054
Furnace oil	$C_{22} - C_{42}$	350 - 570	-	1,002	5,4	-
Heavy residua						

The feedstock contact with ozone-containing air provokes conversion of practically all the components of crude oil or oil products; they participate in reactions of electrophilic O_3 addition or ozone-induce radical oxidation by molecular oxygen. Rates of chemical reactions that can simultaneously proceed during low-temperature ozonolysis vary in a wide range.

Sulphide oxidation to sulphoxides at 20^0C is noticeable for the high reaction rate of 1500-1900 liters/mole·s, it proceeds according to the scheme:

$$\begin{matrix} R_1 \\ & S \\ R_2 \end{matrix} \xrightarrow{+O_3} \begin{matrix} R_1 \\ & C{=}O \\ R_1 \end{matrix} \xrightarrow{+O_3} \begin{matrix} R_1 & & O \\ & S & \\ R_2 & & O \end{matrix}$$

Ozonization destroys thiophene cycles, as a result a part of sulfur eliminates from the liquid phase and another part oxidizes to appropriate sulphoxide and sulphone groups without destruction of the carbon skeleton.

When combined with X-ray irradiation, processing of oil feedstock by ionized air causes increase in concentration of the gasoline fraction by 1.6-2.0 times and corresponding decrease in concentration of mercaptane sulfur by 2.5-4.2 times.

Decrease in total sulfur and its predominant concentration in high-molecular compounds improve crude oil quality and prepare it for the more efficient high-temperature processing.

The next stage of processing was radiation-thermal cracking of ozonized oil. Radiation processing of high-viscous oil with heightened concentrations of pitches and asphaltenes leads to decomposition of polyaromatic structures, including a greater part of sulfur compounds. This process is accompanied by sulfur release and its transfer to light fractions, partially to the gaseous phase.

In the first turn, it can be related to sulfur contained in aliphatic chains serving as links between structural blocks of pitches and asphaltenes. Detachment of alkyl substituents with double C=C bonds from aromatic rings gives rise to concentration of unsaturated hydrocarbons in light fractions of the liquid RTC product (Fig.48).

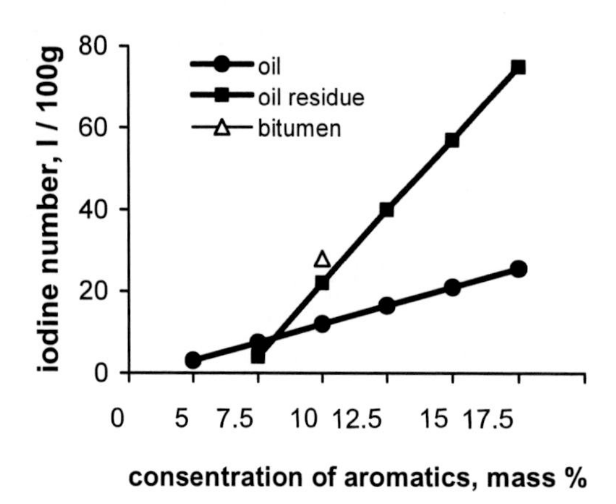

Figure 48. Dependence of iodine number on concentration of aromatic hydrocarbons in gasoline fraction of the liquid product (LP) obtained by RTC of high-sulfuric Tatarstan oil [42].

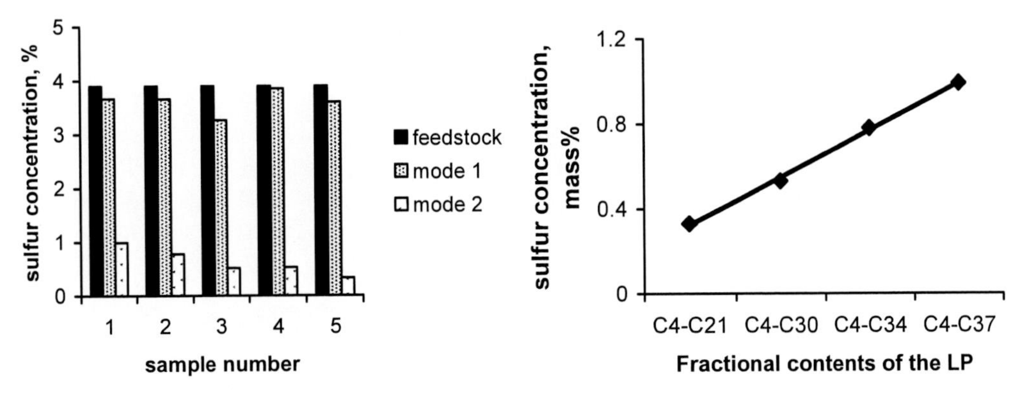

Fractional contents of samples:
1 – C_4-C_{21}; 2,3 - C_4-C_{30}; 4 – C_4-C_{34}; 5 – C_4-C_{37}

Figure 49. Concentration of total sulfur in the liquid product after RTC of high-sulfur Tatarstan oil with bubbling by ionized air in the field of X-ray radiation (mode 1) and after RTC of ozonized oil (mode 2) [42].

Data on alterations of sulfur contents in the liquid product of low-temperature (20^{0}C) and high-temperature (400^{0}C) radiation processing of high-sulfuric Tatarstan oil are summarized in Fig.49. In these experiments, oil was bubbled by ionized air during 20-30 minutes at the rate of 20 mg/s per 1 kg of feedstock in the field of X-ray radiation (P=0.275 Gy/s). Sample 3 was bubbled at the heightened temperature of 60^{0}C.

Deep destructive processing of crude oil with initial sulfur concentrations up to 4 mass % by radiation-thermal method provides liquid RTC products with sulfur concentrations in the range of 0.5-1.2 mass %, that is 2-4 times lower than the sulfur content in the corresponding product of thermal cracking.

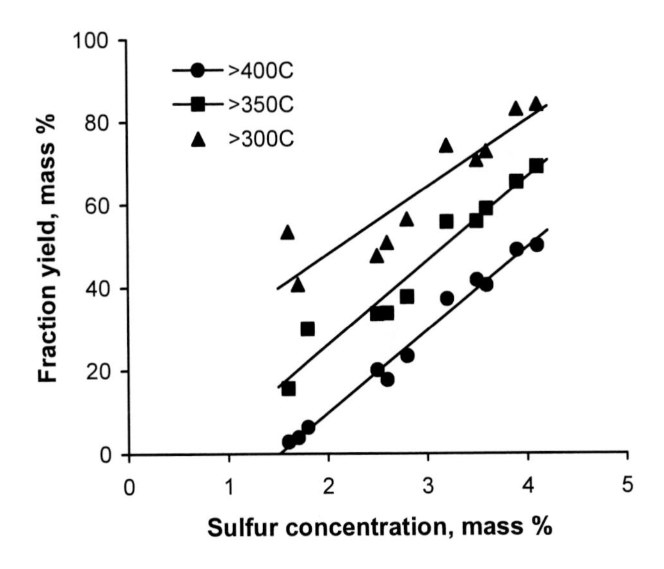

Figure 50. Total sulfur concentration in heavy residue (T_{boil} >300, 350 and 400^0 C) after RTC of high-sulfuric Tatarstan oil versus residue concentration in the overall liquid RTC product (data of gas-liquid chromatography).

RTC of high-sulfuric oil carries away in average 25 mass % of sulfur into gaseous phase; in the case of preliminary oil processing by ozone-containing air the share of sulfur converted to gases increases up to 60 mass %. The coking residue of high-sulfur heavy oil radiation processing makes 10-15 mass %, sulfur concentration in the residue is usually 5-6 mass%.

Fig.50 shows distribution of total sulfur in the heavy fractions of the liquid RTC product, viz in the heavy residua with boiling temperature higher than 300^0 C, higher than 350^0 C, and higher than 400^0 C.

The heavier is the residue and the higher is its concentration in the overall liquid product after radiation processing the higher is sulfur concentration in the residue.

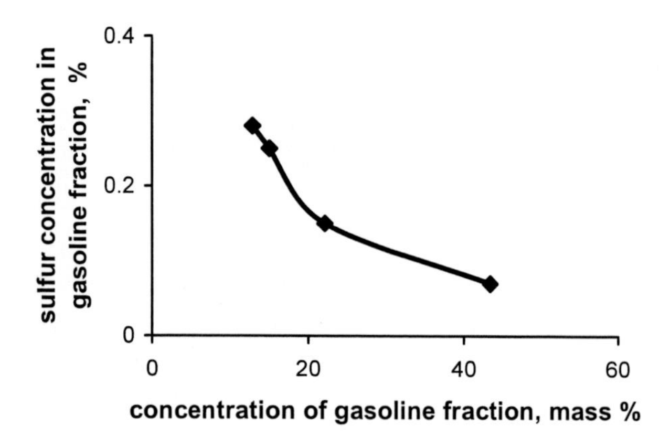

Figure 51. Total sulfur concentration in gasoline fraction versus concentration of gasoline fraction after furnace oil processing in different conditions

The pronounced synergetic effect of radiolysis and ozonolysis enhanced by X-ray irradiation was detected when furnace oil (mixture of light fractions and heavy gas oil with initial gasoline concentration of 17.6 mass %) was processed at the room temperature. The observed low-temperature cracking becomes apparent, first of all, in the considerable increase in the yield of gasoline fraction. Total sulfur concentration in the overall product of furnace oil processing decreases by 23 mass % compared with the unprocessed material. Increase in the yields of gasoline fraction is accompanied by deeper desulfurization of the gasoline fraction (Fig.51).

Generally, processing of high-sulfuric crude oil and petroleum products by ionized ozone-containing air provides deep oxidation of sulfur compounds and improves fractional contents of the products.

The low-temperature cracking initiated by the combined action of ionized radiation and radiation ozonolysis is not only an energy-saving process; in the case of electron accelerator application it uses only by-products of its operation (gamma-background and ionized air). However, the reaction rates of this process are considerably lower compared with radiation-thermal cracking at heightened temperatures, even if special requirements for the feedstock hydrocarbon composition are satisfied.

The search for more universal methods of high-rate oil radiation processing at lowered temperatures has resulted in development of PetroBeam^{TM} technology.

Petrobeam™ Process

PetroBeam™ process [55] is the proprietary technology developed in PetroBeam, Inc., U.S.A.

On the contrast to RTC, the new technology provides the same rates of upgrading and deep processing of heavy oil and bitumen at nearly room temperature and nearly atmospheric pressure. It is a technology of the highest energy saving.

PetroBeam™ process is based on experimental and theoretical analysis of the dependence of the rates and mechanisms of self-sustaining radiation-induced cracking reactions in hydrocarbons on temperature, dose rate of ionizing irradiation and structural state of the feedstock.

Thermally Activated Diffusion of Chain Carriers to Electron-Excited Molecules

According to the classic theory of radiation-thermal cracking [3] the initial rate of hydrocarbon cracking reaction by the radical mechanism under ionizing irradiation is defined by the equation:

$$W = K_0 [R] \exp(-\frac{E}{kT}) \tag{22}$$

where E is activation energy for chain propagation.

It is supposed that dynamically equilibrium radical concentration [R] is formed during a period of time much shorter than that characteristic for cracking reaction and defined by the second-order recombination kinetics:

$$G_R P = K_r [R]^2, \tag{23}$$

where K_r is recombination rate constant; G_R is radiation-chemical yield of radicals that initiate cracking; P is radiation dose rate.

Thus

$$W = K_0 \left(\frac{G_r}{K_r} \right)^{1/2} \exp\left(-\frac{E}{kT} \right) P^{1/2}. \tag{24}$$

The following reaction rates constants that enter into equation (24) are characteristic for hydrocarbons: G_r= 3-5 molecules/100 eV; $E \approx 80$ kJ/mol; $K_0 \approx 4 \cdot 10^{-10}$ cm^3/molecules · s ($\approx 10^{12}$s^{-1}); $K_r \approx 3 \cdot 10^{-13}$ cm^3/molecules · s ($\approx 10^9$s^{-1}).

Formula (24) predicts dependence $W \sim \sqrt{P}$; it comes from the theory that assumes radiation initiation of the chain cracking reaction but does not take into account effect of ionizing radiation on chain propagation. However, analysis of the available experimental data on radiation-thermal cracking of hydrocarbons indicates to a much stronger dependence $W(P)$.

A hydrocarbon molecule can react with a light alkyl radical if the molecule was electronically excited under ionizing irradiation up to the level necessary for the reaction proceeding . In this case, concentration of thermally excited molecules $\exp\left(-\frac{E}{kT}\right)$ in formula (24) should be increased by addition of concentration c^* of the molecules that have got electron excitation up to the reaction level. It should be also taken into account that the probability of the radical contact with an excited molecule proportional to the number of the radical diffusion jumps. Therefore, the energy for chain continuation can be written as a sum

$$E = E_0 + \Delta E\,, \tag{25}$$

where ΔE is the activation energy for light radical diffusion in a hydrocarbon mixture.

Then formula (24) should be changed to

$$W = K_0\left(\frac{G_r}{K_r}\right)^{1/2}\left[\exp\left(-\frac{E_0+\Delta E}{kT}\right)+\frac{K_1}{K_0}c^*\exp\left(-\frac{\Delta E}{kT}\right)\right]P^{1/2}, \tag{26}$$

where constant K_1 includes the frequency and the entropy factors of diffusion.

As an excited molecule looses its energy according to the first order kinetics, concentration c^* should be proportional to the dose rate, i.e.

$$c^* = \frac{G^*}{K^*}P\,, \tag{27}$$

where K^* is the rate of energy losses by the excited electron states.

Radiation-chemical yields G^* of the excited molecules are about 2-3 molecules / 100 eV [3].

Taking into account formula (28), equation (27) can be rewritten as

$$W\cdot\exp\left(\frac{E}{kT}\right)P^{-1/2} = A + B(T)P\,, \tag{28}$$

$$A = K_0\left(\frac{G_r}{K_r}\right)^{1/2},\quad B(T) = K_0\left(\frac{G_r}{K_r}\right)^{1/2}\frac{G^*}{K^*}\exp\left(\frac{E_0}{kT}\right). \tag{29}$$

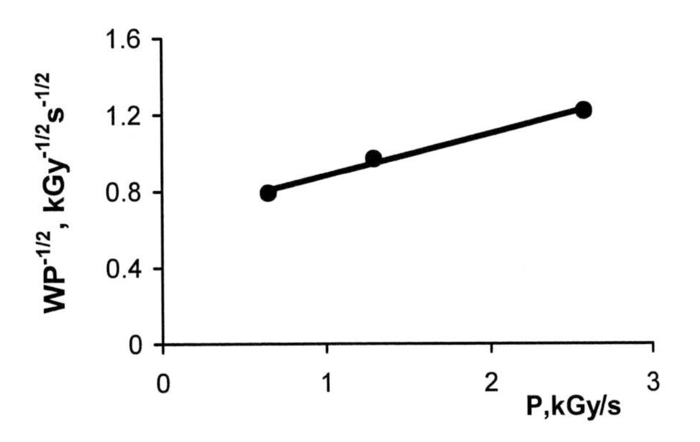

Figure 52. Dependence of the rate of RTC of pentadecane at 500C on dose rate in coordinates $W P^{-1/2}$ - P (recalculated from [8]).

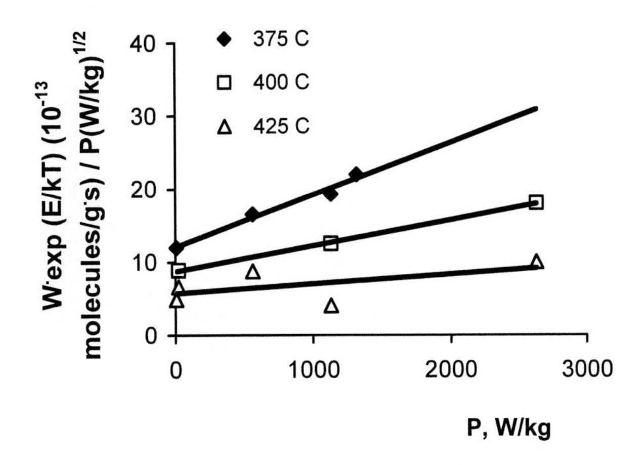

Figure 53. Dependence of the rate of hydrogen formation on radiation dose rate during RTC of pentadecane for different temperatures of RTC in coordinates $W \cdot \exp\left(\frac{E}{kT}\right) P^{-1/2} - P$ (recalculated from [8]).

It is evident that the function *B(T)* can be considered as constant only in the very narrow temperature range.

Equations (24) and (28) were verified using reprocessed experimental data on radiation-thermal cracking available in literature and special experiments on radiation thermal processing of different types of oil feedstock.

Some of the recalculated data of different works represented in Figs.52-54 demonstrate the validity of equation (29).

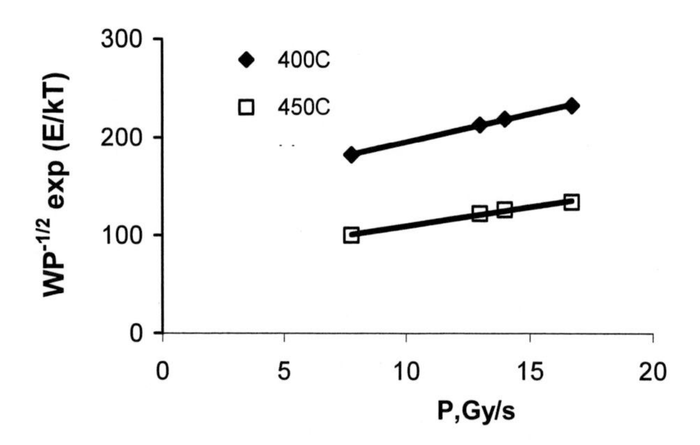

Figure 54. Dependence of the rate of n-hexadecane RTC at different temperatures on dose rate in coordinates $W \cdot \exp\left(\frac{E}{kT}\right) P^{-1/2} - P$ (recalculated from [7]).

Similar results were obtained by processing of our experimental data on RTC of different types of heavy hydrocarbon feedstock [19, 36, 42] (Figs.55-58).

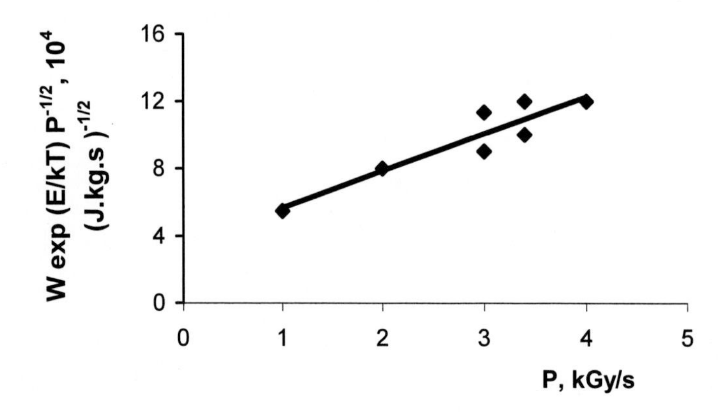

Figure 55. Dependence of the rate of RTC of Karazhanbas crude oil in the temperature range of 370-400C on the dose rate of electron irradiation in coordinates $W \cdot \exp\left(\frac{E}{kT}\right) P^{-1/2} - P$ [19].

Estimation of the activation energy ΔE in equation (26) from the experimental data for different hydrocarbon feedstock yields the average value

$$\Delta E = E - E_0 = 8570 \text{ J / mole}$$

that corresponds to the activation energy for diffusion of light molecules characteristic for liquid hydrocarbons [54].

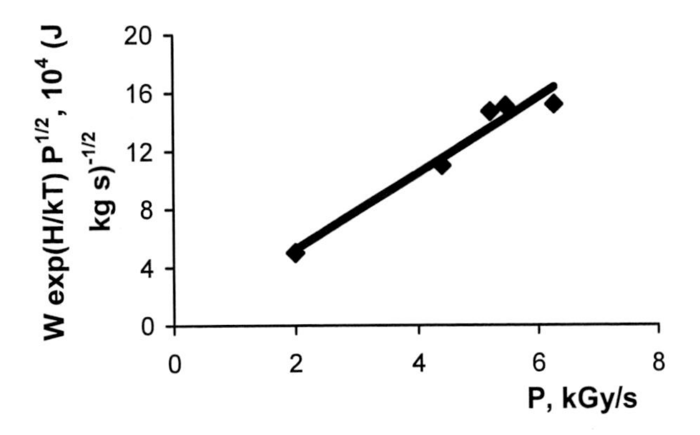

Figure 56. Dependence of the rate of RTC of Bugulma crude oil (Tatarstan, Russia) in the temperature range of 380-400^{0}C on the dose rate of electron irradiation in coordinates $W\cdot\exp\left(\frac{E}{kT}\right)P^{-1/2} - P$

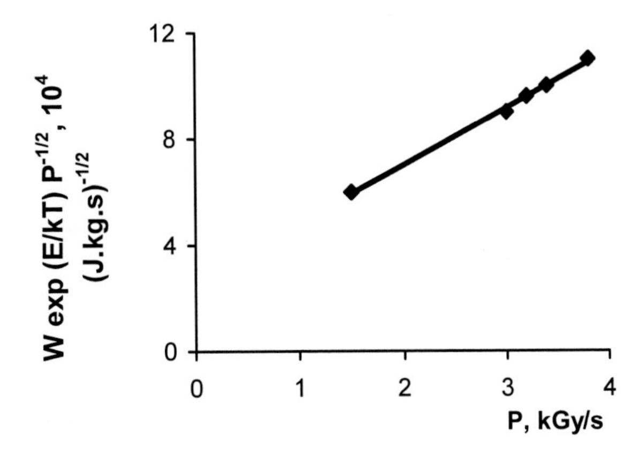

Figure 57. Dependence of the rate of RTC of fuel oil produced by Zuzeeevneft (Tatarstan, Russia) in the temperature range of 400-410^{0}C on the dose rate of electron irradiation in coordinates $W\cdot\exp\left(\frac{E}{kT}\right)P^{-1/2} - P$ [42].

The estimated value of the rate of energy losses by the excited electron molecules is

$$K^* \approx 2\bullet 10^6\, s^{-1}.$$

It corresponds to the life-time of excited molecular states of about $5\cdot 10^{-7}$ s. At the first glance, it is in contradiction with the well-known statement based on experimental observations that the typical rate of molecule excitation energy transfer is ~10^{14}-10^{15} s^{-1} [3, 43] that approximately corresponds to the time between electron-electron collisions.

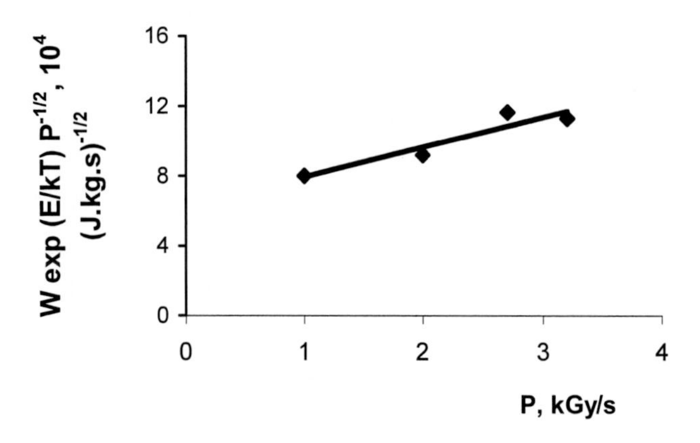

Figure 58. Dependence of the rate of RTC of fuel oil produced by Atyrau refinery (Kazakhstan) in the temperature range of 390-400^{0}C on the dose rate of electron irradiation in coordinates $W \cdot \exp\left(\frac{E}{kT}\right) P^{-1/2} - P$ [36].

The more detailed analysis of the lifetimes of the excited states [47] shows that two types of excited molecular states can appear in hydrocarbons under the action of ionizing radiation: short-living singlet and long-living triplet states.

The triplet excited states were observed in radiolysis studies of alkane and cycloalkane systems both for solute species and the solvent (as a rule, they were observed in aromatic compounds). Such excited states form as a result of ions neutralization. An example of this process is a reaction

$$A^+ + A^- \rightarrow A^* + A \qquad (30)$$

Note that our estimation of the lifetimes of excited molecules corresponds to ion lifetime subject to their pair recombination.

It was noted [47] that triplet excited states can appear not only due to collapse of the coupled ion pairs, they can appear in other process too. One of these processes is the direct excitation of molecules by the electrons. Another process is ion recombination in spurs that contain several ions. In these cases, such processes are possible as, for example, fast neutralization of ions with uncorrelated spins from the neighboring pairs. The remaining charges form a pair in the triplet state.

Lifetimes of excited triplet molecules of aromatic compounds in liquid hydrocarbons at room temperature can amount to tens and even hundreds microseconds if quenchers are absent. Therefore, we can suggest that the observed non-linear dependence of the cracking rate on $\sqrt{P}$ is caused by the contribution of excited molecules in the triplet state.

The singlet excited molecules usually have the lifetime much lower than that of triplet excited molecules; such states are studied using luminescence measurements by the methods of nano- and picosecond radiolysis. The singlet excited molecules solvent form, first, due to the direct interaction of ionizing radiation with hydrocarbons

$$RH \;\text{^^^^^}\!\!\rightarrow RH^* \tag{31}$$

and, second, according to the reaction

$$RH^+ + e_s^- \rightarrow RH^*. \tag{32}$$

Considering their short lifetime, the singlet excited states cannot contribute directly to propagation of the cracking chain reaction proceeding by the radical mechanism. However, as a result of inter-combinational inversion the singlet excited molecules can pass into long-living triplet states [56]. Disintegration of excited molecules into radicals

$$RH^* \rightarrow R' + R'' \tag{33}$$

can contribute to the radical chain initiation.

In different conditions of irradiation of hydrocarbon mixtures, the radical mechanism of radiation cracking can be altered and supplemented by other mechanisms of initiation and propagation of the chain reaction. The description above does not define concretely the cracking mechanism but suggests that the quasi-equilibrium concentration of the chain carriers is controlled by their coupled recombination.

This approach can be justified by observations showing that due to high mobility of light radicals in liquid hydrocarbon mixtures even at room temperature about 70% free radicals diffuse beyond the bounds of spurs [57]. It reduces the role of processes in tracks and spores that cause changes in the order of the reaction of radical recombination and suppress radical reactions as the dose rate increases.

Fig. 59 shows how considerable can be decrease in RTC temperature as dose rates increases. For example, at the temperature of 400^{0}C and the dose rate of electron irradiation of 4kGy/s the cracking rate is 3 s^{-1}.

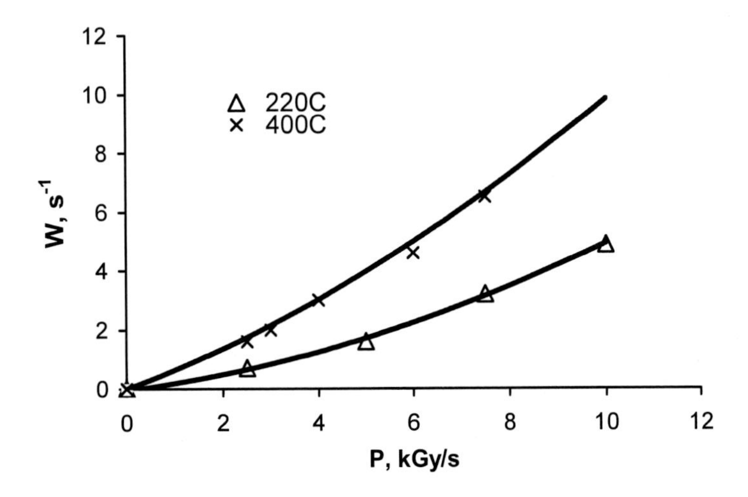

Figure 59. Dependence of the rate of RTC of Karazhanbas crude oil on the dose rate of electron irradiation at 220^{0}C and 400^{0}C.

To provide the same cracking rate at 220^{0}C we would have to increase dose rate up to 5.7 kGy/s, i.e. we should make it higher by 1.5 times. According to formula (3) attainment of the same cracking rate at 220^{0}C would require increase in the dose rate by $\exp\left(\frac{2E(T_1-T_2)}{T_1T_2}\right) \approx 51550$ times, i.e. proceeding of the chain reaction with a noticeable rate would be practically impossible at this temperature.

Thus, the revealed effect of diffusion-enhanced interaction of free radicals with electron-excited molecules allows realization of RTC of hydrocarbon feedstock using technologically acceptable dose rates at temperatures of 150-200^{0}C with the same rate that earlier could be achieved only at the temperature of about 400^{0}C.

Cold Radiation-Induced Cracking

In the previous section it was shown that when the dose rate is high enough, thermally activated diffusion of free radicals to the excited molecules generated by ionizing radiation raises the rate of the chain cracking reaction. At the average dose rate of about 2 kGy/s contributions of molecules exposed to electronic and thermal excitation become comparable in heavy oil. Further increase in dose rate must result in a noticeable probability of athermal electron excitation of molecules up to the level necessary for the chain propagation. At low temperatures, when probability of thermal molecule excitation and probability of thermally activated diffusion of radicals are negligibly low, the transfer from radiation-thermal to "cold" radiation-induced cracking should be observed. In this process, both initiation and continuation of the chain are caused only by the action of radiation.

The rate of cold cracking rate will be defined by expression

$$W = K_0\left(\frac{G_r}{K_r}\right)^{1/2} c^* P^{1/2} = K_0\left(\frac{G_r}{K_r}\right)^{1/2} \frac{G^*}{K^*} P^{3/2} \sim P^{3/2}. \tag{34}$$

Qualitative estimation using formula

$$c^* = \frac{G^*}{K^*} P$$

gives the value of about $5 \cdot 10^{-10}$ for c^* at the dose rate of 40 kGy/s.

At the temperature of 400^{0}C concentration of thermally excited molecules is $c^*_{therm} = \exp\left(-\frac{E}{kT}\right) \approx 3.5 \cdot 10^{-7}$ and the cracking rate is about 2 s^{-1} at the dose rate of 2 kGy/s. Taking into account that concentration of cracking-initiating radicals changes proportionally to $\sqrt{P}$, we shall obtain that at the dose rate of ~40 kGy/s cracking rate should be lower by two orders of magnitude that will make 0.02 s^{-1}. This rate of reaction is high enough, therefore the phenomenon of cold cracking should be quite pronounced beginning from the dose rate values of about several kGy per second.

Experiments have shown that chain cracking reactions with the rate of conversion not less than 0.020 s^{-1} were observed in these conditions (Figs.60, 61). High-viscous crude oil was irradiated by 2 MeV electrons at room temperature at the dose rate of 40 kGy.

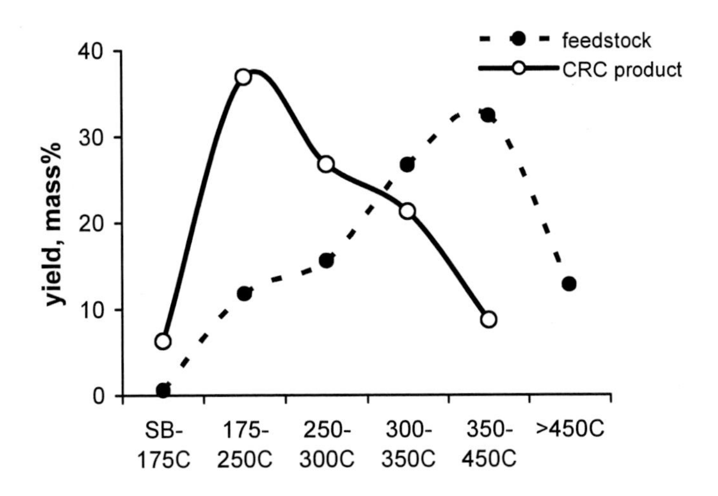

Figure 60. Chromatograms of samples of high-viscous crude oil before and after cold radiation cracking (CRC).

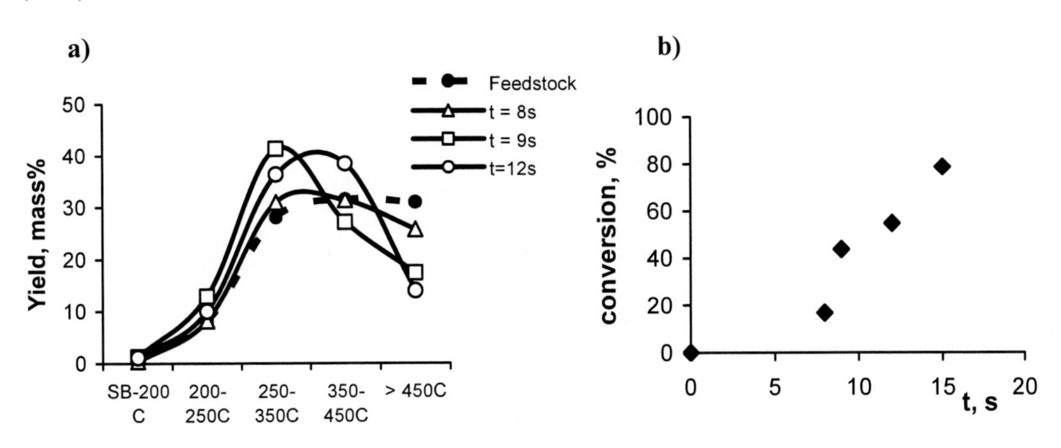

Figure 61. Fractional contents of the products of cold radiation cracking of high-viscous crude oil at the average electron dose rate of 20-40 kGy/s

Fig. 62 demonstrates increase in the reaction rate with the dose rate of electron irradiation characteristic for the cold radiation cracking.

Fig. 62 shows considerable changes in the hydrocarbon contents after cold radiation cracking of bitumen, the heaviest hydrocarbon feedstock.

Increase of the cracking temperature up to 150-200^{0}C provokes intensification of chain carriers diffusion to radiation-excited molecules and considerable increase in the reaction rate.

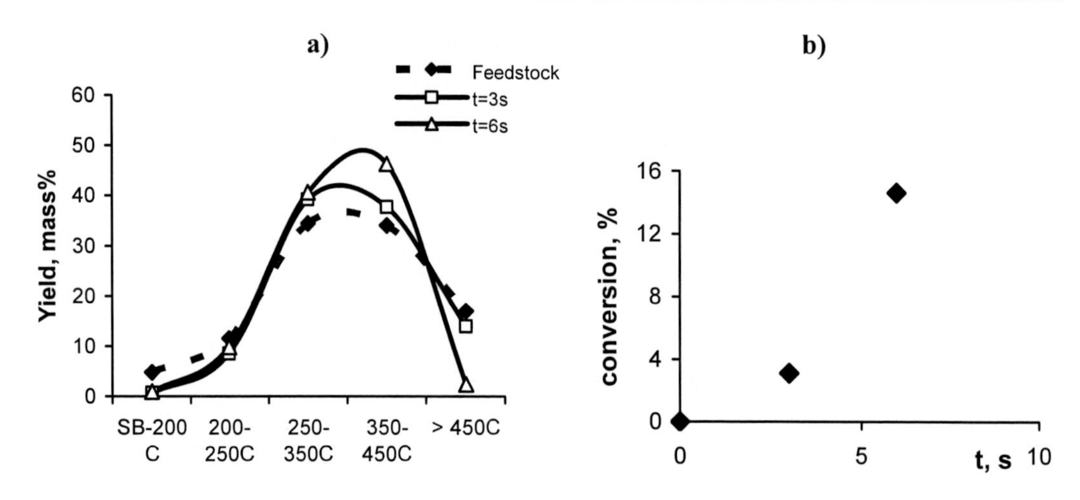

Figure 62. Fractional contents of the products of cold radiation cracking of Karazhanbas oil (well #2) – (a) -and degree of oil conversion after irradiation in the continuous mode (b) at the electron dose rate of 80 kGy/s.

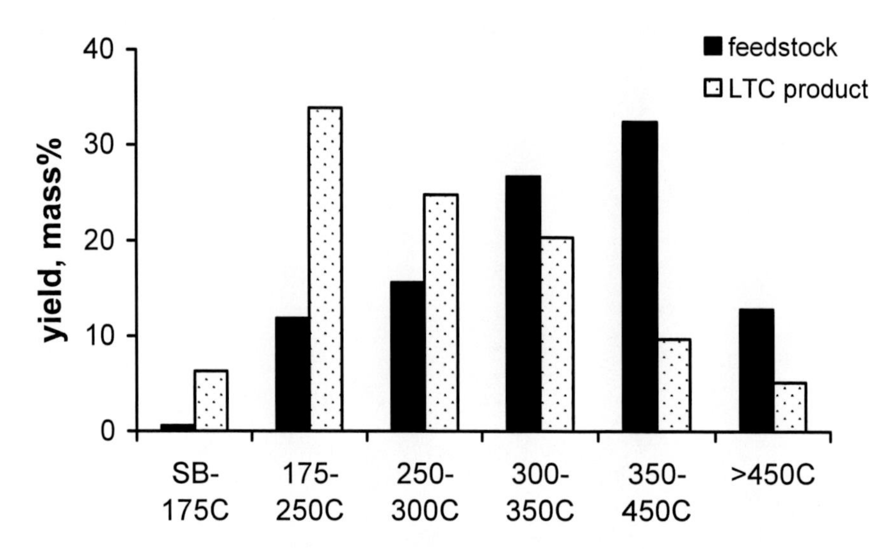

Electron dose – 26 kGy, dose rate – 10 kGy/s, temperature of processing – 220^{0}C.

Figure 63. Fractional contents of Karazhanbas crude oil and the product of its low-temperature radiation cracking (LTC)

Effect of Oil Structural State on the Rate of Radiation Cracking

All the experimental data represented in the previous section were obtained by oil processing in static conditions. They show high yields of light products of radiation cracking at lowered temperatures. However, these data are characterized by lower reaction rates, and therefore, higher doses necessary for the given degree of conversion compared with radiation-thermal cracking at heightened temperatures.

The main reason for it is formation of dense structures in the heavy hydrocarbon feedstock at lowered temperatures. These radiation-resistant clusters facilitate energy losses by radiation-exited molecules and intensify undesirable strong polymerization that limit the yields and deteriorates stability of the radiation-upgraded products.

Application of special methods of oil supply to the radiation-chemical reactor and preliminary oil processing using mechanical, acoustic, or electromagnetic methods provides effective destruction of oil colloid system and allows increase in the cracking rate by nearly two orders of magnitude. As a result, heavy oil feedstock can be processed with the high production rate characteristic for radiation-thermal cracking but at considerably lower temperatures. It dramatically reduces or completely eliminates energy consumption for the feedstock heating.

In particular, Fig. 63 shows that application of the feedstock bubbling with ionized air initiates reaction with the rate of 4,9 s^{-1} at the temperature of 220^0C and the dose rate of 10 kGy/s. It is 63% higher than the cracking rate at the temperature of 400^0C and at the dose rate of 4 kGy/s. Oil processing in this mode has practically resulted in liquidation of the heavy residue with the boiling temperature higher than 450^0C.

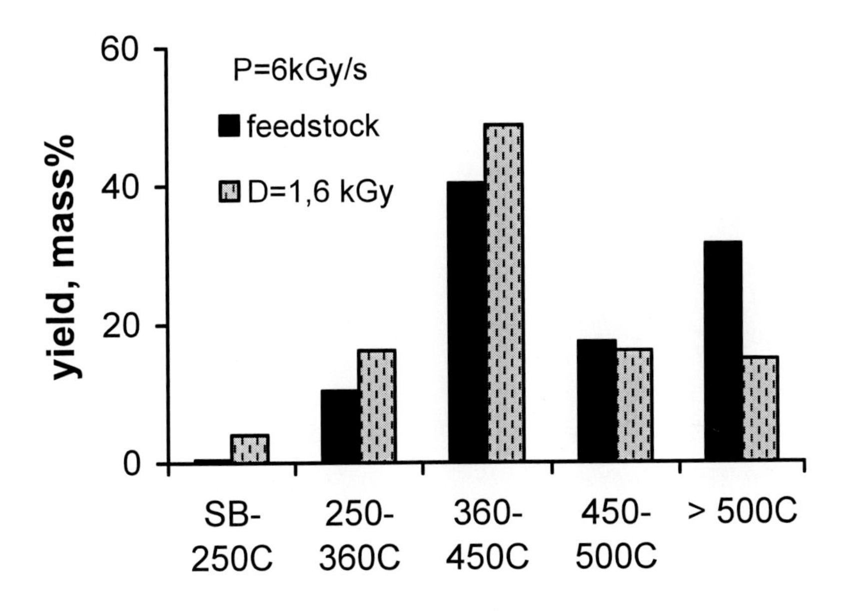

Figure 64. PetroBeam™ processing of high-viscous fuel oil in flow conditions.

Fig. 64 illustrates considerable increase in the reaction rate and in the yields of light fractions after radiation cracking at 40^0C achieved even at the much lower dose rate of electron irradiation by application of flow conditions characterized by high flow rates and considerable shear stresses in the flow.

Generalized Model of PetroBeam™ Process

PetroBeam™ technology is the method for deep destructive processing of hydrocarbons *in a wide temperature range from room temperature to 400⁰C* based on the new generalized concept of self-sustaining radiation cracking of hydrocarbons where the known phenomenon of radiation-thermal cracking enters as a particular case.

The theory of radiation-thermal cracking assumes radiation initiation of self-sustaining cracking of hydrocarbons by radiation generation of chain carriers and thermal chain propagation. The necessity of feedstock heating for chain propagation defines the lower temperature limit for radiation-thermal cracking (about 350^0C for heavy hydrocarbon mixtures).

Additional contributions to the reaction rate appear when interaction of radical with radiation-excited molecules is taken into account. These terms are negligibly small in the usual conditions of radiation-thermal cracking but they can become predominant in the conditions providing high concentrations of radiation-generated excited hydrocarbon molecules.

In the generalized model, radiation cracking of hydrocarbons can proceed in a wide temperature range down to room temperature with thermal, radiation or combined radiation-thermal chain propagation. Generally, dependence of the initial rate of cracking on dose rate and temperature, more complicated compared with the classic model, can be written in the following form:

$$W = \sum_i [A_1(S_i(T))\exp\left(-\frac{E}{kT}\right)P^{1/2} + A_2(S_i(T))\exp\left(-\frac{\Delta E}{kT}\right)P^{3/2} + A_3(S_i(T))\,P^{3/2}] \quad (35)$$

In equation (35), ΔE is the activation energy for chain carrier diffusion; $S_i(T)$- temperature dependent parameters of the initial structural state of oil processed. Equation (35) was confirmed in a number of experiments on oil radiation processing.

At the given temperature, values of the functions $A_j(S_i)$ can be considerably increased by destruction of oil cluster structure using mechanical, acoustic, or electromagnetic methods.

The first term is predominant in the temperature range of about 350-450^{0}C. It describes the known process of *radiation-thermal cracking* (radiation initiation and thermal chain propagation of the self-sustaining cracking reaction) with the correction on the original structural state of the feedstock. This correction becomes essential at the lower end of the temperature interval of radiation-thermal cracking. It is used in PetroBeam™ technology for higher diversity of commodity products and increase in the yields of light fractions at lower temperatures of radiation-thermal cracking by combination of ionizing irradiation and other types of processing.

At the temperature lower than 350^0C the first term of equation (35) becomes negligibly small. The second term describes the *low-temperature cracking* enhanced by thermally activated diffusion of chain carriers to electron-excited molecules. It prevails in the temperature range of 200-350^{0}C characterized by rather high mobility of chain carriers.

The third term predominates at lower temperatures. It describes *cold cracking* of hydrocarbons when chain reaction are both initiated and propagated by radiation. However, it contains the temperature dependent structural factor which considerably affects the rate of cracking at the temperatures below 200^{0}C.

The second and the third terms can be considerable at lowered temperatures when processing conditions provide the high rate of athermal chain propagation in cracking reaction. High processing rates are provided by heightened dose rates of electron irradiation and high values of the structural parameter $A_j(S_i)$. The latter is achieved by different methods, such as feedstock preliminary heating up to moderate temperatures, special forms of feedstock supply into reactor and combination of irradiation with other types of processing.

Examples of PetroBeam^{TM} Technology Application

The technology was tested at the PetroBeam facility simulating high production rates of industrial oil processing. Some of the applications of the PetroBeam^{TM} technology to heavy oil upgrading are given below.

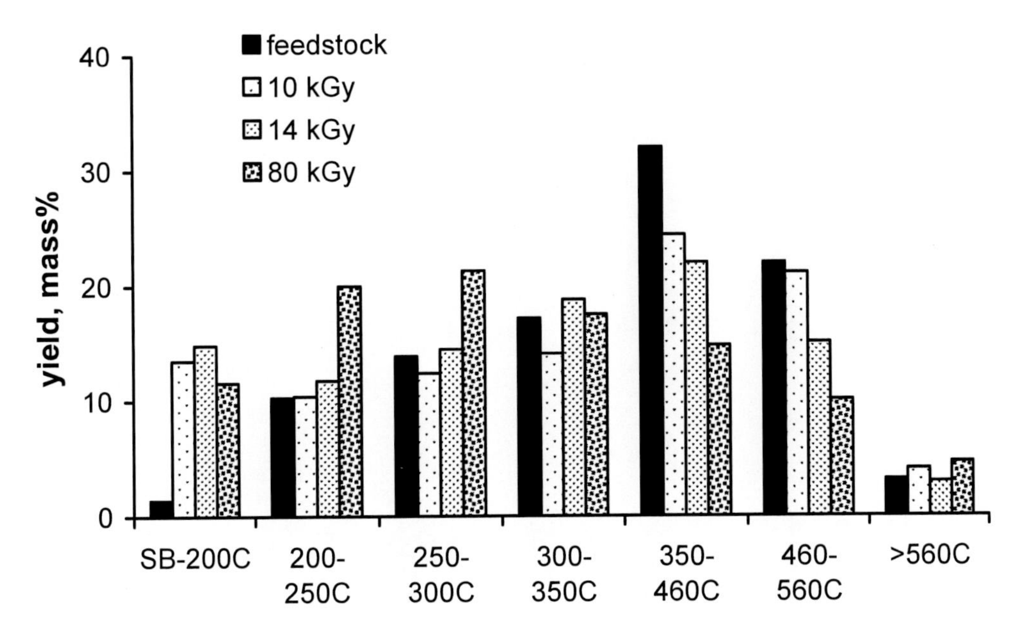

Figure 65. Fractional contents of high-viscous oil after PETROBEAM^{TM} processing.

Fig.65 shows fractional contents of the liquid product obtained by processing of high-viscous oil in PetroBeam flow conditions at 35^{0}C. Conversion of the heavy residue and the yields of light fractions in PetroBeam process at nearly room temperature are close to those obtained by means of RTC at 400^{0}C.

Fig.66 illustrates the results of PetroBeam processing of high-paraffin oil. Due to very strong radiation-induced polymerization, deep processing of this sort of oil by means of RTC is very difficult.

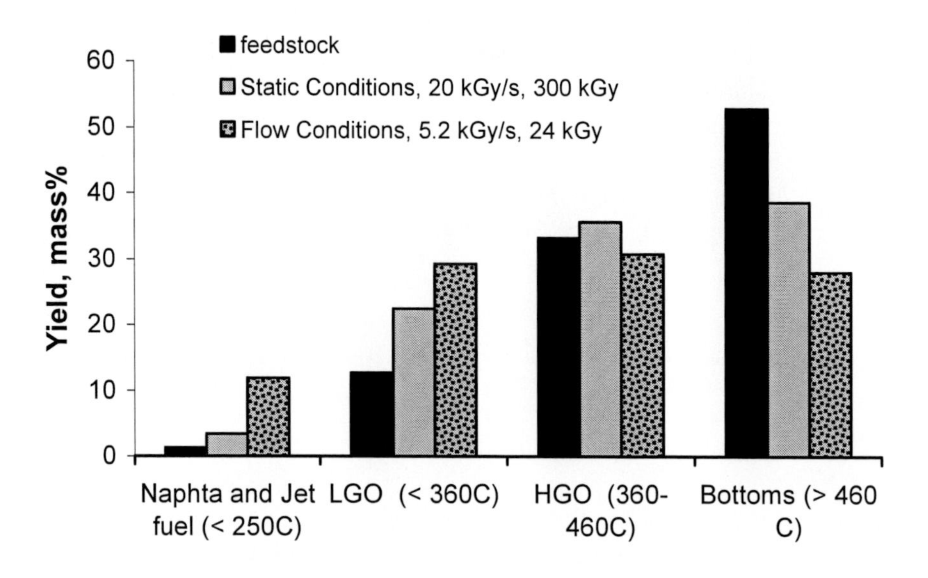

Figure 66. Fractional contents of high-paraffin crude oil and product of its PetroBeam™ processing at 60°C.

Effective solution of the problem was recently found in frames of the PETROBEAM™ process. Radiation processing of high-paraffin oil using special irradiation conditions and special forms of the feedstock supply to the reactor provides high yields of stable light products.

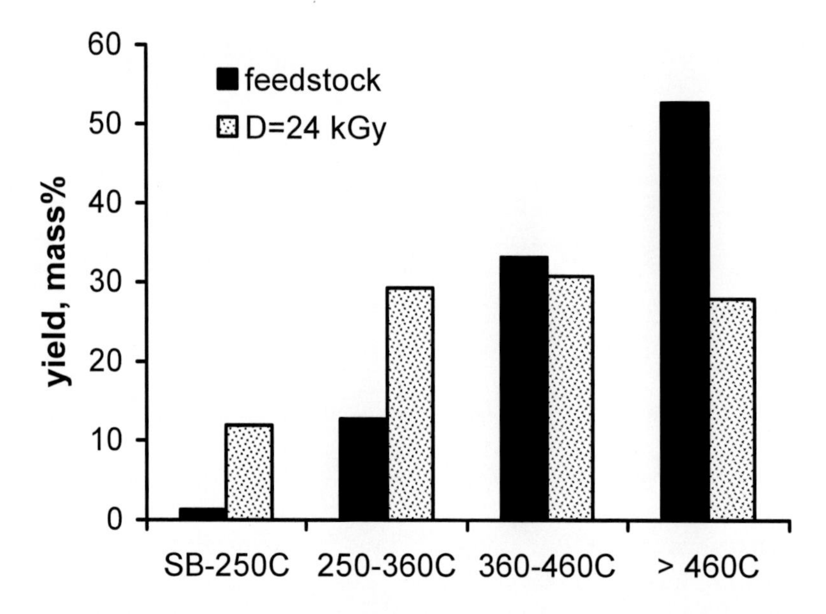

Figure 67. Fractional contents of the high-paraffin fuel oil and the product of its PETROBEAM™ processing in flow conditions.

Successful application of PetroBeam^{TM} technology to high-paraffin fuel oil, a complicated type of oil feedstock with high concentration of heavy aromatics and high-molecular paraffins, is illustrated in Fig.67.

The launch of the PetroBeam facility with the capacity of 1000 barrels a day for production of high-quality synthetic oil from bitumen and heavy oil residua is planned in 2008.

Conclusion

The main problem to be solved by oil industry for the effective processing of any type of hydrocarbon feedstock is the control of cracking reactions in conditions that provide combination of high processing rate and sufficient conversion with maximum simplicity and economic efficiency at minimum energy expense. Technology of oil radiation cracking allows economically, technologically and environmentally overcoming this problem by the most effective way.

The general trend in development of radiation methods for oil refining during last decades was their expansion to upgrading and deep processing of a greater variety of heavy oil feedstock and production of various designed produs together with the radical reduction of the process temperature.

Revelation of the phenomena of cold radiation cracking and chain carrier diffusion to radiation-excited hydrocarbon molecules has created preconditions for development of Petrobeam^{TM} process, new highly economic energy-saving technology for deep processing of heavy oil, bitumen, oil wastes and residua.

Operation of PetroBeam pilot lines for heavy oil and bitumen radiation processing have demonstrated high efficiency of low-temperature radiation processing at heightened dose rates of electron irradiation for different types of oil feedstock, such as aromatic-rich high-viscous crude oil, high-paraffinic oil and bitumen, difficult for traditional refining. Application of this type of oil radiation processing allows combination of the high production rate, the high degree of oil conversion and the highest energy savings compared with any conventional technology of hydrocarbon processing.

Technological approaches developed today can be a real base for the new generation of refineries combining high production rate and economic efficiency with the ability of easy re-orientation to processing of different types of oil feedstock and production of different types of oil products corresponding to the market demands.

References

[1] Berejka A.J. Reactor Design Concepts for Radiation Processing. – *Radiat. Phys. Chem.*, 2003.

[2] Wentworth R.L., Canfield M.P. Patent Literature in Radiation Chemistry, Dynatech Corporation, Cambridge, Massachusetts, 1964.

[3] Topchiev, A.V., Polak L.S. Radiolysis of Hydrocarbons. Moscow: Acad. Sci. USSR, 1962, 208 pp.; Topchiev, A.V. *Radiolysis of Hydrocarbons*, Amst-London-N.Y.: El. Publ. Co.,1964, 232pp.

[4] Lavrovsky K.P., Brodsky A.M., Zvonov N.V., Titov V.B. Radiation-Thermal Conversion of Oil Fractions // *Neftekhimiya (Oil Chemistry),* 1961, v. 3, No. 3, p.370-383.
[5] Lavrovsky K.P. Catalytic, Thermal, and Radiation-Chemical Conversion in Hydrocarbons".- Moscow, *Nauka Publ.*, 1976, p.312-373; 255-263.
[6] Gabsatararova S.A., Kabakchi A.M. Effect of Gamma-Irradiation Dose Rate on Formation of Unsaturated Compounds During Radiation-Thermal Cracking of n-Heptane // *Khimiya Vysokikh Energiy (High Energy Chemistry),* 1969, v.3, No.2, p. 126-128.
[7] Panchenkov G.M., Putilov A.V., Zhuravlev G.I. Study of Basic Regularities of n-decane Radiation-Thermal Cracking // *Khimiya Vysokikh Energiy (High Energy Chemistry),* 1981,v.15, N 5, p.426-430.
[8] Mustafaev I.I. Radiation-Induced Thermal Conversion of Heavy Oil Fraction and Organic Part of Oil-Bitumen Species. // *Khimiya Vysokikh Energiy (High Energy Chemistry)*, 1990, v.24, No.1, p. 22-26.
[9] Zhuravlev G.I., Voznesenskaya S.V., Borisenko I.V., Bilan L.A.. Radiation-Thermal Processing of Heavy Oil Residua // *Khimiya Vysokikh Energiy (High Energy Chemistry)*, 1991, v.25, No.1, p. 27-31.
[10] Zaykin Yu.A., Zaykina R.F., Mirkin G. On Energetics of Hydrocarbon Chemical Reactions by Ionizing Irradiation // *Radiat. Phys Chem.*, 2003, v. 67/3-4, pp. 305-309.
[11] Bugaenko L.P., Kuzmin M.G., Polak L.S. Chemistry of High Energies. – Moscow: *Chimiya*, 1988 – 368 pp.
[12] 12. Nadirov N.K., Zaykina R.F., Zaykin Yu.A. Progress in High Viscosity Oil and Natural Bitumen Refining by Ionizing Irradiation". - *Oil and Bitumen. Kazan,* 1994, V.4, pp.1638-1642.
[13] Nadirov N.K.,Zaykina R.F., Zaykin Yu.A. New High-Efficient Technologies of Heavy Oil and Oil Rests Refining // *Energetics and Fuel Resources of Kazakhstan,* 1995, No.1,p.65-69.
[14] Zaykina R.F., Zaykin Yu.A.,Nadirov N.K." Mechanisms and Kinetics of Radiation Methods of Heavy Oil Fractions Radiation-Induced Thermal Cracking // Oil and Gas, Annex to the "Reports of the Ministry of Science-Academy of Science of the Republic of Kazakstan", 1997, N2, p.83-89.
[15] Nadirov N.K., Zaykina R.F., Zaykin Yu.A., Mamonova T.B., Bakirova S.F. Radiation Method for Sulfur Compounds Conversion in Oil Products // *Oil and Gas of Kazakstan,* 1997, No. 3, p.129-134.
[16] 16. Nadirov N.K., Zaykin Yu.A., Zaykina R.F. Radiation Technologies of Oil Products Refining // *Oil & Gas of Kazakstan*, 1997, No.3, p.43-45.
[17] Zaykin Yu.A., Zaykina R.F., Nadirov N.K. Radiation Processing of the Wastes of High-Paraffinic Oil Extraction // Oil and Gas (Kazakhstan), 1999, No.1 (5); Zaikin Yu.A., Zaikina R.F., Nadirov N.K. Radiation Technologies for Oil Industry. Proc. 2 Internat. Conf. "Nuclear and Radiation Physics", Almaty, 1999, *Institute of Nucl. Phys.*, v.1, p. 209-214.
[18] Zaykina R.F., Alyev B.A. Synergetic Effects in High-Viscous and Paraffinaceous Oil. In: Problems of Open System Evolution. Almaty: Complex Publ., 2001, p. 114-120.

[19] Zaykin Yu.A., Zaykina R.F., Mamonova T.B, Nadirov N.K. Radiation-Thermal Processing of High-Viscous Oil from Karazhanbas Field // *Radiat. Phys. Chem.*, 2001, vol. 60, p.211-221.

[20] Zaykina R.F., Zaykin Yu.A., Mamonova T.B. Radiation Methods for Demercaptanization and Desulfurization of oil Products // *Radiat. Phys. Chem.*, 2002, v.63/2, p.617-619.

[21] Zaykina R.F., Zaykin Yu.A. Radiation Technologies for Production and Regeneration of Motor Fuels and Lubricants // *Radiat. Phys. Chem.*, 2003, v.65, p. 169-172.

[22] Zaikin Yu.A., Nadiriv N.K. On the "Solid-State" Approach to Analysis of Oil Mixtures Viscosity // *Oil & Gas (Kazakstan),* 2004, No.3 (23), .p.116-125.

[23] Zaykin Yu.A, Zaykina R.F.. Radiation Technologies for Priority Branches of Kazakhstan Industry // *News of Kazakhstan Science*, 2004, issue 2 (81), p. 40-44.

[24] Zaykin Yu.A., Zaykina R.F., Silverman J. Radiation-thermal conversion of paraffinic oil // *Radiat. Phys. Chem.*, 2004, v.69/3, p.229-238.

[25] Zaikin Yu.A., Zaikina R.F. On Criteria of Synergetic Effects in Radiation-Induced Transformations of Complex Hydrocarbon Mixtures // *Oil & Gas (Kazakstan)*, 2004, No.2 (22), .p.64-73.

[26] Zaykina R.F., Zaykin Yu.A., Yagudin Sh.G., Fahruddinov I.M. Specific Approaches to Radiation Processing of High-Sulfuric Oil // *Radiat. Phys. Chem.*, 2004, v. 71, p. 467-470.

[27] Zaykin Yu.A., Zaykina R.F. Bitumen Radiation Processing // *Radiat. Phys. Chem.*, 2004, v. 71, p. 471-474.

[28] Zaykin Yu.A., Zaykina R.F., Nadirov N.K. Radiation-Thermal Cracking of Hydrocarbons and Its Application for Deep Conversion of Oil Feedstock // *Oil & Gas (Kazakstan)*, 2004, No.4 (24), .p.47-54.

[29] Zaikin Yu.A., Zaikina R.F., Mirkin G., Nadirov N.K. Radiation technology as a Real Base for the New Generation of Oil Refineries // *Oil and Gas (Kazakhstan)*, 1999, No.3 (7), p.93-99.

[30] Zaykina R.F., Zaykin Yu.A., Mirkin G., Nadirov N.K. Prospects of Radiation Technology Application in Oil Industry // *Radiat. Phys. Chem.*, 2002, v.63/2, p.621-624.

[31] Mirkin G.,. Zaykin Yu.A, Zaykina R.F.. Radiation Methods for Upgrading and Refining of Feedstock for Oil Chemistry. *Radiat. Phys Chem.*, 2003, 67/3-4 pp. 311-314.

[32] Zaikina R., Zaikin Yu., Silverman J., Al-Sheikhly M.. Potentialities of Hydrocarbon Radiation Processing for Reduction of Environmental Pollution by Petroleum Products. Proceedings of the 4th International Conference "Oils and Environment", *Gdansk University of Technology*, 2005, p. 296-303.

[33] Akpabio Em.J. Cand. Diss.Thesis. Baku, 1992, 22 pp.

[34] Kamyanov V.F.., Sivirilov P. P., Lebedev A.K., Shabotkin I.G., Production of Motor Fuels by Iniated Cracking of Natural Bitumen. *Oil and Bitumens*. Kazan, 1994, vol. 5, 1750-1754.

[35] Musaev G.A., Mamonova T.B., Malibov M.S., Musaeva Z.G., A Study of Properties and Oil-Bitumen Rocks of Kazakhstan by Thermocatalytic Method // *Energetics and Fuel Resources of Kazakhstan*, 1994, No.4, 51 – 56.

[36] Development of Experimental Facility for Processing of Hydrocarbon Components of Oil Bitumen. Technical Report on ISTC project K-930. *Almaty*, Kazakhstan, 2006, 44 pp.

[37] Dolivo G., Gaumann T., Ruf A.. Photoinduced Isomerization and Fragmentation of the Pentane Radical Cation in Condensed Phase // *Radiation Physics and Chemistry*, 1986, v.28, No.2, p.195-200.

[38] Stepukhovich A.D., Ulitsky V.A. Kinetics and Thermodynamics of Radical Cracking Reactions - Moscow: *Khimiya Publ.*, 1975.-256 p.

[39] Scheer M.D., McNecby J.R., Klein R. // *J. Chem. Phys.*, 1962, v.36, p.3504-3505.

[40] Hamson R.F., McNecby J.R. // *J. Chem. Phys.*, 1965, v.42, p.2200-2208.

[41] Farkhadova G.T., Rustamov M.I., Agayeva et al. Influence of Temperature on Contents of n-Heptane Conversion Products on Zeolite-Containing Catalysts // *Neftekhimiya (Oil Chemistry),* 1987, No.3.

[42] Development of the Methods for Processing of Oil Products Using Complex Radiation-Thermal Treatment and Radiation Oxonolysis. Final Report on IAEA Project (Research Contract # 11837/RO), *Almaty*, Kazakhstan, 2004, 34 pp.

[43] Radiation Chemistry of Hydrocarbons. Ed. by G. Földiak, Budapest: *Akademiai Kiado*, 1981,-304 p.

[44] Saraeva V.V. Radiolysis of Hydrocarbons in Liquid Phase – Moscow State Univ. Press, 1986 . - 256 p.

[45] Aspects of Hydrocarbon Radiolysis. Ed. By Gäuman *T.* and Hoigne T. - London-New York: Academic Press, 1968. - 273 p.

[46] Kossiakoff A., Rice F.O. // *J.Am. Chem.Soc.*, 1943, 65, p.590-595.

[47] Pikaev A.K.. Modern Radiation Chemistry. Solids and Polymers. Applied Aspects. - Moscow: *Nauka Publ.*, 1987. - 448 p.

[48] Evans, R.C. Introduction to Crystal Chemistry. Cambridge Univ. Press, 1964. - 360 pp.

[49] Kurdumov S.S. et al. Concentration of Heavy Metals and Demetallization of Oil Residua Under Thermal Processing in Presence of Carriers of Different Nature and Polymer Wastes // *Neftekhimiya (Oil Chemistry)*, 1999, v.39, No.4, p. 260-264.

[50] Nadirov N.K., Zaykina R.F., Mamonova T.B. Prospects of Ozone Use to Rise Efficiency of High Viscous Oil Transportation and Refining // *Oil and Gas of Kazakstan*, 1997, No.3, p.159-164.

[51] 51.Zaikin Yu.A., Zaikina R.F. Processing of Oil Products Using Complex Radiation-Thermal Treatment and Radiation Oxonolysis. Proceedings of the Second Eurasian Conference "Nuclear Science and its Applications". Almaty, *Institute of Nuclear Physics*, 2002., v.3, p. 164 – 169.

[52] Zaykin Yu.A., Zaykina R.F. Stimulation of Radiation-Thermal Cracking of Oil Products by Reactive Ozone-Containing Mixtures // *Radiat. Phys. Chem.*, 2004, v. 71, p. 475-478.

[53] Zaikin Yu.A. New Technological Approaches to Cleaning, Upgrading and Desulfurization of Oil Wastes and Low-Grade Oil Products. Proceedings of the 4th International Conference "Oils and Environment". Gdansk University of Technology, 2005, p. 275-282.

[54] Tables of Physical Quantities. Ed. by I.K. Kikoin. – Moscow: *Atomizdat*, 1976, 1008 pp.

[55] Self-Sustaining Cold Cracking of Hydrocarbons. US patent and Trademark Office. Patent application of 12/16/2005, Registration No. 45,587; Docket Number P79993US00GP. Patentee: PetroBeam, Inc, U.S.A.; Authors: Yuriy A. Zaikin , Raissa F. Zaikina.

[56] Egorov G.F., Terekhov G.A., Medvedovsky V.I. – *Khimiya Vysokikh Energiy (High Energy Chemistry*, 1972, v.6, p. 425-429.

[57] Burns W.G., Holroyd R.A., Klein G.V. – *J. Phys. Chem.*, 1966, v.70, p.910.

In: Radiation Physics Research Progress
Editor: Aidan N. Camilleri, pp. 105-154

ISBN: 978-1-60021-988-7

Chapter 2

POSITRON IMPACT DIFFERENTIAL IONIZATION STUDIES

R.D. DuBois[1], O.G. de Lucio[1] and A.C.F. Santos[2]
[1] Department of Physics, University of Missouri-Rolla, Rolla, MO 65409 USA
[2] Instituto de Física, Universidade Federal do Rio de Janeiro, Rio de Janeiro, Brazil

Abstract

Since the discovery of the positron, low to intermediate energy (< 1 keV) positron impact ionization studies have provided insight into how antimatter and matter interact. Plus, by comparing positron and electron impact data, insights and deeper understanding of matter-matter interaction dynamics and mechanisms are achieved. Until recently however, the vast majority of experimental studies for positron impact have been limited to measurements of integral cross sections; only a handful of experiments which provide differential ionization information have been performed. The primary reason for this is associated with the low-intensity positron beams that are available. This results in extremely weak differential signal rates and precludes using standard experimental methods and techniques. This paper gives an overview of differential studies that have been performed to date with emphasis on the methods used and examples of data acquired. This is followed by a discussion of ongoing work at the University of Missouri-Rolla where for the past several years we have been developing techniques which have enabled us to obtain highly differential ionization information using subfemtoamp positron beam intensities. Our basic method involves crossing a positron beam with a simple gas jet. A specially designed spectrometer equipped with a position sensitive detector is used to detect post-collision positrons for a range of scattering angles and energy losses. Ionization events are identified by using a weak electric field to extract target ions from the interaction region. Time of flight techniques determine their charge state, i.e., the degree of target ionization. A second position sensitive channelplate detects electrons ejected into a wide range of angles parallel and perpendicular to the beam direction. Coincidences between these post collision particles have allowed us to study single and multiple ionization as a function of energy loss and scattering angle, to obtain fully kinematic information about single ionization, and to measure the differential electron emission resulting from double and triple ionization by positron impact. Our method has also provided the first look at "ultra inelastic" collisions where the incoming particle transfers all of its available energy to a target electron. Experimental approaches and modifications leading to our present setup plus various examples of data we have obtained are provided. Where appropriate, comparisons with electron and photon impact data are made to illustrate differences and help interpret the positron data. Finally, limitations of our present methods plus possible future studies and needs are discussed.

Introduction

Since their prediction and discovery more than 70 years ago, positrons have been used in a multitude of ways to learn about the basic properties of antimatter, to gain insight into how matter and antimatter interact, to probe the properties of matter, and to obtain a deeper understanding of matter-matter interactions. These studies, and the knowledge gained from them, currently impact a wide variety of fields which range from atomic and condensed matter physics, to astrophysics, chemistry, biology, medicine, materials research, as well as antimatter production and properties. Sometime in the future positrons may even play a role in the development of new energy sources based on antimatter-matter annihilation.

This paper concentrates on just one of these areas, namely how positrons are used to extend our understanding of energetic inelastic atomic interactions and collision dynamics. For positron studies outside the scope of the present article, the reader is referred to two excellent review articles and references therein [1,2]. The present work begins by describing why inelastic collisions are of interest, then briefly outlines the physics that takes place. Next, descriptions and examples of what can be (or has been) learned from total cross section and differential cross section studies is given. Particular emphasis is placed on how comparisons of positron, electron and photon impact data provide information about ionization mechanisms and how charge and mass effects influence the interaction kinematics. In the next section, brief descriptions of experimental methods that have been used to date and examples of differential data that these methods have provided are given. The next sections follow the progress of various methods that we have used at the University of Missouri-Rolla in order to obtain increasingly more complex differential data for positron impact. Examples of published and unpublished data obtained to date are provided and where possible comparisons with electron or photon impact data are made. Finally, we conclude with comments about what lies in the future; e.g., possible and planned improvements with regard to our studies, future experiments that are of interest, and needs with regard to performing improved differential positron impact studies.

Inelastic Collisions

Background

Inelastic atomic interactions lead to energy and momentum transfer from one atomic particle to the bound electrons of another; this results in some of the electrons being excited, ionized, or transferred between particles. For interactions involving molecules, fragmentation can occur; for interactions with solids, electrons, ions, and atoms can be ejected from the surface or bulk. Such processes play important roles in radiation damage to solids and to biological material, in plasma and chemical processes, in various types of lighting sources, etc.

For these reasons, inelastic atomic interactions have been extensively studied for decades. Highly differential, detailed experiments have been performed and very sophisticated theories have been developed. As a result, much has been learned with regard to which mechanisms are important, when various mechanisms dominate or make insignificant contributions. In addition, differential studies provide detailed information about the particle dynamics when atomic particles interact. With regard to theory, although much is known a single standard approach capable of modeling any arbitrary collision system at all collision energies has yet to emerge. This is because even the simplest inelastic atomic process is a time-dependent, many-body problem, i.e., it involves the time evolution of the coulomb forces acting between three particles, namely a "target" consisting of a nucleus and a bound electron, and a "projectile" consisting of a bare charge. (Since this paper will be discussing both positron and electron impact, for clarity, particularly in the case of electron impact ionization, the incoming and scattered particle will be referred to as the projectile.) Since the Schrödinger equation is not analytically solvable for more than two particles, even the simplest inelastic atomic interaction can only be solved using various approximations.

To complicate matters further, most atomic interactions of interest involve many-electron ions, atoms or molecules. Thus any quantal or classical approach must model the time evolution of many, mutually-coupled coulomb forces, plus the correlated behavior among the various particles, plus multi-electron as well as single-electron transitions. Describing multiple ionization of atoms is far from a simple task. The difficulty arises because of the many possible pathways leading to the final target state. For example, multi- as well as single- electron transitions involving just one shell or in some cases both inner and outer shells are possible and must be considered in constructing any theoretical model. But, the statistical distributions of the various available inelastic alternatives, as well as the way the electrons dynamically correlate, significantly change the dependence of the multiple-ionization cross sections with respect to the single ionization for different projectile velocities. In addition, additional effects such as target polarization before, during, and after the collision and electron exchange or transfer must also be modeled.

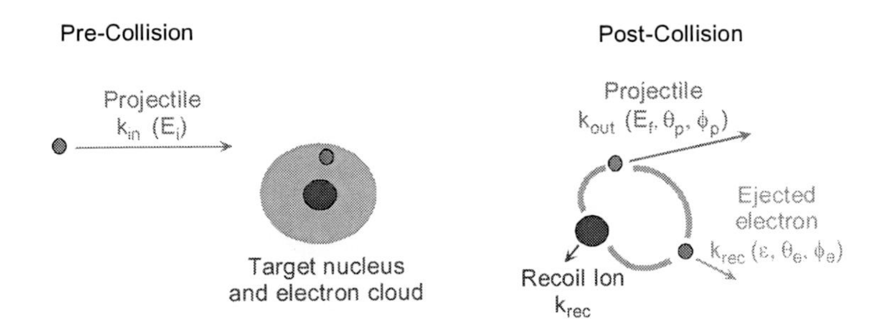

Figure 1. Schematic of a 3-body inelastic collision showing the incoming and scattered projectile plus the ejected electron and recoil ion.

As we will be discussing various types of integral and differential data and making comparisons between positron, electron, and photon impact ionization data, we must first understand how these are connected to the collision dynamics and what types of differential information can be obtained. Figure 1 shows a schematic of a 3-body, inelastic collision where the incoming projectile interacts with the bound electron. For ionizing collisions, the final state consists of a scattered projectile which has lost some energy, an ejected electron, and a recoil target ion. The forces between these particles are indicated. To completely define the interaction dynamics, the energies and angles, or all three momentum components, of two of the three particles need to be known. This, commonly referred to as triply differential cross section (TDCS) data, combined with conservation of energy and momentum, provides complete information about the collision kinematics and therefore the most stringent test of theoretical models.

In TDCS studies, the incoming and scattered projectile momenta define a collision plane, as shown in Figure 2, with the most likely process being the target electron ejected in this plane. As it scatters, the projectile loses energy and therefore transfers energy and momentum to the target electron and nucleus. The momentum transfer ΔK is

$$\Delta\vec{K} = \vec{k}_o - \vec{k}_f \tag{1}$$

$$\left|\Delta\vec{K}\right|^2 = k_o^2 + k_f^2 - 2k_o k_f \cos\theta \tag{2}$$

where k_o is the incident projectile momentum, k_f is the scattered projectile momentum, and θ is the scattering angle. Measuring different scattering angles and energy losses probes different values of momentum transfer and provides detailed information about the interaction and dynamics.

Traditionally, however, the energy-angles or momentum components have been measured for only one of the outgoing particles. This is equivalent to integrating over the parameters of the other particle, i.e., over all collision planes, and represents doubly differential cross section (DDCS) data. Singly differential (SCDS) experiments have also been performed where only angle or energy information has been measured for one particle. Finally, total ionization cross sections, σ_T, are obtained in experiments which integrate over all ejected electron or scattered projectile parameters or measure the total ion production. In cases where the ion production is measured, time-of-flight or magnetic selection can be used to provide partial cross sections for different degrees of ionization, σ_q, q being the degree of ionization.

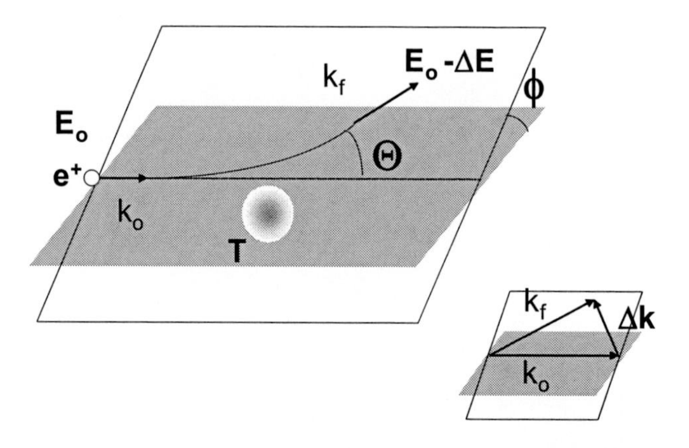

Figure 2. Schematic view of the scattering of a positron by a target. Scattering is in a collision plane defined by the incoming and scattered projectile directions; the plane is tilted at an angle φ is this representation. The small inset shows the initial, final, and momentum transfer vectors.

When the projectile velocity is large compared to the bound target electron velocity and the projectile is scattered at angles near zero, the momentum transfer is small. In this case, the dipole regime dominates and a connection with photon impact can be made. The connection can be summarized as follows. As shown in Figure 3, a particle with charge z and velocity v moving past a bound electron at an impact parameter b (part A) produces a time dependent electric field (part B). The energy loss by the projectile can be calculated by integrating the electric force over the collision time and is given by

$$\Delta E = \frac{\Delta p^2}{2m} = \frac{2z^2e^4}{mv^2b^2} \tag{3}$$

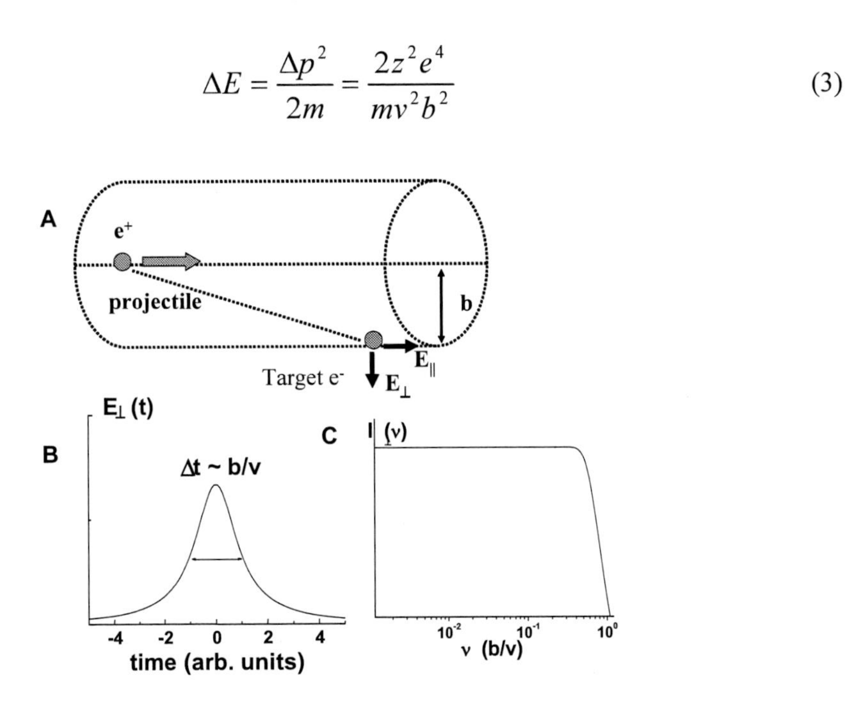

Figure 3. A) A glancing collision between the projectile and a target electron at rest. B) The transverse component of the electric field due to the projectile as seen by the target electron. C) The spectrum of virtual photons as a function of the photon frequency.

The connection with photoionization is that the sharply pulsed electric field in the time domain Δt~b/v for charged particle impact is equivalent to that produced by a beam of virtual photons. In the frequency domain the probability that these virtual photons have a particular frequency is a flat continuum with the maximum frequency being approximately v/b, as illustrated in part C of the figure. Thus the perturbation experienced by the target due to the interaction with a charged particle may be regarded as tantamount to a photon pulse whose frequency components are given by the Fourier transform of the electric field produced by the charged particle. Each individual component is equivalent to the electric field experienced by the target electron when it absorbs a photon of that frequency. Charged particle and photon impact data can be compared if information about the projectile energy loss is know since a particular energy loss is equivalent to the energy of a particular photon absorbed by the target. In making this comparison, it is important to remember that due to the high frequency cutoff to the virtual photon field such comparisons are most valid when a) the energy loss is small compared to the initial energy and b) the momentum transfer is also small (i.e., for small scattering angles) [3,4].

It must also be kept in mind that there is still a cardinal and significant difference between inelastic charged particle-atom and photon-atom interactions. Namely, for photon impact only the ejected electron leaves after the collision whereas for charged particle impact, the ejected electron plus a lower energy scattered particle leave. Thus, for charged particle impact the energy loss minus the ionization energy, ΔE - IP, is the energy imparted to the target electron. This is in pronounced contrast to the photon-impact case where, except for Compton scattering, the entire photon energy minus the ionization energy is imparted to the target electron.

Ionization Mechanisms and Total Cross Section Studies

Accounting for and modeling all of the ionization processes and effects mentioned above is a daunting task; so experimental guidance as to when various processes should be included or excluded is essential. This information can often be obtained from total cross section measurements and positron impact data has proven to be a powerful tool for this. For example, ionization induced by positron rather than by electron impact results in certain interaction channels being turned on or off, e.g., the inclusion of the positronium, Ps, formation channel and the exclusion of the electron exchange channel for positron impact being prime examples. Thus, information about the relative importance of these various channels and whether it is necessary to include them in theoretical models can be obtained. As an illustration, see Figure 4. The Ps channel is seen to dominate at low impact energies (in the few tens of eV range) whereas at higher impact energies the ionization-excitation channels, the dashed curve, are the most important. Data shown are from references 5-7.

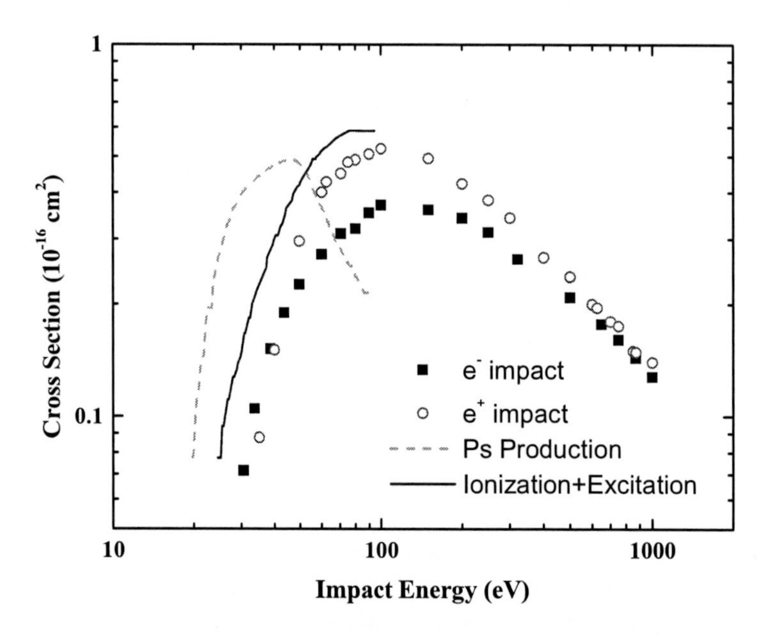

Figure 4. Cross sections for single ionization of helium by positron and electron impact. Green dashed curve: Ps formation channel from ref. 5; blue solid curve: Ionization + excitation channel from ref 5; open circles: e^+ impact data from Knudsen et al. (ref. 6); solid squares: e^- impact data from Shah et al. (ref. 7)

Another important use of positron data is to investigate how the projectile charge and mass influences inelastic collisions. For large impact velocities, perturbation theories such as the Born approximation predict that single ionization cross sections are proportional to the square of the projectile charge and therefore should be independent of the sign of the charge. But, at other energies or if one looks more closely, differences are found. Therefore cross sections for equal velocity positron, electron, and proton impact are extremely useful. For example, projectile charge differences such as target polarization and post-collision interactions can be investigated by comparing positron and electron impact data. With regard to target polarization, one would expect that for positron impact the target electron cloud is attracted toward the positron as it moves past the target atom or molecule and that this would enhance the interaction cross section. In contrast, for electron impact the electron cloud is pushed away and the cross section should decrease. These expectations are confirmed by the other curves shown in Figure 4. Note that the target polarization effects are largest around the cross section maximum and are still observable at impact energies of several hundred eV. The reader is referred to reference 8 for additional discussions along this line. It should be mentioned that electron exchange, which is present for electron impact but absent for positron impact, will also lead to a smaller electron impact cross section. However, and again the reader is referred to reference 8, it is generally assumed that electron exchange is not responsible for the large differences seen in Figure 4.

Figure 4 also shows that these effects become unimportant at high impact energies and the positron and electron impact cross sections merge. This is in accordance with the Born approximation which predicts identical single ionization cross sections for high energy positron and electron impact. In contrast, many experiments have shown that for double ionization the cross sections are significantly different, roughly a factor of 2. Similar differences have been found for proton and antiproton impact, always with the negative projectile yielding the higher cross section. This also means that the ratio of double to single ionization will be larger for negative projectiles, as demonstrated in Figure 5 for positron and electron impact [9,10]

Many years ago McGuire [11] provided an explanation for these charge related differences. He explained that they arise because double ionization can result either from *i*) a single interaction between the projectile and one of the target electrons which results in the emission of two electrons or *ii*) from two independent interactions during the same collision, each leading to the ejection of one electron. In the first case, the cross section amplitude is proportional to $[Ze/v]$; in the second it is proportional to $[Ze/v]^2$, Z being the projectile charge and v being its velocity. Adding amplitudes and squaring yields a double ionization cross section that has a cross term proportional to $[Ze/v]^3$ which is positive for negatively charged projectiles and negative for positively charged projectiles. This explanation is generally accepted and often referred to as the TS-1 and TS-2 mechanisms. Note that at low impact energies the TS-2 mechanism will dominate and the ratio of double to single ionization will have a v^{-1} dependence while at very high energies the TS-1 mechanism dominates and a constant value for the ratio is expected. According to the data in Figure 2 the TS-2 mechanism never truly dominates whereas additional high energy data are needed to confirm whether a pure TS-1 mechanism is responsible for the double ionization of argon at 1 keV.

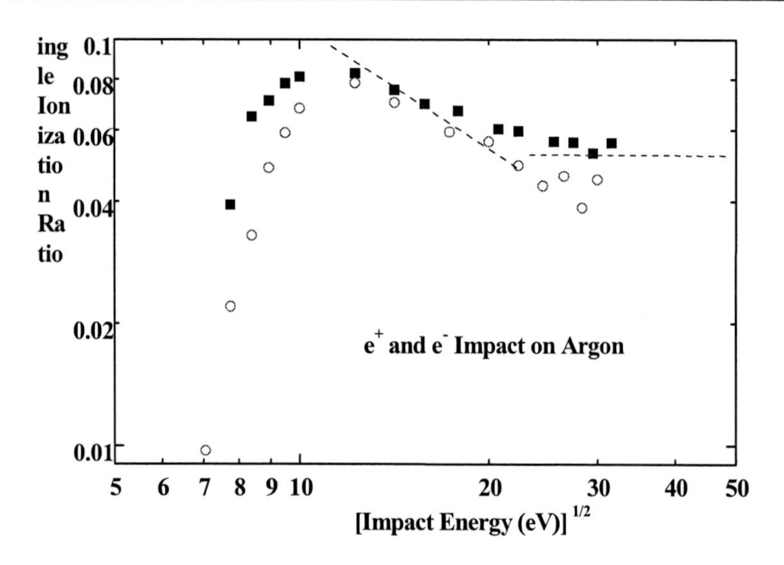

Figure 5. Double to single cross section ratios for positron and electron (open and solid squares) impact ionization of argon. Data are from Bluhme et al. (ref. 9) and McCallion et al. (ref 10). The sloped dashed curve represents a v^{-1} dependence of the ratio.

It should be mentioned that other conceptual arguments such as changes in the binding energies of the target electrons, changes of the projectile trajectory [12], momentum transfer to the target, inner shell contributions or differences coming from post-collision interactions have also been invoked to explain differences resulting from electron and positron impact [13,14].

Thus, by comparing positron and electron impact total cross section data important information about ionization mechanisms can be obtained. More importantly, these comparisons can be used to isolate and identify charge dependent effects. Likewise, by comparing positron and proton impact data, effects associated solely with the projectile mass can be isolated. It is well known that mass effects influence the amount of momentum and energy that can be transferred in a collision as well as the magnitudes of the cross sections at low collision velocities. However, there may be additional effects as implied by recent studies [15] where differences between electron and proton impact single ionization cross sections were noted for fast collisions with heavier atoms. If correct, this implies that simple, one-electron perturbation theories are inappropriate as they cannot account for these effects. More sophisticated theoretical treatments or models are required. It should be noted that in making comparisons in the search for charge or mass effects, it is better to use ratios of cross sections, e.g., double to single ionization, or comparisons of cross section shape rather than absolute magnitude as this will avoid uncertainties associated with absolute normalization of different sets of data. The reader is referred to ref. 8 for a good discussion of charge and mass effects.

With regard to these types of studies, total ionization cross sections are readily available for electron and proton impact. For positron impact, the beam intensities and hence signal strengths are much lower because of available radioactive sources and energy moderator efficiencies. However, total cross section data are available from several sources. Hence, positron-electron and positron-proton comparisons for all of the rare gases have been made. See ref. 8 and references therein.

Interaction Dynamics and Differential Cross Section Studies

As demonstrated, total or integral cross section measurements can provide information about which ionization mechanisms are important plus some information about charge and mass effects. However, for details about how these influence the collision dynamics and reaction kinematics, differential information is required. In particular, kinematic effects associated with altering either the projectile charge or mass provide severe tests of theoretical models. In order to advance towards a full description of the dynamics dominating the collision and to test the theoretical models available, fully differential studies for different projectile charges and masses are required. Differential measurements involving positrons also allow us to probe regimes where electron impact cannot provide information. For instance, the dynamics of large energy loss ionizing collisions cannot be investigated using electron impact since it is impossible to determinate if the outgoing particle was ejected from the atom or was the scattered projectile.

Changing the sign of the projectile charge changes attractive forces to repulsive and visa versa in Figure 1. This influences the scattered projectile and ejected electron kinematics, and hence the differential cross section, in two ways. First, it alters the pre- and post-collision target polarization. This influences both the impact parameter and the impact energy slightly depending on whether the projectile sees an attractive or repulsive force due to the target nucleus and its electron cloud. Second, the post-collision kinematics are influenced due to the reversal of the post-collision force between the scattered projectile and ejected electron. As will be shown, these effects have been observed in doubly differential data. To date no comparisons using TDCS have been possible, because prior to our efforts at UMR no TDCS positron impact data were available except for very limited kinematic parameters. However, as shown in Figure 6 theory predicts differences in the magnitudes, shapes, and orientations of the binary and recoil lobes measured for positron and electron impact. With respect to electron impact, for positron impact the recoil lobe is predicted to be larger whereas the binary lobe is smaller and oriented at a smaller angle. Obviously, TDCS data are needed for both electron and positron impact to test these predictions.

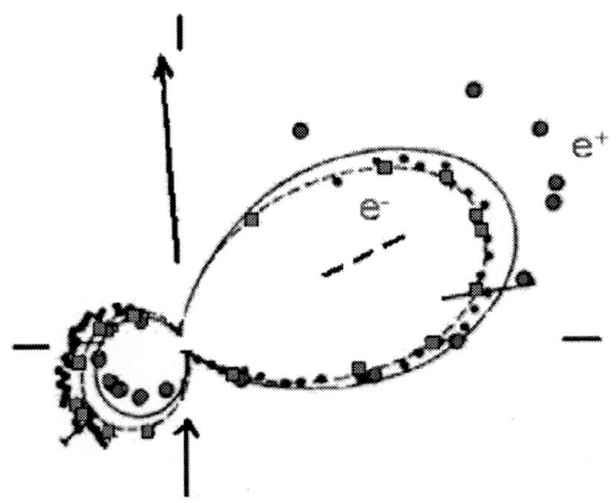

Figure 6. Theoretical predictions of TDCS for 10 eV electron emission resulting from 600 eV e^+ and e^- impact on helium. Large solid circles, 2nd Born predictions for e^+ impact; solid squares, solid and dashed lines, 2nd Born predictions for e^- impact; small dots, e^- impact data. The arrows show the directions of the incoming and scattered projectile while the dashed line shows the direction of momentum transfer. The short lines indicate emission at 0° and ± 90°. See ref. 16 for details.

Another very important aspect of changing the electron charge, i.e., using positron rather than electron impact, is that the scattered projectile and ejected electron can be unequivocally identified. This means that their kinematics can be investigated independently and individual components of theoretical models can be tested. Also, positron impact permits studying highly inelastic collisions for lepton impact. This cannot be done using electrons because only for small energy losses or specific situations is one rather confident as to which was the scattered and which was the ejected particle.

With regard to mass effects, changing the projectile mass, but not its velocity, drastically alters the projectile scattering angle and the percentage of the initial kinetic energy that can be lost in the collision. This influences the dynamics and determines whether a classical or a quantum mechanical model is required for describing the process. An example of a mass effect is that certain processes such as electron capture to the continuum appear as sharp peaks for proton impact [17] but as broad bumps for positron impact [18-20].

Experimental Methods for Differential Studies

Overview

For electron impact, singly and doubly differential cross section data have been available almost from the advent of atomic physics. In the 1960's the final step in probing interaction dynamics was made when Ehrhardt and collaborators [21] used two rotatable energy analyzers to measure the angular distributions for low-energy "target" electron emission in coincidence with fast, forward-scattered, "projectile" electrons. Figure 7 shows a schematic of their apparatus and an example of their TDCS data. Using effectively the same experimental techniques, TDCS studies have advanced significantly and a variety of collision systems have been studied for a broad range of impact energies. Ionization of specific subshells has even been measured [22].

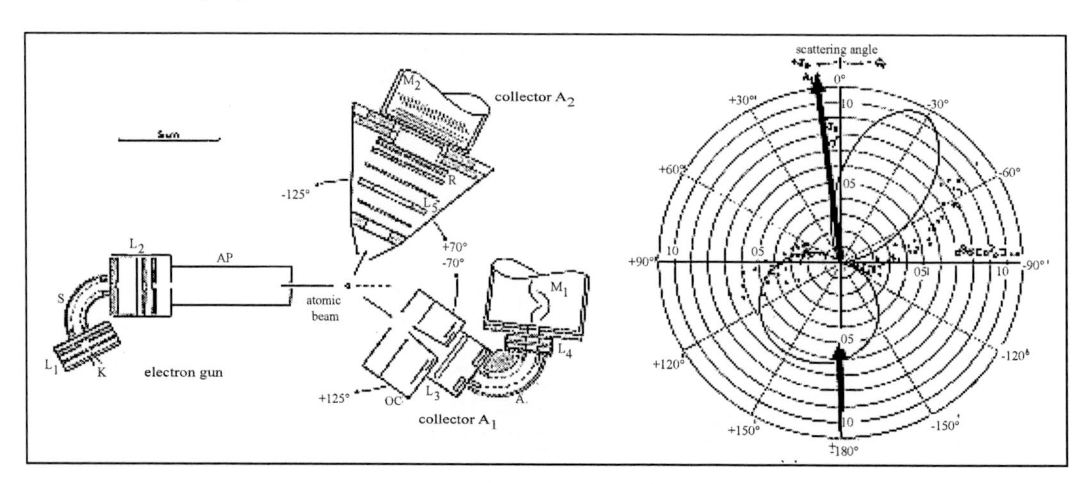

Figure 7. Apparatus used for TDCS for electron impact. [from Ehrhardt et al. [21]. The apparatus has two rotatable energy analyzers, one for the high energy scattered projectile and one for the low-energy ejected electron. Sample data are for 114 eV electron impact on helium. The arrows indicate the directions of the incident and scattered projectile. The dots are for emission of 15 eV electrons; the solid lines are theoretical expectations.

In comparison, positron impact studies are far behind. This is because the typical positron beam intensities that can be obtained using radioactive sources and energy moderators are femtoamps or smaller rather than nanoamps or microamps as is the case for electron impact. This is a major limitation. As a result, the vast majority of positron impact studies have concentrated on total cross section measurements. However, in the past couple decades a few pioneering positron impact differential ionization studies have been performed.

To give a feel for the difficulties involved, for total cross section measurements the signal rates are typically a few percent of the beam rate. Thus for a 1 femtoamp beam, the signal rates are several hundred Hz; such rates are easily measured. However, in differential studies only a portion of phase space is sampled. For example, using a single energy analyzer for a DDCS study limits the acceptance solid angle, $\Delta\Omega$, and the energies detected, $\Delta\varepsilon$. Typical values are $\Delta\Omega$ =1-10 millisteradian and an energy resolution of approximately 10%. Here $\Delta\varepsilon$ is equal to the energy resolution times the energy detected. For electron beam intensities on the order of 1 microamp, the DDCS signal rates typically range from a few tens of Hz to many hundred Hz. This means that the expected DDCS rates for a 1 femtoamp positron beam are much too small to be measured using standard electron impact techniques.

For TDCS measurements, the signal rates are even smaller since the second spectrometer samples only a portion of the DDCS signal. To illustrate, Figure 8 shows the solid angle sampled by an ejected electron detector positioned at some large angle with respect to the incoming projectile plus the scattered projectile intensity and solid angle sampled by a projectile detector positioned at some small forward angle. For intermediate or higher energy electron impact ionization and low energy electron emission, the scattered projectile intensity decreases rapidly between 0° and 90° and then increases slightly. [23,24] For a DDCS measurement the signal rate is proportional to this scattered projectile intensity integrated over all scattering angles; for a TDCS measurement it is proportional to the number of projectiles scattered within the projectile detector solid angle, $I_p(\Delta\theta_p,\Delta\varphi_p)$, e.g., within the box shown in Figure 8.

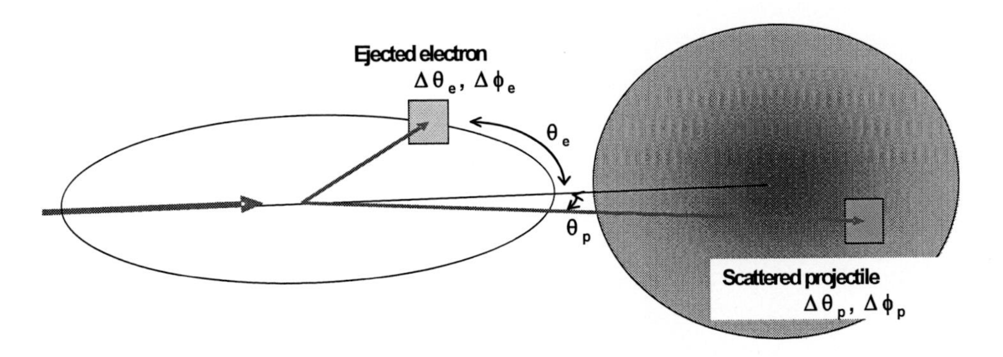

Figure 8. Schematic of the scattered projectile intensity and solid angles for detecting the ejected electron and the scattered projectile. The open circle represents the scattering plane defined by the incoming and outgoing projectile vectors (the red arrows). Not shown are the out-of-plane angles, φ_e.

To estimate how much smaller the TDCS rates are as compared to the DDCS rates let us assume 500 eV impact, a small energy loss, a 5° scattering angle, and that we wish to define the ejected electron and scattered projectile energies to ±½ eV, the ejected scattering angle, θ_e, to ±3°, and the collision plane to ±5°. Let us also assume we use hemispherical spectrometers for detecting both the scattered projectile and the ejected electron and that their entrance apertures are 5 cm from the scattering center. The energy resolution, $\Delta\varepsilon/\varepsilon$, for hemispherical spectrometers is roughly $\Delta r/R$ where Δr is the entrance aperture width and R is the mean radius of the inner and outer hemispheres. Here, we shall assume R = 5 cm. Thus, for 1 eV resolution at 500 eV, Δr_p must be approximately 0.1 mm, which in turn means that $\Delta\theta_p$ in Figure 3 is approximately 0.12°. The projectile scattering intensity corresponding to a DDCS measurement for a particular electron emission energy is given by $\int_0^{2\pi} d\phi_p \int_0^{\pi} I_p(\Delta\theta_p,\Delta\phi_p)\, d\theta_p$, whereas for a TDCS measurement the scattered particle intensity sampled by the projectile solid angle is $\int_0^{\Delta\phi_p} d\phi_p \int_{\theta_p}^{\theta+\Delta\theta_p} I_p(\Delta\theta_p,\Delta\phi_p)\, d\theta_p$. Fitting data from references 23 and 24 with a polynomial, these integrals were calculated for 5° scattering and the values of $\Delta\theta_p$, $\Delta\varphi_p$ estimated above. Doing so yields a TDCS signal that is smaller than the DDCS signal by roughly a factor of 35. This explains the limited number of data points and error bars for electron impact TDCS data plus why innovative, special techniques are required to make such measurements possible for positron impact.

Various methods can be used to greatly enhance the signal collection efficiency in order to make differential measurements possible for positron impact. Options include using large solid angle detectors which may or may not be equipped with position-sensitive anodes or using biased grids or time-of-flight rather than traditional spectrometers for energy analysis. Where high signal rates are expected, specially designed spectrometers can be used. In combination with these options, the use of multi-particle coincidence techniques helps considerably since background contributions are virtually eliminated. This allows reliable data to be extracted under the low-signal rate conditions for positron impact. Another option to greatly improve the data collection efficiency is the recoil ion momentum spectroscopic technique (RIMS), where the momenta of several products of a single collision are measured simultaneously. However, as will be discussed below, different RIMS methods than those currently used are needed for positron experiments based on radioactive sources.

Previous Methods and Results

As an example of one of these techniques used to obtain SDCS data, Falke et al. [25] used a relatively simple setup consisting of biased grids and rotatable channel electron multiplier detectors to measure single and double ionization of argon, positronium formation (electron capture), and transfer ionization, all as a function of observation angle. These data and a schematic of the apparatus are shown in Figure 9. Here, the curves have been displaced vertically for display purposes. Note that the single electron processes, namely single target ionization and Ps formation, have similar angular distributions while two-electron processes, e.g., double and transfer ionization, have nearly identical angular distributions.

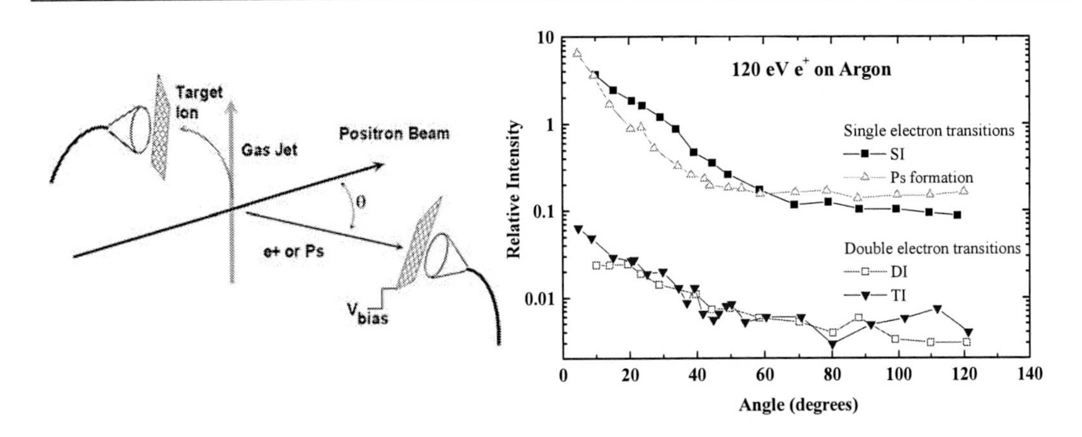

Figure 9. Schematic of apparatus (left) and SDCS data (right) from Falke et al. [25]. Data show angular dependences for one and two electron processes in 120 eV e^+ – Ar collisions. Final state products are: single ionization (SI): e^+,e^-,Ar^+; Ps formation: Ps,Ar^+; double ionization (DI), e^+,$2e^-$,Ar^{2+}; transfer ionization (TI): Ps^-,Ar^{2+}. The curves have been shifted vertically and normalized for display purposes.

Another SDCS measurement was performed at University College, London, (UCL) where a hemispherical collector system was used to measure the energy spectrum for electron emission due to positron impact. [26]

With regard to DDCS, a few positron impact measurements have been made. Schmitt et al [27] used a rotatable energy analyzer to measure and compare 15 eV electron emission between 20 and 90 degrees for 100 eV positron and electron impact. Using this method, accumulation times for each for positron impact data point were on the order of 10^5 seconds. Thus, to reduce background effects, the emitted electrons were measured in coincidence with singly ionized argon ions. These data, shown in Figure 10, demonstrate that a larger number of electrons are ejected in the forward direction for positron impact than for electron impact. This enhancement for positron impact, suppression for electron impact, in the forward electron emission is due to the attractive/repulsive post-collision forces that were mentioned earlier.

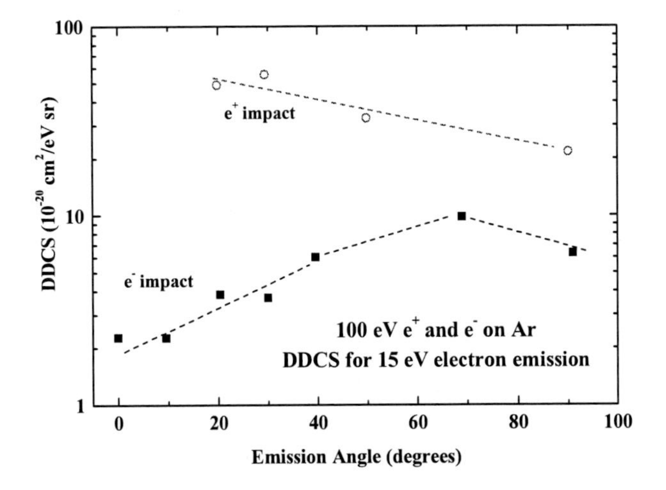

Figure 10. DDCS electron emission data from Schmitt et al. [27]. Data are for 100 eV positron and electron impact ionization of argon and 15 eV electron emission. Dashed lines are to guide the eye.

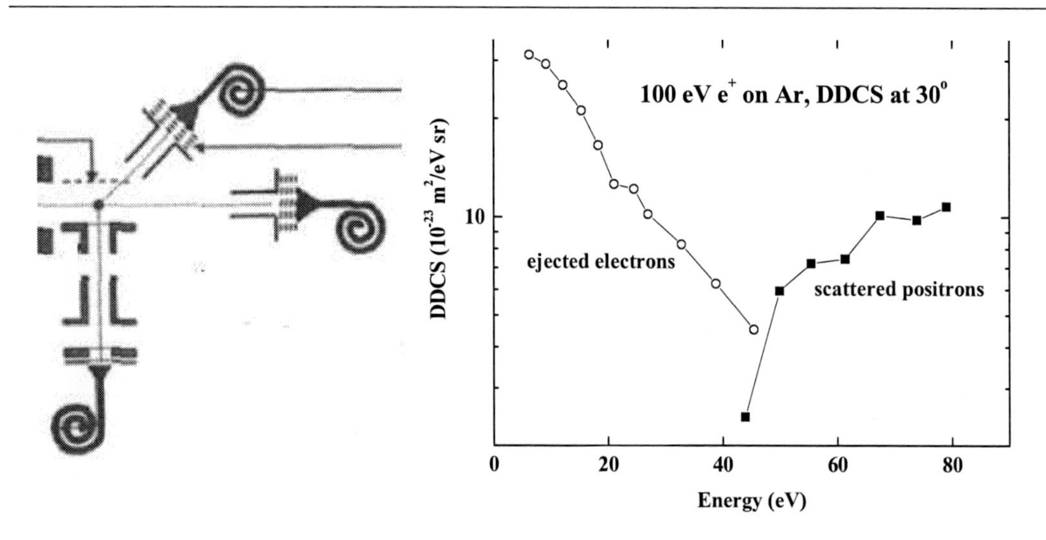

Figure 11. UCL DDCS apparatus and measurements of the ejected electrons and scattered positrons for 100 eV e^+ impact on argon. The observation angle is 30°. Data are from ref. 28.

At University College in London, Kövér and Larricchia and coworkers [28,29] have performed a couple doubly differential electron emission studies. Using a rotatable channel electron multiplier and biased grids for energy analysis, DDCS for the ejected electrons and scattered positrons were measured at 30° and 45°. Coincidences with target ions were used to eliminate background contributions. This method is similar to that shown in Fig. 9, the difference being that ejected electrons and scattered positrons are collected by the rotatable detector. In this work, the coincidence signal rates for positron impact ranged from 10^{-2} to 10^{-4}/s; for electron impact, the rates were considerably higher. Figure 11 shows a schematic of their apparatus and a comparison of their 30° positron and electron impact data. An obvious difference in shapes is seen. However, interpreting these differences depends on how the data are normalized with respect to each other. We do note that as shown, for 15 eV ejected electrons, the DDCS is larger for positron impact than for electron impact, just as was seen in Fig. 10. Another important aspect of these data is that for positron impact the projectile and target contributions to the DDCS are measured separately, something that cannot be done in "standard" electron impact e-2e measurements. Thus individual parts of theoretical descriptions can be tested.

The main efforts of the UCL group have been to investigate electron capture to the continuum (ECC) [18-20]. In this process, the post-collision Coulomb interaction between the ejected electron and the scattered positron can cause the two particles to emerge from an ionizing event with closely matched velocities and directions, i.e., the electron may be considered to have been transferred to a continuum state of the scattered positron. This process is most effective at lower impact energies which provide more time for the outgoing particles to interact with each other. Due to the equal masses of the positron and electron, ECC effects are maximum when each particle has the same kinetic energy which is $(E_0 - IP)/2$ where E_0 is the incident energy and IP is the ionization potential of the target. For example, for 100 eV positron impact, the 0° spectra for electron emission or scattered positrons would both have a peak at 43 eV. The reader should keep in mind that for electron impact, the post-collision interaction should lead to a minimum in the scattered plus ejected energy distributions at 0°.

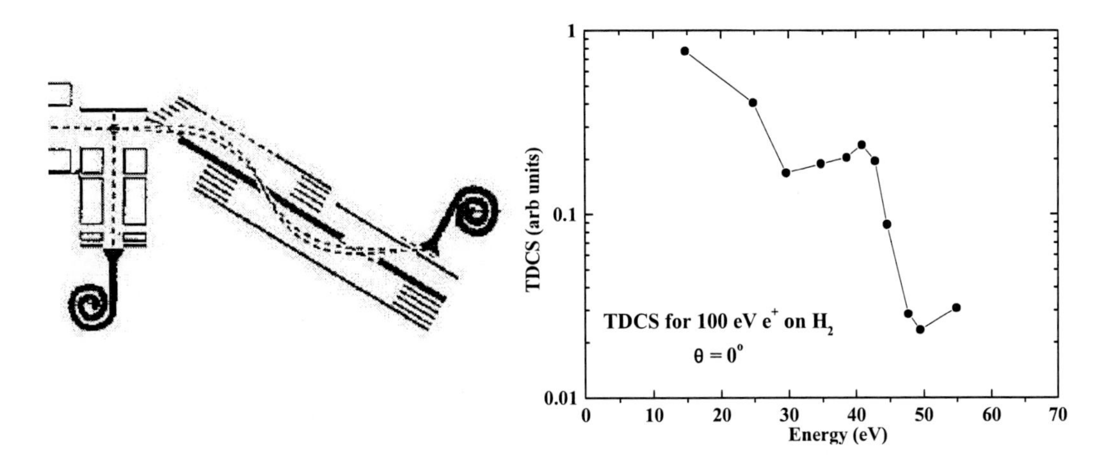

Figure 12. UCL zero degree apparatus and TDCS data for 100 eV e^+ impact ionization of H_2. Data are for electron emission in coincidence with positron scattering, both at 0°. From ref. 20.

For this they used a specially designed spectrometer to measure both the scattered positrons and the ejected electrons at 0°. As the signal rates are relatively large at this angle, no special techniques such as large solid angle detectors or special energy analysis techniques were needed. A schematic of the zero degree spectrometer for these measurements is shown in Figure 12. These zero degree data represent the first TDCS for positron impact but because of the angle chosen provide only limited tests of TDCS theories. Also, no electron or proton impact data direct are available for making direct comparisons. However as mentioned previously, with regard to mass effects, the broad cusp seen for positron impact is in sharp contrast to the narrow cusp found for proton impact.

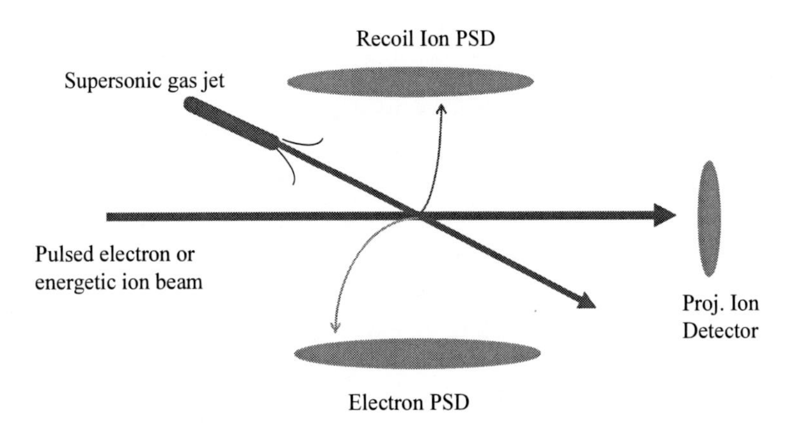

Figure 13. Schematic diagram of a recoil ion momentum spectroscopy apparatus. Here a weak electric field is used to extract recoil ions and electrons produced at a point interaction region where the beam and a supersonic jet cross. Time-of-flight and position-sensitive detection of the ejected electron and recoil ion provide information about their momentum components; coincidences between these particles provide full kinematic information for single ionization events.

As mentioned above, RIMS techniques greatly improve the data collection efficiency. This has resulted in many new insights into the collision dynamics for electron, ion, and photon impact.[30] For electron impact, highly detailed information has been obtained for both single and double ionization of atoms. [31 and references therein] A schematic of the RIMS technique is shown in Figure 13. The basic principle of this technique is to use a weak electric field to extract target ions produced in a collision. Since the ion energies are milli-eV, the detection solid angle is 4π. Using a position sensitive detector to measure the x,y positions and time-of-flight techniques, all three momentum components of the ion can be determined. By adding a magnetic field and a second position sensitive detector, similar data provide the momentum components for a large percentage of the ejected electrons. These six momentum components plus conservation of energy and momentum equations are sufficient to completely define the collision dynamics, i.e., to measure TDCS. Resolution conditions, particularly for the recoil ions, require a cold, point target. This is achieved by crossing the projectile beam with a supersonic gas jet where the gas has been precooled to cryogenic temperatures.

For positron studies based on radioactive sources, several difficulties must be overcome in order to use the RIMS method. The first has to do with measuring the flight times. This requires knowing when the collision took place, i.e., knowing time zero. For ion impact where the projectile scattering angles and energy losses are small, this can be done by detecting the post-collision projectile ion and projecting backwards in time to when the collision took place. For lepton impact the scattering angles can be many tens of degrees and the projectile can lose a large percentage of its initial velocity. Thus, it is not possible to accurately predict when the collision took place. In the case of electron impact, a pulsed beam can be used to overcome the time zero problem. However, for the weak positron beams that can be produced using radioactive sources, pulsed beams cannot be used due to the low duty cycle (beam on/beam off times). So a different approach is needed. One method is to use position sensitive detection of all three particles. In this case, measuring the three x,y positions with no time-of-flight information would be sufficient to completely define the kinematics. This could be achieved either by using a "standard" RIMS setup where two of the post-collision particles are measured by one of the detectors or by adding a third position sensitive detector.

The second problem associated with the RIMS technique that is important for positron impact studies is that the achievable resolution requires using a cold, point target. Thus, the beam diameter is typically 1 mm and a supersonic gas jet with cryogenically cooled gas source is used. For positron impact, larger beam diameters must be used due to the weaker beams that can be produced. Plus, target atom densities achieved using a supersonic jet positioned many centimeters away from the collision region are generally less than those for differentially pumped gas cells or simple gas jets. This leads to less beam-target overlap and lower signal rates, plus lower resolution data than for electron impact.

UMR Differential Ionization Experiments

Initial Methods and Results

Approximately ten years ago, at UMR we started a program with the goal to perform highly differential ionization studies for positron impact. The initial concept was to use RIMS techniques, simply because many different kinematic parameters can be collected simultaneously and with high efficiency. Our initial plans were to inject the positron beam along the longitudinal axis of the spectrometer, as shown in Figure 14. This requires passing the beam through the center of the entrance and exit detectors. To do this, we used a triangular configuration of three 50 mm diameter channelplates and a delay line anode with gaps in the horizontal and vertical windings. To hide the beam from the large detector and anode voltages, grounded tubes extended through the entire detector. As seen, in this configuration the exit detector would be sensitive to forward scattered positrons (shown by the solid red line) and, by using double hit electronics, also to the recoil ions (shown by the solid blue arrow) arriving a few microseconds later. Depending on the electric field strength and size of the detector, a percentage of the ejected electrons (shown by the dashed green lines) could be directed to and detected by the entrance detector. Detection of all three particles and their positions would be sufficient for TDCS information. In addition, this method would simultaneously provide singly, and doubly, differential information about projectile scattering and about the electron emission. Subject to signal rates, coincidences with recoil ions would generate these data for different degrees of target ionization.

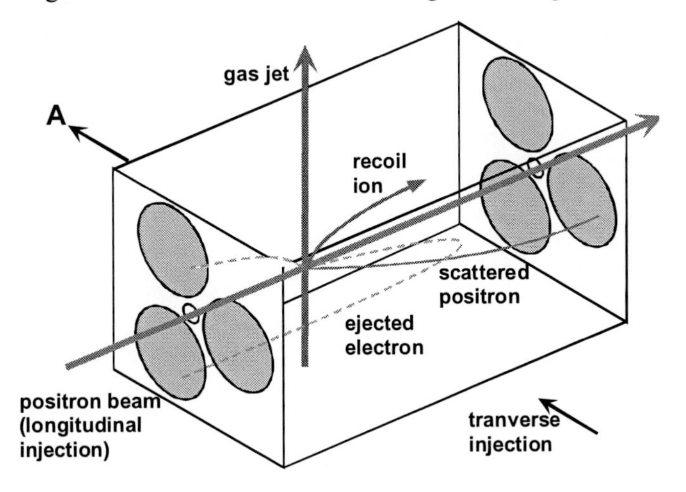

Figure 14. Schematic of RIMS setup originally planned for UMR positron studies. Plans were to inject the beam longitudinally and record forward scattered positrons and recoil ions at the spectrometer exit. Electrons emitted in the backward direction or turned around by the electric field would be detected by the entrance detector. The transverse line A indicates the direction used for the initial studies.

Although highly efficient, the major experimental difficulty with this method lies with passing a low-energy beam though the entrance detector. To minimize loss of solid angle, a 2 mm diameter was planned. Unfortunately, we were unable to focus enough of our positron beam intensity to a small enough diameter to pass it through the tube, so this method was abandoned and the following approach was pursued.

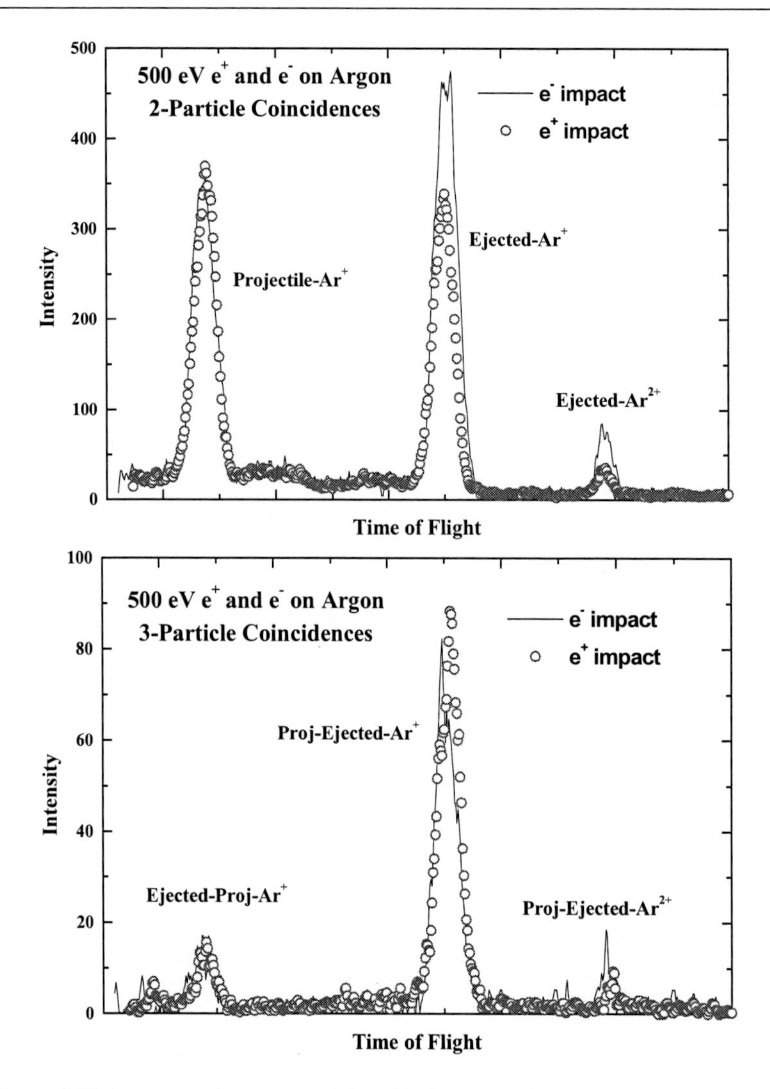

Figure 15. Time-of-flight spectra for two-particle (left side) and three-particle (right side) coindidences for 500 eV positron impact ionization of argon. The peaks are identified as to which of the post-collision particles are detected. See text for details.

Rather than injecting the beam longitudinally, it was injected transverse to the electric field. This eliminated the problem with beam focus and by adding another channelplate detector at point A, scattered projectiles could still be detected rather than by using the spectrometer recoil ion detector as shown in Figure 14. Standard RIMS techniques using recoil ion-ejected electron coincidences could still be used to generate doubly and triply differential ionization data. Figure 15 shows an example of initial data obtained using this method. For these data, the projectile detector was sensitive to projectile scattering between 0° and ±15°. The upper portion of the figure shows 2-particle coincidences for both positron and electron impact, i.e., scattered projectile-recoil ion or ejected electron-recoil ion coincidences, while the lower portion shows 3-particle coincidences, i.e., scattered projectile-elected electron-recoil ion coincidences. For experimental reasons, the "scattered projectile" and "ejected electron" spectra were combined by using a time delay to shift them with respect to each other.

These data provide several interesting pieces of information. Most importantly, they demonstrate that the signal rates are sufficient to allow TDCS measurements for positron impact; not just for single, but also for multiple, ionization processes. The data also provide information about charge effects. For example, when the projectile-Ar^+ single ionization intensities are normalized together, there is more intensity for electron impact in the 2-particle electron emission single ionization peak but not in the 3-particle coincidence peak. Perhaps this is due to large angle scattered events which are also counted by the ejected electron detector. Also note the larger intensity for electron impact double target ionization, consistent with our earlier double ionization discussions. The other interesting information seen in the lower spectrum, which was obtained using double hit electronics, is the large peak in the center which indicates that the scattered projectile detector recorded the first hit with the "ejected electron" being counted later, i.e., the projectile was fast, the ejected electron was slow. In contrast, the small peak at the left side of the spectrum is for events where just the opposite happened, i.e., the electron detector signal was first, the projectile detector signal was second. This could indicate that the projectile was scattered through a large angle and the electron was ejected in the forward direction.

After initial efforts, using a supersonic jet was abandoned because the signal rates were very low making it difficult to establish an overlap with the positron beam. The low signal rates were due to a) low beam intensity, e.g., approximately 2-3 kHz, b) large beam diameter, approximately 4 mm, which reduced the amount of overlap with the gas jet, and c) low target atom densities for the supersonic jet positioned many centimeters away from the collision region. Note that for electron impact experiments, the beam intensity can be increased to compensate for low target density; for positron impact, this is not possible.

Because of the low signal rates, simple monitoring of the recoil ion rate could not be used to adjust the jet position and establish a beam-target overlap. The 2D recoil ion signal could be used but was slow and inefficient. Much higher rates could be obtained by reversing all voltages in the beamline in order to create an electron beam using secondary electrons produced at the energy moderator. However, the transverse electric field in the spectrometer deflected the electron beam slightly and in the opposite direction as the positron beam. That, and because of slight changes in the injected beam trajectories, finding the correct overlap for positron impact proved to be extremely difficult and uncertain.

To overcome limitations of low target densities, the supersonic jet was replaced with a simple gas jet emerging from a needle. The needle was placed as close to the beam as possible and could be positioned from outside the chamber. For this setup, the signal rates were higher and an overlap could be quickly established for electron impact using the recoil ion 2D images. For positron impact, this could still be done, but was slower. By using a higher electric field, these 2D images allowed i) maximizing the beam-target overlap, ii) estimating the target density and overlap profile by comparing the overlap signal intensity with the background intensity along the beam path (the absolute background pressure was known), and iii) "smoothing distortions" in the electric field near the needle by making sure that the spot produced by ions extracted from the beam-jet overlap region was located on the line produced as the beam interacted with background gases during its passage through the spectrometer. The electric field distortions were because the needle was inserted near the center of the spectrometer. This introduced an asymmetric potential near the beam-target overlap volume. Monitoring the 2D recoil ion images showed when the distortion in the vicinity of the interaction volume could be minimized by applying small voltages to the needle.

The final two changes that were made were to 1) design and construct a large acceptance angle projectile energy analyzer equipped with a position-sensitive detector sensitive to a wide range of projectile energies and 2) reduce the size of the spectrometer. For the energy analyzer, many designs were investigated with the final design consisting of a modified cylindrical spectrometer, as shown in Figure 16. For the large acceptance angles we needed, SIMION tracings showed that various energies focused on the exit plane shown, thus the deviations from the normal cylindrical spectrometer configuration. In order to "collect" a wider range of vertical (out of the plane of the figure) scattering angles and project them on the channelplate, the inner cylinder has a shorter length than the outer cylinder. In its original configuration, horizontal scattering angles between ±17° were focused at the exit plane while vertical scattering angles between approximately ±20° were detected. By deflecting the incoming beam in a known fashion, it was shown that the beam position on the detector was linear as a function of the vertical scattering angle for angles less than ±12°. Reducing the size of the recoil ion spectrometer was done to minimize the beam deflection as it passed through the electric field and to reduce the recoil ion flight times to be compatible with available electronics.

Figure 16 illustrates the essential features of the apparatus. As seen, projectiles having different post-collision energies are focused to different locations on the projectile position-sensitive-detector (PSD). For small energy losses, the main beam also is detected. It should be noted that sweeping the main beam across the detector by changing the spectrometer voltages showed "lensing" effects resulting in an enhancement/loss of intensity at certain points. This happened when the localized beam swept across individual wires of the one-dimensional grid that was placed at the exit of the projectile spectrometer in order to smooth the electric field. However, these lensing effects are strongly suppressed for a broad beam, such as our continuum spectrum of energies for scattered projectiles.

In order to intersect the gas jet at the center of the extraction region, a set of biased plates deflected the incoming beam in the opposite direction than the extraction did. These deflections are greatly exaggerated in the figure. Recoil ions were extracted and time focused onto a second PSD. Using Roentdek delay line anodes and COBOLDPC list-mode data acquisition software [32], time-of-flight coincidence spectra were collected using the recoil ion channelplate signal as a start and the ion and projectile anodes signals for stops. The time-of-flight spectra provided information about the recoil ion charge states, i.e., the degree of target ionization, and the 2D projectile spectra provided information about the energy lost in the collision. The 2D recoil ion spectra provided information about beam-target overlap, background contributions, and where the collisions occurred. Further details and examples of data obtained using this setup are provided in the next section.

First, however, a brief discussion about our positron beam production and transport is in order. A ^{22}Na source, mounted on a linear manipulator so it could be positioned close to a tungsten mesh moderator, was used for positron production. The moderator had a 6 mm diameter. The extracted positrons were transported to the scattering chamber using a simple transport system which consisted of an anode, electrostatic lenses, plus vertical and horizontal steerers. The anode and first lens were followed by a 15° deflection in order to prohibit line-of-sight high energy positrons and photons from entering the scattering chamber. After deflection, two more lenses plus horizontal and vertical steering plates were used to focus the positron beam through the entrance aperture located a couple centimeters before the RIMS spectrometer. The source, moderator, and transport system were shielded by lead and surrounded by magnetic shielding.

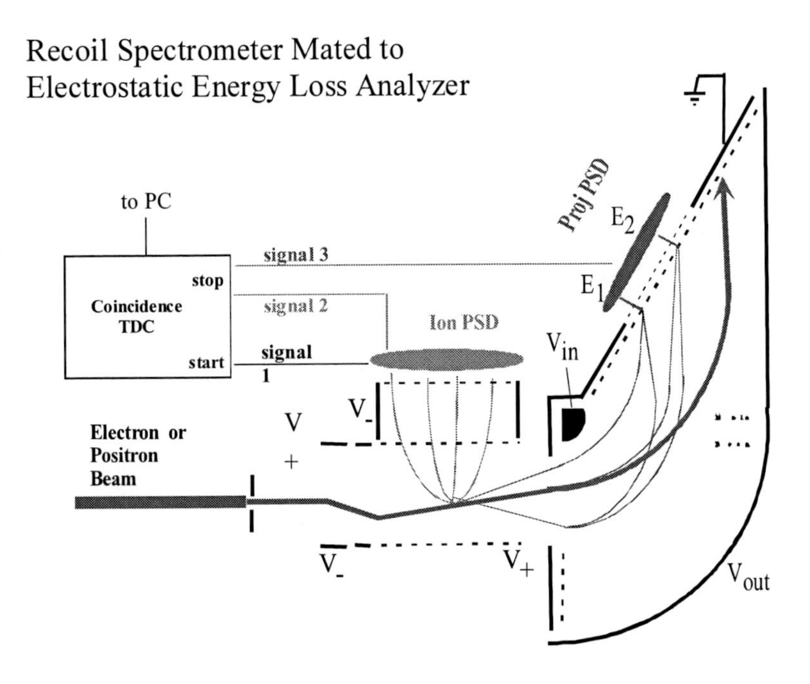

Figure 16. Apparatus used at UMR for energy loss and DDCS studies. Red indicates the recoil ion trajectories and PSD; blue are the trajectories and PSD for scattered projectiles.

Initially, the beam transport components had inner diameters of 12 mm in order to minimize the probability of high energy photons generating background counts on the channelplate detectors. As background counts did not present a problem, the inner diameters were increased to 25 mm in order to increase the transported beam intensity. SIMION ray tracings indicated that most 0-1 eV positrons extracted from the moderator could be focused on the chamber entrance to a spot size approximately the same diameter as when they were produced and that the beam was fairly parallel at the interaction volume. Unfortunately, the beam intensity on target has always been much smaller than anticipated, e.g., approximately 3 kHz for a 50 mCi source. This may be, in part, due to moderator efficiency but is certainly due to beam focus at the chamber entrance, e.g., measurements of the beam intensity just before the entrance aperture plus monitoring the injected beam with a position-sensitive detector indicates considerable beam loss at the entrance aperture. When the maximum extracted positron energy was increased to approximately 3 eV in the SIMION simulations, a much poorer focus was found, more in agreement with our observations. Various methods that we have used to compensate for low positron beam intensities include increasing the beam diameter from 4 to 6 mm, using the highest target pressure possible, increasing various solid angles, and maximizing the recoil ion detection efficiencies. Note that, unlike for total cross sections measurements where magnetic fields are used for very efficient beam transport, such methods cannot be used for our differential studies unless the magnetic field can be terminated in a manner which allows efficient extraction of a parallel beam.

For electron impact, two methods were used. First, and simplest, was to reverse all voltages in the transport system and use the secondary electron emission from the moderator. This provided beam intensities roughly one order of magnitude larger than the positron intensity. However, the energy width of the beam was rather broad. The other method we have used is to insert a small electron gun in the beamline, just before the entrance aperture. The energy width in this case is comparable to that of the positron beam. In either case, the intensity is sufficient such that a smaller entrance aperture can be used. Note that for electron impact TDCS measurements, we found that limiting the beam intensity to less than 10 kHz yielded the cleanest data.

Energy Loss and DDCS Measurements

Using the setup shown in Fig, 16, our first experiments employed recoil ion-projectile coincidences to study single and multiple target ionization as a function of projectile energy loss. This was done for both positron and electron impact and, in order to enhance the beam intensity and signal rates, relatively high impact energies were used. Doing so allowed us to detect a broader range of energy losses plus decreased the scattering angles.

The processes studied were

$$e^{\pm}(E_o) + A \rightarrow e^{\pm}(E_o - \Delta E, \Delta\theta_p, \Delta\phi_p) + A^{n+} + ne_{\mathrm{T}}^{-} \tag{3}$$

Where E_o is the initial projectile energy and ΔE is the energy loss, $\Delta\theta_{\mathrm{p}}$ and $\Delta\phi_{\mathrm{p}}$ are the azimuthal and polar scattering angular ranges sampled with θ_{p} and ϕ_{p} both being 0°. n is the number of target electrons, e_{T}, removed.

Experimentally, coincidences between scattered projectiles and singly or multiply charged recoil ions were measured. In practice this means either setting windows on the projectile 2D spectra which correspond to various energy losses and sorting the data to generate recoil ion time-of-flight spectra for specific energy losses, Figure 17, or setting windows on the various charge state peaks in the time-of-flight spectra and sorting the data to generate projectile 2D spectra for single, double, and triple ionization. An example for single ionization is shown in Figure 18. In both cases, different windows are used for determining random background events, which are then subtracted and a window was placed on the recoil ion 2D spectra to only accept events from where the beam-gas jet overlapped.

The time-of-flight spectra clearly demonstrate that the relative percentage of multiple ionization increases significantly with increasing energy loss. Note that the intensities have not been corrected for recoil ion detection efficiencies. Doing so would decrease the multiple ionization peak intensities to the order of 30–40% with respect to the single ionization intensity. Contributions from N_2 background gas ions are also visible. The bottom right part of Figure 17 indicates more 'noise' due to lower statistics since their yields are several orders of magnitude smaller than the top left part of Figure 17.

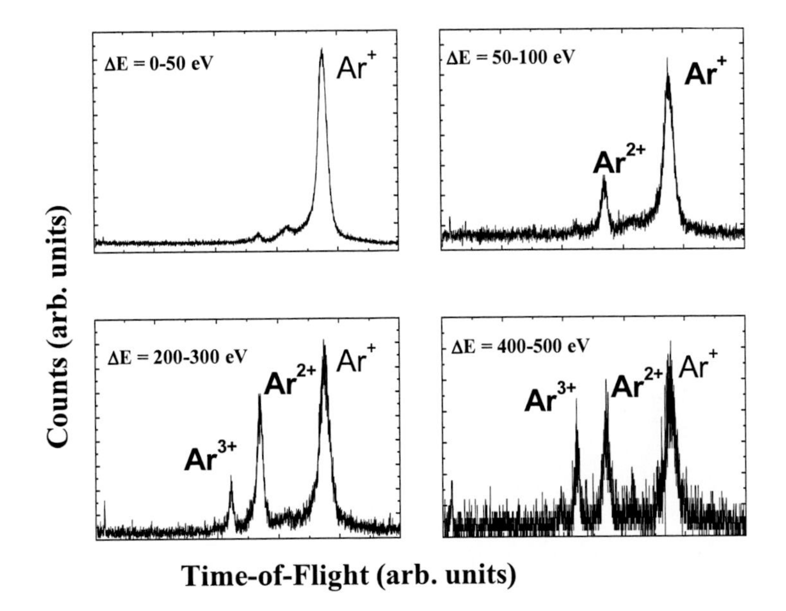

Figure 17. Time-of-flight spectra for 1 keV electron impact on argon. The spectra are for different energy losses, ΔE, and for projectile scattering angles between $\pm$ 17°.

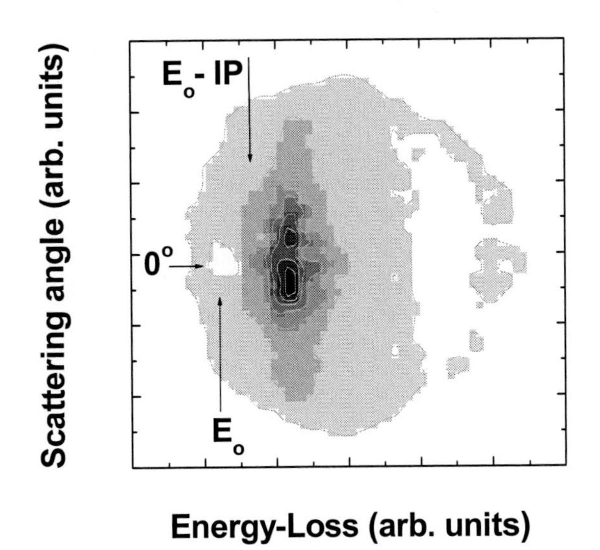

Figure 18. 2D plot of scattered projectiles as a function of the energy-loss (horizontal scale) and scattering angle (vertical scale) for single ionization of argon by 1 keV electron impact. E_o indicates the position of the main beam and E_o – IP the threshold for single ionization of argon.

These data are for 1 keV electron impact on argon. In the top right spectrum, the energy loss is sufficient to remove one or two electrons from the M shell of argon. Here, single electron removal dominates. The bottom right spectrum shows energy losses sufficient for removing several M-shell electrons or an L-shell electron. Here, subsequent Auger processes can then lead to double, triple or higher degrees of target ionization. For the bottom right spectrum the projectile has lost approximately nearly half of its initial kinetic energy and triple ionization has increased to the few per cent level.

Turning our attention to the projectile 2D spectrum in Figure 18, where positive and negative projectile scattering angles are the upper and lower vertical axis and projectile energy loss is the horizontal axis. In this figure, several things are apparent. First, the "hole" seen at location 0°, E_o is where the main beam was detected. The hole comes from subtracting a background that is slightly too large, since negative numbers are not visible in the log intensity display used in the figure. The next thing that is noted is the high intensity (dark region) that appears for energy losses slightly larger than the ionization potential. This intensity, which corresponds to the single ionization yield, decreases with increasing energy loss or scattering angle.

Since this 2D spectrum is for single ionization, the intensity within a box centered at a particular energy loss and scattering angle is related to the doubly differential cross section for single ionization at that angle and energy loss by

$$\frac{d^2\sigma^{q+}}{d\Omega dE} = \frac{N_{RP}^{q+}(\Omega, E_o - \Delta E)}{\eta_{q+}(N_P/\eta_P)N_T\Delta\Omega\Delta E} \tag{4}$$

Here $N_{RP}^{q+}(\Omega, E_o - \Delta E)$ is the number of recoil ions of charge state q measured in coincidence with projectiles having a final energy E_o - ΔE that are scattered into angle Ω, i.e., the intensity within the box. N_T is the number of target atoms, N_P is the beam intensity (number of projectiles), η_{q+} and η_P are the recoil and scattered projectile detection efficiencies, $\Delta\Omega$ and ΔE are the projectile solid angle and the energy range accepted, respectively.

N_P or N_T were not measured directly, instead they were obtained from noncoincidence measurements of the number of recoils, N_R, and time-of-flight spectra where the peak areas, N_R^{q+}, are integrated and charge state fractions, $F_{q+}= N_R^{q+}/N_R$, are calculated. N_R is the sum of all charge state peak areas. These quantities are related to the gross ionization cross section, σ, as follows

$$\sum_q q\frac{N_R^{q+}}{\eta_{q+}} = N_P N_T \sigma \tag{5}$$

The left side can be rewritten as

$$\sum_q q\frac{N_R^{q+}}{\eta_{q+}} = N_R \sum_q q\frac{F_{q+}}{\eta_{q+}} \tag{6}$$

The recoil ion and scattered projectile detection efficiencies were taken from references 33 and 34. In both cases, the efficiencies were adjusted for transmission through the entrance grids of the detectors. Thus, using these procedures absolute doubly differential cross sections (DDCS) for single and multiple ionization of atoms by positron and electron impact were determined.

If all scattering angles were detected, our energy loss measurements would be equivalent to SDCS as calculated using the 'modified Rutherford formula' [35]

$$\frac{d\sigma}{dE} = q^2 \frac{2\pi e^4}{mv^2(\varepsilon + IP)^2} \tag{7}$$

Here ε is the ejected electron energy and IP is the ionization potential, i.e., $\Delta E = \varepsilon + IP$. Experimentally, integrating over all scattering angles is done by setting vertical windows on the projectile 2D spectra. For small energy losses, all relevant scattering angles are collected in this setup and we should expect our energy loss data to decrease as ΔE^{-2}. As the energy loss increases, the scattering angles also increase which means that comparisons with the above formula increasingly become less accurate. It should be kept in mind that in the case of electron impact, even though one cannot distinguish whether the detected electron is the scattered primary or ionized target electron, kinematic arguments imply that in the forward direction the first half of the energy loss range should consist primarily of scattered projectile electrons with the second half being primarily composed of ejected electrons.

As an example of the absolute DDCS measured using this setup and techniques, Figure 19 shows data obtained at 0° for 500 and 1000 eV electron impact on Ar, the solid and open symbols respectively. For comparison, the solid line is obtained by extrapolating the 500 eV absolute measurements of DuBois and Rudd [36] to 0°. The good agreement between the two independent sets of data, even at larger energy losses where the cross sections are several orders of magnitude smaller, demonstrates the reliability of the apparatus and the absolute normalization procedure. Note that the final projectile energy has been scaled in order that data for all impact energies are between 1 and 0. Thus, the 1 keV data demonstrates that as the projectile energy increases the distribution becomes more peaked near small energy losses and decrease in magnitude, in accordance with equation 7.

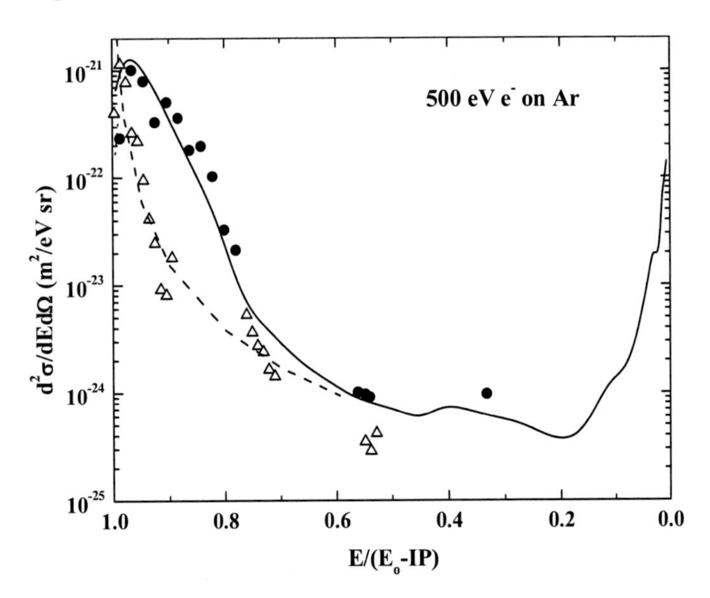

Figure 19. DDCS at 0° for single ionization of argon by electron impact. Solid line: DDCS for 500 eV electron impact extrapolated to 0° from DuBois and Rudd [36]; closed circles and open triangles: present data for 500 and 1000 eV electron impact. Dashed line is ΔE^{-2} dependence where ΔE is the energy loss.

For positron impact, the signal rates were smaller. So to improve the statistics the single and multiple ionization yields were integrated for vertical scattering angles between ± 17° and were binned and averaged over a small range of ΔE energies which ranged from 5 to 20 eV for the smallest and large energy losses, respectively. The results for single, double, and triple ionization of argon are shown in Figure 20 for 750 eV positron impact. These data are compared to similar data obtained for 800 eV electron impact by using a scaled energy axis. As seen, for positron impact the doubly differential energy loss single ionization yields decrease rapidly for the first 50 eV of the energy loss and then more slowly out to the highest energy losses measured. In comparison, double ionization decreases more slowly. For triple ionization, due to the extremely small intensity, approximately five orders of magnitude smaller that the maximum intensity for single ionization, about the only thing that can be said is that within the experimental uncertainties, no triple ionization was observed until ionization of the L-shell was energetically possible. The reader is reminded that for energy losses less than 250 eV the L shell cannot be ionized and multiple ionization in this range implies the removal of multiple M-shell electrons.

In comparing the positron and electron impact data, several major differences are noted. First of all, for electron impact the single ionization yields are much larger for the first 25% of the energy loss whereas they are much smaller for energy losses larger than 50%. Kinematic arguments imply that these zero degree data are dominated by the scattered projectile signal (i.e., the scattered electron in the case of electron impact) with perhaps the final data point indicating the contributions from the ejected electron. For double ionization, the yields are always larger for electron impact for small energy losses and again are smaller at larger energy losses. For triple ionization, after the L-shell is opened the yields for electron and positron impact are identical within experimental uncertainties. However, for energy losses below those capable of ionizing the L-shell, triple ionization by electron impact is observed whereas for positron impact no triple ionization was observed. We attribute this as being due to detection of either the Auger electron or the other electrons emitted from the target; events that can't be detected when we are measuring scattered positrons. Finally, it is important to note that a) the single and double ionization yields observed below their respective thresholds are due to our limited projectile energy resolution and that b) the energy resolution for the electron data was poorer since the beam was produced by using secondary electrons emitted from a surface.

In later experiments, the entrance aperture of the projectile spectrometer was changed such that the horizontal scattering angle resolution was improved to 0° ± 6.5°. In addition, a longer inner spectrometer cylinder was installed in order to preserve the linearity of the vertical scattering angles over a broader range. With these changes, better defined doubly differential data could be collected. Figure 21 shows an example for single and double ionization of argon by 1 keV electrons [37]. Note that, in contrast to single ionization, which maximizes at 0° and then falls off fairly rapidly with increased scattering, the lower portion indicates that double ionization has a broader and less peaked distribution. No positron data are available for comparison.

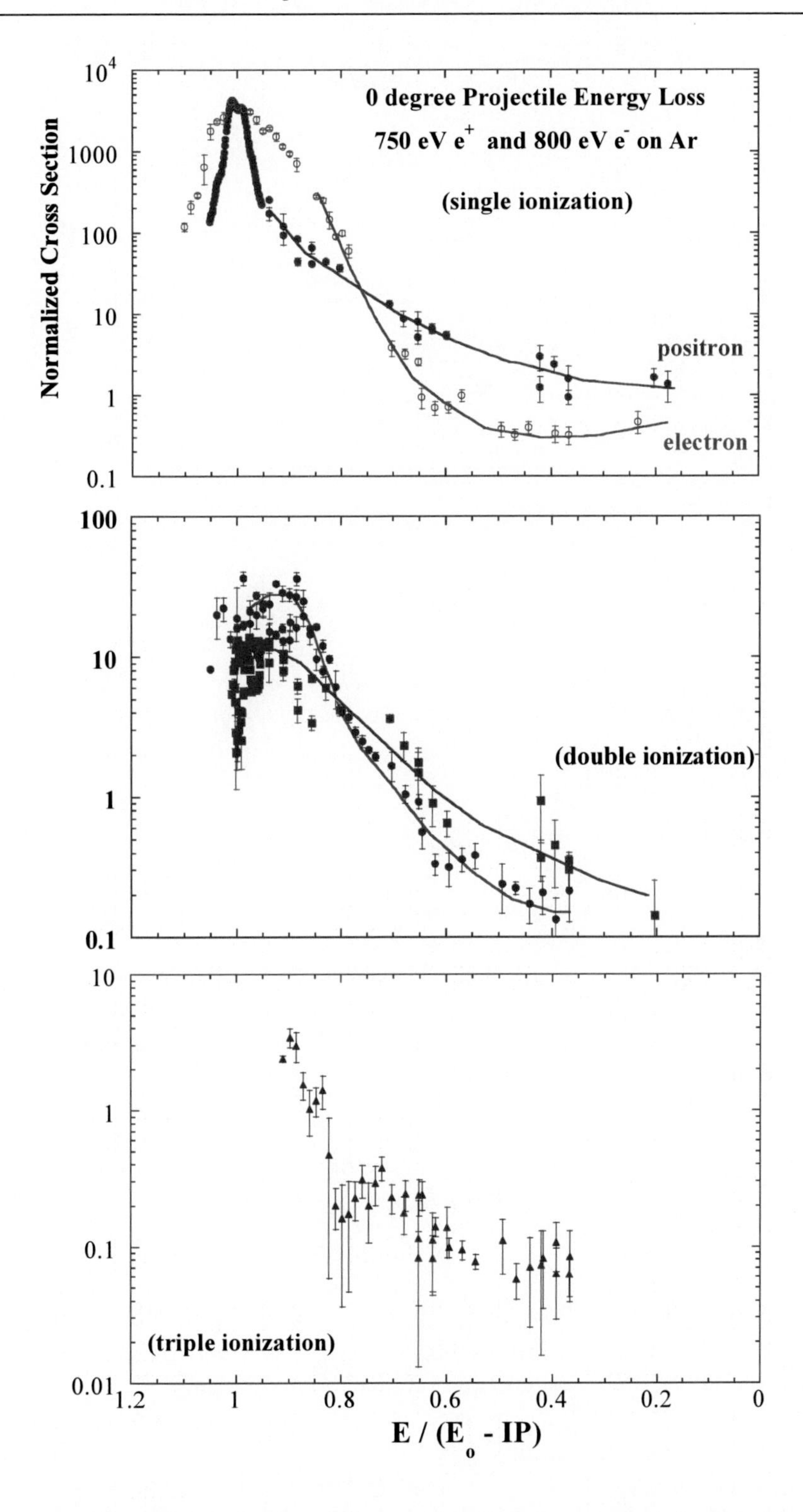

Figure 20. Single and multiple ionization yields, as a function of energy loss, for 750 eV positron (red data points and curves) and 800 eV electron (blue data points and curves) impact on argon. Data are for projectile scattering into horizontal and vertical angles of $0^{\circ} \pm 17^{\circ}$. The curves are to guide the eye. Note the vertical axis changes between single, double, and triple ionization.

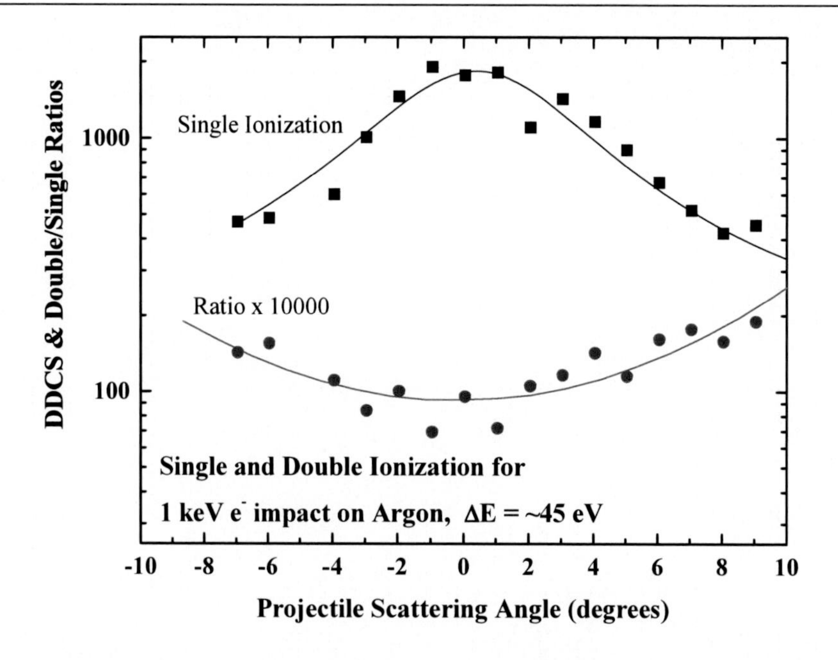

Figure 21. Upper data points: DDCS for single ionization of argon by 1 keV electron impact. Lower data points: Ratio of DDCS for double with respect to single ionization. In both cases, the projectile scattering energy loss was 55 eV. From ref. 37

Differential Multiple Ionization

Comparison of Positron and Electron Impact

To address the question about differences between positive and negative projectiles, rather than comparing our doubly-differential cross sections directly, the relative fractions for single and multiple ionization were calculated as a function of energy loss and compared. The relative yields were obtained from

$$f_{q+} = \frac{N^{q+}/\eta_{q+}}{\sum_q N^{q+}/\eta_{q+}} \tag{8}$$

where f_{q+} is the fraction of q times ionized argon, η_{q+} is the detection efficiency for Ar^{q+}, and N^{q+} is the number of recoil ions with charge state q recorded for a particular energy loss. Figure 22 shows results for 750 eV positron [38] and electron impact.

Figure 22 shows that for both positron and electron impact, single target ionization dominates over the entire range of energy loss and that its contribution decreases slowly as a function of the projectile energy loss. The amount of double ionization increases rapidly for the first 100 eV of energy loss and then seems to plateau. Above 250 eV where the L-shell can be ionized, another increase possibly followed by another plateau is seen. Similar features are observed for triple ionization. These effects were also found for other impact energies (not shown). No significant differences are observed between the single-ionization fractions resulting from positron and electron impact up to 350 eV. Above 350 eV, the positron data are 18% larger than the corresponding electron impact data. For double ionization, the relative amount is systematically larger for electrons than for positron impact. This charge effect was discussed for total double-ionization cross section data; here we see that it occurs independent of how much energy is transferred in the collision, i.e., independent of the impact parameter, at least for the forward-scattering Again note that no triple ionization was observed below 200 eV for positron impact.

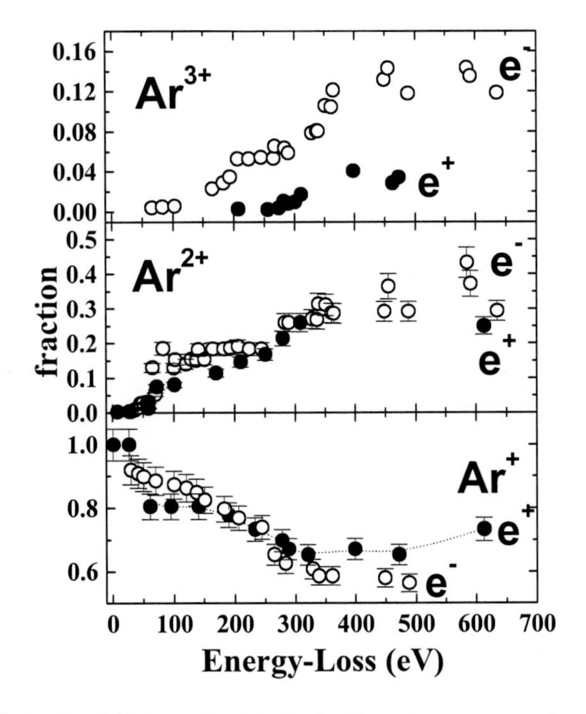

Figure 22. Fractions of single, double, and triple ionization of argon as a function of the projectile electron loss for 750 eV positron (closed circles) [38] and electron (open circles) impact.

In order to investigate these effects further, similar data were collected for ionization of Kr [39]. In this case, 750 eV impact energies were also used, but the positron data are for φ_p and θ_p scattering angles of 0° ± 17° whereas for electron impact a narrow slit was placed at the entrance of the spectrometer. Thus, for electron impact the scattering angles were φ_p = 0° ± 6.5° and θ_p = 0° ± 17°. Also, the energy loss range for positron impact was not as extensive as for electron impact. The charge state fractions as a function of energy loss are compared in Figure 23. Single ionization again dominates over the entire range of energy loss. Its contribution decreases from 100% near the threshold down to 60% just below the M shell edge, then remains constant. For large energy losses, where the coincidence rate was very low, there is scatter in the measured values but the overall trend is obvious. The percentage of double ionization increases rapidly for the first 100 eV, reaches a maximum of roughly 15% after the M shell is opened, after which it remains constant. Triple and quadruple exhibit similar effects, the differences being that the increase is slower and the plateau values are smaller.

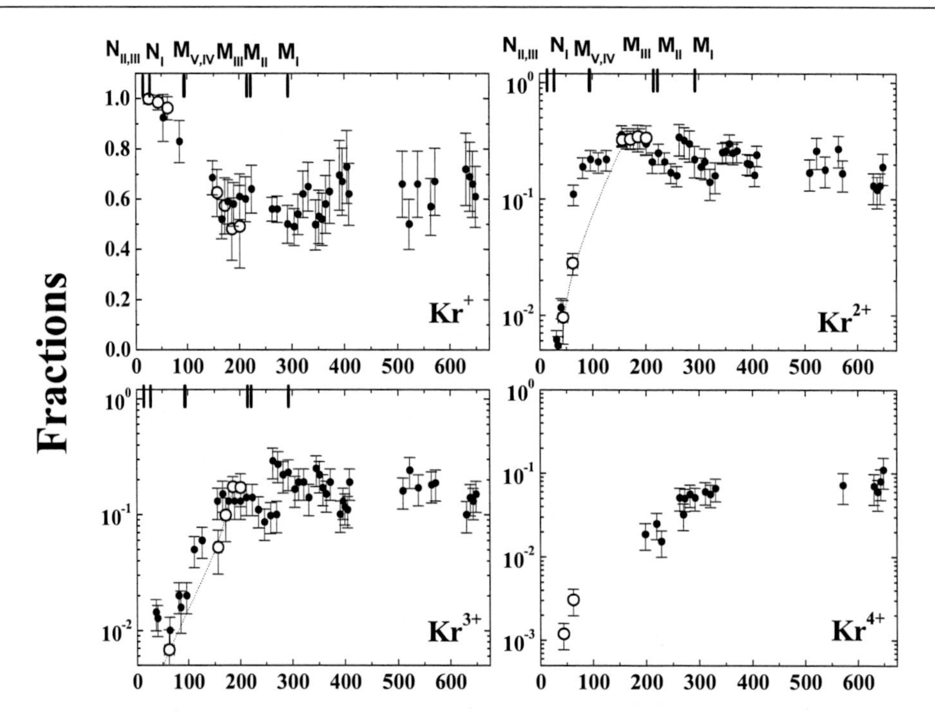

Figure 23. Fractions of single, and multiple ionization of Kr by 750 eV positron and electron impact (open and closed circles). Data are as a function of the projectile energy loss, the horizontal axis [39]. Vertical lines indicate the binding energies of N and M shells. Horizontal arrows are fractions determined using total cross sections from refs. 40-42.

Because of the experimental uncertainties and the limited positron data, comparing positron and electron impact is more difficult than for ionization of argon. However, we again see that the relative amount of double ionization is systematically higher for electron impact than for positron impact, at least for small energy losses. Above 200 eV, both fractions appear to be the same. The same is observed for triple ionization. Included in the figures are values (indicated by the horizontal arrows) obtained from total multiple ionization cross section measurements [40-42]. With respect to our differential measurements, the integral values for multiple ionization are smaller than our plateau values and the electron impact multiple ionization values are all larger than the positron values, e.g., electron (positron) impact: single 0.90 (0.95), double 0.064 (0.033), triple 0.026 (0.017), quadruple 0.071 (no data).

Differences between multiple ionization by positron and electron impact are better observed by means of the average charge state of the recoil ions, which is given by

$$\langle q \rangle = \frac{\sum_q qN^{q+}/\eta_{q+}}{\sum_q N^{q+}/\eta_{q+}} \tag{9}$$

where the quantities have been defined previously. Average recoil ion charge states, as a function of energy loss, are compared for ionization of argon by 750 eV positron and electron impact in Figure 24. The data show that the average charge state steadily increases up to 400 eV after which it remains constant. They also demonstrate that electron impact is more efficient at removing more than one electron. The asymptotic values are 1.55 for electron impact and 1.25 for positron impact.

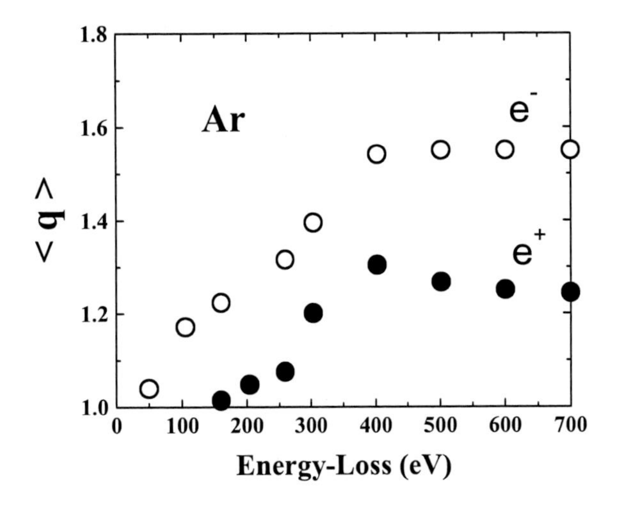

Figure 24. Average ionization charge state for 750 eV electron and positron impact on argon as a function of the projectile energy loss.

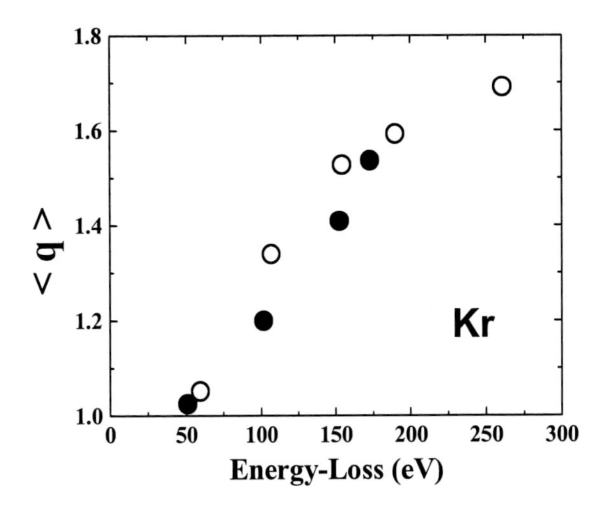

Figure 25. Right figure. Average charge state for 750 eV electron and positron impact on krypton as a function of the projectile energy loss.

Figure 25 shows corresponding data for ionization of krypton. The average charge states for electron impact are systematically above the values for positron impact, but the differences are much smaller than for ionization of argon. The reasons for the smaller differences might be interpreted as saturation effects for increasing target atomic number. Another explanation is trajectory effects, since as stated previously the positively and negatively charged particles follow different trajectories as they pass through the target field

[8,9,12]. Since ionization probabilities are larger for close compared to distant collisions, the positron multiple-ionization cross sections are reduced due to the Coulomb effect. This prevents positrons from penetrating deep into the target atom. On the other hand, electrons are attracted to the target nucleus, sinking in deeper compared to positrons, so that the ionization takes place at reduced impact parameters closer to the maximum of the ionization probability.

Comparison to Photoionization

As discussed in the introduction, charged particle and photon impact data can be compared if information about the projectile energy loss is known. This is because a particular energy loss is equivalent to the energy of a particular photon absorbed by the target. In making this comparison, it is important to remember that due to the high frequency cutoff to the virtual photon field such comparisons are most valid when a) the energy loss is small compared to the initial energy and b) the momentum transfer is also small (i.e., for small scattering angles) [3,4]. Both conditions are met for our small energy loss data; the second condition is met for all of our measurements to date.

Therefore, in Figures 26 and 27 double-to-single ionization ratios for ionization of argon and neon by 750 eV positron and/or electron impact are compared to the corresponding photoionization ratios taken from refs. 43-50. In both cases, the cross-section ratio increases rapidly as a function of the projectile energy loss (photon energy) until the transferred energy equals the average kinetic energy of the outermost target electrons (78 eV, for argon and 116 eV for neon) [44], after which it remains constant. Also note that in Figure 26, the photoionization data agree better with the electron impact data than with the positron impact data. The reader should keep mind when referring to Figures 26 and 27, that we have extended the comparison between charged particle impact and photoionization past the limit of its validity.

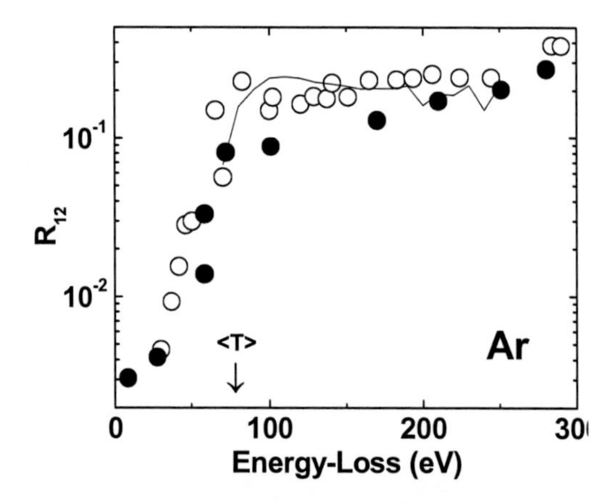

Figure 26. Ratio of double-to-single ionization of argon by positrons (closed circles), electrons (open circles), and photons (line) as a function of the projectile energy loss or photon energy. The photon data are average values between refs. 43-45. <T> indicates the average kinetic energy of the argon 3p electrons (78 eV) [44].

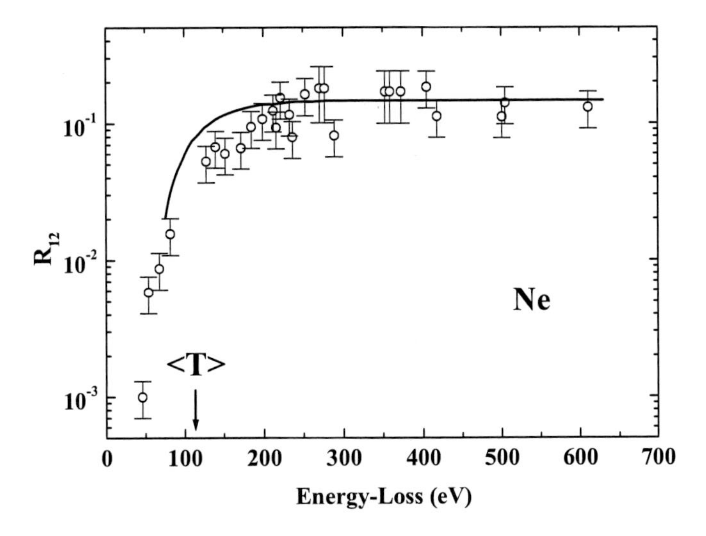

Figure 27. Ratio of double-to-single ionization of neon electrons (open circles), and photons (line) as a function of the projectile energy loss or photon energy. The photon data are average values between refs. [43-50]. <T> indicates the average kinetic energy of the neon 2p electrons (116 eV) [44].

Information about the Interference Term

Single ionization of an atom occurs when the projectile interacts with one of the target electrons. Thus the transition amplitude is proportional to the projectile charge, q, and the cross section is proportional to q^2.

$$\sigma_1 \propto q^2 \tag{10}$$

For double ionization, as pointed out in the previous sections, the TS-1 and shake-off mechanisms mean that the projectile collides with one of the target electrons and the second electron is ejected due to subsequent processes. Thus means the transition amplitude is again due to one interaction with the projectile so is again proportional to the projectile charge.

$$a_{TS-1,SO} \propto q \tag{11}$$

For the TS-2 double ionization process, the projectile interacts independently with both target electrons so the transition amplitude is proportional to the projectile charge squared, i.e., it is proportional to q for each interaction.

$$a_{TS-2} \propto q^2 \tag{12}$$

Note that at large impact parameters which correspond to small momentum transfer and small energy losses, the shake-off mechanism is expected to dominate over the two-step mechanism which peaks at smaller impact parameters.

Thus, the double ionization cross section is given by

$$\sigma^{2+} = \left| a_{TS-1,SO} + a_{TS-2} \right|^2 \tag{13}$$

or, using a different terminology, by

$$\sigma^{2+} = \left| a_{1st-order} + a_{2nd-order} \right|^2 \tag{14}$$

where 1st-order and 2nd-order refer to the number of interactions with the projectile.

Therefore for charged particle impact and considering only outer shell ionization, the double ionization cross section can be written as

$$\sigma_{e-}^{2+} = \sigma_{1st-order}^{2+} + \sigma_{2nd-order}^{2+} \pm \sigma_{\text{int}}^{2+} \tag{15}$$

where $\sigma^{2+}_{\text{1st-order}}$ is the first-order contribution which includes the shake-off and the TS-1 mechanisms, $\sigma^{2+}_{\text{2nd-order}}$ is the 2nd-order TS-2 mechanism, and σ_{int}^{2+} is the term due to the interference between the first- and second-order mechanisms [46].

Note that the interference term is proportional to the cube of the projectile charge which makes it positive for negative projectiles and negative for positive projectiles. Therefore, it is possible to estimate the relative importance of the various terms leading to double ionization by dividing equation 15 by the single ionization cross section, equation 10. Doing so yields

$$R_2(e^{\pm}) = R_{2,1st\ order} + R_{2,2nd\ order} \mp R_{2,\text{int}} \tag{16}$$

where $R_2(e^{\pm})$ is our measured double ionization ratio for positron (electron impact). From equation 16, note that $R_{2,\text{int}}$ can be written as

$$R_{2,\text{int}} = \frac{R_2(e^-) - R(e^+)}{2} \tag{17}$$

Thus, this somewhat simplistic approach provides a method of obtaining information about quantal interference between the first and second order interaction terms simply via a comparison of ratios of cross sections where the ratios have been measured as a function of energy loss.

Next, we use the well established fact that for 1st-order interactions, the double ionization cross section, $\sigma^{2+}_{\text{1st-order}}$, is proportional to the single ionization cross section, σ^{+}_{1}, with the constant of proportionality being essentially the same for both charged particle and photon impact. This means that eq. 16 can be rewritten as

$$R_2(e^{\pm}) = R_2(h\nu) + R_{2,2nd\ order} \mp R_{2,\text{int}} \tag{18}$$

where $R_2(h\nu)$is the double ionization ratio measured for photoionization. Keep in mind that in equation 18 the ratios must be for particular energy losses which are related to the photon energy by $h\nu = \Delta E$. Thus, using equation 18 and values for $R_{2,int}$ obtained using equation 17, the relative importance of $R_{2,2nd\text{-}order}$ can be estimated.

The TS-1 mechanism can be estimated as follows. In the TS-1 *particle impact* process, the first *slow* electron that is ejected receives an energy equal to the energy loss minus the ionization potential. As it leaves, it interacts with and knocks out one of the remaining target electrons. This model resembles electron impact ionization of an ion A^+. In the TS-1 *photoionization* process, a target electron absorbs an energy E = hν - IP and then interacts with the remaining target electrons to eject a second electron. Thus, the TS-1 double ionization contribution can be estimated by normalizing the electron impact single ionization cross sections to the positron (electron) double ionization fractions at 30 eV where the TS-1 mechanism is expected to dominate (see ref. 47).

These steps have been done with the relative contributions of the interference, shake-off, and TS-1 terms being plotted in Figure 28 as a function of the projectile energy loss for 750 eV positron and electron impact on argon. Here, in order to estimate the interference term, we have made smooth curve fits to experimental data for positrons and electrons. We then calculated $R_{2,int}$ using eq. 17.

As seen, the contribution of the TS-2 term is minimal for forward scattering angles at 750 eV. Comparing the recommended data from ref. 51 with the present double-to-single ionization data, at higher energy losses both the TS-1 and the TS-2 mechanisms are negligible and hence shake-off is the dominant mechanism for producing doubly ionized neon. For very small energy losses, the difference between positron and electron data is small, meaning the first order mechanisms dominate and there is no contribution of the second order TS-2 mechanism. For larger energy losses, the difference increases and is 20 % at 200 eV. At 75 eV, both the TS-1 and SO mechanisms compete with 50% contribution each. Above 100 eV the shake-off mechanism dominates for forward scattering angles.

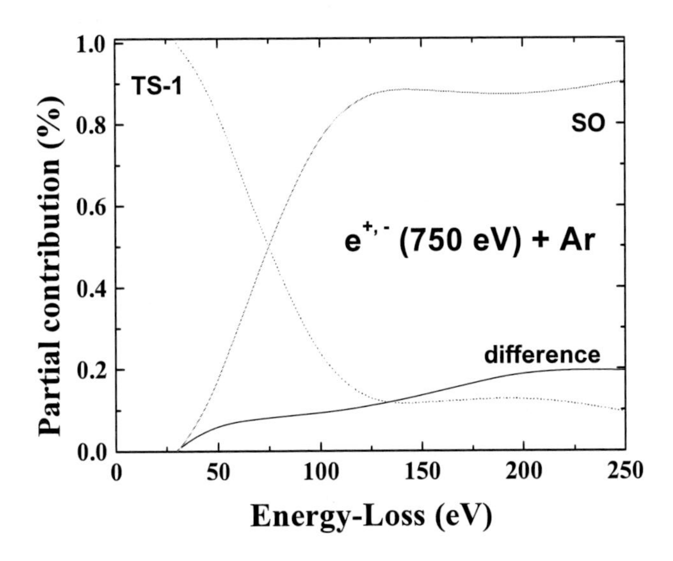

Figure 28. Partial contributions of the interference (Equation 15), TS-1, and shake-off (SO) to double-to-single ionization cross-sections as a function of projectile energy-loss for 750 eV positron (electron) impact on argon.

Ultra Inelastic Collision Studies

Positrons also provide the only method capable of studying collision dynamics at its most extreme limit, namely where the incoming projectile transfers all of its energy and momentum to the target. As is well known by first year physics students, in an inelastic collision between a stationary target of mass m and an incoming projectile of mass M and velocity v_o, if the projectile scatters at an angle $+\theta$ and has a final velocity v_f, the target particle leaves with velocity v at an angle $-\varphi$, both angles being measured with respect to the direction of the incoming projectile. Conservation of energy and momentum require that

$$v = \frac{M}{m+M} 2v_o \cos\varphi \qquad (19)$$

Classically, for headon collisions, $\varphi = 0$ and the target velocity is maximum. For a light target and a heavy projectile, e.g., an incoming ion and a "free" target electron, M >> m and $v_{max} = 2v_o$. For lepton impact, M = m which means that $v_{max} = v_o$ and $v_f = 0$, i.e., the target electron leaves with the entire initial energy and the incoming lepton stops. For ionization of an atom, a recoil ion is produced but because of its large mass it carries away only a few meV of energy. Therefore, the ejected electron should leave with the initial energy minus the ionization potential energy.

Thus, studying the absolute extreme limit for which ionizing collisions can happen is possible by measuring the electron emission at zero degrees and energies approaching the initial energy. As stated, this cannot be done using ion-atom data since such ultra-inelastic ionization processes cannot occur for ion impact. For electron impact, the zero degree small energy loss data are overwhelmingly dominated by scattered projectiles. Therefore, again ultra-inelastic processes cannot be studied. At first glance, one might think of using polarized beams and targets in order to differentiate between the scattered projectile and ejected electron. However, because headon events are rare, one has to worry about spin flip processes. More important, however, techniques that measure the electron spin for individual particles, rather than looking at differences between a large number of particles, don't currently exist.

Therefore, positrons offer the only current method for studying ultra-inelastic processes. However, as stated, these headon events are rare. An estimate of the total interaction probability would be the area of a circle whose diameter is the classical radius of an electron, namely 2.5 x 10^{-26} cm^2. The differential signal would be smaller still. It is possible that this cross section is too small to be measured.

To investigate this, we reversed the voltages applied to our projectile spectrometer and measured the electron emission in the forward direction for 750 eV positron impact single ionization of argon. Figure 29 shows these data. The upper (red) curve is the scattered positron intensity shown previously. The lower (blue) data points are the ejected electron signal. Note the signal decreases monotonically and reaches a constant background value at approximately the beam energy. From the absolute scaling of the scattered positron signal shown in Fig. 18, the background level here represents a differential cross section of approximately 10^{-29} cm^2/eV sr. Note that the collection time for these data was approximately 2 months with half of that devoted to establishing the background level. These data clearly demonstrate that even using sub-femtoamp beams, extremely small differential cross sections can be cleanly measured using our techniques.

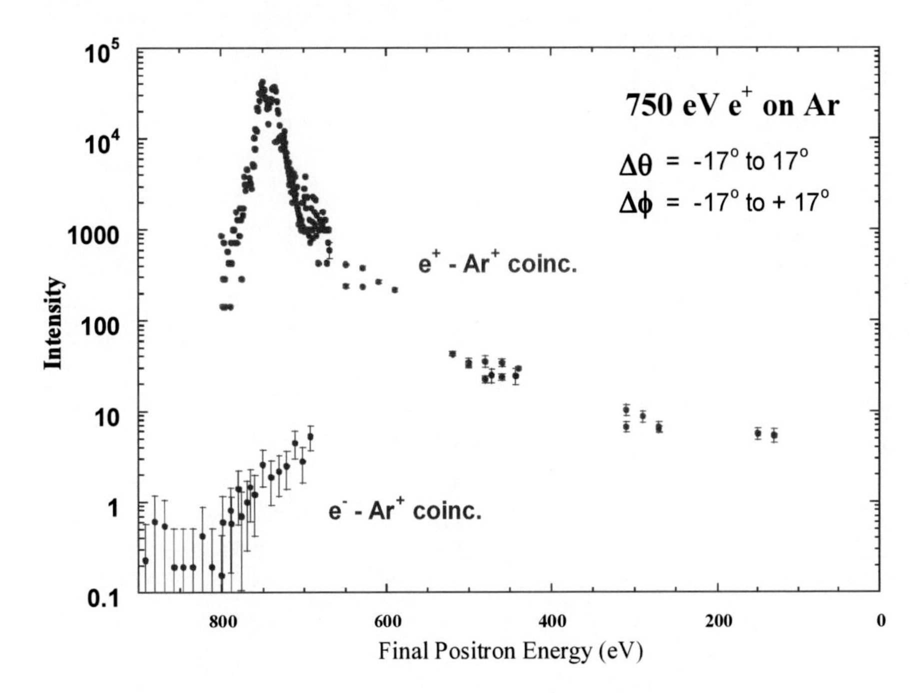

Figure 29. Scattered positrons and ejected electrons (red and blue data points) measured at zero degrees for 750 eV positron single ionization of argon.

Unfortunately, because of the achievable energy resolution we cannot clearly establish if 100% of the available energy was transferred to the electron in these collisions. In other words, we cannot state how valid the approximation is that in hard collisions the electron can be considered to be totally free. Other interesting questions associated with this type of experiment include: a) since a third body (the target nucleus) is present, what is the maximum momentum that can be transferred to the electron? b) what happens in double ionization? Can one of the electrons carry away all of the available energy? Future studies might address such questions.

TDCS Measurements

As stated, when we initiated our positron studies at UMR our goal was to perform triply differential ionization studies and to obtain kinematically complete information about the interaction dynamics in positron-atom collisions. As seen, after much trial and error and numerous modifications to the experimental apparatus and our techniques, we have been able to generate increasingly more complex differential data. The latest modifications to the apparatus include: 1) replacing the recoil ion position-sensitive channelplate with a channel electron multiplier, 2) moving the position-sensitive detector to just outside the electron extraction plate of the spectrometer in order to detect electrons ejected into a large range of θ_e, φ_e angles, 3) adding apertures at the ion and electron extraction plates to restrict coincidence events to those occurring at the beam-gas jet overlap, 4) rotating the recoil ion spectrometer by 90° so that its electric field is in the θ_p direction, and 5) reducing the width of the entrance slit of the projectile spectrometer in order to restrict the range of φ_p angles in order to define a collision plane. A schematic of the apparatus after these changes were made is shown in Figure 30.

Switching from a channelplate to a channel electron multiplier improves the ion detection efficiency [33], thus increasing the coincidence rate slightly. Adding the aperture at the ion extraction plate meant that only recoil ions produced at the beam gas intersection are recorded. These changes allowed the position-sensitive detector and electronics to be used for the ejected electron channel. Placing the electron detector as close as possible allowed electrons emitted between 90° ± 50° in both the θ_e, φ_e directions to be detected. These angles, of course, are subject to the strength of the electric field and the emitted electron energy. This will be discussed further at a later time.

The aperture added at the electron extraction plate helped eliminate electrons produced at points other than the desired interaction region. The slit at the projectile spectrometer entrance restricted projectile scattering angles to φ_p = 0° ± 2.5° while the size and distance of the projectile channelplate restricted the maximum θ_p angles to ± 7.5°. Note that by aligning the electric field of the recoil ion spectrometer with the θ_p scattering angle means that pure in-plane scattering should produce a well-defined stripe on the electron detector whereas out-of-plane scattering for a particular value of θ_e would form a circular pattern, as illustrated in Figure 31.

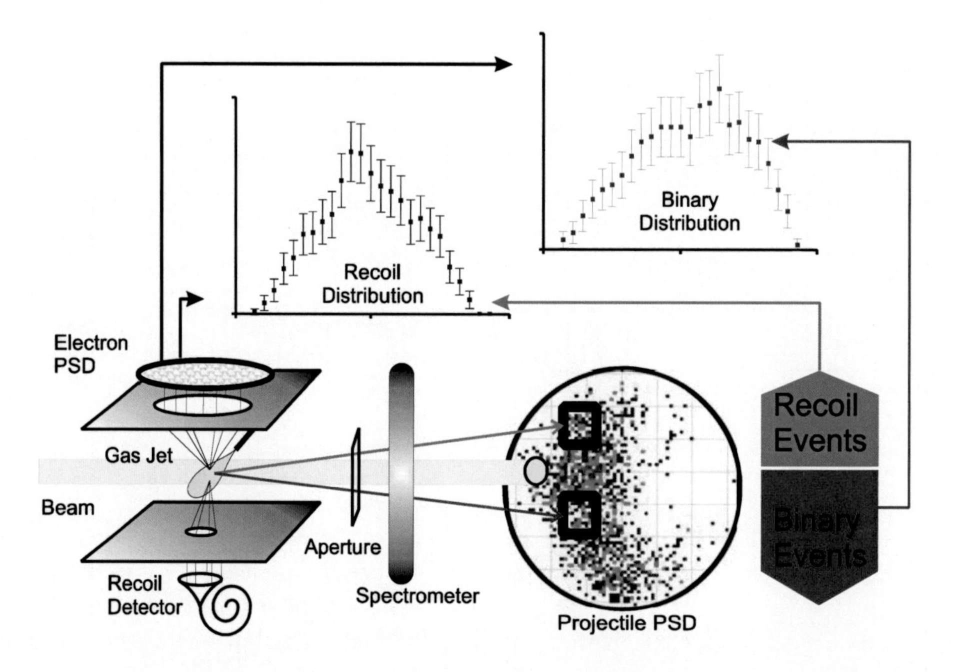

Figure 30. Experimental apparatus used for TDCS data acquisition. The two upper spectra show an example of the ejected electron distributions for binary and recoil interactions. The lower circular 2D spectra is an example of the 2D projectile spectra, the circle corresponds to the position of the beam on the detector and regions corresponding to binary and recoil events have been distinguished.

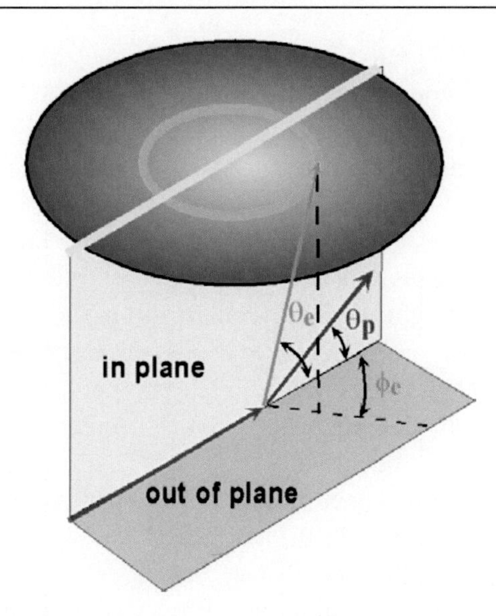

Figure 31. Schematic of signal on electron detector for TDCS setup. In-plane electron emission is the vertical plane and results in a stripe along the beam direction. Out-of-plane emission is the horizontal plane and results in a circular pattern where the diameter of the circle depends on θ_e.

With this setup, TDCS data were acquired using a computer and an 8-channel TDC where the four delay-line anode signals from the projectile and ejected electron detectors served as starts and the recoil ion signal served as a common stop. At 500 eV, the positron beam intensity was about 2000 counts/sec, while the electron rates were about 200 counts/sec and recoil rates on the order of 1 count/minute. This gave a triple coincidence rate for all ejected electron energies and angles of approximately 15/hr. At 200 eV, the triple coincidence rate was approximately 4/hr. Roentdek CoboldPC software was used for list-mode data accumulation and analysis. The data consisted of "projectile TOF spectra" using recoil ion-scattered projectile coincidences and "electron TOF spectra" recoil ion-ejected electron coincidences. These provided information as to whether the scattered projectiles and ejected electrons which were detected resulted from single, double, triple, etc., target ionization or from background interactions.

Simultaneously, 2D spectra were collected for the scattered projectiles and for the ejected electrons. The projectile 2D spectra provided information about scattering angles and energy loss, just as was done in our DDCS and energy loss measurements. The electron spectra provided information about the emission angles. In both cases, by setting windows on the various peaks and on the background in their respective TOF spectra, the 2D data could be sorted to provide the above information for single, double and triple ionization events where background effects have been subtracted.

Projectile TDCS Information

By combining these pieces of information it is possible to construct TDCS spectra and extract information from them. For example in the projectile channel, information about the relative intensities for binary and recoil interactions as a function of the energy loss and scattering angle of the projectile are easily displayed. This is done by setting windows on the single ionization peak in both the projectile and electron TOF spectra and then generating 2D projectile spectra. The windows ensure that single ionization occurred and that a scattered projectile and an ejected electron were detected. Note that setting a window on the single ionization peak in the electron TOF spectrum is equivalent to integrating over all emission angles that were detected. Figure 32 shows examples of TDCS measured in the projectile channel for positron and electron impact at 500 eV.

As mentioned before, the horizontal axes correspond to the energy loss of the projectile with the main intensities (brightest/yellow colors) located for energy losses slightly larger than the ionization potential. The vertical axes denote the scattering angle with zero located at the beam position height. For 500 eV impact each channel corresponds to an energy loss of approximately 1 eV and a scattering angle of 0.2 degrees. The solid circles denote the position and size of the beam in both cases.

For positrons we see a well defined ridge in the lower part of the detector. We also see an asymmetry between the intensities in the upper and lower regions, particularly for larger scattering angles and energy loss. Note that the loss of intensity at the upper portion of the spectrum occurs before the edge of the detector is reached. For small energy losses the upper and lower portions are almost symmetric. Referring back to the apparatus schematic in Fig. 30, note that the lower portion of the projectile TDCS 2D spectrum corresponds to projectiles which have scattered "down" but projectiles which are in coincidence with electrons which were emitted "up". (Towards the electron detector which is above the target.) Thus, momentum conservation tells us that the lower ridge in Fig. 32 is due to binary interactions. On the other hand, the projectile TDCS intensities above the beam location occur for "up" scattered positrons. As they are also in coincidence with "up" ejected electrons, these intensities are related to recoil events, since the only way both could leave in the same direction is if one interacts with a third body. Thus, the 2D projectile spectrum provides information about the relative importance of binary and recoil interactions as a function of energy loss and scattering angle.

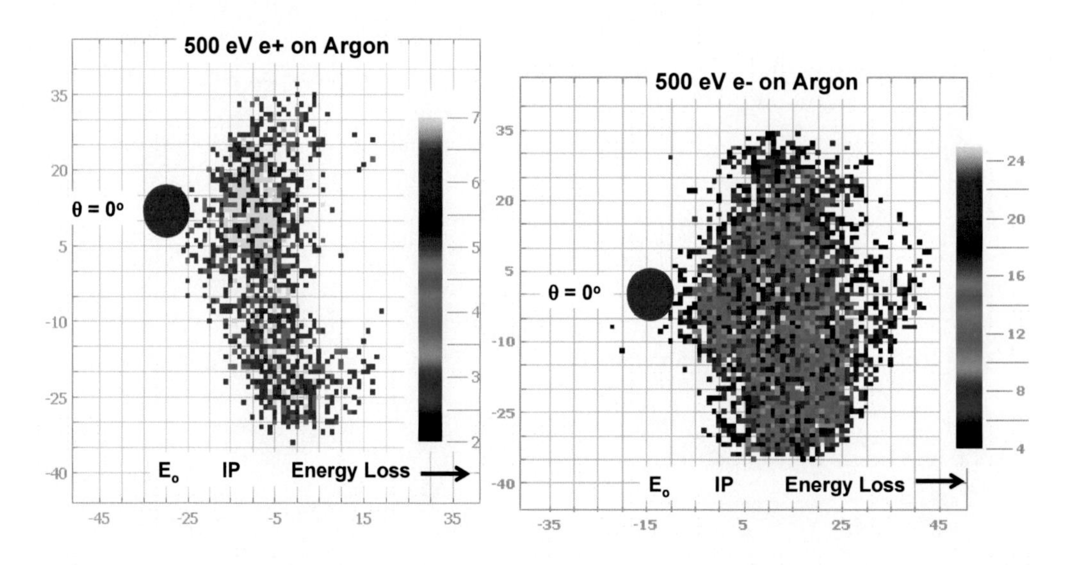

Figure 32. Projectile TDCS plot for 500 eV positron (left figure) and electron (right figure) impact on Ar. See text for details.

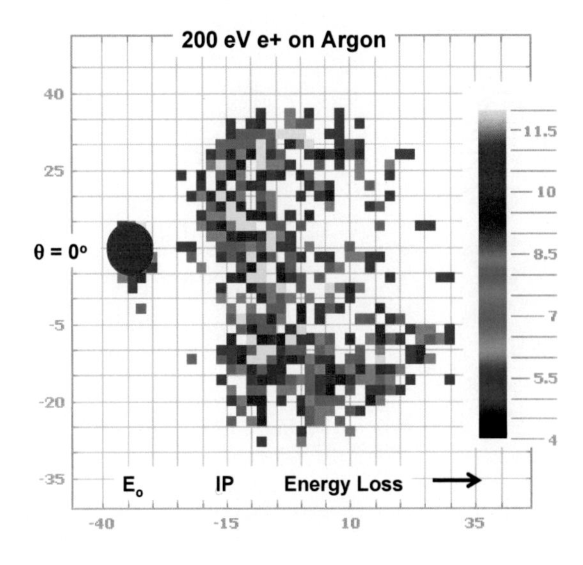

Figure 33. Projectile TDCS for 200 eV e^+ on Ar. The solid circle shows the beam.

For electron impact the major differences are that the ridge is broader and less defined and the data appears to extend further along the zero scattering angle. Note that this is in accordance with what we observed in our early energy loss measurements (see Fig. 20).

Figure 33 shows projectile TDCS data for 200 eV positron impact. Here, each horizontal pixel corresponds to approximately 0.45 eV per channel, while each vertical pixel is again 0.2°. With respect to what was observed at 500 eV, the data are much more diffuse in both energy loss and scattering angle, i.e., the binary/recoil ridge is smeared out. However, again the TDCS spectrum indicates an extra intensity for the binary region, relative to the recoil intensity.

Ejected Electron TDCS Information

The traditional way of displaying TDCS data is to show the angular dependences of the ejected electrons as a function of the electron energy and projectile scattering angle. This is done in our experiment by selecting a range of energy losses and projectile scattering angles, i.e., the boxes in the 2D projectile spectra shown in Figure 30, and sorting the electron 2D spectra. This generates 2D electron emission spectra where the initial and final energies of the projectile plus its scattering angle are known. As discussed above, if the scattering plane is well defined and only in-plane scattering takes place the 2D electron spectra will consist of only a well defined line and for the "down" scattered projectiles, the 2D electron spectra measure the angular distributions for binary interactions; for the "up" scattered projectiles, the angular distributions are for recoil interactions. Examples of these TDCS viewed in the ejected electron channel [53] are shown in Figure 34. It should be mentioned that to improve the statistics the 2D electron emission data have been summed over a range of φ_e angles and binned for θ_e angles, i.e., along the direction of the beam.

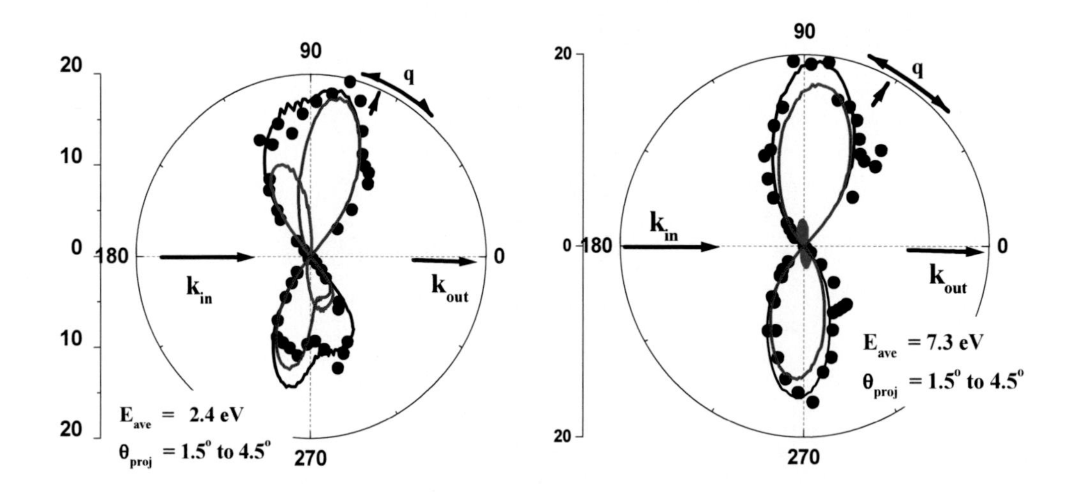

Figure 34. Triply differential electron emission intensities for single ionization of argon by 500 eV positrons. From ref. 52. Experimental data are represented by the solid circles, horizontal arrows represent the initial and scattered positron directions; radial and curved arrows show momentum transfer direction and range. The solid lines are calculated curves for the binary and recoil emission which has been convoluted over the different experimental parameters. Left: small ejected electron energies; Right: intermediate ejected electron energies.

These data, the filled circles, are for single ionization of Ar by 500 eV positron impact and presented as a polar plot which illustrates the intensities and angular dependences of the binary and recoil lobes. For these polar plots, the angles are the geometric angles, not the original electron emission angles. This distinction plus the various curves will be explained below. Included in the figures are the mean values plus directions and ranges of the expected momentum transfer and horizontal arrows indicating the incoming and final projectile directions. The mean ejected electron energies, E_{ave}, for the left and right figures are 2.4 eV and 7.3 eV, respectively. The projectile scattering angle is between 1.5° and 4.5°.

In order to understand these features, many factors have to be considered. First, because of the extended beam and target dimensions their overlap forms an interaction volume. This translates into a range of scattering angles and scattering planes, particularly for the small scattering angles we have measured. Note that for a particular scattering angle the range of scattering planes is determined by the entrance slit width to the spectrometer and the width of the interaction volume whereas the range of scattering angles is determined by the height of the interaction volume and the height of the box shown in the projectile 2D spectrum in Fig. 30. The height of the interaction volume also translates into a range of starting points for the ejected electrons, i.e., a range of electric field potentials in the recoil spectrometer.

Second, our large beam diameter generated a large spot size on the projectile detector, see figs. 32-33. The beam profile was measured to be Gaussian and by definition all projectiles in the profile have the same energy, namely E_0. However, our projectile spectrometer and position-sensitive projectile detector mean that each horizontal pixel corresponds to a particular energy loss. Therefore, because of the beam width and box size width used to sort the electron TDCS data, a range of energy losses, i.e., of ejected electron energies, must be considered. This range is given by the convolution of the beam profile over the range of energy losses for each horizontal window used on the projectile 2D spectra. This convolution is also a Gaussian defined by a centroid energy, E_{cent}, and has a width determined by the spectrometer resolution and beam spot size, approximately ±3 eV for 500 eV impact. Note that for low energy losses, after convolution a portion of the lower energy tail lies between E_o and IP, i.e., it corresponds to "negative energy" electron emission. Therefore it does not contribute to our ejected electron spectrum. In this case, we define E_{ave} which is the average energy of all observed electrons which were emitted. Observe that because the Gaussian has a large width when E_{ave} is small, E_{cent} can be negative, e.g., it was -4.4 eV for the left figure above. However, for larger values of the electron emission, E_{ave} and E_{cent} are approximately the same, e.g., 7.3 and 6.4 eV for the right figure above.

Third, the ejected electrons must pass through two grids, one at the exit plate of the recoil spectrometer and one just before the electron channelplate, before being detected. As one moves away from a point directly above where the electron was produced, the grid transmission decreases. This decrease depends upon the angle of approach which in turn depends upon various electric fields plus the original values of the electron energy and emission angle. Also, as one moves away from a point directly above the interaction volume, the solid angle for a fixed surface area on the electron detector becomes larger. Neither of these has been taken into account for the experimental data shown in Fig. 34.

Fourth, the electric field used to extract target ions seriously influences the emitted electron trajectories. Before performing a complete analysis it was expected that the electric field effects would consist of a) turning around low energy electrons which were originally emitted in toward the recoil ion detector, b) slightly "narrowing" the widths of the binary and recoil lobes, and c) enhancing the magnitudes of the lobes and shifting the lobes toward 90 degrees. All of these being important for low-energy electron emission, e.g., for electron energies less than or on the order of the voltages applied to the extraction plates of the recoil ion spectrometer. For higher energy emission, it was assumed they would be of minor importance. However, our broad Gaussian energy profile means that even for $E_{cent} \sim$ 10-15 eV, our windows still include electrons having very low energies and these electrons contribute to our measured binary and recoil lobes. Also, as shown in Figure 35 the combination of electric field effects plus solid angle effects don't simply narrow the distributions, for isotropic emission peaks at the ends of the distributions are generated because a large range of emission angles all arrive at the same point on the detector.

This provided a qualitative explanation of effects originally thought to be unimportant. For quantitative information, a computer program was written where the "exact" binary and recoil electron emission lobes were represented by cosine squared functions with their relative magnitudes and absolute directions left as adjustable variables. These "exact" lobes were convoluted over all experimental parameters and compared with experimental measurements. As a starting point, the binary lobe was placed in line with the momentum transfer direction and the recoil lobe was directly opposite. Using the dimensions and fields in our recoil spectrometer, the trajectories and detection positions at the electron channelplate were calculated at the emission angle was varied in one degree steps. This was done for a grid of starting points covering the interaction volume and for a grid of projectile final points covering the box sizes used on the projectile detector. This was also done for a Gaussian profile of electron energies as defined by our projectile energy loss conditions. In other words, a complete convolution over all experimental parameters was performed and the results compared to our measurements. Then, the relative maxima of the binary and recoil lobes plus their absolute positions were adjusted until the best agreement with the low and high energy electron emission data were achieved.

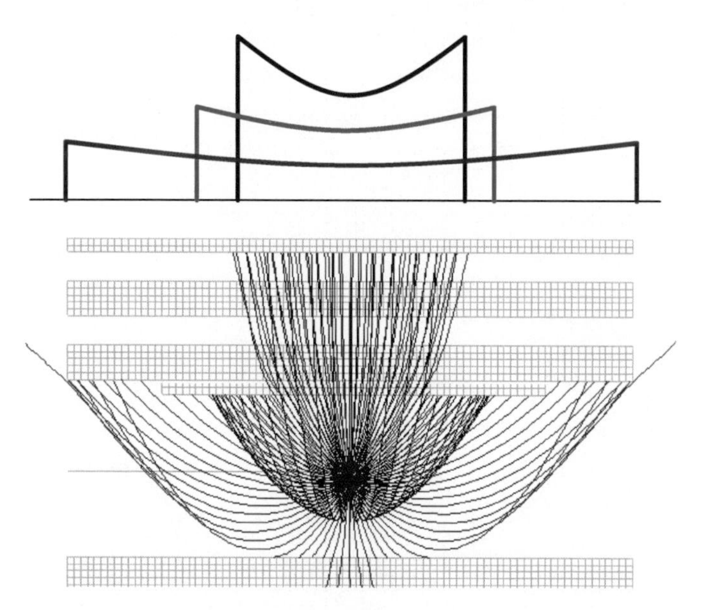

Figure 35. Bottom: SIMION simulation for the trajectories isotropic electron emission under the influence of the extraction electric field. Shown are trajectories for low (dark blue), intermediate (light blue) and high (red) emission energies. The cross hatched rectangles represent the top and bottom plates of the recoil spectrometer with an aperture attached to the top plate, plus the electron channelplate and the plate it is attached to. Top: Integration of the number of events hitting the electron detector.

The results are shown by the red, blue and black lines in figure 34. Here the red lines (the lobes centered at angles smaller than 90° and 270°) represent "real" binary and recoil events whereas the blue lines (the smaller lobes at angles larger than 90° and 270°) represent "false" events, i.e., events where low-energy binary and recoil electrons are initially emitted away from the electron detector but are turned around by the electric field. The black curve nearly matching the data points is the sum of these two. As seen, for low energy emission, the contribution of "false" electrons is large and asymmetric or double lobed features appear. For larger energy emission, few electrons are turned around and the "false" lobes are small. Our simulations also demonstrated that the directions and widths of the lobes are altered due to decreases in grid transmission with increasing angle of incidence plus detection solid angle and electric field effects. These lead to increasing truncation of the observed electron intensities at forward and backward angles, i.e., altered peak width and direction. Finally, in order to best reproduce our experimental data, we found that the relative recoil to binary intensities needed to be approximately equal at 1 eV and slowly decrease to 1/3 at 20 eV. In addition, to match the 2.4 eV data it was necessary to shift the positions of the binary and recoil peaks to smaller angles by 10° and 40° whereas to match the 7.3 eV data a 10° shift to larger angles was required. Values intermediate to these were found for energies between 2.4 and 7.3 eV.

Angular Distributions of the Ejected Electrons

As discussed previously, for charged particle impact, double ionization can result from either a single interaction between the projectile and the target followed by the ejection of the second electron (TS-1) or from two independent interactions with the colliding projectile where each interaction removes one electron (TS-2). In the Born approximation, TS-1 corresponds to the first order term while TS-2 corresponds to the second order term. Because of their different dependences on the impact velocity, at high impact energies the TS-1 mechanism will dominate. Therefore, since single ionization also involves only a single projectile-target interaction, using total cross sections the ratio of double to single ionization should be a constant whenever the TS-1 mechanism dominates, i.e., the ratio of double to single ionization should be a constant value at high impact energies. At lower impact energies, the TS-2 mechanism becomes more and more important and deviations from a constant value should be seen. This is demonstrated in figure 5.

However, with regard to differential cross sections the situation is different. In the TS-1 double ionization mechanism, the electron emission spectrum consists of a "single ionization spectrum" accompanied by a "second collision between the ejected electron with another bound electron spectrum" or a "shakeoff of the second electron" spectrum. In contrast, in the TS-2 double ionization mechanism, the electron emission spectrum consists of two identical "single ionization spectra". Thus, in this simplistic picture we would expect a constant for the differential electron emission ratios calculated for double with respect to single ionization. For the TS-1 mechanism, we would expect that the ratio is not constant.

Therefore, in Figure 36 ratios of the angular distributions, $\sigma(\theta_e)$, for single and multiple ionization by positron and electron impact are calculated and compared. Note that these data only involve recoil ion-ejected electron coincidences, i.e., no projectile information is used. This corresponds to integrating over all possible final state projectile parameters.

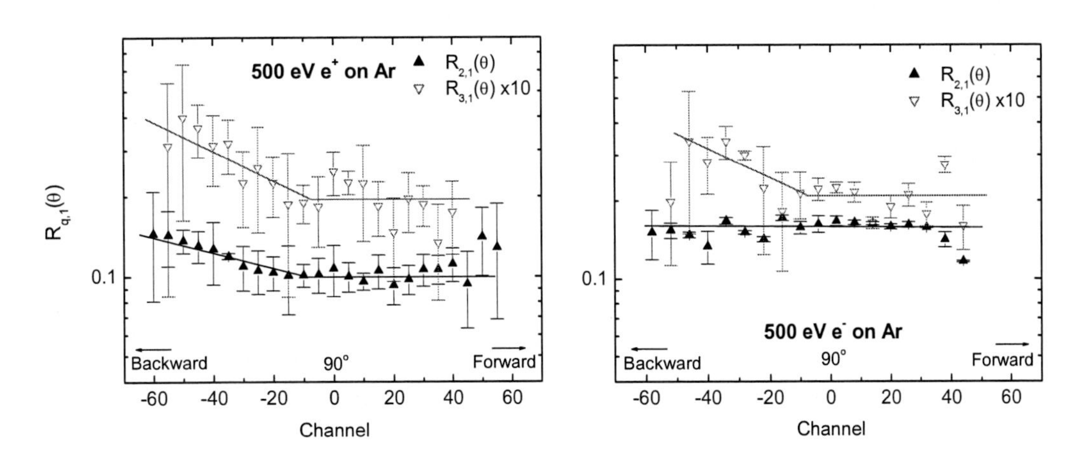

Figure 36. Ratios of singly differential electron emission cross sections, σ_q / σ_1, resulting from ionization of Ar by 500 eV positrons (left figure) and electrons (right figure). Positron data from ref. 52.

In the case of the double ionization (solid points) differences between the ratios for electrons and positrons can be noticed. For electron impact the ratio is flat within statistics. In contrast, the positron ratios are flat for forward electron emission but in the backward direction they steadily increase. For triple ionization (open circles), the ratios for both electron and positron exhibit the same shape and this shape is similar to that observed for double ionization by positron impact. Notice that triple ionization ratio has been scaled by a factor of 10.

According to our arguments above, this can be interpreted as meaning that for 500 eV electron impact the TS-2 double ionization mechanism dominates whereas for the same energy positron impact the increase in the backward direction implies that the TS-1 mechanism is important. However, another interpretation of the increase in the backward direction is that while single ionization is purely a two body interaction, higher degrees of ionization require the presence of more bodies. Therefore the slope on the ratios for backward electron emission could indicate a many body interaction, i.e., collisions in which the nucleus or another electron is involved. This would be consistent with the stronger increase observed for the triple ionization ratios. Obviously, additional studies at additional energies are required to answer these questions.

Conclusions

Future Plans, Improvements and Studies

To summarize, we have outlined the various processes that take place in inelastic collisions and have discussed how positron studies can provide information about which processes are active plus details about the interaction dynamics. Various methods that have been used by others and examples of integral and differential positron impact data obtained to date were presented. This was followed by outlining the experimental approaches we have used at UMR to generate various forms of differential ionization data. After many improvements, a long sought goal has finally been achieved, namely our latest series of studies are generating TDCS data for single ionization. More importantly, these data can probe a range of collision parameters. In addition, our work has gone beyond single ionization and has generated the first differential electron emission information for multiple ionization by antiparticles. A point that we would like to emphasize is that we are able to obtain TDCS data and differential data for multiple ionization using beam intensities as low as 1 kHz or less. Data collection is slow and statistics are low but using multi-coincidence techniques makes it possible to collect clean, background free data.

Our short term goals are to continue to improve our apparatus and to collect more and better TDCS data. Two improvements that have already taken place are a) perturbations on the ejected electron trajectories have been reduced by using a smaller electric field used to extract recoil ions and b) the range of ejected electron angles that can be detected has been increased by removing the aperture at electron extraction plate and using a strong field to project the image of ejected electrons passing through the grid at the electron plate directly on the channelplate. Experiments in progress or planned for the near future include new measurements at a higher impact energy and repeating our 200 eV measurement using a lower extraction field, plus TDCS electron impact measurements so direct comparisons can be made.

Improvements that we hope to apply later are to install a larger diameter projectile detector and to better define the collision plane. A larger detector would primarily allow us to investigate larger scattering angles; a secondary benefit would be that a wider range of energy losses could be collected for a single setting of spectrometer voltages. To better define the collision plane, several parameters need to be considered. Namely, the entrance slit width of the spectrometer needs to be narrower, the scattering angle needs to be larger, and the beam diameter needs to be smaller. All of thee can easily be changed, but at a major cost in signal rate. Therefore, this improvement requires more positron beam intensity. It should also be noted that reducing the beam diameter and increasing the beam intensity would also lead to an improved energy resolution in our experiments as currently the ejected electron energy is defined as the overlap of the Gaussian beam profile measured at the detector and the energy window that we use for sorting the TDCS. Decreasing the beam spot size would help, increasing the beam intensity would allow a narrower energy window to be used.

At present, there are several efforts around the world to construct intense positron sources, with desired intensities approaching 10^9 slow positrons/second. It should be evident from examples presented here that the availability of these sources would revolutionalize positron research and allow studies that can only be dreamed of at present. This would allow TDCS studies to be made for much larger scattering angles, plus would allow RIMS techniques to be used. Regarding the RIMS technique, using a synchrotron-based intense positron source would provide the time zero information needed for determining the third momentum component via time-of-flight measurements.

With some effort, intense positron sources should also make fully differential double ionization studies possible. As demonstrated throughout this paper, double ionization provides many interesting opportunities with regard to studying projectile charge effects.

An intense positron source would also provide the possibility of extending our knowledge into unknown areas. One of these we have already discussed, namely ultra inelastic collisions. To reiterate, only for positron impact ionization can we investigate the ultimate extreme in collision conditions, namely where 100% of the available energy and all of the momentum is transferred. Studying this regime will tell us when a target electron can really be assumed to be a free electron or when does it appear to be bound. What happens if it is the only electron in the atom, or in a particular shell? What happens when two electrons are ionized? What happens to the incoming positron? Does it get captured and annihilated? There are many interesting questions with regard to these types of studies.

The other interesting study would be to study Ps impact ionization. The UCL group has studied Ps production and collisions for many years [53] but Ps impact ionization at higher energies has never been looked at. Many years ago, one of us (RDD) measured total cross sections for hydrogen atom impact ionization [54]. Those measurements showed how the bound electron played an active role (where it interacts directly with a target electron) and a passive role (where it partially screens the projectile nuclear charge and this screened nuclear charge interacts with the target electron). For Ps impact, the projectile consists of a positron and an electron "circling around each other", i.e., there is no massive central charge as in an atom. Therefore for Ps impact ionization, do the positron and electron act independently with the target electron?, do they tend to mutually screen each other?, what happens? If they act independently one would expect the Ps impact total ionization cross section to be twice that as for positron or electron impact. If they mutually screen each other, one would expect the coulomb field and hence the cross section to be much smaller than for positron or electron impact alone. An intense positron source would allow a reasonable number of Ps to be produced, even at higher energies. Hence, these questions could be answered. Obviously, such data would test theoretical predictions of the structure of Ps.

As a final suggestion, this paper has only talked about ionization of atoms. Future studies will obviously also extend to positron-molecule interactions. For example, positron binding to molecules and how this influences their reactivity could be probed in detail. Positron trapping, either inside or outside, Fullerene cages could be studied. The list is endless BUT without more intense positron sources than currently available, most of these future studies and questions will remain in the category "interesting but unanswered."

Acknowledgements

Work supported by the National Science Foundation and CNPq (Brazil).

References

[1] Bromley, M. W.; Lima, M. A. P.;Laricchia, G. *Phys. Scr.* 2006, 74, C37.
[2] Surko, C. M.; Gribakin, G. F.; Buckman, S. J. *J. Phys. B: At. Mol. Opt. Phys.* 2005, 38, R57.
[3] Jackson, J. D. *Classical Electrodynamics*, Willey,New York, 1975.
[4] Brion, C. E. *Comments At. Mol. Phys.* 1985, 16, 249.
[5] Sueoka, O.; Mori, S. *J. Phys. B: At. Mol. Opt. Phys.* 1994, 27, 5083.
[6] Knudsen, H.; Brun-Nielsen, L.; Charlton, M.; Poulsen, M. R. *J. Phys. B: At. Mol. Opt. Phys.* 1990, 23, 3955.
[7] Shah, M. B.; Elliott, D. S.; McCallion, P.; Gilbody, H. B. *J. Phys. B: At. Mol. Opt. Phys.* 1988, 21, 2751-2761.
[8] Paludan, K.; Laricchia, G.; Ashley, P.; Kara, V.; Moxom, J.; Bluhme, H.; Knudsen, H.; Mikkelsen, U.; Møller, S. P.; Uggerhøj, E.; Morenzoni, E. *J. Phys. B: At. Mol. Opt. Phys.* 1997, 30, L581–L587.
[9] Bluhme, H.; Knudsen, H.; Merrison, J. P.; Nielsen, K. A. *J. Phys. B: At. Mol. Opt. Phys.* 1999, 32, 5835-5842.
[10] McCallion, P.; Shah, M.B.; Gilbody, H. B. *J. Phys. B: At. Mol. Opt. Phys.* 1992, 25, 1061-1071.
[11] McGuire, J. *J. Phys. Rev. Lett.* 1982, 49, 1153.
[12] Helms, S.; Brinkmann, U.; Deiwokis, J.; Hippler, R.; Schneider, H.; Segers, D.; Paridaens, J. *J. Phys. B: At. Mol. Opt. Phys.* 1994, 27, L557.
[13] Schultz, D. R.; Olson, R. E.; Reinhold, C. O. *J. Phys. B: At. Mol. Opt. Phys.* 1991, 24, 521.
[14] Knudsen, H.; Reading, J. F. *Phys. Rev.* 1992, 212, 107.
[15] Cavalcanti, E.G.; Sigaud, G. M.; Montenegro, E. C.; Schmidt-Böcking, H. *J. Phys. B: At. Mol. Opt. Phys.* 2003, 36, 3087.
[16] Sharma, S.; Srivastiva, M. K. *Phys. Rev. A* 1988, 38, 1083.
[17] Fiol, J.; Rodríguez, V. D.; Barrachina, R. O. *J. Phys. B: At. Mol. Opt. Phys.* 2001, 34, 933-944.
[18] Kövér, Á.; Laricchia, G.; Charlton, M. *J. Phys. B: At. Mol. Opt. Phys.* 1993, 26, L575.

[19] Kövér, Á.; Laricchia, G. *Phys. Rev. Lett.* 1998, 80, 5309.

[20] Kövér, Á.; Paludon, K.; Laricchia, G. *J. Phys. B: At. Mol. Opt. Phys.* 2001, 34, L219-L222.

[21] Ehrhardt, H.; Schulz, M.; Tekaat, T.; Willhelm, K. *Phys. Rev. Lett.* 1969, 22, 89.

[22] Waterhouse, D. K.; McCarthy, I. E.; Williams, J. F. *Phys. Rev. A* 1998, 57, 3565.

[23] DuBois, R. D. Ph.D. thesis, University of Nebraska-Lincoln 1975.

[24] Santos, A. C. F.; Hasan, A.; DuBois, R. D. *Phys. Rev. A* 2005, 71, 034701.

[25] Falke, T.; Brandt, T.; Kühl, O.; Raith, W.; Weber, M. *J. Phys. B: At. Mol. Opt. Phys.* 1997, 30, 3247.

[26] Moxom, J.; Laricchia, G.; Charlton, M.; Jones, G. O.; Kövér, Á. *J. Phys. B: At. Mol. Opt. Phys.* 1992, 25, L613.

[27] Schmitt, A.; Cerny, U.; Möller, H.; Raith, R.; Weber, M. *Phys. Rev. A* 1994, 49, R5.

[28] Kövér, Á.; Laricchia, G.; Charlton, M. *J. Phys. B: At. Mol. Opt. Phys.* 1994, 27, 2409.

[29] Kövér, Á.; Finch, R. M.; Charlton, M.; Laricchia, G. *J. Phys. B: At. Mol. Opt. Phys.* 1997, 30, L507-L512.

[30] *Ten Years of COLTRIMS and Reaction Microscopes* 2004 Ed by J. Ullrich book and electronic version Max-Planck-Institute für Kernphysik Heidelberg.

[31] Dorn, A.; Moshammer, R.; Schröter, C. D.; Zouros, T. J.; Schmitt, H.; Kollmus, H.; Mann, R.; Ullrich, J. *Phys. Rev. Lett.* 1999, 82, 2496.

[32] http://www.roentdek.com/

[33] Krems, M.; Zirbel, J.; Thomason, M.; DuBois, R. D. *Rev. Sci. Instr.* 2005, 76, 093305.

[34] Müller, A.; Djurić, N.; Dunn, G. H.; Belić, D. S. *Rev. Sci. Instr.* 1986, 57, 349.

[35] Rudd, M.E.; Kim, Y.–K.; Madison, D. H.; Gay, T. J. *Rev. Mod. Phys.* 1992, 64, 441.

[36] DuBois, R. D.; Rudd, M. E. *Phys. Rev. A* 1978, 17, 843.

[37] Santos, A. C. F.; Hasan, A.; DuBois, R. D. *Phys. Rev. A* 2004, 69, 032706.

[38] DuBois, R. D.; Doudna, C.; Lloyd, C.; Kahveci, M.; Khayyat, K.; Zhou, Y.; Madison, D. H. *J. Phys. B At. Mol. Opt. Phys.* 2001, 34, L783-L789.

[39] Santos, A. C. F.; Hasan, A.; DuBois, R. D. *Phys. Rev. A* 2004, 69, 032706.

[40] Paludan, K.; Laricchia, G.; Ashley, P.; Kara, V.; Moxon, J.; Bluhme, H.; Knudsen, H.; Mikkelsen, U.; Møller, S. P.; Uggerhøj, E.; Morenzoni, E. *J. Phys. B: At. Mol. Opt. Phys.* 1997, 30, L581.

[41] Helms, S. *J. Phys. B: At. Mol. Opt. Phys.* 1995, 28, 1095.

[42] Rejoub, R.; Lindsay, B. G.; Stebbings, R. F. *Phys. Rev. A* 2002, 65, 042713.

[43] Holland, D. M. P.; Codling, K.; West, J. B.; Marr, G. V. *J. Phys. B: At. Mol. Opt. Phys.* 1979, 12, 2465

[44] Bizau, J. M.; Wuilleumier, F. J. *J. Electron Spectrosc. Relat. Phen.* 1995, 71, 205.

[45] Van der Wiel, M.J.; Wiebes, G. *Physica (Amsterdam)* 1971, 54, 411.

[46] Saito, N.; Suzuki, I.H. *Int. J. Mass. Spectrom. Ion Processes* 1992, 115, 157.

[47] Carlson, T. A. *Phys. Rev.* 1967, 156, 142.

[48] Wright, G. R.; Van der Wiel, M. J. *J. Phys. B: At. Mol. Opt. Phys.* 1976, 9, 1319.

[49] Samson, J.A. R.; Angel, G. C. *Phys. Rev. A* 1990, 42, 5328.

[50] Schmidt, V.;Sandner, N.; Kuntzemuller, H.; Dhez, P.; Wuileumier, F.; Kallne, E. *Phys. Rev. A* 1976, 13, 1748.

[51] Andersen, L. H.; Hvelplund, P.; Knudsen, H.; Møller, S. P.; Sørensen, A. H.; Elsener, K.; Rensfelt, K.-G.; Uggerhøj, E. *Phys. Rev. A* 1987, 36, 3612.

[52] de Lucio, O. G.; Gavin, J.; DuBois, R. D. *Phys. Rev. Lett.* 2006, 97, 243201.

[53] Murtagh, D. J.; Szluinska, M.; Moxom, J.; Van Reeth, P.; Laricchia, G. *J. Phys. B: At. Mol. Opt. Phys.* 2006, 39, 1251.
[54] DuBois, R. D.; Kövér, Á. *Phys. Rev. A* 1989, 40, 3605.

In: Radiation Physics Research Progress
Editor: Aidan N. Camilleri, pp. 155-191
ISBN 978-1-60021-988-7

Chapter 3

RADIATIVE IONIZATION: THE LINK BETWEEN RADIATIVE ELECTRON CAPTURE AND BREMSSTRAHLUNG

D. H. Jakubassa-Amundsen
Mathematics Institute, University of Munich,
Theresienstrasse 39, 80333 Munich, Germany

Abstract

This chapter is devoted to the theory of radiative ionization of target atoms in energetic collisions with highly stripped projectiles. It is set into context with the simultaneously occurring processes of nonradiative electron capture to continuum and radiative capture to bound projectile eigenstates. Among other processes linked by inverse kinematics, particular emphasis is laid on the relation between radiative ionization and electron-nucleus bremsstrahlung. Specific features of the electron and photon spectra and their angular distributions as well as the photon linear polarization are reviewed. In addition, new results are presented and are compared with experiments using relativistic uranium beams. The validity of the theoretical model is also inferred from a comparison with accurate partial-wave calculations for bremsstrahlung and radiative electron capture.

1. Introduction

The interaction of charged particles by means of the Coulomb field as well as their coupling to weak photon fields are basically well-understood processes. Nevertheless, the atomic collision physics has kept its fascination all over the years. A particularly intriguing aspect is the close relation between specific processes which usually are treated isolated from each other. This interrelation allows for experimental tests of the theory from a different point of view, maybe even at conditions which never could be met in the isolated process. On the other hand, previously unexplained experimental features suddenly become clear if viewed from another frame of reference.

There exist several possible relations between atomic processes in collision physics. The first is that the physical process induced by the interaction between the active particles is the same, but the initial or final states differ from each other. Restricting ourselves to radiation physics, an example is radiative recombination (where the electron is initially free) in relation to radiative electron capture (REC, where the electron is quasifree, i.e. in a loosely bound initial state) [1]. As another case we have REC (with a bound electronic final state) and radiative ionization (RI, sometimes also termed RECC, where the electron is released), respectively. A second type of interrelation concerns different possible processes under the restriction of the same initial and final states. Such processes usually occur simultaneously and often require coincidence experiments to separate them. As examples may serve the two processes RI and the nonradiative electron capture to continuum (ECC) [2, 3], but also REC and Coulomb capture to bound states (EC) [4], where either the Coulomb field or the radiation field induces the transition. Like REC and RI, also EC and ECC are linked by means of the continuity across the projectile's ionization threshold [5].

A very important type of interrelation is the one by inverse kinematics. In processes related in this way the initial and final states are interchanged; one might also view one process as the time-reversed second process. For instance we have photoionization (incoming photon, emitted electron) and radiative recombination (incoming electron, emitted photon), respectively [6, 7]. Relaxing the requirement that the initial state of one process be completely the same as the final state of the second process, one may also view photoionization (with a free outgoing electron) and REC (with a quasifree incoming electron) as processes linked by inverse kinematics [7]. Likewise, photoionization (with an initially bound electron) and bremsstrahlung (with a free outgoing electron) have been considered as inverse processes [8]. However, the phrase inverse kinematics is also applied to processes which are linked by a frame transformation. For these not only the physics is the same, but also their observation provided the reference frame is changed accordingly. The show-piece is target ionization and electron loss, respectively, in H + H collisions but, again relaxing the condition of identical initial states, RI and bremsstrahlung can be viewed as inverse processes too [9]. In particular, this is true for the short-wavelength limit of bremsstrahlung and the radiative electron capture to the continuum threshold which leads to cusp electrons, respectively [10, 11].

Once two reactions are identified as being linked by one of the above-mentioned relations, the same theoretical model can be applied to their description. If, for instance, a process initiated by a free electron is well described by some theory T, the impulse approximation (IA) will give an excellent description for the same process initiated by a quasifree electron, provided the collision is sufficiently energetic. In fact, the frame-transformed RI for relativistic collision velocities and a hydrogen target leads to results which are very close to those for bremsstrahlung under the condition that the underlying theory T is the same [12]. Similarly, the radiative recombination provides a very good approximation to the impulse approximation for REC at relativistic collision velocities [7]. There remains, however, a basic difference between processes initiated by a free and a quasifree electron if multiply differential cross sections are considered. In the quasifree case, the momentum provided by the parent nucleus adds to the momentum balance and changes a sharp photon line (in the case of a free electron) to a broad energy distribution governed by the Compton profile. That becomes evident in the REC spectra (see e.g. [13]).

This chapter is devoted to the relativistic formulation of the theory for radiative ionization, nonradiative electron capture to continuum and radiative electron capture to bound states using the formal scattering theory as a common starting point. The competition between RI and ECC in the electron spectra is reviewed in section 2, supplemented with the interpretation of new experimental results for the forward peak. The close relation between RI and the elementary process of bremsstrahlung is discussed in section 3. Particular emphasis is laid on the influence of the collision parameters, such as projectile and target nuclear charge as well as the collision velocity, on the angular distribution of the emitted photons and their degree of polarization. Furthermore, in section 4, the relativistic theory for REC is derived in some detail. This theory is compared to an existing REC theory that is based on the inverse photoeffect, and the calculated photon yield is contrasted to that from RI near the continuum threshold of the projectile. A summary of all results is given in section 5.

2. Radiative Ionization: The Model and its Comparison with ECC

RI, interpreted as capture of a target electron into the continuum of an ionized, ideally bare, projectile with the simultaneous emission of a photon, is the dominant background process in photon spectra from fast, asymmetric collisions ($Z_P \gg Z_T$, Z_P and Z_T being the nuclear charges of the projectile and target, respectively) for photon energies $\hbar\omega$ below the threshold T_0. This threshold is related to the fact that an electron at rest in the target frame cannot radiate more than its complete kinetic energy relative to the projectile frame. One has $T_0 = (E_0' - mc^2)/[\gamma(1 - \frac{v}{c}\cos\vartheta_k)]$ where $\boldsymbol{v}$ is the collision velocity, mc^2 the electron's rest energy, $\gamma = (1 - v^2/c^2)^{-\frac{1}{2}}$, $E_0' = \sqrt{(\gamma mvc)^2 + m^2c^4}$ the electronic collision energy in the projectile reference frame, and ϑ_k the photon emission angle. As such RI was first identified by Kienle et al [14] and correctly interpreted soon afterwards [15, 16]. Later, RI became directly visible as a ridge near T_0 in the continuous photon spectra in very fast collisions ([17]; for recent work see e.g. [10]). The breakthrough for RI came, however, just now with a coincidence experiment where U^{88+} was collided with nitrogen and where the photon and electron momenta were recorded simultaneously [11]. There, a photon spectrum, due entirely to RI, was measured for the first time and interpreted by theory.

2.1. The Impulse Approximation for Relativistic Collisions

Within the semiclassical independent-particle approximation, the general expression for the transition amplitude from an initially bound target state to a projectile continuum state $\overline{\psi_{f,P}^{(\sigma_f)'}(x')} = \psi_{f,P}^{(\sigma_f)'+}(x')\,\gamma_0$ of momentum $\boldsymbol{k}_f'$ and spin σ_f is given by (in atomic units, $\hbar = m = e = 1$)

$$a_{fi} = -\frac{i}{c}\int d^4x'\;\overline{\psi_{f,P}^{(\sigma_f)'}(x')}\;d_\lambda^+\;\left(\hat{S}\;\mathbb{A}(x)\;\hat{S}^{-1}\right)\;\hat{S}\;\Psi_i^{(\sigma_i)}(x). \tag{2.1}$$

The RI and ECC processes differ only in the choice of the electromagnetic transition field $\mathbb{A}(x)$ and in the presence (RI) or absence (ECC) of the photon creation operator d_λ^+ in (2.1). Projectile-frame related quantities are denoted by a prime, and we have chosen this frame as our frame of reference. $\Psi_i^{(\sigma_i)}(x)$ is the exact electronic scattering state which relates asymptotically to a target eigenstate $\psi_{i,T}^{(\sigma_i)}(x)$ with spin σ_i and space-time vector $x = (ct, \boldsymbol{x})$. x is connected to x' by a Lorentz transformation, $x = \Gamma x' + b$, where b is the impact parameter. The scattering state as well as the field $\mathbb{A}(x)$ have to be transformed into the projectile frame by means of the Lorentz boost operator $\hat{S}$. If the z-axis (respectively the unit vector $\boldsymbol{e}_z$) is chosen along $\boldsymbol{v}$, then

$$\hat{S}(v) = \sqrt{\frac{1+\gamma}{2}} \left(1 - \frac{\gamma v/c}{1+\gamma}\,\alpha_z\right) \tag{2.2}$$

with its inverse $\hat{S}^{-1}(v) = \gamma_0 \hat{S}(v)\,\gamma_0 = \hat{S}(-v)$. α_z, γ_0 are Dirac matrices [18].

For asymmetric collisions with $Z_T \ll Z_P$ one may expand the scattering state $\Psi_i^{(\sigma_i)}(x)$ in terms of the weak target field while retaining the correct asymptotics. The lowest-order term in this expansion leads to the strong potential Born approximation [19, 60],

$$\hat{S}\,\Psi_i^{(\sigma_i)}(x) = \frac{1}{c}\sum_{s=1}^{4}\int d\boldsymbol{q}\; dE\; \psi_{q,off}^{(s)\prime}(x')\,\left(q_s'(x'), \hat{S}\,\psi_{i,T}^{(\sigma_i)}(x)\right). \tag{2.3}$$

Here, $q_s'(x')$ is a relativistic plane wave of energy E characterized by the four-spinor $u_q^{(s)}$ [21]. $s = 1, 2$ denotes the spin directions of the particle states, and $s = 3, 4$ those of the antiparticle states. Furthermore, $\psi_{q,off}^{(s)\prime}(x')$ is an off-shell projectile continuum state with momentum $\boldsymbol{q}$. The deviation from the energy shell is determined by the binding energy of the initial target state. Therefore the target potential is included to some extent in the transition operator. The off-shell effects vanish for $Z_T = 0$.

For sufficiently high collision velocities ($v \gg Z_T/n_i$ with n_i the initial-state main quantum number) an on-shell approximation can be made. This replacement of $\psi_{q,off}^{(s)\prime}(x')$ by a projectile continuum eigenstate $\psi_{q,P}^{(s)\prime}(x')$ leads to the impulse approximation. Since the differences between the on-shell and off-shell approximation relates to the target field, they can be used as an indicator of validity of the strong potential Born theory itself. In the nonrelativistic case where the IA was tested against the strong potential Born theory, the difference between the two theories was on the level of 10 percent in the cusp region, getting smaller when v increases [20].

In the case of radiative ionization the interaction in (2.1) arises from the photon field, defined in the projectile frame, $\gamma_0\,\hat{S}\,\mathbb{A}(x)\,\hat{S}^{-1} = -\boldsymbol{\alpha}\boldsymbol{A}_\lambda' e^{ik'x'} d_\lambda^+$ with $\boldsymbol{A}_\lambda' = c\,\boldsymbol{e}_\lambda/(2\pi\omega'^{1/2})$ where $k' = (\omega'/c, -\boldsymbol{k}')$ is the 4-momentum of the photon and $\boldsymbol{e}_\lambda$ its polarization direction. $\boldsymbol{\alpha} = (\alpha_x, \alpha_y, \alpha_z)$ is the vector of the Dirac matrices. In the IA, the transition amplitude is thus governed by the radiation matrix element

$$\boldsymbol{W}_{rad}(\sigma_f, s, \boldsymbol{q}) = \int d\boldsymbol{x}'\; \psi_{f,P}^{(\sigma_f)\prime +}(\boldsymbol{x}')\,\boldsymbol{\alpha}\, e^{-i\boldsymbol{k}'\boldsymbol{x}'}\; \psi_{q,P}^{(s)\prime}(\boldsymbol{x}') \tag{2.4}$$

and is given by [22]

$$a_{fi,\lambda}^{RI} = \frac{2\pi i}{\gamma}\sqrt{\frac{1+\gamma}{2}}\;\boldsymbol{A}_\lambda' \sum_{s=1}^{4}\int d\boldsymbol{q}\; e^{i\boldsymbol{q}_\perp \boldsymbol{b}}\; \boldsymbol{W}_{rad}(\sigma_f, s, \boldsymbol{q})$$

$$\cdot \left[u_q^{(s)+} \left(1 - \frac{\gamma v/c}{1+\gamma} \alpha_z \right) \varphi_{i,T}^{(\sigma_i)}(\boldsymbol{q}_0) \right] \delta\left(E_f' + \omega' - E_i^T/\gamma + q_z v\right) \tag{2.5}$$

where $\varphi_{i,T}^{(\sigma_i)}(\boldsymbol{q}_0)$ is the initial-state wavefunction in momentum space, $\boldsymbol{q}_0 = (\boldsymbol{q}_\perp, q_{0z})$ with $\boldsymbol{q}_\perp$ perpendicular to $\boldsymbol{v}$ and $q_{0z} = E_i^T v/c^2 + q_z/\gamma$, E_i^T and E_f' being the electron energy in its initial and final state, respectively[1].

For the nonradiative capture, the interaction in (2.1) results from the target Coulomb potential $V_T(\boldsymbol{x})$, viz. $\gamma_0 \hat{S}\, \mathbb{A}(x)\, \hat{S}^{-1} = \gamma V_T(\boldsymbol{x}) \left(1 + \frac{v}{c} \alpha_z\right)$, which is conventionally decomposed into its Fourier components with weight factor $1/p^2$. Using the IA, we define the transition matrix element T_{00}/p^2 with

$$T_{00}(s, \boldsymbol{q}, \boldsymbol{p}) = \int d\boldsymbol{x}'\, \psi_{f,P}^{(\sigma_f)'+}(\boldsymbol{x}')\, e^{i\boldsymbol{p}_\perp \boldsymbol{x}'_\perp + i p_z \gamma z'} \left(1 + \frac{v}{c} \alpha_z\right) \psi_{q,P}^{(s)'}(\boldsymbol{x}'). \tag{2.6}$$

Then the transition amplitude for ECC is given by [3]

$$a_{fi}^{ECC} = \frac{i Z_T}{\pi} \sqrt{\frac{1+\gamma}{2}} \sum_{s=1}^{4} \int d\boldsymbol{q} \int \frac{d\boldsymbol{p}}{p^2}\, e^{i(\boldsymbol{p}_\perp + \boldsymbol{q}_\perp)\boldsymbol{b}}\, T_{00}(s, \boldsymbol{q}, \boldsymbol{p}) \tag{2.7}$$

$$\cdot \left[u_q^{(s)+} \left(1 - \frac{\gamma v/c}{1+\gamma} \alpha_z \right) \varphi_{i,T}^{(\sigma_i)}(\boldsymbol{q}_0) \right] \delta(E_f' + p_z \gamma v + q_z v - E_i^T/\gamma),$$

where $\boldsymbol{q}_0 = (\boldsymbol{q}_\perp, q_{0z})$ is defined as above.

The differential cross section for the emission of an electron with energy E_f into the solid angle $d\Omega_f$ is obtained by means of integrating over impact parameter and by averaging and summing, respectively, over the initial and final spin states. For ECC, one has

$$\frac{d^2\sigma^{ECC}}{dE_f d\Omega_f} = \frac{k_f E_f'}{2c^2} \sum_{\sigma_i, \sigma_f} \int d^2\boldsymbol{b}\, \left|a_{fi}^{ECC}\right|^2. \tag{2.8}$$

Concerning RI, this prescription leads to the fourfold differential cross section for a photon of frequency ω ejected into the solid angle $d\Omega_k$ with a fixed polarization direction $\boldsymbol{e}_\lambda$ and a simultaneously emitted electron,

$$\frac{d^4\sigma_\lambda^{RI}}{dE_f d\Omega_f d\omega\, d\Omega_k} = \frac{k_f E_f' \omega \omega'}{2c^5} \sum_{\sigma_i, \sigma_f} \int d^2\boldsymbol{b}\, \left|a_{fi,\lambda}^{RI}\right|^2. \tag{2.9}$$

For the comparison with ECC, (2.9) has to be summed over the two polarization directions and integrated over the photon momentum degrees of freedom,

$$\frac{d^2\sigma^{RI}}{dE_f d\Omega_f} = \int d\omega\, d\Omega_k \sum_\lambda \frac{d^4\sigma_\lambda^{RI}}{dE_f d\Omega_f d\omega d\Omega_k}. \tag{2.10}$$

For the numerical evaluation, a semirelativistic approximation is used for the wavefunctions in order to deal with analytic, closed expressions for the transition matrix elements

[1] In previous work [3, 22] this spin sum was erroneously truncated at $s = 2$. However, the contribution of $s = 3, 4$ is negligibly small in the weakly relativistic regime considered.

(2.4) and (2.6). It involves Darwin functions for the bound states [23] and Sommerfeld-Maue functions for the continuum states [24]-[26] which are accurate up to first order in Z/c, where Z is the respective nuclear charge (for their explicit form, see section 4.1). They coincide with the exact Coulomb eigenstates in the nonrelativistic limit. Tests of their validity by comparing the results with those obtained from accurate relativistic wavefunctions are provided in sections 3 and 4.

In order to understand the features in the momentum distribution of the photons and electrons given below we will analyze the fourfold differential RI cross section in somewhat more detail. To do so, we apply for the sake of demonstration a peaking approximation. This approximation relies on the fact that the bound-state wavefunction $\varphi_{i,T}^{(\sigma_i)}(\boldsymbol{q}_0)$ is (for s-states) strongly peaked at $q_0 = 0$. According to the δ-function in (2.5), the z-component q_{0z} vanishes if the photon energy takes the value

$$\omega_{peak}(\vartheta_k) = \frac{\omega'_{peak}}{\gamma(1 - \frac{v}{c}\cos\vartheta_k)}, \qquad \omega'_{peak} = \gamma E_i^T - E'_f. \tag{2.11}$$

For photons with $\omega \approx \omega_{peak}$ we set $\boldsymbol{q} = (\boldsymbol{0}, -\gamma E_i^T v/c^2)$ corresponding to $q_0 = 0$ everywhere in the transition amplitude (2.5) except in $\varphi_{i,T}^{(\sigma_i)}(\boldsymbol{q}_0)$ and in the δ-function. Splitting the Darwin function into a scalar part $\tilde{\varphi}_{i,T}(\boldsymbol{q}_0)$ times a spinor function which can be taken at $q_0 = 0$ (see the expression above (4.2)), the differential cross section becomes proportional to

$$v \int d\boldsymbol{q}\, |\tilde{\varphi}_{i,T}(\boldsymbol{q}_0)|^2\, \delta(E'_f + \omega' - E_i^T/\gamma + q_z v) = J_i\left(E_i^T/v - \frac{1}{\gamma v}(E'_f + \omega')\right) \tag{2.12}$$

which is the target Compton profile. Hence one obtains a peak in $d^4\sigma_\lambda^{RI}/dE_f d\Omega_f d\omega d\Omega_k$ at $\omega = \omega_{peak}(\vartheta_k)$, shaped by this Compton profile.

While in the actual RI calculations no such peaking approximation is made we have, however, resorted to a transverse peaking approximation in the case of ECC. This approximation is applied to handle the multiple integral in the transition amplitude (2.7). For fast collisions it is well justified, the more so, the higher the collision velocity [3].

2.2. RI Photon Spectra and Angular Distributions in Comparison with Experiment

Fig.1 shows the photon spectrum from collisions of U^{90+} with N_2 at a fixed photon emission angle in comparison with theory. The experimental RI spectrum is obtained from the singles photon spectrum by subtracting the REC spectrum (which was recorded in coincidence with U^{89+} ejectiles) as well as the background radiation [10]. Its absolute scale results from a fit of the measured L-REC spectrum to an accurate REC theory [27]. In order to obtain the theoretical doubly differential RI cross section, (2.9) is summed over the polarization directions and integrated over the electron degrees of freedom. Here and in the following, if not stated otherwise, partly stripped projectiles are treated as bare projectiles with the ionic charge. Also, for the states of a multielectron target atom, an average over the subshells, a Slater-screened charge and experimental binding energies are used. The molecular character of N_2 is disregarded (i.e. the result for N is multiplied by 2). In the figure the one-electron contributions from the target K- and L-shell are shown separately.

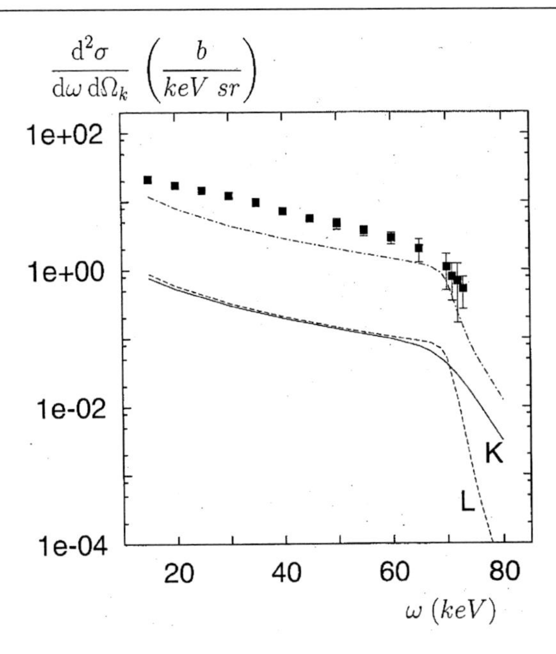

Figure 1. Doubly differential RI cross section from 223.2 MeV/amu $U^{90+} + N_2$ ($v = 80.77$ a.u.) for an emission angle $\vartheta_k = 132°$ as a function of photon energy ω. Shown are RI from the K-shell (——) and L-shell (– – –) of N as well as the total RI from N_2 (– · – · –) as described in the text. Experimental data (■) from Ludziejewski et al [10].

The threshold ridge at $T_0 = 70.4$ keV is clearly identified in experiment and theory. While there is qualitative agreement, theory underestimates experiment by a global factor of 2 which we ascribe to the use of semirelativistic wavefunctions for the heavy uranium projectile (see Figs.15 and 16).

In Fig.2 the angular distribution of the photons is shown. The frequency $\omega = 65$ keV was chosen such that the threshold ridge is present ($T_0 = 65$ keV for $\vartheta_k = 150.1°$). In comparison with experiment the RI theory provides cross sections which are a factor of 3 too low, and the slope is also not well reproduced. The origin of this discrepancy remains unclear.

2.3. Characteristics of the RI and ECC Electron Spectra

For electrons emitted close to the beam direction there are three prominent structures in the electron spectra. One is the binary encounter peak at a kinetic electron energy of $E_{f,kin} = (2v^2 \cos^2 \vartheta_f - 2E_B)/(1 - \frac{v^2}{c^2} \cos^2 \vartheta_f)$ where $E_B = mc^2 - E_i^T$ is the binding energy of the target state [28, 29]. This energy is the maximum energy which can be transferred from the projectile nucleus to an electron at rest. Having a classical origin, the peak rises the higher above the background the larger v. Its shape is determined by the target Compton profile, the width increasing with γv. The second structure is the cusp-shaped forward peak near $E_{f,kin} = (\gamma - 1)mc^2$ which is the energy of the projectile continuum threshold measured in the target frame of reference. The forward peak was first observed by Rudd et al [30] and originates from the long-range Coulomb interaction between the ejected electron and the

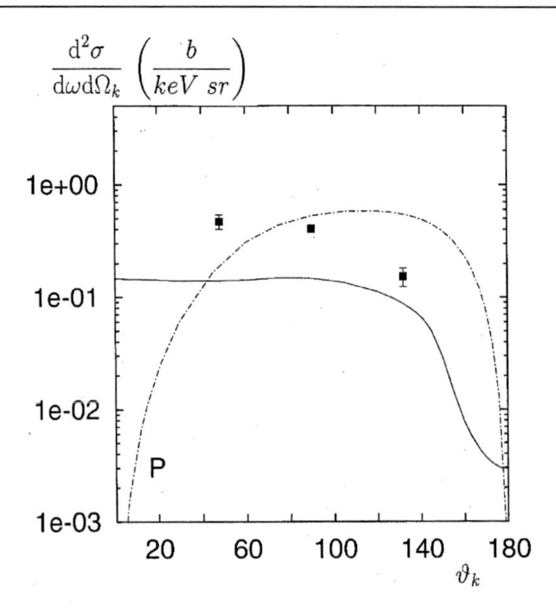

Figure 2. Doubly differential RI cross section per electron from 223.2 MeV/amu U^{90+}+N_2 at ω = 65 keV as a function of photon emission angle ϑ_k. Theory: ——— . Experiment: ■, Ludziejewski et al [10]. Also shown is the photon linear polarization P relating to $d^2\sigma/d\omega d\Omega_k$ for this process (left-hand scale without dimension, $- \cdot - \cdot -$).

projectile which manifests itself in the normalization constant of the final-state electronic wavefunction. This normalization constant introduces a divergence $\sim 1/k'_f$ into the target-frame doubly differential cross section [31, 32]. If $\vartheta_f = 0$ and $k_f = v$, the shape of the measured peak is governed by the detector resolution (the consideration of which renders the cusp maximum finite). The underlying background with its discontinuity at $\boldsymbol{k}_f = \boldsymbol{v}$ (which is only correctly described in a higher-order theory with respect to the electron-projectile interaction [33, 34]) leads to an asymmetry of the forward peak. Finally, the third structure in the electron spectra is the increase of intensity towards $E_{f,kin} \to 0$, the so-called soft-electron peak. Such electrons with a small kinetic energy can be released in distant collisions and usually represent the main portion of the total ionization cross section.

In Fig.3a the spectra of the forward electrons ($\vartheta_f = 1.5°$) resulting from Ar^{18+}+ H collisions are shown. The cross sections for RI and ECC were calculated from (2.10) and (2.8), respectively. The collision velocity ($v = 33.85$ a.u., $\gamma = 1.03$) is on the border to the nonrelativistic regime. The three structures are clearly seen in both processes, the binary encounter peak at $E_{f,kin} = 66.3$ keV, the cusp near $E_{f,kin} = 16.3$ keV and the rise when $E_{f,kin} \to 0$. At the chosen collision energy ECC is dominant for the soft electrons and the binary encounter electrons, but the cusp intensity is larger for RI than for ECC. While the ECC cusp is skewed to the low-energy side and is rather weak, the RI cusp rises much higher above the background, and its high-energy wing is enhanced.

The reason for the different cusp intensities lies in the fact that ECC requires a high momentum transfer ($q_{min}^{ECC} = |\frac{1}{\gamma v}(E'_f - \gamma E_i^T) + p_z|$) to the target nucleus, while in RI the emitted photon carries away the excess energy ($q_{min}^{RI} = \frac{1}{\gamma v}|E'_f + \omega' - \gamma E_i^T|$). Therefore the dominant cusp contribution for RI comes from the maximum of the target Compton

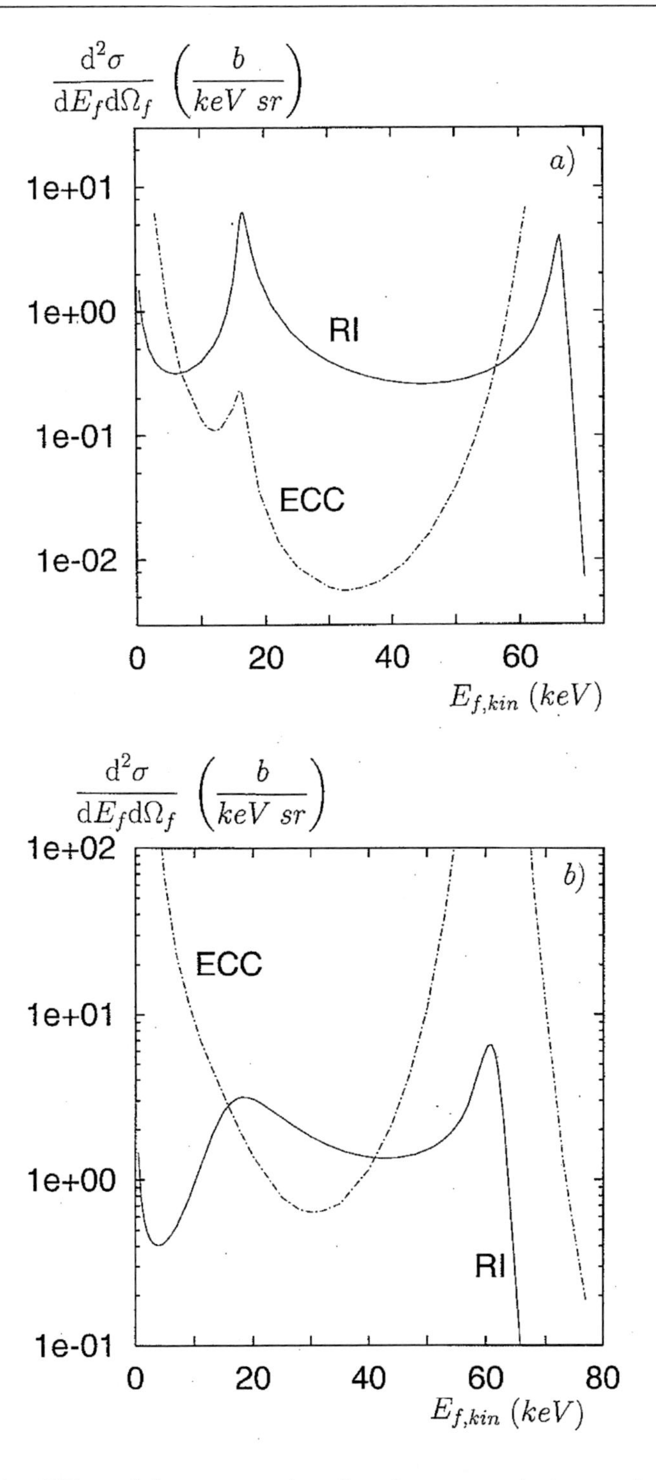

Figure 3. Doubly differential cross section for electron emission at $\vartheta_f = 1.5°$ in 30 MeV/amu Ar^{18+}+ H collisions (a) and at $\vartheta_f = 15°$ in 30 MeV/amu Kr^{36+}+ He collisions (b) as a function of kinetic electron energy $E_{f,kin} = E_f - mc^2$. ——, RI; $- \cdot - \cdot -$, ECC.

profile. On the other side, q_{min}^{ECC} increases with collision velocity (for $p_z = 0$) such that for ECC mainly the outer wing of the Compton profile contributes. As a consequence, ECC decreases much faster with v than RI. We remark that for ECC the matter is actually more complicated because the interaction potential V_T can help to supply the necessary momentum ($p_z \neq 0$). The inference given above remains correct, however.

The analysis given above may also help to explain the different cusp asymmetries of the two processes. For ECC, in the projectile frame of reference, the incoming electron moves close to the target nucleus because it has to exchange a large momentum. It is then dragged along with the target nucleus after colliding quasielastically with the projectile. Thus it is ejected predominantly along $-\boldsymbol{v}$. In contrast, in the RI process the electron is quasifree and thus only subject to the projectile field. The loss of nearly all its energy to the photon requires a deeply inelastic scattering in the vicinity of the nucleus. The electron is thereby guided around the projectile by the attractive potential and emerges in the direction of $\boldsymbol{v}$. Therefore, in the target frame of reference, the ECC and RI electrons tend to appear slightly below and above the cusp maximum, respectively.

Fig.3b depicts the (single-electron) spectra from Kr^{36+}+ He collisions at the same velocity but for a larger angle ($\vartheta_f = 15°$). At this angle the ECC cusp has disappeared while the RI forward peak has broadened but is still clearly visible above the background. For the heavier He target, ECC largely dominates the soft-electron and binary encounter peak (the latter having shifted to 61.6 keV). In the forward peak region, ECC and RI are now of comparable importance. The reason is that ECC from the target K-shell strongly increases with Z_T because the momentum distribution is broadened. On the other hand, the doubly differential RI cross section is approximately independent of Z_T since (2.10) involves an integration over the target Compton profile.

Fig.4 shows the RI and ECC spectra from U^{88+}+ N at $\vartheta_f = 3°$ for a higher collision velocity ($v = 56.24$ a.u., $\gamma = 1.1$). For this heavy projectile the cusp asymmetry is clearly visible in both processes since this asymmetry (as well as the peak intensity) increases with Z_P [3, 22]. Moreover, ECC is now also dominating in the cusp region because of the larger Z_T. The RI contribution from the nitrogen L-shell is approximately 5 times that for one K-shell electron while for ECC, the L-shell contribution is only about 5 percent of the K-shell yield at this velocity (due to the stronger fall-off of the bound-state momentum space wavefunction for the less tightly bound L-electrons). Therefore, the curves in Fig.4 were obtained by calculating the capture of an electron from the target K-shell and multiplying it with a factor of 7 and 2 for RI and ECC, respectively.

A comparison with (relative) experimental data for this collision system [11] is made in Fig.5. At an electron emission angle of 0°, the RI spectrum was directly recorded in coincidence with photons emitted at $\vartheta_k = 90°$. The ECC spectrum was obtained from the singles electron spectrum by subtracting the simultaneously measured electron loss to continuum (ELC, in coincidence with U^{89+} ejectiles) and RI spectra (the latter extrapolated to a 4π solid angle for the emitted photon [35]). In the calculations the target subshells were treated separately, using Hartree-Fock states (and experimental binding energies) in the case of RI. The difference to the results from Slater-screening is small, however. Theory is averaged over the spectrometer resolution (in order to facilitate the calculations, ECC is calculated for $\vartheta_f = 1°$ instead of averaging over the angular resolution of 1.9°, and is only averaged over the energy resolution). In the case of RI, the fourfold differential

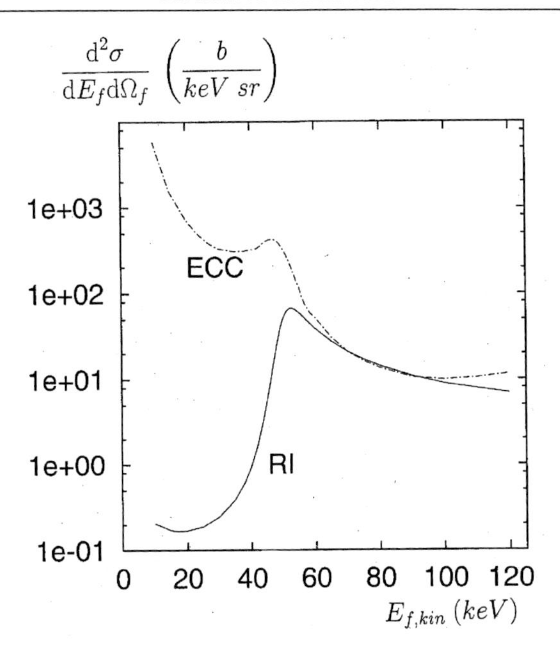

Figure 4. Doubly differential cross section for electron emission at $\vartheta_f = 3°$ in 90 MeV/amu U^{88+}+ N collisions as a function of $E_{f,kin}$. ———, RI; $- \cdot - \cdot -$, ECC.

cross section is in addition integrated over the energy and angular acceptance of the photon detector. Experiment is normalized to theory in the peak maximum. It is seen that the theoretical peak positions and widths compare fairly well with these pioneer data.

As mentioned above the shape of the cusp is strongly influenced by the spectrometer resolution. In Fig.6 the RI cusp cross section (2.10) for one electron from the K-shell in U^{88+}+ N collisions at $\vartheta_f = 0°$ is shown for several different angular resolutions θ_0. There is a shift of the peak maximum to higher electron energies and the peak becomes broader, the more, the larger θ_0. A similar effect is observed when θ_0 is kept fixed at a small value whereas the energy resolution ΔE_f of the spectrometer is increased. The peak shift is exclusively due to the strong asymmetry of the cusp.

2.4. Crossing Velocity of RI/ECC

For low collision energies ECC is strongly dominant since the coupling to the radiation field is suppressed because of the smallness of the fine structure constant. However, due to the slower decrease of RI with velocity as compared to ECC, RI eventually gains importance in the cusp region. Therefore there exists a velocity where the RI and ECC processes provide equal electron intensities in the cusp maximum. This crossing velocity v_{cr} marks the change of shape in the singles cusp spectra (for bare projectiles where ELC is not present) from left-hand skewed to right-hand skewed. Moreover, when $v \gg v_{cr}$, the singles cusp spectra will exclusively be due to RI. The crossing velocity was first estimated within a nonrelativistic approach by Shakeshaft and Spruch [2] for a hydrogen target. Scaling properties of v_{cr} were derived in [3] using the relativistic theory. The crossing velocity is independent of the electron emission angle (for ϑ_f in the forward region). Also, it is only weakly dependent

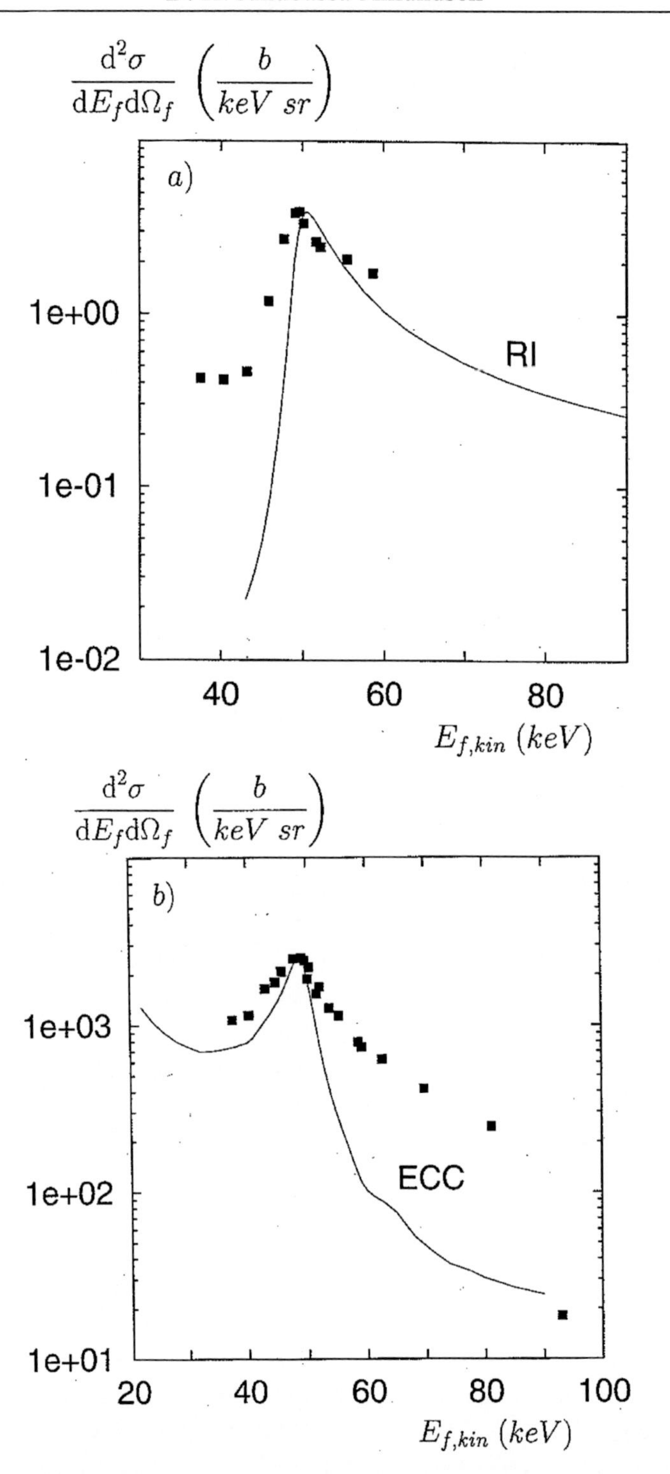

Figure 5. Doubly differential cross section for RI (a) and ECC (b) from 90 MeV/amu $U^{88+} + N_2$ at $\vartheta_f = 0 \pm 1.9°$. Experiment (■, from Nofal et al [11, 35]) is normalized to theory in the maximum.

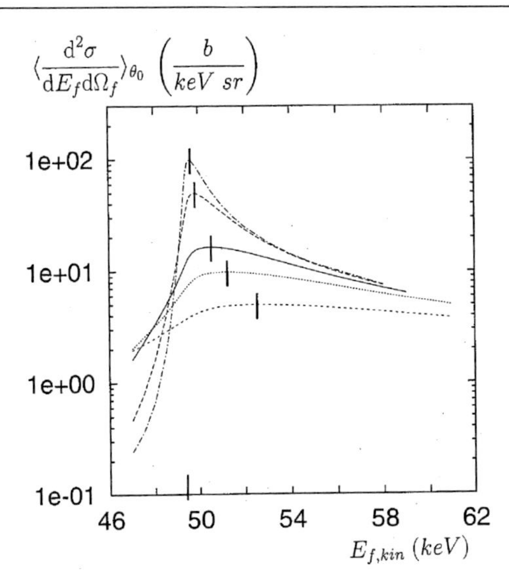

Figure 6. Doubly differential RI cross section for one target K-shell electron in 90 MeV/amu U^{88+}+ N collisions at $\vartheta_f = 0$ averaged over the detector resolution θ_0 as a function of kinetic electron energy. $\theta_0 = 0.5°$ $(- \cdot - \cdot -)$, 1° (– – – –), 3° (———), 5° $(\cdots\cdots)$ and 10° (- - - -). The peak maxima are marked by vertical lines, and the vertical line on the abscissa marks the cusp energy $(\gamma - 1)c^2 = 49.37$ keV.

on the projectile charge because RI and ECC increase with approximately the same power of Z_P (lying between 2 and 3 [3]).

In Fig.7 the doubly differential cross section in the peak maximum at $\vartheta_f = 3°$ for ECC respectively RI from U^{88+}+ N collisions is shown as a function of collision momentum γv. The calculations are done in the same way as described in the discussion of Fig.4. One obtains $v_{cr} = 72.64$ a.u. (from $\gamma v_{cr} = 85.67$ a.u.). Since at $\gamma v = 61.67$ a.u. the ratio of the ECC and RI peak intensities is known experimentally, we have in Fig.7 normalized the experimental RI peak intensity to theory and derived from this an experimental ECC peak intensity which lies a factor of 1.5 above the ECC theory. This is in good accord with the estimated accuracy of the present model: Taken into consideration that for uranium the RI theory is approximately a factor of 2 too low (see Fig.1 and Fig.16) ECC is underpredicted by a factor of 3. In fact, the Sommerfeld-Maue functions lack the relativistic spatial contraction such that the high momentum tails of the target bound state are underestimated more severely than the centroid of the Compton profile. Therefore, these functions are not so good for ECC than for RI.

The crossing velocity depends strongly on the target. This is displayed in Fig.8 for one-electron capture by Xe^{54+} from the target K-shell. The increase of γv_{cr} with Z_T is approximately linear in the weak-relativistic regime ($\gamma \lesssim 1.5$). A similar behaviour is found if the total capture from neutral targets is considered (in the figure, the RI $1s$-capture cross section is multiplied by the number N of target electrons, and the ECC $1s$-capture cross section by a factor of 2). The crossing velocity is, however, lower if all target electrons are considered. This is so because RI gains the factor $N/2$ as compared to ECC[2].

[2]Due to an error in the RI code (only present for relativistic velocities) the numerical results given in the

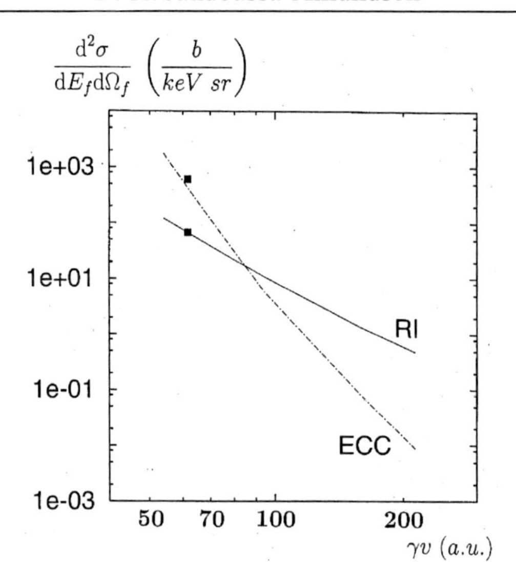

Figure 7. Doubly differential cross section for electron emission at $\vartheta_f = 3°$ from U^{88+}+ N collisions in the respective peak maximum for RI (———) and ECC (− · − · −) as a function of collision momentum γv. The experimental data for $\gamma v = 61.67$ a.u.(■) are from Nofal [35]. The RI datum point is normalized to theory; ECC is calculated from the measured ratio.

3. Radiative Ionization in Terms of Inverse Bremsstrahlung

The elementary process of bremsstrahlung occupies a very important place in physics [36]. Due to the simultaneous observation of the decelerated electron and the emitted photon, a stringent test of the underlying theory, describing the coupling of the radiation field with the field of electrons and nuclei, becomes possible. There exists a vast literature on the (electron-nucleus) bremsstrahlung theory in comparison with experiment, starting with the work of Bethe and Heitler [37] using the Born approximation, and followed later by Bess [25], Maximon and Bethe [26] and Elwert and Haug [38] who employed the Sommerfeld-Maue wavefunctions. Nowadays, calculations with accurate relativistic wavefunctions have become feasible, based on the work of Tseng and Pratt [39].

The measurements of the triply differential cross section $d^3\sigma^{brems'}/d\Omega'_f d\omega' d\Omega'_k$ for a free electron scattering from a (screened) nucleus, which we identify with the heavy projectile, are theoretically well understood. However, they do not cover the short-wavelength limit (SWL) of bremsstrahlung, $\omega' = (\gamma - 1)mc^2$, where the electron has given all its kinetic energy to the photon. The SWL is of particular interest since it requires a large momentum transfer to the nucleus which necessitates close collisions. This allows for a test of the electronic wavefunction in the vicinity of heavy nuclei. The detection of electrons with near-zero kinetic energy in the rest frame of the nucleus is a very difficult task. It can, however, be made feasible by performing the experiment in a moving reference frame. Here lies the importance of radiative electron capture to near-threshold continuum projectile states.

three earlier papers [3, 9, 22] are in part incorrect.

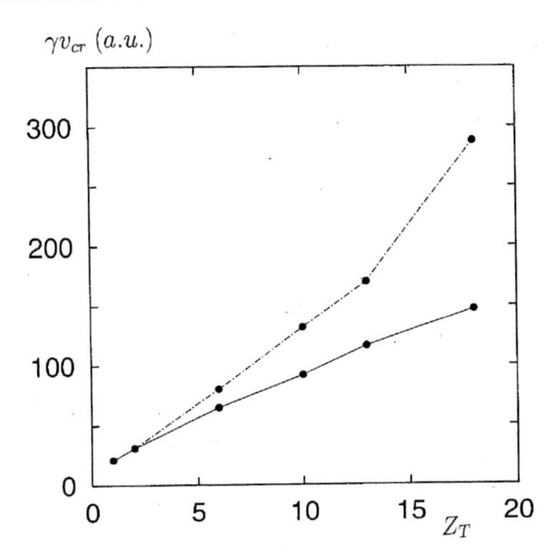

Figure 8. Crossing momentum γv_{cr} as a function of target nuclear charge Z_T. Shown is γv_{cr} as obtained from the doubly differential RI and ECC cross sections evaluated in the RI-peak maximum at $\vartheta_f = 5°$ for one-electron capture from the target K-shell $(- \cdot - \cdot -)$ and for the summed capture from all shells of a neutral target (———) by a Xe^{54+} projectile. The black dots are the calculated values and the lines are eye-guides.

3.1. Radiative Ionization in Inverse Kinematics

The change of reference frame for a given process proceeds in two steps. First we note that the particle momenta in a reference frame moving with a constant velocity $\boldsymbol{v}$ are subject to a Lorentz transformation. Switching from the (unprimed) target reference frame (where RI is observed) to the (primed) projectile reference frame (which is the natural choice for bremsstrahlung), the energies and polar angles of electron and photon are transformed according to

$$E'_f \;=\; \gamma\,(E_f - vk_f\cos\vartheta_f), \qquad \omega' \;=\; \gamma\omega\,(1 - \frac{v}{c}\cos\vartheta_k) \tag{3.1}$$

$$k'_f\cos\vartheta'_f \;=\; \gamma\left(-\frac{vE_f}{c^2} + k_f\cos\vartheta_f\right), \qquad \cos\vartheta'_k \;=\; \frac{\cos\vartheta_k - \frac{v}{c}}{1 - \frac{v}{c}\cos\vartheta_k}.$$

The inverse transformation (from primed to unprimed quantities) is obtained by replacing in (3.1) v with $-v$ throughout.

In the second step we have to account for the reversal of the direction of $\boldsymbol{v}$, i.e. of the z-axis, since a projectile moving with $\boldsymbol{v}$ corresponds to the target moving with $-\boldsymbol{v}$ when the reference frame is changed. The polar angles θ'_f and θ'_k of the decelerated electron and the emitted photon, respectively, in the projectile frame of reference are connected to ϑ'_f and ϑ'_k from the Lorentz transformation (3.1) by means of

$$\theta'_f \;=\; \pi - \vartheta'_f, \qquad \theta'_k \;=\; \pi - \vartheta'_k. \tag{3.2}$$

Finally, the cross section has to be transformed. We use the relativistic invariance of the phase space elements $c^2 d\boldsymbol{k}_f/E_f$ and $c^2 d\boldsymbol{k}/\omega$ of electron and photon, respectively [18,

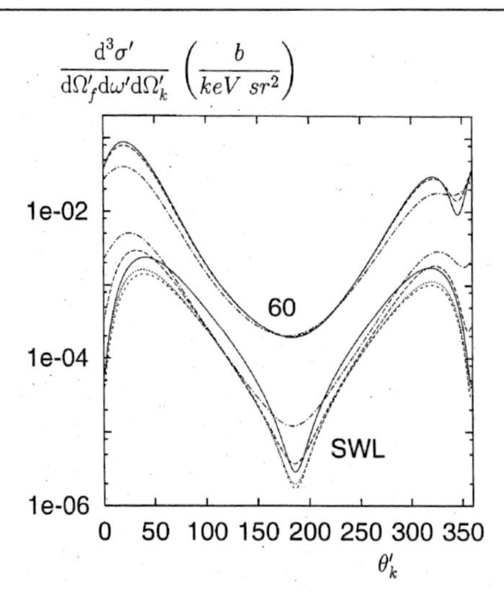

Figure 9. Projectile-frame triply differential cross section for K-shell RI from Xe^{54+} + T in coplanar geometry for $\tilde{E}_0 = (\gamma - 1)mc^2 = 100$ keV, $\theta'_f = 10°$, $\varphi = 0$ as a function of photon angle θ'_k. $\theta'_k > 180°$ corresponds to $2\pi - \theta'_k$ for $\varphi = 180°$. φ is the relative azimuthal angle between photon and electron. T = Ar (— · — · —), C (— — — —), H (· · · · · ·), $Z_T = 0.3$ (- - - -) and the bremsstrahlung result ($Z_T = 0$, ———). The upper bunch of curves is for $\omega' = 60$ keV, the lower bunch is at the SWL ($\omega' = 94.71,\ 99.4,\ 99.97,\ 99.984,\ 99.985$ keV for $Z_T = 18,\ 6,\ 1,\ 0.3,\ 0$, respectively).

p.124], when changing between the primed and unprimed reference frames. Then the fourfold differential cross section (2.9) is transformed into the projectile frame of reference by means of

$$\frac{d^4\sigma_\lambda^{RI'}}{dE'_f d\Omega'_f d\omega' d\Omega'_k} = \frac{\omega' k'_f}{\omega k_f} \frac{d^4\sigma_\lambda^{RI}}{dE_f d\Omega_f d\omega d\Omega_k}. \tag{3.3}$$

3.2. The Bremsstrahlung Limit of RI

In the bremsstrahlung process the initial electron is free and therefore the energy conservation requires

$$E'_0 = \gamma c^2 = E'_f + \omega' \tag{3.4}$$

which fixes the electron energy E'_f once ω' and v are given. In contrast, one has for RI

$$E'_f = \gamma E_i^T - \omega' - q_{0z}\gamma v \tag{3.5}$$

with q_{0z} distributed according to the bound-state target Compton profile (2.12). Therefore, the comparison with bremsstrahlung necessitates the integration of (3.3) with respect to E'_f. This leads to the correspondence, valid for ω' below the SWL [12],

$$\frac{d^3\sigma_\lambda^{brems'}}{d\Omega'_f d\omega' d\Omega'_k} = \lim_{Z_T\to 0} \frac{d^3\sigma_\lambda^{RI'}}{d\Omega'_f d\omega' d\Omega'_k} = \lim_{Z_T\to 0} \int_{c^2}^{\infty} dE'_f \frac{d^4\sigma_\lambda^{RI'}}{dE'_f d\Omega'_f d\omega' d\Omega'_k} \tag{3.6}$$

and in this limit, the RI theory from section 2.1 agrees with the Elwert-Haug theory [38] for bremsstrahlung.

In the following we display the transition from RI to bremsstrahlung by varying the target nuclear charge. In Fig.9 the differential cross section for K-shell RI in $\mathrm{Xe}^{54+} + T$ collisions at $\tilde{E}_0 \equiv E_0' - mc^2 = 100$ keV as a function of photon emission angle θ_k' is shown. The nuclear charge of the one-electron target T is varied from 18 to 0. If one is far from the SWL (e.g. at $\omega' = 60$ keV) the bremsstrahlung limit is approached monotonically when Z_T is decreased, and for a hydrogen target, RI is already indistinguishable from bremsstrahlung. At the SWL, however ($\omega' \approx 100$ keV), the integral on the r.h.s. of (3.6) for any small, but finite Z_T is a factor of 2 below the bremsstrahlung result. This is due to the fact that for non-zero Z_T only one half of the peak shaped by the Compton profile lies in the integration regime (i.e. above c^2) whereas for $Z_T = 0$ one has a δ-type singularity at c^2 in the fourfold differential RI cross section [12]. From Fig.9 it is also evident that the radiation is stronger when electron and photon are emitted to the same side of the beam axis ($0° < \theta_k' < 180°$), a feature which is generally true for bremsstrahlung [36].

Fig.10 provides the dependence of RI on the electron emission angle θ_f' for the same collision parameters and a (one-electron) hydrogen, carbon and argon target. The photon angle $\theta_k' = 30°$ is chosen close to the maximum of the photon distribution (which is located in the forward hemisphere because of the relativistic retardation). It is seen that for the lower frequencies, the electron distribution is also strongly peaked at small angles, and its shape is only weakly influenced by the target. This sharp peak for the lower frequencies is well-known from the bremsstrahlung experiments [40], whereas the angular variations are considerably weakened, but do not vanish, when the SWL is approached [36]. In fact, at the SWL the maximum for forward angles changes to a maximum at backward angles when Z_T is decreased from 18 to 1. We note that, by the inverse kinematics, the SWL backward-to-forward intensity ratio $A \equiv d^4\sigma'(\theta_f' = 180°)/d^4\sigma'(\theta_f' = 0°)$ determines the asymmetry of the RI cusp [12]. In contrast to the ratio calculated from the triply differential cross section as given in Fig.10, A is larger than one (at $\theta_k' = 30°$, translating to a cusp skewed to the high-energy side) and approximately target independent.

Let us now consider the case when the electron is not ejected into the collision plane spanned by $\boldsymbol{v}$ and the photon momentum, but forms an angle φ with the collision plane. In Fig.11 the triply differential RI cross section is shown for this noncoplanar geometry, using again the same collision parameters (and one-electron targets) as before. For the Xe projectile, the variation with φ is quite smooth and the strong target dependence for electrons emitted into the forward hemisphere persists. The φ-dependence of the photon intensity near $\varphi = 180°$ increases, however, considerably when lighter projectiles are used [12].

Figs.10 and 11 can help to understand the dependence of the doubly differential RI cross section, obtained by integrating over the electron angles, on the target nuclear charge (or on the target shells, respectively). The large difference between Ar and H at small angles θ_f' is suppressed by the weight factor $\sin\theta_f'$ when performing the integration. Well below the SWL these differences diminish at the larger θ_f' such that the doubly differential cross section is nearly independent of the initial target state (see Fig.1). In contrast, at the SWL, the intensity for the larger θ_f' is much lower for Ar than for H and therefore also the integrated cross section (see again Fig.1). This is due to the fact that the electrons corresponding to the SWL are very sensitive to the target Compton profile (the tip of which decreases with increasing binding energy).

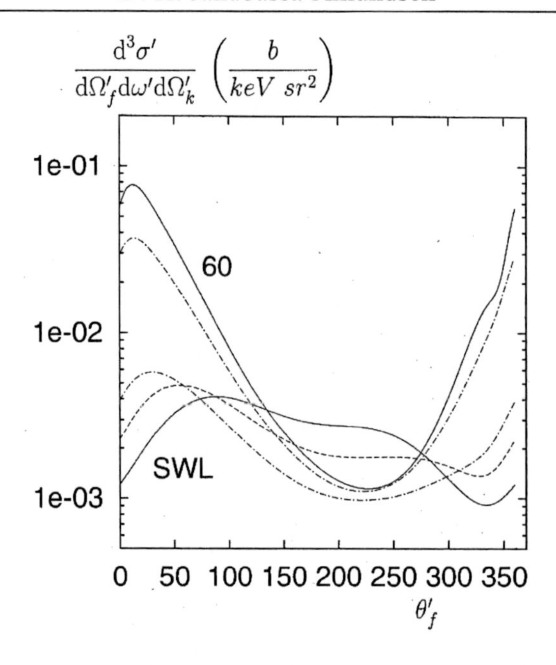

Figure 10. Projectile-frame triply differential cross section for K-shell RI from Xe^{54+}+ T in coplanar geometry ($\varphi = 0$) for $\tilde{E}_0 = 100$ keV, $\theta'_k = 30°$ as a function of electron angle θ'_f. The steep upper curves correspond to $\omega' = 60$ keV, the lower curves to the SWL. T = Ar (— · — · —), C (- - - - -) and H (————).

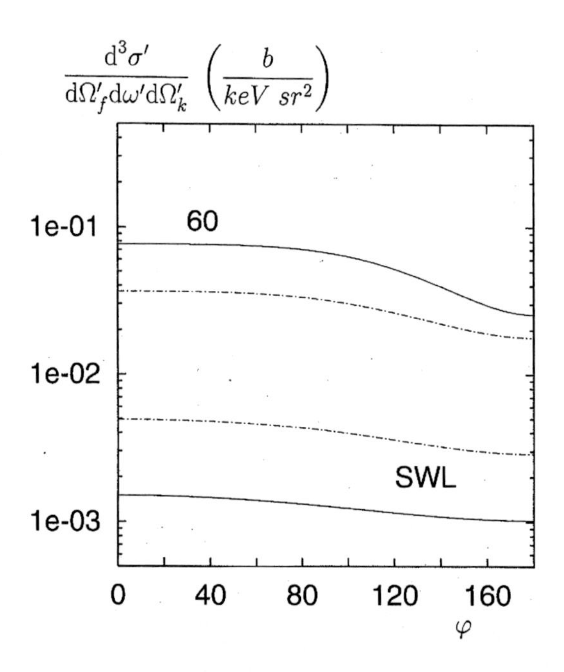

Figure 11. Projectile-frame triply differential cross section for K-shell RI from Xe^{54+}+ T in noncoplanar geometry for $\tilde{E}_0 = 100$ keV, $\theta'_f = 10°$, $\theta'_k = 30°$ as a function of φ. The upper curves correspond to $\omega' = 60$ keV, the lower curves to the SWL. T = Ar (— · — · —) and H (————).

3.3. Photon Linear Polarization

The most detailed observable quantity in the radiative electron capture to continuum process is the fourfold differential cross section including the polarization correlations. This corresponds to a coincidence experiment where in addition to the momentum distributions of electron and photon, also the spin polarization of the electron in its initial and final state as well as the photon polarization are measured. Such a so-called complete experiment provides the most stringent test of theory. In the case of electron-nucleus bremsstrahlung this goal has not entirely been achieved, although there exist measurements on the photon linear polarization induced by unpolarized electrons as well as on the photon emission asymmetry for polarized electron beams (for an overview, see [41]).

Here we will only consider electrons which are unpolarized in their initial state which is the usual situation for electron capture from neutral targets in their ground state. Then the emitted photons can be linearly (but not circularly) polarized [42, 43]. Taking the (x,z) plane as the collision plane, the photon momentum is given by $\boldsymbol{k}' = k'(\sin\vartheta_k', 0, \cos\vartheta_k')$. The two polarization directions of the photon, which have to be perpendicular to $\boldsymbol{k}'$, are chosen as

$$\boldsymbol{e}_{\lambda_1} = (0,1,0), \qquad \boldsymbol{e}_{\lambda_2} = (-\cos\vartheta_k', 0, \sin\vartheta_k'). \tag{3.7}$$

$\boldsymbol{e}_{\lambda_2}$ lies in the collision plane while $\boldsymbol{e}_{\lambda_1}$ is perpendicular to it. For any multiply differential cross section, abbreviated by $d\sigma_\lambda$, the degree P of the photon linear polarization is defined by

$$P = \frac{d\sigma_{\lambda_2} - d\sigma_{\lambda_1}}{d\sigma_{\lambda_2} + d\sigma_{\lambda_1}} \tag{3.8}$$

which coincides with the Stokes parameter C_{03} [42]. In the bremsstrahlung literature, P is usually defined with a negative sign (see e.g. [36, 43, 44]). The definition (3.8) holds also for the projectile reference frame since the transformation (3.1) - (3.3) does not affect the polarization degree of freedom.

It is well-known that for the coplanar geometry (where $\boldsymbol{v}$, $\boldsymbol{k}'$ and $\boldsymbol{k}_f'$ lie in the (x,z) plane) the nonrelativistic bremsstrahlung theory predicts an in-plane polarization ($P = 1$, [45]). Any deviation from $P = 1$ is thus based on relativistic effects [46] or, in the case of RI, on the binding of the initial electron.

Fig.12 compares the polarization relating to the projectile-frame doubly differential cross section for RI from Au^{79+}+ H and one-electron Ar with the bremsstrahlung result [47] for $e+$ Au at $\tilde{E}_0 = 500$ keV and $\omega' = 450$ keV as a function of photon angle θ_k'. The RI result for Au^{79+}+ H is indistinguishable from the bremsstrahlung theory for $e+$ Au^{79+}. The deviations between this theory and experiment may partly be ascribed to the screening by the passive electrons in the neutral Au target used in the experiment, and partly to the inaccuracy of the Sommerfeld-Maue functions for Au^{79+}. When a heavier target is used in the RI calculations, P decreases. This is true in most cases [12].

The determination of the degree of photon polarization when the outgoing electron is not observed can be interpreted as a kind of averaging procedure. Thereby some information on the elementary process of bremsstrahlung is lost [36]. In fact, the polarization of the photons which are detected in coincidence with the outgoing electrons is much different from the one obtained by integrating over the electron emission angles [48]. Fig.13a depicts P resulting from the projectile-frame triply differential RI cross section, defined in (3.6),

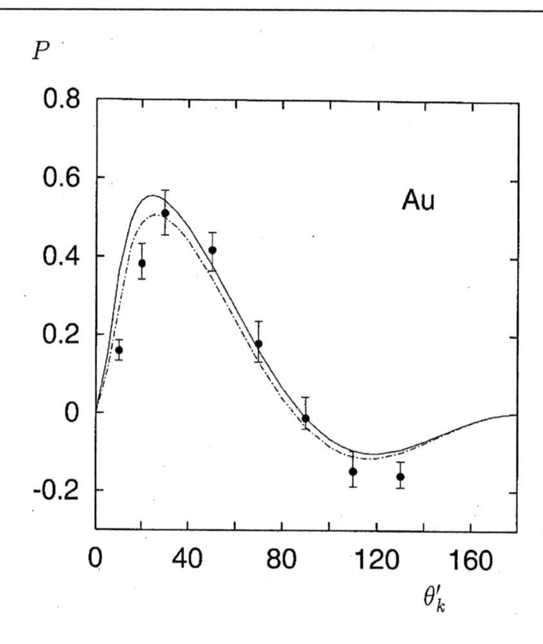

Figure 12. Polarization corresponding to $d^2\sigma'_\lambda/d\omega' d\Omega'_k$ for K-shell RI from collisions of Au^{79+} with H (———) and Ar (—·—·—) at $\tilde{E}_0 = 500$ keV, $\omega' = 450$ keV as a function of θ'_k. Comparison is made with the polarization of bremsstrahlung from equivelocity ($v = 118.25$ a.u.) $e+$ Au^0 collisions (•, [47]).

for the same collision parameters (and one-electron targets) as in Fig.12. The target dependence of P is largest for photons emitted close to the beam direction (θ'_k near 0° or 360°) and it gets stronger when the SWL is approached. In comparison with $d^3\sigma'/d\Omega'_f d\omega' d\Omega'_k$ (summed over λ, Fig.13b) one notes that a large depolarization coincides with the cross section minima. The explanation is simple. When the momenta of the outgoing particles are chosen such that a large momentum transfer to the (projectile) nucleus is required, this can only be achieved by a close collision for which the cross section is small. In close collisions, on the other hand, relativistic effects become particularly important, such that P is lowered. This behaviour is different for the noncoplanar geometry. There is a strong variation of P with φ such that the correspondence between the cross section minima and the minima in P is lost [12]. In fact, for $\varphi \neq 0$ one can have strong deviations from $P = 1$ even in the nonrelativistic case. A striking feature in Fig.13a is the asymmetry of P (with respect to reflection at $\theta'_k = 180°$) because the electron momentum $\boldsymbol{k}'_f$ is slightly tilted away from the beam axis. This asymmetry is much larger than that of the underlying cross section and emphasizes the supplementary information contained in a polarization measurement.

Let us now turn to the polarization related to the fourfold differential RI cross section, and to its dependence on the projectile and the target. In Fig.14 the polarization for RI from Ag^{47+} colliding with H, C, Ar and from U^{92+}, C^{6+} colliding with H is given. The electron energy (at the SWL) is fixed, as is ϑ_f (in the forward direction). At each photon emission angle ϑ_k, the frequency ω is chosen to coincide with the peak frequency (2.11) determined from the tip of the target Compton profile. Under this condition the polarizations corresponding to the triply and fourfold differential cross sections, respectively, show a

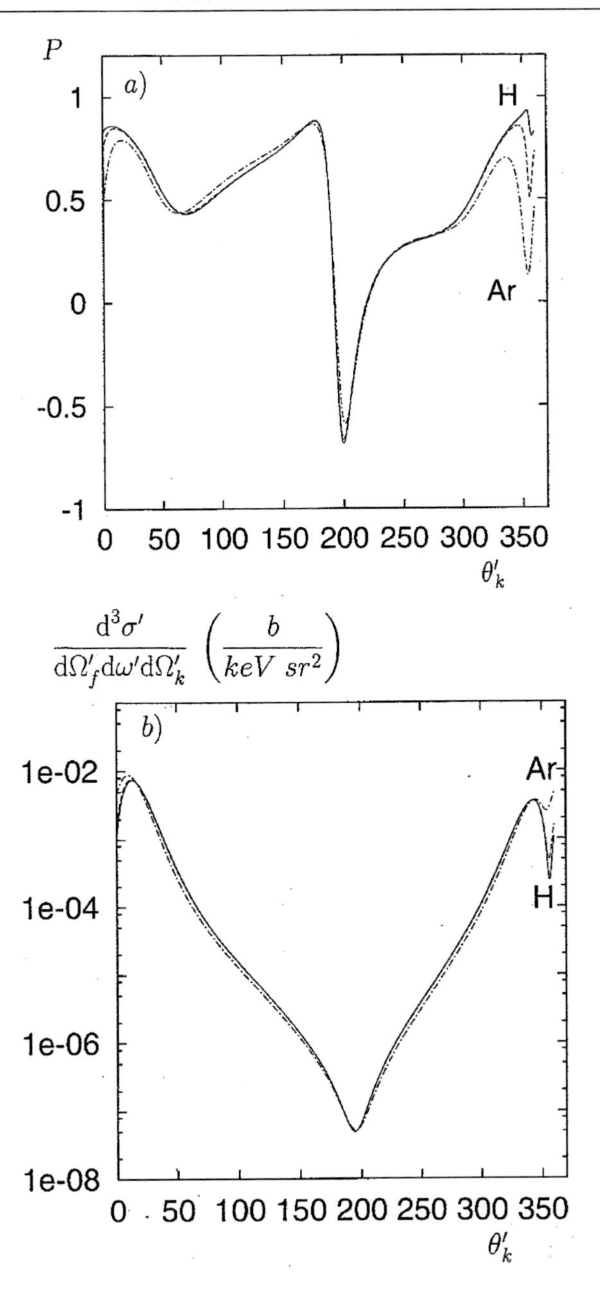

Figure 13. Polarization P (a) and projectile-frame triply differential cross section $d^3\sigma'/d\Omega'_f d\omega' d\Omega'_k$ (b) for K-shell RI from collisions of Au^{79+} with H (——), C(− − −) and Ar (− · − · −) at $\tilde{E}_0 = 500$ keV, $\omega' = 450$ keV, $\theta'_f = 10°$ and $\varphi = 0$ as a function of θ'_k.

similar dependence on the photon angle, and for a hydrogen target there is even complete agreement if ω is not too close to the SWL. The dependence of P on the target-frame emission angle ϑ_k changes, however, again drastically if an additional integration over the electron angles is made (see e.g. Fig.2 and Fig.20).

From Fig.14 it follows that the depolarization and the ϑ_k-variation of P is reduced when the projectile charge increases, a fact confirmed by bremsstrahlung experiments [49]. When the target gets heavier the depolarization increases, as mentioned earlier. If $\vartheta_f = 0°$, P drops to zero at $\vartheta_k = 0°$ and $180°$. In that case of collinear particle emission one has cylindrical symmetry with respect to the beam axix, causing $d\sigma_{\lambda_1} = d\sigma_{\lambda_2}$ in (3.8). We note that the shape of the angular distribution of the photons and of their degree of polarization changes strongly when one switches between the target and the projectile frame of reference [12]. This frame dependence arises from the fact that the threshold T_0 and the peak frequency ω_{peak} depend on ϑ_k (in contrast to the nonrelativistic case) whereas these quantities are angular independent in the projectile frame of reference.

3.4. Sommerfeld-Maue Results in Comparison with Accurate Calculations

In section 3.2 we have established the close relation between RI from a hydrogen target and the elementary process of bremsstrahlung. The bremsstrahlung results using accurate relativistic wavefunctions can therefore serve as a test for the applicability of the semirelativistic Sommerfeld-Maue functions for the unbound projectile electron. There is a considerable number of publications where the two theoretical approaches are compared with experiment (see e.g. [39, 50, 51]). Below we compile some representative bremsstrahlung literature results for very heavy nuclei which we have supplied with RI calculations (that extend the published Elwert-Haug bremsstrahlung results [38]).

In Fig.15 the singly differential bremsstrahlung cross section $d\sigma'/d\omega'$ from 50 keV and 500 keV $e+$ Au^{79+} is shown as a function of photon energy. The calculations from Lee et al [52] employ a partial-wave expansion of the relativistic electronic states in the field of Au^{79+}. Comparison is made with the experiments by Motz [53]. We have included the frame-transformed RI results from Au^{79+}+ H at the same collision energies (which are indistinguishable from the Elwert-Haug results given in [54] for this system). It is seen that at the lower collision energy ($\gamma = 1.1$) the Sommerfeld-Maue functions do quite well, but they get worse when the SWL is approached. At the higher energy ($\gamma = 2$) the deviations reach a factor of 2. For a systematic investigation of the different models at the SWL, see [55].

Now we proceed from the singly differential cross section to the doubly differential cross section for the same system (Fig.16). The accurate bremsstrahlung calculations [39] which compare well with the experimental data from Aiginger are, nearly independently of the photon emission angle, underestimated in our model by roughly a factor of 2 for $\omega' = 480$ keV (which is slightly below the SWL). In Fig.16 are included the results for an Al^{13+} nucleus. For this smaller charge, the Sommerfeld-Maue functions provide a good approximation even at this high collision energy.

Finally, in Fig.17 the triply differential bremsstrahlung cross section for 300 keV $e+$ Au at a photon energy well below the SWL ($\omega' = 150$ keV) is displayed [51]. For the forward electrons considered, there is quite good agreement between the Elwert-Haug theory and

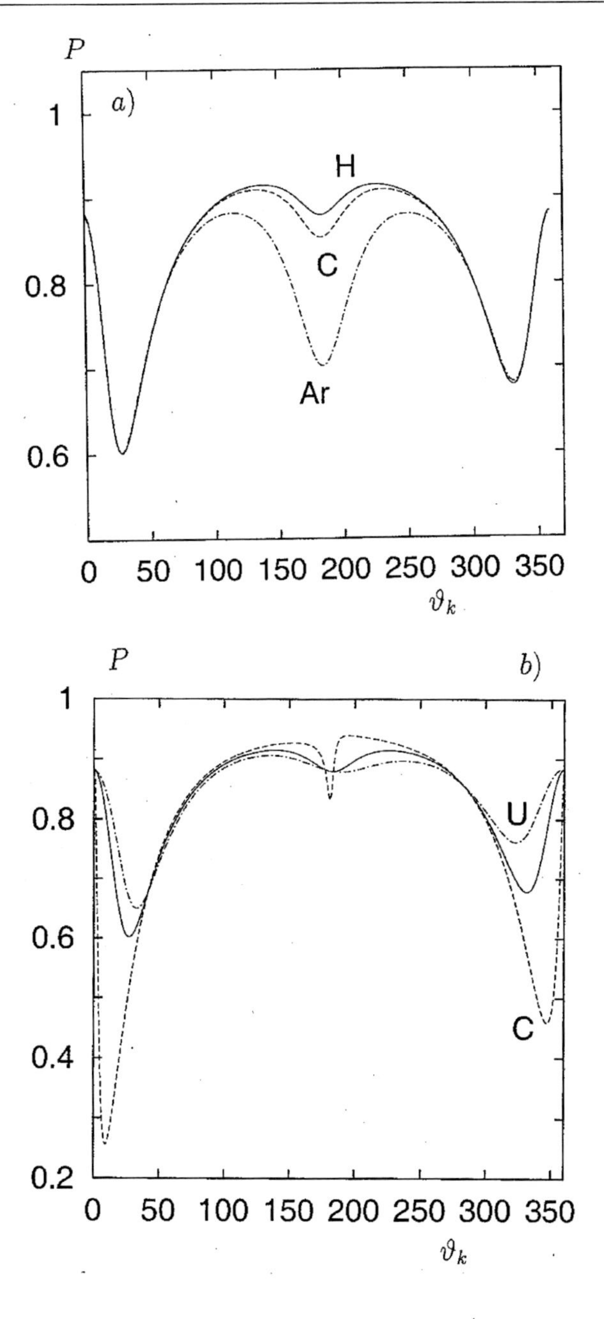

Figure 14. Polarization corresponding to (target-frame) $d^4\sigma_\lambda/dE_f d\Omega_f d\omega d\Omega_k$ for RI at the SWL from collisions of Ag^{47+} with one electron targets H (——), C (– – –) and Ar (– · – · –) (a) and of C^{6+} (– – – –), Ag^{47+} (———) and U^{92+} (– · – · –) with H (b) as a function of ϑ_k. The parameters are $\tilde{E}_0 = 300$ keV $= E_{f,kin}$, $\vartheta_f = 1°$, $\varphi = 0$ and $\omega = \omega_{peak}(\vartheta_k)$ from (2.11).

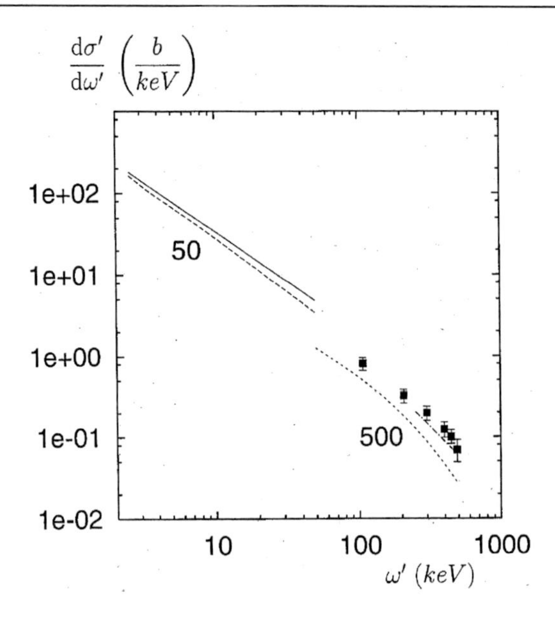

Figure 15. Singly differential projectile-frame cross section $d\sigma'/d\omega'$ for bremsstrahlung from Au as a function of ω'. Partial-wave results for 50 keV (——) and 500 keV (– · – · –) electrons colliding with Au^{79+} [52, 54]. Experimental data for 500 keV $e+$ Au^0 (■, [53]). Frame-transformed RI results for Au^{79+}+ H at $\tilde{E}_0 = 50$ keV (– – – –) and 500 keV (- - - -).

the accurate calculations for an Au^{79+} nucleus if θ'_k is not too large. From this figure it is also seen that accounting for the presence of the passive electrons in Au^0 lowers the cross section in the peak region. To our knowledge, a comparison of the two theoretical models for the triply differential cross section near the SWL covering the whole angular range has not yet been made.

4. Radiative Electron Capture and its Relation to RI

The radiative electron capture, in which a bound target electron is captured into a bound state of a fast, highly stripped projectile with the simultaneous emission of a photon, plays an important role in spectroscopic studies of highly stripped heavy ions and has been investigated in great detail. The early work on REC started with a close collaboration between experimentalists and theoreticians [1, 14, 56]. Relativistic kinematics was first considered by Spindler et al [57]. Their predictions concerning the approximate cancellation between the relativistic retardation and the frame transformation (3.1) in the target-frame photon angular distribution was further verified experimentally by Anholt et al [58] in the case of highly relativistic projectiles.

A consistent relativistic prescription of REC was put forth by Eichler and coworkers [7, 27] who extended the nonrelativistic Kleber model [1]. The differential cross section for photoionization serves as their starting point from which, employing inverse kinematics, the differential cross section for radiative recombination, $d\sigma'_{RR}/d\Omega'_k$, is obtained. This cross section has eventually to be convoluted with the momentum distribution of the target state,

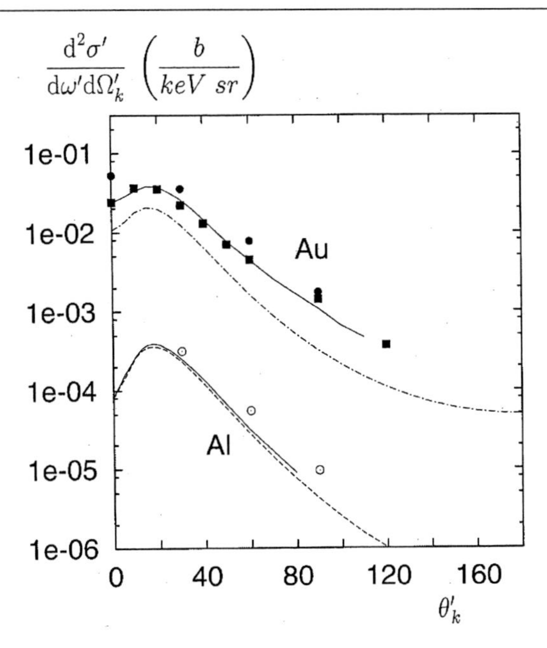

Figure 16. Doubly diffferential projectile-frame cross section $d^2\sigma'/d\omega' d\Omega'_k$ for bremsstrahlung from Au (upper curves) and Al (lower curves) at a collision energy of $\tilde{E}_0 = 500$ keV and $\omega' = 480$ keV as a function of photon emission angle. Partial-wave results (——) for Au^{79+} and Al^{13+}, respectively [39]. The experimental data for Au^0 are from Aiginger (■, taken from [39]) and from Motz (•, [53]). The data for Al^0 are from Motz (○, [53]). Frame-transformed RI results for equivelocity Au^{79+}+ H (– · – · –) and Al^{13+}+ H (– – – –).

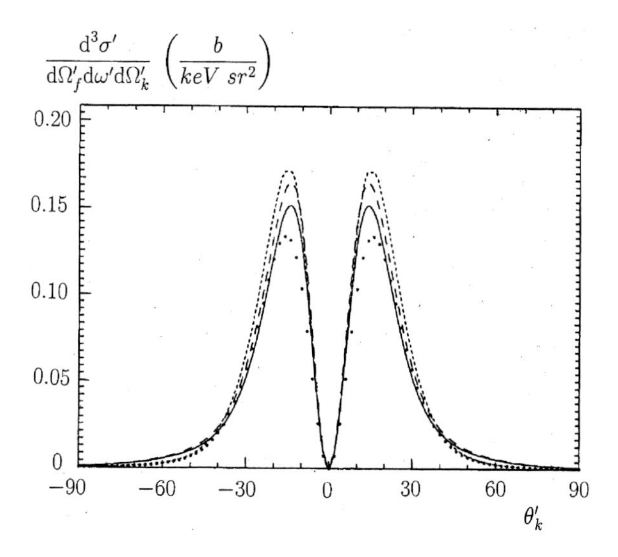

Figure 17. Triply differential bremsstrahlung cross section $d^3\sigma'/d\Omega'_f d\omega' d\Omega'_k$ from 300 keV $e+$ Au collisions at $\omega' = 150$ keV and $\theta'_f = 0$. Shown are the partial-wave results (– – –) and Elwert-Haug results (- - - -) for Au^{79+} as well as for a screened target (———, partial-wave result; · · · · · ·, modified Elwert-Haug theory). Taken from [51].

as dictated by the impulse approximation. Relativistic kinematics leads to [7]

$$\frac{d^2\sigma_{REC}}{d\omega d\Omega_k} = \frac{\omega}{\omega'\gamma} \int d\boldsymbol{q}\, \frac{d\sigma'_{RR}(\boldsymbol{q}'_0)}{d\Omega'_k}\, |\tilde{\varphi}_{i,T}(\boldsymbol{q}_0)|^2\, \delta(E'_f + \omega' - E_i^T/\gamma + q_z v) \tag{4.1}$$

where the prefactor arises from $\frac{d^2\sigma}{d\omega d\Omega_k} = \frac{\omega}{\omega'} \frac{d^2\sigma'}{d\omega' d\Omega'_k}$ and $dq_{0z} = dq_z/\gamma$ (recall the relation $q_{0z} = E_i^T v/c^2 + q_z/\gamma$). $\tilde{\varphi}_{i,T}(\boldsymbol{q}_0)$ is the nonrelativistic target function and $\boldsymbol{q}'_0$ the Lorentz transform of $\boldsymbol{q}_0$. By means of this step-by-step method, accurate relativistic calculations of the photoionization process [59] are sufficient for the determination of REC.

4.1. Relativistic Theory for REC

In the following a relativistic REC theory is outlined which is strictly derived from the formal scattering theory ([18, 22], without employing the photoelectric effect) and which extends the nonrelativistic formalism developed earlier [1, 60]. The Kleber-Eichler formula (4.1) is then recovered by means of additional approximations.

Starting point of the theory is the transition amplitude (2.1), where now $\psi_{f,P}^{(\sigma_f)'}(x')$ represents a bound projectile eigenstate. For heavy projectiles the off-shell approximation (2.3) is made which retains some influence of the target potential that is not accounted for in the Kleber-Eichler method. The impulse approximation is recovered by going on-shell, and the deviations between the (nonrelativistic) on-shell and off-shell results are somewhat larger than for RI [20]. For relativistic velocities they are not expected to play any significant role, however.

The IA transition amplitude $a_{fi,\lambda}^{REC}$ derived in this way has the form (2.5) where now E'_f is the energy of the bound final state and $\boldsymbol{W}_{rad}^{REC}(\sigma_f, s, \boldsymbol{q})$ describes the free-bound transition mediated by the photon field. For light targets we can describe the initial-state momentum-space function by a Darwin function [23]. For spherically symmetric states it has a simple product form [22], $\varphi_{i,T}^{(\sigma_i)}(\boldsymbol{q}_0) = N_i^T\, a_i^{(\sigma_i)}(\boldsymbol{q}_0)\, \tilde{\varphi}_{i,T}(\boldsymbol{q}_0)$, where

$$N_i^T = \left[1 + \left(\frac{Z_T\mu}{n_i}\right)^2\right]^{-\frac{1}{2}}, \quad a_i^{(1)}(\boldsymbol{q}_0) = \begin{pmatrix} 1 \\ 0 \\ \mu q_{0z} \\ \mu q_+ \end{pmatrix}, \quad a_i^{(2)}(\boldsymbol{q}_0) = \begin{pmatrix} 0 \\ 1 \\ \mu q_- \\ -\mu q_{0z} \end{pmatrix} \tag{4.2}$$

with $q_\pm = q_x \pm i q_y$, $\mu = c/(E_i^T + c^2)$.

Then the doubly differential cross section for REC follows from

$$\frac{d^2\sigma^{REC}}{d\omega d\Omega_k} = \frac{\omega\omega'}{2c^3} \sum_{\lambda,\sigma_i,\sigma_f} \int d^2\boldsymbol{b}\, |a_{fi,\lambda}^{REC}|^2$$

$$= \frac{\omega}{\omega'\gamma} \int d\boldsymbol{q}\, F(\boldsymbol{q})\, |\tilde{\varphi}_{i,T}(\boldsymbol{q}_0)|^2\, \delta(E'_f + \omega' - E_i^T/\gamma + q_z v) \tag{4.3}$$

where

$$F(\boldsymbol{q}) = \frac{(2\pi)^4\, \omega'^2}{2c^3\, v} \sum_{\lambda,\sigma_i,\sigma_f} \left| N_i^T \sum_{s=1}^{4} \boldsymbol{A}'_\lambda\, \boldsymbol{W}_{rad}^{REC}(\sigma_f, s, \boldsymbol{q}) \right. \tag{4.4}$$

$$\cdot \left[u_q^{(s)+} \sqrt{\frac{1+\gamma}{2\gamma}} \left(1 - \frac{\gamma v/c}{1+\gamma}\,\alpha_z\right) a_i^{(\sigma_i)}(\boldsymbol{q}_0) \right] \Bigg|^2 .$$

For the sake of comparison we furnish the differential cross section for the radiative recombination of a free electron with momentum $\boldsymbol{q}_0'$,

$$\frac{d\sigma'_{RR}}{d\Omega'_k}(\boldsymbol{q}_0') = \frac{(2\pi)^4\,\omega_0'^2}{2c^3\,v} \sum_{\lambda,\sigma_i,\sigma_f} \left| \boldsymbol{A}'_\lambda\,\boldsymbol{W}^{REC}_{rad}(\sigma_f,\sigma_i,\boldsymbol{q}_0') \right|^2 . \tag{4.5}$$

Here, $\omega_0' = \gamma_0 c^2 - E_f'$ follows from energy conservation and $\gamma_0 c^2 = \sqrt{(q_0' c)^2 + c^4}$ is the collision energy of the electron.

In the limit of vanishing target field, where $\tilde{\varphi}_{i,T}(\boldsymbol{q}_0)$ turns into a δ-function and $q_{0z}' = \gamma\,(-vE_{q_0}/c^2 + q_{0z}) \to -\gamma v$, the integral of (4.1) with respect to ω' provides, as expected, the differential cross section for radiative recombination,

$$\lim_{Z_T\to 0} \int_0^\infty d\omega'\, \frac{d^2\sigma'_{REC}}{d\omega'\,d\Omega'_k} = \frac{d\sigma'_{RR}}{d\Omega'_k}(-\gamma \boldsymbol{v}). \tag{4.6}$$

A straightforward calculation, following the lines of the derivation of the bremsstrahlung limit of RI [12], shows that the r.h.s. of (4.6) is also the $Z_T \to 0$ limit of $F(\boldsymbol{q})$ as required. In this limit one has $\omega' = \omega_0' = \gamma c^2 - E_f' > 0$. When Z_T is finite, the deviations between (4.1) and (4.3) depend on γ and increase with Z_T.

In the evaluation of the radiation matrix element $\boldsymbol{W}^{REC}_{rad}(\sigma_f, s, \boldsymbol{q})$ in (4.4) we use the semirelativistic wavefunctions to obtain a closed expression. The Darwin function for the final $1s_{1/2}$ ground state is given by

$$\psi^{(\sigma_f)'}_{f,P}(x') = N_f^P\, a_f^{(\sigma_f)}\, \tilde{\psi}'_{f,P}(\boldsymbol{x}')\, e^{-iE_f' t}, \tag{4.7}$$

$$a_f^{(1)} = \begin{pmatrix} 1 \\ 0 \\ -i\lambda\partial_{z'} \\ -i\lambda\partial_+ \end{pmatrix}, \qquad a_f^{(2)} = \begin{pmatrix} 0 \\ 1 \\ -i\lambda\partial_- \\ i\lambda\partial_{z'} \end{pmatrix}, \qquad N_f^P = \frac{1}{\sqrt{1 + (Z_P\lambda)^2}}$$

where $\lambda = \frac{c}{E_f'+c^2}$, $E_f' = c^2\sqrt{1-(Z_P/c)^2}$ and $\partial_\pm = \partial_{x'} \pm i\partial_{y'}$. $\tilde{\psi}'_{f,P}(\boldsymbol{x}') = \pi^{-1/2}\, Z_P^{3/2}\, e^{-Z_P r'}$ with $r' = |\boldsymbol{x}'|$, is the nonrelativistic bound-state function.

The Sommerfeld-Maue function for the intermediate unbound state is defined in terms of the derivative of a confluent hypergeometric function [38],

$$\psi^{(s)'}_{q,P}(\boldsymbol{x}') = N_q e^{i\boldsymbol{q}\boldsymbol{x}'}(1 - \frac{ic}{2E_q}\boldsymbol{\alpha}\boldsymbol{\nabla})\, {}_1F_1(i\eta_q, 1, i(qr' - \boldsymbol{q}\boldsymbol{x}'))\, u_{\boldsymbol{q}}^{(s)}, \tag{4.8}$$

$$u_{\boldsymbol{q}}^{(1)} = \sqrt{\frac{E_q + c^2}{2E_q}} \begin{pmatrix} 1 \\ 0 \\ \nu q_z \\ \nu q_+ \end{pmatrix}, \qquad u_{\boldsymbol{q}}^{(2)} = \sqrt{\frac{E_q + c^2}{2E_q}} \begin{pmatrix} 0 \\ 1 \\ \nu q_- \\ -\nu q_z \end{pmatrix},$$

with $N_q = (2\pi)^{-\frac{3}{2}}\, e^{\pi\eta_q/2}\,\Gamma(1 - i\eta_q)$, $E_q = \sqrt{q^2c^2 + c^4}$, $\nu = c/(E_q + c^2)$ and the Sommerfeld parameter $\eta_q = Z_P E_q/(qc^2)$.

In consistency with the accuracy of the semirelativistic functions, only terms up to $O(\alpha^2)$ are kept in the radiation matrix element (like in the bremsstrahlung theory [38]). This means that for the small components of $\psi_{f,P}^{(\sigma_f)'}$, the $\boldsymbol{\alpha}\boldsymbol{\nabla}$ term in (4.8) is disregarded.

We make use of the relations $r'e^{-Z_P r'} = -\frac{\partial}{\partial Z_P} e^{-Z_P r'}$ and $\boldsymbol{\nabla}_1 F_1(i\eta_q, 1, i(qr' - \boldsymbol{q}\boldsymbol{x}')) = -\frac{q}{r'} \left[\boldsymbol{\nabla}_{s_0\,1} F_1(i\eta_q, 1, i(s_0 r' - \boldsymbol{s}_0 \boldsymbol{x}'))\right]_{\boldsymbol{s}_0 = \boldsymbol{q}}$. Then the radiation matrix element can be based on a single integral [61]

$$I_0(Z_P, \boldsymbol{p}, \boldsymbol{s}_0) \equiv \int d\boldsymbol{x}' \, \frac{1}{r'} \, e^{-Z_P r'} \, e^{i\boldsymbol{p}\boldsymbol{x}'} \, {}_1F_1(i\eta_q, 1, i(s_0 r' - \boldsymbol{s}_0 \boldsymbol{x}'))$$

$$= 4\pi \, \frac{1}{[Z_P^2 + p^2]^{1-i\eta_q}} \, \left[(\boldsymbol{p} - \boldsymbol{s}_0)^2 - (s_0 + iZ_P)^2\right]^{-i\eta_q}. \tag{4.9}$$

We write the 4-spinor $a_f^{(\sigma_f)}$ from $\psi_{f,P}^{(\sigma_f)'}$ in the following way,

$$a_f^{(\sigma_f)} = \begin{pmatrix} \chi^{(\sigma_f)} \\ 0 \end{pmatrix} + \begin{pmatrix} 0 \\ \frac{1}{r'} g^{(\sigma_f)} \end{pmatrix} \quad \text{with } \chi^{(1)} = \begin{pmatrix} 1 \\ 0 \end{pmatrix}, \quad \chi^{(2)} = \begin{pmatrix} 0 \\ 1 \end{pmatrix}, \tag{4.10}$$

where $g^{(\sigma_f)}$ contains the derivatives (which act only on $e^{-Z_P r'}$) according to (4.7). One obtains from (2.4) with (4.7) and (4.8),

$$\boldsymbol{W}_{rad}^{REC}(\sigma_f, s, \boldsymbol{q}) = \frac{Z_P^{3/2}}{\sqrt{\pi}} \, N_q \, N_f^P \, (\boldsymbol{M}_1 + \boldsymbol{M}_2 + \boldsymbol{M}_3), \tag{4.11}$$

$$\boldsymbol{M}_1 = -\left[\begin{pmatrix} \chi^{(\sigma_f)} \\ 0 \end{pmatrix}^+ \boldsymbol{\alpha} \, u_{\boldsymbol{q}}^{(s)}\right] \frac{\partial}{\partial Z_P} I_0(Z_P, \boldsymbol{q} - \boldsymbol{k}', \boldsymbol{q})$$

$$\boldsymbol{M}_2 = \frac{icq}{2E_q} \left\{\left[\begin{pmatrix} \chi^{(\sigma_f)} \\ 0 \end{pmatrix}^+ \boldsymbol{\alpha} \, (\boldsymbol{\alpha}\boldsymbol{\nabla}_{s_0}) \, u_{\boldsymbol{q}}^{(s)}\right] \, I_0(Z_P, \boldsymbol{q} - \boldsymbol{k}', \boldsymbol{s}_0)\right\}_{\boldsymbol{s}_0 = \boldsymbol{q}}$$

$$\boldsymbol{M}_3 = \left\{\left[\begin{pmatrix} 0 \\ g^{(\sigma_f)} \end{pmatrix}^+ \boldsymbol{\alpha} \, u_{\boldsymbol{q}}^{(s)}\right] \, I_0(Z_P, \boldsymbol{\varrho} - \boldsymbol{k}', \boldsymbol{q})\right\}_{\boldsymbol{\varrho} = \boldsymbol{q}}.$$

The evaluation of $\boldsymbol{M}_3$ is done by means of $\partial_{z'} e^{-Z_P r'} = -\frac{Z_P}{r'} z' e^{-Z_P r'}$ followed by $z' e^{i\boldsymbol{\varrho}\boldsymbol{x}'} = -i\frac{\partial}{\partial \varrho_z} e^{i\boldsymbol{\varrho}\boldsymbol{x}'}$ and

$$\left[-\frac{\partial}{\partial \boldsymbol{\varrho}} I_0(Z_P, \boldsymbol{\varrho} - \boldsymbol{k}', \boldsymbol{q})\right]_{\boldsymbol{\varrho} = \boldsymbol{q}} = \frac{8\pi}{[Z_P^2 + (\boldsymbol{q} - \boldsymbol{k}')^2]^{2 - i\eta_q}} \left[k'^2 - (q + iZ_P)^2\right]^{-i\eta_q}$$

$$\cdot \left\{\boldsymbol{q}(1 - i\eta_q) - \boldsymbol{k}' \left(1 - i\eta_q + i\eta_q \, \frac{Z_P^2 + (\boldsymbol{q} - \boldsymbol{k}')^2}{k'^2 - (q + iZ_P)^2}\right)\right\}. \tag{4.12}$$

This leads to the doubly differential REC cross section for the emission of a photon with polarization direction $\boldsymbol{e}_\lambda$,

$$\frac{d^2\sigma_\lambda^{REC}}{d\omega d\Omega_k} = \frac{(2\pi)^3 (1 + \gamma) \omega \omega' Z_P^3}{2c^3 \gamma^2 v} \, |N_i^T \, N_f^P|^2 \int d\boldsymbol{q} \; |\tilde{\varphi}_{i,T}(\boldsymbol{q}_0)|^2 \; \delta(E_f' + \omega' - E_i^T/\gamma + q_z v)$$

$$\cdot |N_q|^2 \sum_{\sigma_i,\sigma_f} \left| A'_\lambda (\tilde{M}_1 + \tilde{M}_2 + \tilde{M}_3) \, (1 - \frac{\gamma v/c}{1+\gamma} \, \alpha_z) \, a_i^{(\sigma_i)}(q_0) \right|^2 , \qquad (4.13)$$

where $M_l \equiv \tilde{M}_l u_q^{(s)}$, $l = 1, 2, 3$ and the completeness relation for $u_q^{(s)}$ was used.

The present theory has to be contrasted against the Sauter formula [62, 63] for the photoionization in the Kleber-Eichler model. Sauter also employs the Sommerfeld-Maue wavefunctions, but in addition makes an expansion in η_q to lowest order which invalidates his theory for the heavy projectiles.

4.2. Sommerfeld-Maue Results in Comparison with Accurate Calculations

There exists a systematic investigation of the angular dependence of the REC photons using accurate relativistic wavefunctions [7]. At the relativistic collision energies considered the binding of the initial-state electron plays only a minor role (if an integration over the photon energies is performed [7, Fig.32]). Therefore these calculations are actually done for the radiative recombination instead of REC.

In Fig.18 REC induced by 100 MeV/amu bare projectiles of charge $Z_P = 50$ and 70 is shown. For $Z_P = 50$, the difference between the Sommerfeld-Maue result and the accurate partial-wave calculations is very small except for the forward photons. At the velocity $v = 58.68$ a.u. ($\gamma = 1.1$) the deviations from the nonrelativistic impulse approximation [20] are already quite strong. For $Z_P = 70$ (Fig.18b) the Sommerfeld-Maue functions become poor for $\vartheta_k < 20°$. In particular they fail to correctly reproduce the strong spin-flip transitions which are responsible for the entire REC photon intensity at $\vartheta_k = 0°$ [7]. The high forward cross sections predicted by the partial-wave theory were recently verified experimentally for K-shell REC by U^{91+} and U^{92+} projectiles [64, 65]. The failure of the Sommerfeld-Maue functions in correctly predicting the spin-flip contributions for REC into strongly bound projectile eigenstates is due to their missing relativistic spatial contraction. Therefore the electron is not adequately described when being close to the nucleus. In this context we remark that spin-flip transitions during RI are of lesser importance than for REC because in RI the final state is unbound and therefore not localized so close to the nucleus.

4.3. REC in Comparison with Radiative Ionization

In Fig.19 the photon angular distribution for RI in 100 MeV/amu Xe^{54+}+ H collisions, resulting from an integration over the electronic degrees of freedom, is compared to the REC results. This system was chosen because, according to Fig.18a, the Sommerfeld-Maue functions provide reliable REC results for all ϑ_k except in a small forward cone. Since the relativistic effects are more important for REC than for RI, it follows that the Sommerfeld-Maue wavefunctions are then also appropriate for RI.

The plotted RI results are for photon energies ω_{peak} corresponding, respectively, to the SWL and to electrons with a kinetic energy of approximately half the Xe^{53+} ground-state binding energy ($E'_{f,kin} = 21.8$ keV). For the doubly differential REC cross section we have also chosen $\omega = \omega_{peak}^{REC}$. This REC peak position is obtained from (4.13) – in analogy to

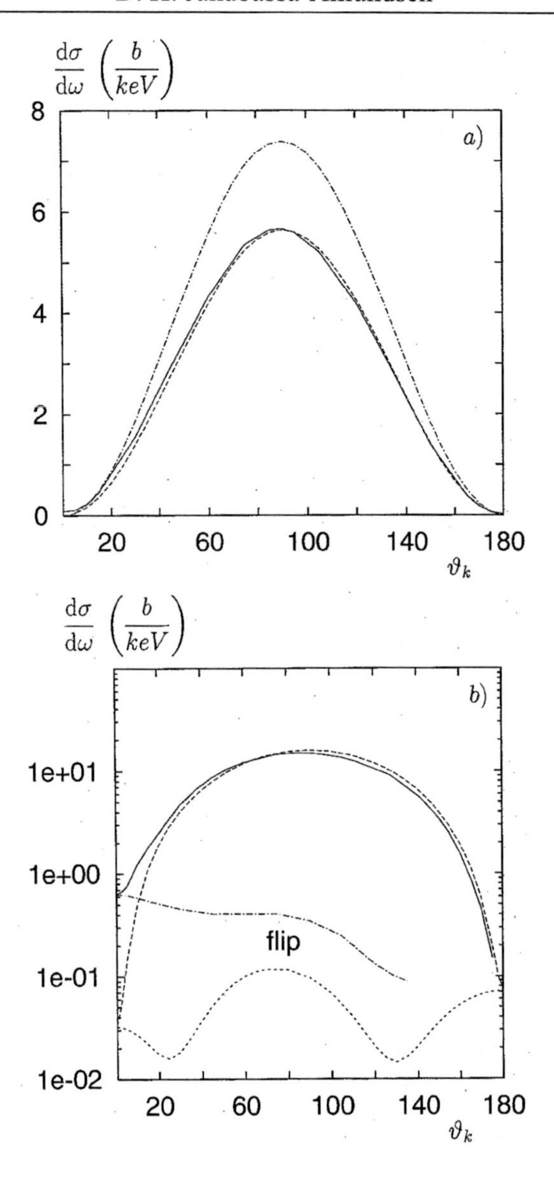

Figure 18. Singly differential cross section for REC into the K-shell of bare projectiles with $Z_P = 50 \quad (a)$ and 70 (b) at 100 MeV/amu as a function of photon angle ϑ_k. The partial-wave results from [7] are for a free initial electron (———), the present REC results use a hydrogen target (– – – –). For comparison, the nonrelativistic result for $Z_P = 50$ colliding with H (– · – · –) is included in (a). In (b), the spin-flip contributions to the singly differential cross section are shown separately (– · – · –, partial-wave expansion; - - - -, Sommerfeld-Maue functions).

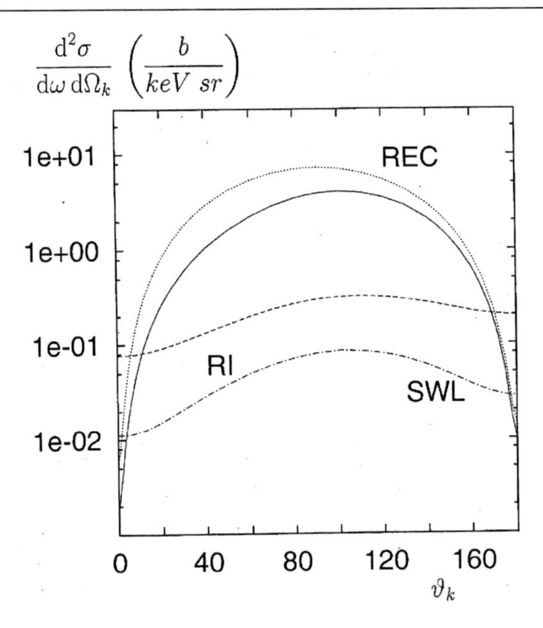

Figure 19. Doubly differential cross section for photons from the K-shell REC in 100 MeV/amu Xe^{54+}+ H collisions at the REC peak frequency (4.14) (———) and for RI photons at threshold (ω from (2.11) for $E'_{f,kin} = 10^{-3}$ keV, $-\cdot-\cdot-$) and for $E'_{f,kin} = 21.8$ keV (– – – –). Included is the singly differential REC cross section $d\sigma/d\Omega_k$ (left-hand scale in b/sr, $\cdots\cdots$). All calculations use semirelativistic wavefunctions.

(2.11) – as

$$\omega_{peak}^{REC}(\vartheta_k) = \frac{\gamma E_i^T - E_f'}{\gamma(1 - \frac{v}{c}\cos\vartheta_k)} \tag{4.14}$$

with $E_f' < mc^2$ the ground-state energy. It is seen that the angular distribution of the REC photons in the peak maximum differs very much from the RI results at the ionization threshold. In contrast, the RI into higher-lying continuum states has a photon distribution similar to the one at the SWL. This shows that it is not the final-state energy E_f' but the final-state momentum distribution which is the decisive quantity in determining the shape of the cross section. For the sake of comparison we have included the REC angular distribution obtained from the photon-energy integrated cross section. The similarity between $d\sigma^{REC}/d\Omega_k$ and the angular distribution from the doubly differential REC cross section in the peak maximum is obvious. This behaviour corresponds to the previously discussed similarity between the triply and fourfold differential RI cross sections below the SWL.

4.4. Photon Linear Polarization

For the photoionization process the polarization correlations were studied in great detail [66], using the relativistic partial-wave formalism [67]. These investigations were motivated by realizing that the photoeffect can serve as a polarizer of electrons, a transmitter of polarization from photons to electrons, or an analyzer of polarized radiation. It is the latter effect which translates to the observation of linearly polarized REC photons caused by unpolarized quasifree electrons. Within the Kleber-Eichler model, the correlation parameter C_{10} of photoionization [66] corresponds to the degree P of linear polarization in REC.

Compared to photoionization the REC process has the advantage that the photon escapes from the target more or less unperturbed whereas in the inverse process the emitted electron may undergo successive collisions inside the target. As an application of REC it was suggested [68] to use the polarization of the emitted photon to gain information on the spin polarization of ion beams. In a recent pilot experiment, P was measured for K-shell REC in 400 MeV/amu $U^{92+}+ N_2$ collisions [69] and was found to be in accord with the predictions from accurate REC calculations [70, 71].

Aiming at a comparison of the angular dependence of P which results from REC and threshold-RI, respectively, we recall a common property of P which renders the deviations between the two processes smaller than in the case of the photon distributions. For symmetry reasons the RI polarization vanishes when the photon is emitted parallel or antiparallel to the beam axis ($\vartheta_k = 0°, 180°$) provided the emitted electron – if observed – is ejected into the beam direction too. For the photoeffect it was shown that a consistent relativistic theory also leads to a vanishing P for $\vartheta_k = 0°$ and $180°$ [66], and the same is true for REC [68, 71]. When Sommerfeld-Maue functions are used, P decreases strongly near $0°$ and $180°$ but does not vanish, whereas for RI, one does get $P = 0$. This is another indication that the semirelativistic functions are more appropriate for RI than they are for REC.

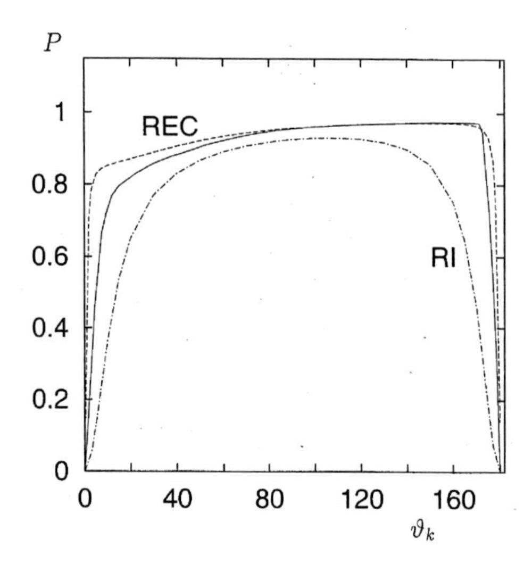

Figure 20. Polarization of photons from collisions with a 300 MeV/amu Ar^{18+} projectile as a function of photon emission angle. Shown are accurate REC results for capture of a free electron into the argon K-shell (———) and REC results using Sommerfeld-Maue functions and a hydrogen target (– – – –). The difference between P relating to $d^2\sigma/d\omega d\Omega_k$ at ω_{peak}^{REC} and to $d\sigma/d\Omega_k$ is indistinguishable. Included are RI results relating to $d^2\sigma/d\omega d\Omega_k$ for ω from (2.11) corresponding to the SWL ($E'_{f,kin} = 10^{-3}$ keV, $- \cdot - \cdot -$).

From Fig.20 where P resulting from K-shell REC and threshold-RI in 300 MeV/amu $Ar^{18+}+$ H collisions is shown, it is seen that at photon angles between $40°$ and $130°$ the photon degree of polarization is indeed quite similar for the two processes. However, its decrease to zero towards $0°$ and $180°$ is much steeper in the case of REC. For a heavier projectile the angular dependence of P from REC and threshold-RI is similar to the one in Fig.20 (see [7, Fig.54]), but the two processes differ somewhat more from each other [12]. A comparison with REC results using accurate wavefunctions [7] shows that the semirelativistic wavefunctions are appropriate for an argon projectile (except for photons ejected into the forward direction), whereas they start to deteriorate for $Z_P > 50$.

5. Conclusion

We have discussed the significant features of the momentum distributions of the outgoing photon and electron in the process of radiative electron capture to the projectile continuum. The underlying model was the relativistic impulse approximation, derived from scattering theory, with the use of semirelativistic Sommerfeld-Maue functions for the electronic states as the only additional approximation. Relying on the fact that bremsstrahlung is the inverse process of RI for a vanishing target field, the accuracy of our model, agreeing with the Elwert-Haug theory in this limit, could be tested against available theoretical results which use an accurate relativistic partial-wave representation of the wavefunctions. We conjecture that for collision velocities in the weak-relativistic regime ($\gamma \lesssim 2$) our model is accurate for not too heavy projectiles ($Z_P \lesssim 50$) in the region where the cross sections are large, but it may underestimate the cross sections when they are very small. For uranium projectiles the structures in the momentum distributions are qualitatively correctly predicted, but the absolute values come out a factor of 2 too low. This underprediction for uranium is confirmed by the comparison with experimental singles photon spectra which have been put on an absolute scale by normalizing to the well-established REC peak maxima.

Concerning the process occurring simultaneously with RI, the Coulomb capture into the projectile continuum, we have given predictions for which collision parameters both processes provide comparable electron intensities in the forward peak (cusp) region. The importance of this particular region of the electron spectrum is, on the one hand, its sensitivity to relativistic effects resulting from the close electron-projectile collisions. On the other hand, it seems to be the only region where radiative ionization can be the dominating process. We have established the strong dependence of the ratio between RI and ECC on the target species and on the collision velocity by using the same theoretical model for both processes. Recent coincidence experiments on 90 MeV/amu $U^{88+} + N_2$, allowing for the detection of the RI and ECC cusp electron spectra in one measurement, have confirmed the strong dominance of ECC predicted by theory for this collision system. In addition, they have established the shape of the RI cusp with its skewness to the opposite side of the ECC cusp. They have thus proven the feasibility of measuring the short-wavelength limit of bremsstrahlung with the tool of inverse kinematics.

If comparison is made between the frame-transformed RI and the bremsstrahlung results, there are considerable differences in the photon and electron angular distributions as well as in the photon linear polarization for the heavier targets. These differences increase when the photon frequency approaches the short-wavelength limit. When the target field is decreased to zero and the photon frequency well below the SWL, there is a smooth transition to the bremsstrahlung limit. At the SWL, on the other hand, the RI limit for vanishing target charge falls a factor of 2 below the bremsstrahlung theory. A related phenomenon concerns the deviations in the (photon or electron) angular distributions when calculated, respectively, from the triply and fourfold differential RI cross sections (the latter taken at the electron energy that provides the peak value of the cross section) which are present at the SWL but absent at lower frequencies. All these SWL-pecularities arise from the fact that the cusp-like forward peak structure is superimposed on a background which is shaped by the Compton profile [22].

For the radiative electron capture to bound states we have derived a relativistic formulation of the impulse approximation and have again employed the Sommerfeld-Maue wavefunctions for the electron. We have compared it to results from the nonrelativistic impulse approximation and have found similar photon angular distributions, but a lower global intensity if the proper relativistic prescription is used. We have also contrasted the REC photon angular distributions to those from the doubly differential RI cross section at the short-wavelength limit. Whereas for the collision system investigated the RI angular distribution is comparatively flat with shallow minima near 0° and 180°, the REC distribution has a pronounced maximum near 100° with very deep minima at 0° and 180°. We ascribe this dissimilarity to the different momentum distributions of the outgoing electron in the ground state and near the ionization threshold, respectively. The supplementary comparison of our model with available partial-wave results for REC is not so conclusive for RI as in the case of bremsstrahlung. We attribute the considerable deviations between our model and the literature results for the forward-emitted REC photons in collisions with very heavy projectiles to the necessity of very close collisions when the electron is captured into the K-shell. For electrons in unbound final states, on the contrary, the small-distance part of the wavefunction, where the semirelativistic functions fail, is expected to be not so important.

We conclude that the radiative ionization presents itself indeed as a link between bremsstrahlung and radiative capture to bound states. However, RI exhibits a good deal of peculiarities which make it worth while an object of study for its own sake. Further experiments, in particular on an absolute scale which is independent of theory, are highly desirable.

Acknowledgments

It is a pleasure to thank S.Hagmann and M.Nofal for the close collaboration on the experimental side. I whould also like to thank J.Ullrich for supporting contacts with the physical community.

References

[1] Kleber, M.; Jakubassa, D.H. (1975). *Nucl. Phys.* **A252**, 152-162.

[2] Shakeshaft, R.; Spruch, L. (1978). *J. Phys.* **B11**, L621-L627.

[3] Jakubassa-Amundsen, D.H. (2007). *Eur. Phys. J.* **D41**, 267-274.

[4] Briggs, J.S.; Dettmann, K. (1974). *Phys. Rev. Lett.* **33**, 1123-1125.

[5] Lucas, M.W.; Steckelmacher, W.; Macek, J.; Potter, J.E. (1980). *J. Phys.* **B13**, 4833-4844.

[6] Bethe, H.A.; Salpeter, E.E. Quantum mechanics of one-and two-electron systems. In *Encyclopedia of Physics* Vol. 35; Flügge, S.; Ed.; Springer: Berlin, 1957; p.381,406.

[7] Eichler, J.; Stöhlker, Th. (2007). *Phys. Rep.* **439**, 1-99.

[8] McVoy, K.W.; Fano, U. (1959). *Phys. Rev.* **116**, 1168-1184.

[9] Jakubassa-Amundsen, D.H. (2006). *Radiat. Phys. Chem.* **75**, 1319-1329.

[10] Ludziejewski, T. et al (1998). *J. Phys.* **B31**, 2601-2609.

[11] Nofal, M. et al (2007). Submitted to Phys. Rev. Lett.

[12] Jakubassa-Amundsen, D.H. (2007). *J. Phys.* **B40**, in print

[13] Anholt, R. et al (1986). *Phys. Rev.* **A33**, 2270-2280.

[14] Kienle, P. et al (1973). *Phys. Rev. Lett.* **31**, 1099-1102.

[15] Schnopper, H.W. et al (1974). *Phys. Lett.* **47A**, 61-62.

[16] Jakubassa, D.H.; Kleber, M. (1975). *Z. Phys.* **A273**, 29-35.

[17] Yamadera, A.; Ishii, K.; Sera, K.; Sabata, M.; Morita, S. (1981). *Phys. Rev.* **A23**, 24-33.

[18] Bjorken, D.; Drell, S. D. *Relativistic Quantum Mechanics*; BI: Mannheim, 1964

[19] Jakubassa-Amundsen, D.H.; Amundsen, P.A. (1980). *Z. Phys.* **A297**, 203-214.

[20] Jakubassa-Amundsen, D.H. (1987). *J.Phys.* **B20**, 325-336.

[21] Rose, E.M. *Relativistic Electron Theory* Vol.1; BI: Mannheim, 1971; Sect.III.

[22] Jakubassa-Amundsen, D.H. (2003). *J. Phys.* **B36**, 1971-1989.

[23] Davidović, D.M.; Moiseiwitsch, B.L.; Norrington, P.H. (1978). *J. Phys.* **B11**, 847-864.

[24] Sommerfeld, A.; Maue, A.W. (1935). *Ann. Physik* **22**, 629

[25] Bess, L. (1950). *Phys. Rev.* **77**, 550-556.

[26] Maximon, L.C.; Bethe, H.A. (1952). *Phys. Rev.* **87**, 156.

[27] Ichihara, A.; Shirai, T.; Eichler, *J. (1994). Phys. Rev.* **A49**, 1875-1884.

[28] DePaola, B.D. et al (1995). *J. Phys.* **B28**, 4283-4290.

[29] Jakubassa-Amundsen, D.H. (1997). *J. Phys.* **B30**, 365-385.

[30] Rudd, M.E.; Sautter, C.A.; Bailey, C.L. (1966). *Phys. Rev.* **151**, 20-27.

[31] Salin, A. (1969). *J. Phys.* **B2**, 631-636.

[32] Macek, J. (1970). *Phys. Rev.* **A1**, 235-241.

[33] Dettmann,K.; Harrison, K.G.; Lucas, M.W. (1974). *J. Phys.* **B7**, 269-287.

[34] Jakubassa-Amundsen, D.H. (1983). *J. Phys.* **B16**, 1767-1781.

[35] Nofal, M. (2007). PhD Thesis, University of Frankfurt, and private communication

[36] Haug, E.; Nakel, W. *The Elementary Process of Bremsstrahlung*, World Scientific Lecture Notes in Physics vol. 73; World Scientific Publications: Singapore, 2004

[37] Bethe, H.A.; Heitler, W. (1934). *Proc. Roy. Soc.* (London) **A146**, 83

[38] Elwert, G.; Haug, E. (1969). *Phys. Rev.* **183**, 90-105.

[39] Tseng, H.K.; Pratt, R.H. (1971). *Phys. Rev. A* **3**, 100-115.

[40] Hub, R.; Nakel, W. (1967). *Phys. Lett.* **24A**, 601-602.

[41] Nakel, W. (2006). *Radiat. Phys. Chem.* **75**, 1164-1175.

[42] Tseng, H.K.; Pratt, R.H. (1973). *Phys. Rev. A* **7**, 1502-1515.

[43] Haug, E. (1969). *Phys. Rev.* **188**, 63-75.

[44] Gluckstern, R.L.; Hull, M.H. Jr. (1953). *Phys. Rev.* **90**, 1030-1035.

[45] Gluckstern, R.L.; Hull, M.H. Jr.; Breit, G. (1953). *Phys. Rev.* **90**, 1026-1029.

[46] Fano, U.; McVoy, K.W.; Albers, J.R. (1959). *Phys. Rev.* **116**, 1159-1167.

[47] Motz, J.W.; Placious, R.C. (1960). *Nuovo Cimento* **15**, 571

[48] Behnke, H.-H.; Nakel, W. (1978). *Phys. Rev.* **A17**, 1679-1685.

[49] Bleier, W.; Nakel, W. (1984). *Phys. Rev. A***30**, 607-609.

[50] Shaffer, C.D.; Tong, X.-M.; Pratt, R.H. (1996). *Phys. Rev.* **A53**, 4158-4163.

[51] Tseng, H.K. (2002). *J. Phys.* **B35**, 1129-1142.

[52] Lee, C.M.; Kissel, L.; Pratt, R.H.; Tseng, H.K. (1976). *Phys. Rev.* **A13**, 1714-1727.

[53] Motz, J.W. (1955). *Phys. Rev.* **100**, 1560-1571.

[54] Tseng, H.K.; Pratt, R.H. (1974). *Phys. Rev. Lett.* **33**, 516-518.

[55] Pratt, R.H.; Tseng, H.K. (1975). *Phys. Rev.* **A11**, 1797-1803.

[56] Schnopper, H.W. et al. (1972). *Phys. Rev. Lett.* **29**, 898-901.

[57] Spindler, E.; Betz, H.-D.; Bell, F. (1979). *Phys. Rev. Lett.* **42**, 832-835.

[58] Anholt, R. et al. (1984). *Phys. Rev. Lett.* **53**, 234-237.

[59] Pratt, R.H.; Ron, A.; Tseng, H.K. (1973). *Rev. Mod. Phys.* **45**, 273-325.

[60] Jakubassa-Amundsen, D.H.; Höppler, R.; Betz, H.-D. (1984). *J. Phys.* **B17**, 3943-3949.

[61] McDowell, M.R.C.; Coleman, J.P. *Introduction to the Theory of Ion-Atom Collisions*; North-Holland: Amsterdam, 1970; p.366

[62] Sauter, F. (1931). Ann. Physik **9**, 217; *Ann. Physik* **11**, 454

[63] Fano, U.; McVoy, K.W.; Albers, J.R. (1959). *Phys. Rev.* **116**, 1147-1156.

[64] Stöhlker, Th. et al. (2001). *Phys. Rev. Lett.* **86**, 983-986.

[65] Bednarz, G. et al. (2003). *Hyperfine Interact.* **146**, 29

[66] Pratt, R.H.; Levee, R.D.; Pexton, R.L.; Aron, W. (1964). *Phys. Rev.* **A134**, 916-922.

[67] Pratt, R.H.; Levee, R.D.; Pexton, R.L.; Aron, W. (1964). *Phys. Rev.* **A134**, 898-915.

[68] Surzhykov, A.; Fritzsche, S.; Stöhlker, Th.; Tashenov, S. (2005), *Phys. Rev. Lett.* **94**, 203202, 1-4.

[69] Tashenov, S. et al. (2006). *Phys. Rev. Lett.* **97**, 223202, 1-4.

[70] Surzhykov, A.; Fritzsche, S.; Stöhlker, Th.; Tashenov, S. (2003), *Phys. Rev.* A**68**, 022710, 1-7.

[71] Eichler, J.; Ichihara, A. (2002). *Phys. Rev.* **A65**, 052716, 1-5.

In: Radiation Physics Research Progress
Editor: Aidan N. Camilleri, pp. 193-217
ISBN: 978-1-60021-988-7

Chapter 4

PULSED LASER ABLATION AND DEPOSITION OF THIN FILMS OF RARE EARTH IONS-DOPED FLUORIDES

P. Bicchi, M. Anwar-ul-Haq and S. Barsanti
Department of Physics - University of Siena - Via Roma, 56 – 53100 Siena – Italy

Abstract

This chapter describes the production of thin films of Nd^{3+}-doped fluorides of high optical quality, to be used as devices in non linear optics, via the pulsed laser deposition technique. The innovative aspect of this research is the use of the same monocrystalline undoped fluoride as the substrate for the deposition. The technique is briefly recalled and the experimental apparatus is described in some detail. We describe how two different kinds of Nd^{3+}-doped fluoride films, namely Nd^{3+}:YF_3 and Nd^{3+}:$LiYF_4$, can be obtained on a pure $LiYF_4$ substrate from the ablation of a Nd^{3+}:$LiYF_4$ bulk crystal by changing some ablation/deposition parameters such as the substrate temperature and the presence or absence of a buffer gas in the ablation/deposition chamber. The onset of a film on the substrate is checked by interferometric measurements. The optical characterization of the films includes both polarized laser induced fluorescence spectroscopy analysis, which testifies of the kind of fluoride film produced, and the $^4F_{3/2}$ Nd^{3+} manifold lifetime measurement, which is related to the Nd^{3+} ions concentration. Both the laser induced fluorescence spectra and the lifetime measurements are compared with the corresponding ones obtained in the bulk crystal. Some data on the films thickness and on their morphology are also presented together with some checks on the ablation plume constituents and on their expansion dynamics.

1. Introduction

1.a. General Remarks

The development of solid-state devices is, nowadays, oriented towards a strong integration and miniaturization and for this reason the research is more and more oriented towards the study of materials with proper optical properties and smaller and smaller dimensions. An answer to this quest is given by the development of micrometric and sub micrometric crystalline films of different classes aimed to the development of high efficiency active and passive optical devices. In particular the realization of devices in film shape would make possible the development of micro-lasers. The latter magnify all the advantages of the lasers as the film shape favours the removal of the heat in excess from the active media and the confinement of the radiation, which maximizes the interaction zone between the pump and the active medium with a reduction of the lasing threshold [1] and the possibility of obtaining wave guiding systems [2].

The research on nano-crystalline structures uses several techniques for thin films deposition [3]. They are essentially divided in two classes: thermal deposition techniques and ion-beam techniques. The first class includes the thermal evaporation [4] (TE – evaporative deposition of a target material thermally heated), the molecular beam epitaxy [5] (MBE – a technique in which molecular beams, produced by thermally evaporated elemental sources, are deposited onto a heated crystalline substrate to form thin epitaxial layers), the liquid phase epitaxy [6] (LPE – a technique in which the elements to be deposited come from liquid solutions), the chemical vapour deposition [7] (CVD - the deposition comes from the reaction of gases above a heated substrate) and the metal organic chemical vapour deposition [8] (MOCVD – the film deposition on a substrate is obtained by selecting the vapours of a volatile compound). In the second class the most commonly used are the sputtering [4, 9] (film deposition on a substrate due to the emission of atoms and clusters from a target bombarded by energetic particles, usually ions) and ion implantation [10] (a doping technique via absorption of ions from a target). Most of these techniques were developed and improved to optimize the deposition of a particular material. After the mid eighties a new technique took the lead. It is based on the ejection of materials from a target due to the interaction with highly energetic photons produced by a laser source, with subsequent deposition on a suitable substrate. Its name is pulsed laser deposition (PLD).

1.b. Pulsed Laser Deposition

The PLD process [11] is in principle very simple. The essential elements of the technique are illustrated in fig. 1. A high intensity pulsed laser beam is focused on a target. When the power density exceeds a threshold value, the interaction of the photons with the target causes material ejection via thermal and/or electronic processes [12]. This ejected material, the "ablated plume", consists of a mixture of energetic particles such as atoms, molecules, electrons, ions, clusters, sub micrometric or micrometric solid particulates and fused particles. It expands away from the interaction volume into the vacuum to form a nozzle jet mainly maintaining a symmetric distribution about the target surface normal. The ejected flux then recondense on an appropriate, properly heated substrate to form the film. The whole process takes place in a vacuum chamber, where both the target and the substrate are housed, in which a known quantity of buffer gas may be introduced. This aspect guarantees the possibility of creating a growth area without contamination and of operating in a controlled background gas atmosphere, which is a necessary condition for the growth of some special material films. For the flexibility of the experimental set-up, PLD is considered one of the most efficient and versatile techniques to get thin films of any kind of material to produce optoelectronic devices. In addition PLD is well known as a technique able to ensure in the film the same stoichiometric ratio of the elemental components of the bulk [13], to control the film growth rate [13], to allow film deposition also on substrates kept at low temperature [14] and to facilitate the realization of multiple layer films with variable refractive index [15], so to function as waveguides for the laser wavelengths [16]. PLD has also disadvantages to cope with, mainly the presence of micro sized particulates in the plume and its narrow forward angle distribution which makes the realization of large area films quite difficult. The growth and quality of the deposition depend on a number of parameters such as the choice of the substrate, its temperature, the substrate/target geometry, the deposition rate and the energy of the plume constituents which in turn are affected by the wavelength, duration and energy of the laser pulses, and on the percentage of inert or reactive gas inside the vacuum chamber [11-22].

The versatility of PLD is affirmed beyond any doubt by the fast growing list of deposited materials. A comprehensive review updated to 1994 is given in ref. [23]. Since then the list has grown considerably. We report some examples without any intention of completeness. Several results are reported on the PLD preparation of thin films of metallic oxides with ferroelectric properties [24], due to the interest that bidimensional structures of these compounds arise in the field of signal elaboration, electro-optics and data storage. Among microwave devices PLD has been applied to produce ferrite thin films while nanostructured films of tin oxides have been deposited on an alumina polycrystalline substrate [25]. In the field we are concerned with, non linear optics, active planar waveguides have been produced by depositing sub micrometric films of chalcogenide and tellurite, doped with different compounds, on silica substrates [26]. The PLD technique has been applied to produce guided lasers using thin films of Ti:Sapphire on sapphire substrate [27], of Nd-doped $YAlO_3$ both on $YAlO_3$ and sapphire substrates [28] and of Nd-doped $Gd_3Ga_5O_{12}$, on YAG substrates [1, 2]. Quite recently Eu-doped YVO_4 [29] and $GdVO_4$ [30] films grown by PLD have been reported together with a PLD produced waveguiding film of Nd-doped Sc_2O_3 [31].

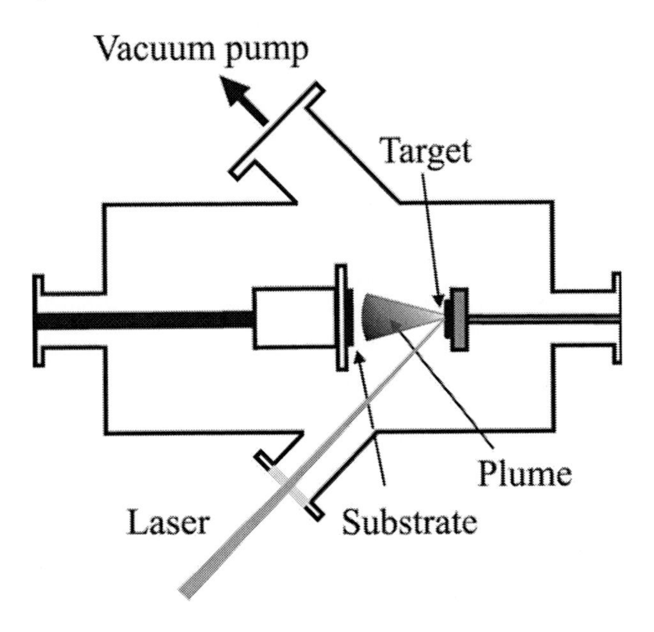

Figure 1. Schematic diagram of a typical PLD set-up.

1.c. Rare Earth Ions-Doped Fluorides

Fluoride crystals have been considered for decades good candidates as hosts for rare-earth (RE) trivalent ions to realize efficient laser sources in the IR region due to their remarkable optical characteristics such as high transparency, phonon energies lower than the corresponding ones in oxides associated with comparable emission cross sections and high resistance to optical damage [32]. In spite of this, fluorides find much fewer applications than oxides because their growth process requires a very careful methodology as a very small contamination (even few tens of ppm) of the OH^- radical inside the crystals can irremediably compromise their photonic quality. For this reason the starting materials must undergo a fluoridation process to make them free from any OH^- contamination and consequently they must be stored and manipulated in anhydrous atmosphere. In addition fluorides are more vulnerable to fractures and to thermal shocks than oxides, which add technical problems in their manipulation. These difficulties may be among the reasons of the limited production and study of thin films of these materials even if RE-doped fluoride films have been obtained with LPE [33, 34], vacuum deposition and sol-gel technique [35] and PLD [36-38] mainly on amorphous or Si substrates.

In the following we describe the results of the research carried on in our laboratory which aims to the production of monocrystalline RE-doped fluoride sub-micrometric films on monocrystalline substrates of the same undoped fluoride. These films could be operated as planar waveguides for the laser emission, so to enhance the lasing efficiency as detailed in the precedent section. In particular we mean to develop films suitable for active devices in the infrared wavelength range, towards which the bio-medical, micro-machining and nanotechnology areas show an increasing interest. The technique that emerges as the most appropriate is PLD essentially for the congruent transfer from the bulk target to the film and the possibility of working in a high vacuum or a background gas controlled atmosphere. The most innovative aspect of this research is the use of the same monocrystalline undoped bulk crystals as the substrates. This aspect insures very good accordance for the crystal lattice constants in the interface between the film and the substrate so favouring the developing of real crystalline depositions [39]. The fluoride we start with is $LiYF_4$ (YLF) used pure as the substrate and doped with Nd^{3+} ions as the bulk target. After a description of the experimental apparatus, we report about the study done on the ablated plume in two different experimental environments. Next we show how by acting on the ablation/deposition parameters we can get two different kinds of Nd-doped fluoride films, namely Nd^{3+}:YF_3 (Nd:YF) and Nd^{3+}:$LiYF_4$ (Nd:YLF). The morphological analysis of the films and their optical characterization follow. The former is obtained by analyzing the deposition with a scanning electron microscope (SEM). The films are spectroscopically characterized by recording the room temperature polarized fluorescence with E || or E⊥ to the c- axis of the YLF substrate in the two IR regions around 900 nm and 1050 nm after the film has been irradiated with a diode laser radiation around 800 nm. A check of the $^4F_{3/2}$ Nd^{3+} manifold lifetime is also performed by exciting the film with the same radiation mechanically chopped. The excitation corresponds to the fundamental manifolds transition $^4I_{9/2} \rightarrow {}^4F_{5/2} + {}^2H_{9/2}$ of the Nd^{3+} ion while the fluorescences around 900 nm and 1050 nm are due to the $^4F_{3/2} \rightarrow {}^4I_{9/2}$ and $^4F_{3/2} \rightarrow {}^4I_{11/2}$ manifolds transitions as shown in fig. 2 where a simplified scheme of the lowest energy levels of the Nd^{3+} ions [40] is plotted together with their absorption spectrum [41] when they are hosted in a YLF matrix. The diode laser excitation used in the experiment is also indicated.

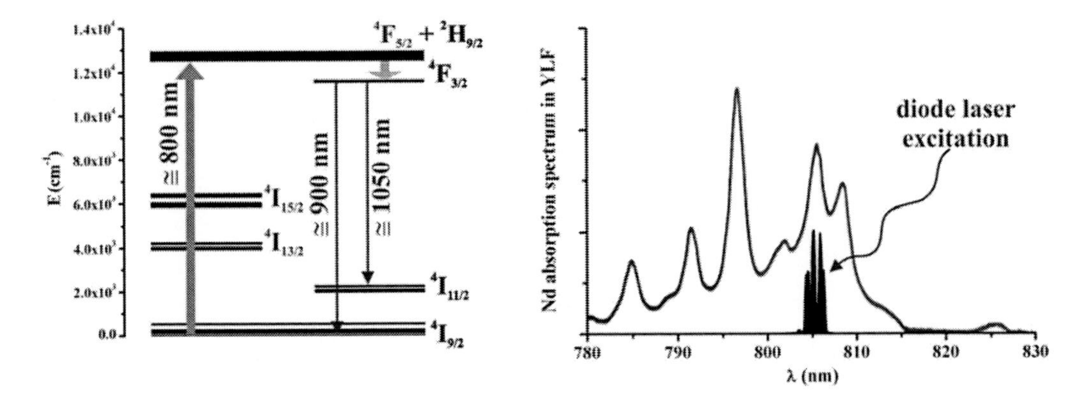

Figure 2. Simplified diagram of the lowest energy levels of the Nd^{3+} ions in YLF. The transitions of interest for this chapter are evidenced. The absorption spectrum of the Nd^{3+} ions embedded in YLF is also plotted together with the diode laser excitation used in this experiment.

2. Experimental Apparatus

2.a. YLF and Nd:YLF Crystals

Both the target (Nd:YLF) and the substrate (YLF) crystals were grown with a home made Czochralski furnace [42]. The growth process starts from purified LiF and YF_3 powders for the YLF crystal and from purified LiF, YF_3 and NdF_3 powders for the Nd:YLF crystal. The relative ratios were chosen so to obtain a 1.5% atomic concentration (at.) of Nd^{3+} ions in the bulk target crystal. The single crystalline nature and the axis orientation of both crystals were checked by means of the X-ray Laue technique so to cut oriented parallel flat faces samples whose wider faces were optically polished by alumina powder. The Nd:YLF target has a radius of 7.2 mm with 3 mm thickness. The YLF substrate has similar dimensions with the thickness reduced to 2 mm.

2.b. Ultra High Vacuum Chamber

The ultra high vacuum (UHV) chamber, whose section parallel to the base plane is sketched in fig. 3, consists of a cylindrical stainless steel vessel of internal diameter 440 mm and height 260 mm. The chamber is evacuated by a Leybold Turbovac V-150 turbo molecular pump (base pressure $\approx 10^{-5}$ Pa) connected to its base via a gate valve to control the pumping speed. An additional flange is connected to the base to mount electrical and thermal feedthroughs. All along the lateral surface there are other 9 flanges to mount view ports, pressure gauges, a gas inlet, the laser input window and the target and substrate holders. The latter are mounted on stainless steel rods aligned along one of the chamber axis. There mutual distance can vary from a minimum of 10 mm up to 75 mm. Along the perpendicular axis two 80 mm diameter view ports of 3.5 mm thickness are used for spatial laser analysis of the ablated plume and for

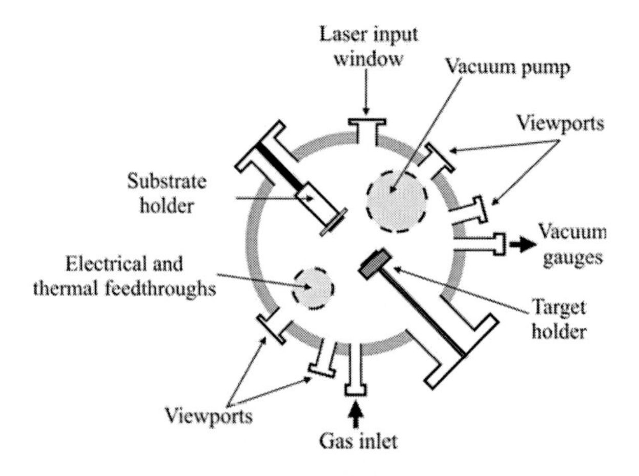

Figure 3. Schematic top view of the UHV chamber for PLD.

Figure 4. Photos of: a) the whole UHV chamber with details of the fibre coupled plume emission detection system, b) the UHV chamber inside with the substrate and target holders, c) a side view of the UHV chamber.

in situ analysis of the deposited film. Two extra view ports of 30 mm diameter and 2.5 mm thickness are added at 30° with the target/substrate axis for possible ellipsometric measurements. All these view ports are made out of KODIAL glass whose optical transmission is 90% or above in the range 360 nm – 2.0 μm. Only the ablation laser input window is a quartz disk of 50 mm diameter and 3.3 mm thickness, which allows the use of ablation pulses also in the far UV. The window is mounted so that the laser can impinge on the target at 45° to the normal. The distance from the center of the chamber is 34 mm. With this geometry the window can be maintained in its place up to about 3 hours without any measurable attenuation of the incoming laser radiation. The attenuation results from the accumulation of materials unavoidably deposited also on the laser input window during the ablation process. The pressure inside the chamber is measured via a Leybold Combitron CM350 which is a microprocessor-controlled vacuum meter that enables continuous measurement and monitoring of the vacuum pressure between atmospheric pressure and $5x10^{-6}$ Pa by combining two measurements done by a Thermovac TR301 Pirani gauge and a Penningvac PR-35 gauge. The chamber is also equipped with a gas inlet valve through which a high purity gas can be introduced for deposition in a controlled atmosphere. During the evacuation process the chamber can be baked up to 200°C to remove humidity and accidental contamination. The top of the vacuum chamber consists of a large flange that houses another KODIAL 80 mm diameter view port which, connected with a fibre coupled detection system, allows optical emission measurements of the ejected plume. This flange can be removed to access the inside of the chamber. Photos of the whole chamber assembly are shown in fig. 4.

2.c. Film Growth System

The film deposition apparatus consists of two main parts: the ablation set-up and the deposition system. They are sketched in fig. 5 (a, b).

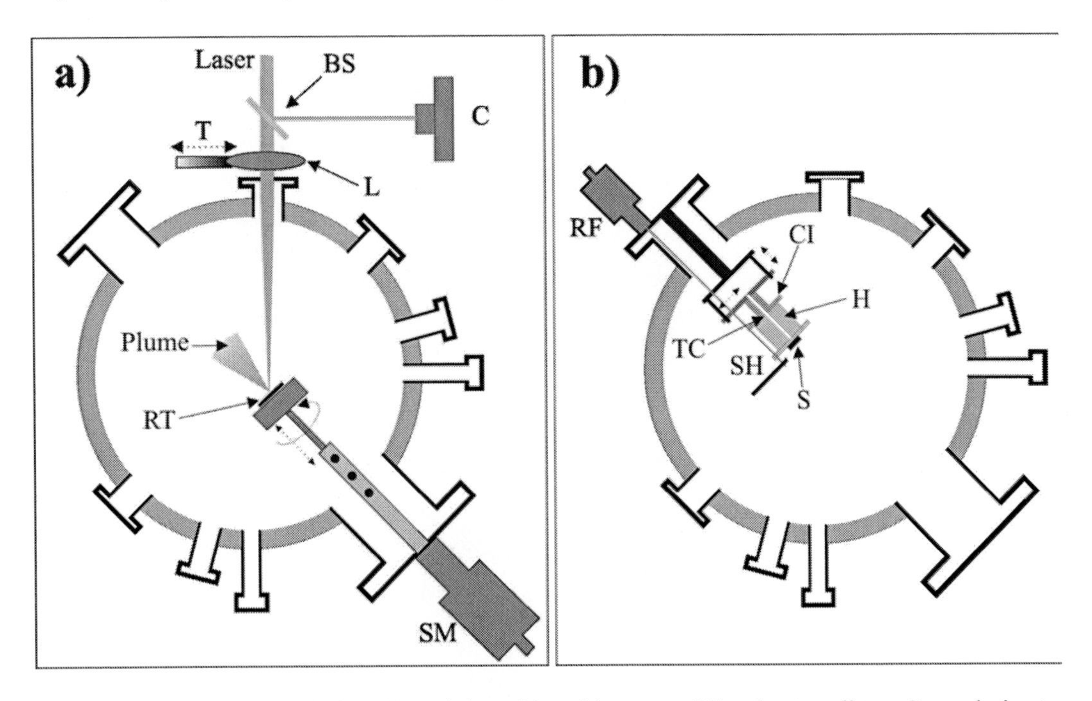

Figure 5. Sketch of the ablation (a) and deposition (b) set-up. BS = beam splitter; C = calorimeter (power meter); T = translator; L = lens; RT = rotating target; SM = step motor; RF = rotating feedthrough; CI = ceramic insulator; TC = thermocouple; H = heater; SH = shutter; S = substrate

2.c.1. Ablation Set Up

The target bulk Nd:YLF crystal is mounted on the flat target holder which is kept rotating by a step motor placed outside the vacuum chamber with a maximum speed of 1.25 rpm, to prevent craterization of the crystal surface. The stainless steel rod ending with the target holder can be translated along the chamber axis to adjust the mutual target/substrate distance. The target is cleaned by typically 1500 laser shots prior to the deposition. The third harmonic output at 355 nm of a QUANTEL YG 580 Nd-YAG laser provides the ablation pulses at a repetition rate of 10 Hz. The maximum pulse energy is 130 mJ and its duration is 13 ns. The ablation beam is focused by a 300 mm focal quartz lens mounted on a piezo-translator in front of the input window, to an 800 μm diameter spot and hits the target surface at 45° to the normal. The ablation fluencies range from 4 to 10 J/cm^2. The laser fluency is constantly controlled by measuring the laser power via a beam splitter and a power meter placed just outside the input window, after the system calibration done with the open chamber. The translator moves the ablation beam 650 μm horizontally every 7 minutes. This has the advantage of providing fresh polished target surface to the ablating beam and at the same time favours the stability of the plume axis orientation.

2.c.2. Deposition System

The YLF substrate is mounted at the center of the hot plate of the heater holder which comprises two basic parts: the heater itself and the thermally insulating ceramic piece on which the heater is fixed. This assembly can translate along one stainless steel rod aligned with the target holder one to adjust the substrate/target distance. The substrate holder can also translate 30 mm perpendicularly to this direction for off-axis deposition. The commercially available heater Heat Wave Labs Inc., model # 101491-01 can reach temperatures up to 1200°C in absence of oxygen and can hold samples up to 20 mm diameter. A type K thermocouple is fixed at direct contact at the center of the hot face and the temperature controller EUROTHERM 2116 is used to control the heating/cooling cycles and to stabilize the temperature within ±1°C during the deposition. Due to the elevated fluoride crystals thermal shock sensitivity, the heating/cooling cycles are arranged in such a way to increase/decrease the substrate temperature linearly at a rate of 1°C/min. Typical YLF substrate temperatures during the deposition range between 650°C and 750°C. The substrate heater/holder assembly is equipped with a stainless steel shutter to cover the substrate during the target cleaning laser shots. The shutter can be rotated via a mechanical feedthrough housed in the substrate holder flange.

2.d. Plume Analysis

The optical emission of the ablated plume is studied with the set-up sketched in fig. 6. The optical emission arising from the plume is collected at 90° to the plume axis via a lens/iris combination (iris diameter = 3 mm, lens focal length = 3 mm) from the plume image in the focal plane of a 300 mm lens placed above the view port on the topside of the ablation chamber. Housing for filters is also provided. An optical fibre transfers this radiation to the input slit of a Jobin Yvon (JY) 320 monochromator equipped with a 1200 grooves/mm grating and used with a 1.2 nm resolution. A time integrated frequency resolved spectrum is recorded via a photomultiplier and a boxcar averager system along the plume axis at different distances, d, from the plume origin. The available spectral range is 375 – 700 nm. When the time resolved analysis of the single lines in the plume emission spectrum is needed, the monochromator is fixed to the selected wavelength and the output of the photomultiplier is sent to a digital oscilloscope (Tektronix TDS 520, 500 MHz bandwidth) for analysis. A controlled pressure of buffer gas can fill the ablation chamber to check its influence on the plume emission and expansion.

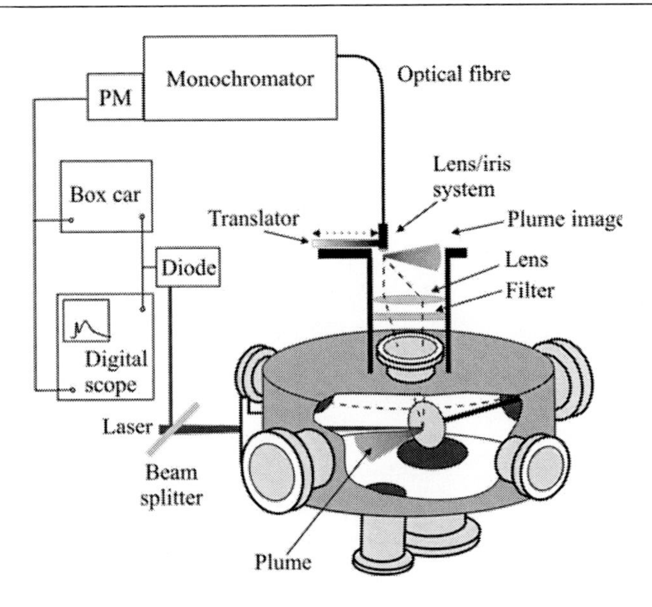

Figure 6. Sketch of the plume emission detection set-up. PM = photomultiplier.

2.f. Film Morphological Analysis

The film surface properties were analyzed with a SEM XL20 PHILIPS microscope with a maximum resolution of 3 nm. The microscope is equipped with a standard secondary electron detector.

2.e. Film Optical Characterization

After the deposition, the substrate is cooled down to room temperature and some preliminary checks are done *in situ*. The effective realization of a film growth is monitored by interferometric techniques. The radiation from a He-Ne laser is sent at grazing incidence on to the substrate. The presence of a layer on it is testified by the formation of interference fringes collected on a screen as illustrated in fig. 7. An average estimate of the film thickness can also be derived from the fringes pattern. We have up to now deposited films of 100 - 400 nm average thickness. The check of the presence of the dopant Nd^{3+} ions in the film is done by recording the laser induced fluorescence (LIF) spectrum from the layer after excitation at grazing incidence with the same ablation laser beam at 355 nm, attenuated to 3 mJ/cm^2. The LIF fluorescence in the range 375 - 500 nm is collected at 30° to the excitation direction via a quartz lens, filtered by a 385 nm long pass filter, dispersed by the mentioned JY 320 monochromator with 1.7 nm resolution and recorded via a box car averager with 100 ns gate delay with respect to the laser pulse to prevent the scattered exciting laser radiation from masking the UV side of the spectrum. The arrangement is also sketched in fig. 7. This spectrum is compared with that obtained from the bulk crystal following the same excitation with a similar set-up. In this case the monochromator resolution is 0.6 nm.

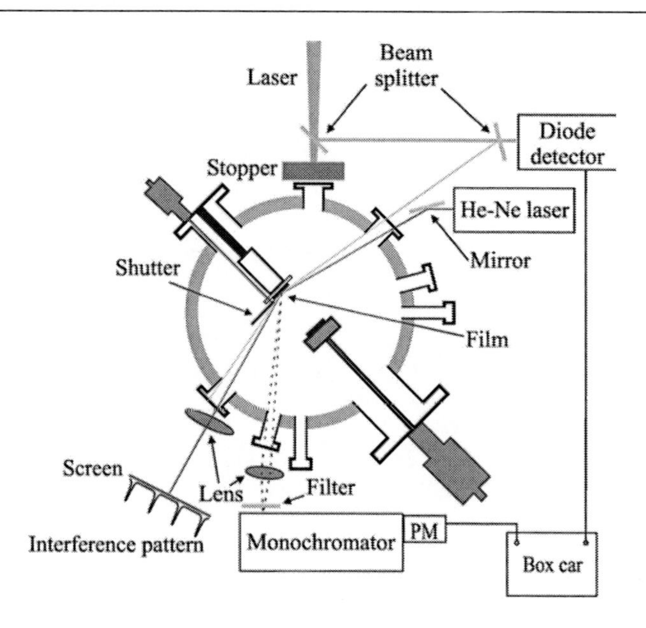

Figure 7. Experimental apparatus for the *in situ* analysis of the deposited film. PM = photomultiplier.

After these preliminary checks the films are optically characterized *ex situ* to determine their kind and quality. The room temperature fluorescence spectra polarized with E || or E⊥ to the c- axis of the YLF substrate are recorded after excitation of the film with a cw diode laser radiation focused to a 600 μm spot with a two lenses system. The two investigated regions extend around 900 nm and 1050 nm corresponding to the emission from the $^4F_{3/2}$ manifold to the $^4I_{9/2}$ and $^4I_{11/2}$ manifolds respectively, as shown in fig. 2 of section *1.c.* The exciting radiation wavelength is set at 806.6 nm by checking for the maximum fluorescence intensity from the Nd:YLF bulk crystal. The former, after being modulated at 380 Hz, incides upon the film at ~ 70° and the fluorescence is collected parallel to the film face normal. The JY 320 monochromator, equipped with an 830 nm long pass filter to minimize the scattered laser light, disperses the fluorescence. The signal is detected with an S-1 response curve photomultiplier and processed via a lock-in amplifier. The detection range is 400 - 1200 nm. The $^4F_{3/2}$ manifold lifetime measurement is done by using the same excitation source chopped at 100 Hz in such a way to give laser pulses with an extinction time less than 5 μs. The diode laser intensity is properly attenuated so to avoid non-linear effects. The transient fluorescence signal is detected via the same monochromator/photomultiplier assembly and is memorized in the TDS 520 digital oscilloscope for analysis. The whole system has a response time of about 20 μs. The basic elements of both frequency resolved and time resolved spectra detection are sketched in fig. 8. Both the fluorescence polarized spectra and the lifetime measurements were compared with the equivalent ones obtained in the bulk crystal with a similar set-up after the same excitation. For all the *ex situ* spectra the monochromator resolution is always chosen so to maximize the signal to noise ratio, without affecting the linewidths of the resolved lines.

Throughout all the experimental steps, both data acquisition and analysis are computerized. Anytime frequency or time resolved LIF spectra are recorded, the whole experiment is triggered by a small fraction of the exciting laser radiation. The monochromator calibration is done by recording and measuring the spectra from two r.f. excited Hg or Cd spectral lamps.

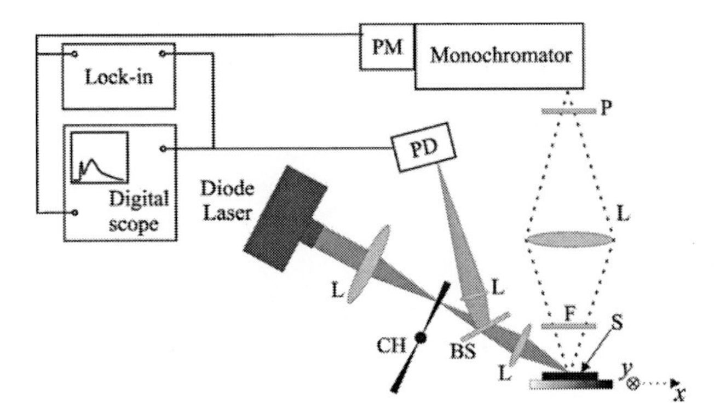

Figure 8. Experimental apparatus for the *ex situ* optical characterization of the deposited film. PM = photomultiplier; P = polarizer; L = lens; F = filter; S = substrate with the grown film; BS = beam splitter; CH = chopper; PD = photodiode; (y, x) = vertical and horizontal translator.

Before proceeding to the deposition on YLF substrates, the whole ablation/deposition apparatus was checked with a preliminary deposition from an Nd:YLF bulk crystal on a Si (100), perpendicular to the plane, crystal substrate as detailed in ref. [43]. The first check of deposition on an YLF substrate heated to 650°C was realized with this apparatus, by ablating a Nd:YLF bulk crystal (1.5 % at.) with a 4 J/cm^2 ablation fluency in high vacuum. The result, detailed in ref. [44], was a Nd:YF film showing, together with some deficiencies, some hints of crystalline nature.

3. The Ablated Plume

The ablation laser pulse induces a rapid heating of the target material which boils off and vaporizes. In this way some target material is ejected. This process occurs within some picoseconds [12]. When nanoseconds laser pulses are used, as in our case, the later part of the pulse is absorbed by the plume with subsequent excitation and ionization of its different elemental components. The result of this chain of events is the formation of a plasma and of optical emission. The plume with its neutral or charged components, whose excitations decay with their characteristic times, then expands away from the laser pulses/target interaction volume. During the expansion the plume can so be analyzed either by detecting the emitted radiation or by monitoring the produced ions/electrons.

We performed both a frequency and a spatial/temporal resolved emission spectroscopy of the plume ejected from the bulk Nd:YLF target for identifying the different species and getting some information on their dynamics. The studies on laser ablated plumes of fluoride or RE-doped fluoride crystals are scarce if not absent. This study is a self standing complicated subject and there are dedicated experiments just begun in our laboratory [45]. The observations we report are preliminary and limited, but they gave us some inputs in order to be able to determine the geometry and some ablation parameters to start off with the PLD film deposition we pursue. The plume was studied in two different environments: in vacuum, typical pressure 8x10^{-5} Pa, and in presence of 1Pa of background inert gas.

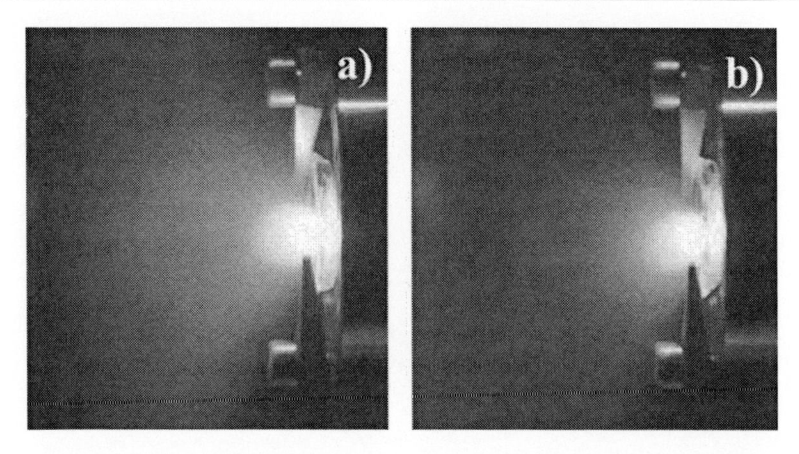

Figure 9. Photos of the plume ablated from the Nd: YLF bulk crystal with 10 J/cm^2 laser fluency in vacuum (8x10^{-5} Pa) (a) and in presence of 1 Pa of He pressure (b).

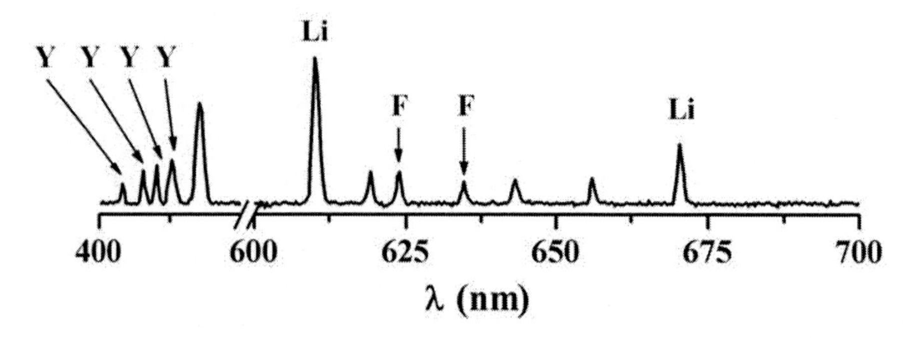

Figure 10. Portion of the emission spectrum of the plume ablated from the Nd: YLF bulk crystal with laser pulses at 355 nm, recorded along the plume axis at d = 6 mm from the bulk. The crystal is ablated in vacuum with a laser fluency of 10 J/cm^2.

The plume is easily seen with the naked eye and extends a few centimeters from the target. Two photos of the plume originated by 10 J/cm^2 laser pulses in vacuum and in presence of 1 Pa He pressure are shown in fig. 9 (a,b). The confinement of the plume when in presence of the background gas is evident. The emission spectrum in the UV-visible range is very rich in features and contains elemental lines of all the components of the host lattice. An example is given in fig. 10 for d = 6 mm along the plume axis. Some lines belonging to the different elemental components of the bulk could be assigned and a few of them were used for the spatial/temporal analysis. No clear evidence of the dopant could be extracted from this spectrum. The reason is most probably due to the very low concentration of Nd^{3+} ions in the YLF matrix.

We determined the ablation fluency threshold to be around 3 J/cm^2 by measuring the intensity of the signal at 610.3 nm, corresponding to the Li transition $3\ ^2D_{3/2} \rightarrow 2\ ^2P_{1/2}$, as a function of the laser fluency at d = 0. The time resolved analysis of some identified lines, enlightens in part the dynamics of the correspondent species. We recorded the time resolved signals at different distances from the target along the plume axis for the mentioned Li* line as well as for other lines at 410.2 nm (Y*) and 623.9 nm (F*). Figure 11 shows the intensity

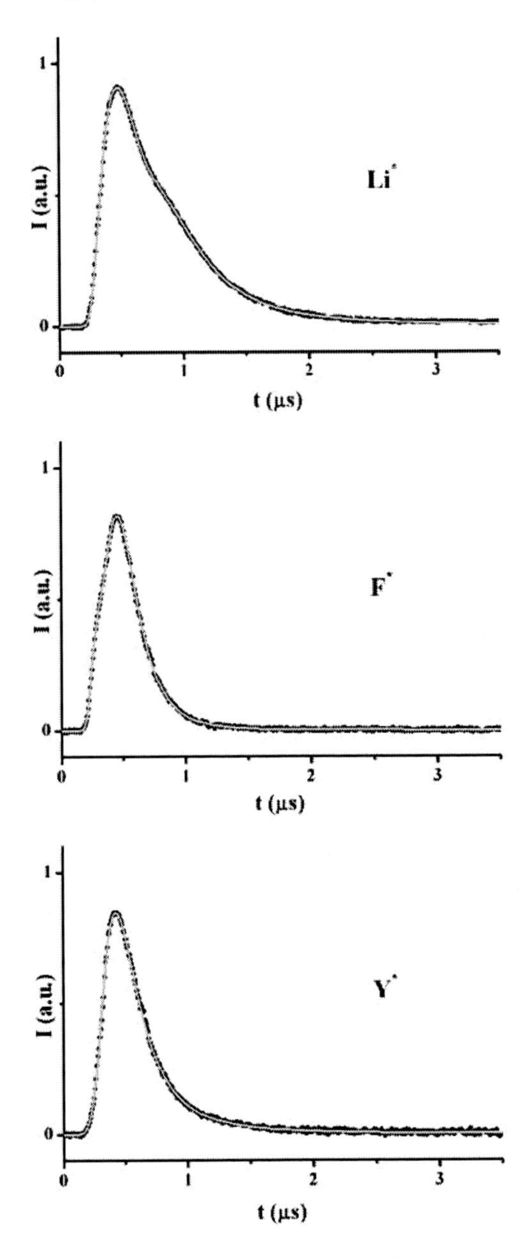

Figure 11. Emission profiles of Li*, F* and Y* atoms recorded versus time at d = 6 mm from the Nd: YLF bulk crystal, following ablation with 355 nm laser pulses at 10 J/cm^2 fluency in vacuum, plotted together with the respective MB distribution fits (grey lines).

profiles of the mentioned lines recorded in vacuum at d = 6 mm. They can all be fitted by a shifted center of mass Maxwell Boltzmann (MB) distribution [46] of the form:

$$I(t) = \sum_{j=1}^{n} A_j \left(\frac{d}{t}\right)^3 e^{\frac{m(d - v_j t)^2}{2KT_j t^2}} \quad (1)$$

Where I(t) is the signal intensity detected at distance d from the target, A is a normalization parameter, m is the mass of the species and K is the Boltzmann constant. All the lines could be recorded and fitted by the same MB distribution for d up to 30 mm. The fit gave velocities ranging from $5x10^3$ m/s up to 2 $x10^4$ m/s, values in accordance with those of several other ablated materials [46].

When the same time resolved fluorescences were detected in presence of 1 Pa of He, the MB distribution gave values in the range $3x10^3$ m/s to less than $1x10^4$ m/s. It is worth mentioning that at longer distances from the bulk, the presence of the gas atmosphere favoured the evidence of a second fast peak in the time resolved fluorescence profile of the different species. A similar circumstance was already reported in the literature for laser ablated oxides [47].

4. Films and Characterization

The films obtained so far were all deposited by placing the target bulk Nd:YLF crystal (1.5% at.) and the pure YLF substrate crystal at 35 mm distance along the chamber axis. We used different ablation/deposition parameters as described in the following, but in all cases from the *in situ* analysis we got evidence of the layer deposition and of the Nd^{3+} ions presence in it. The interference pattern of the He-Ne light provided evidence of the formation of a layer on the substrate and of its average thickness. The LIF spectrum recorded from the films at room temperature after excitation with the attenuated ablation laser radiation, as detailed in section *2.e*, evidenced the presence of the Nd^{3+} ions. An example of such spectrum is shown in

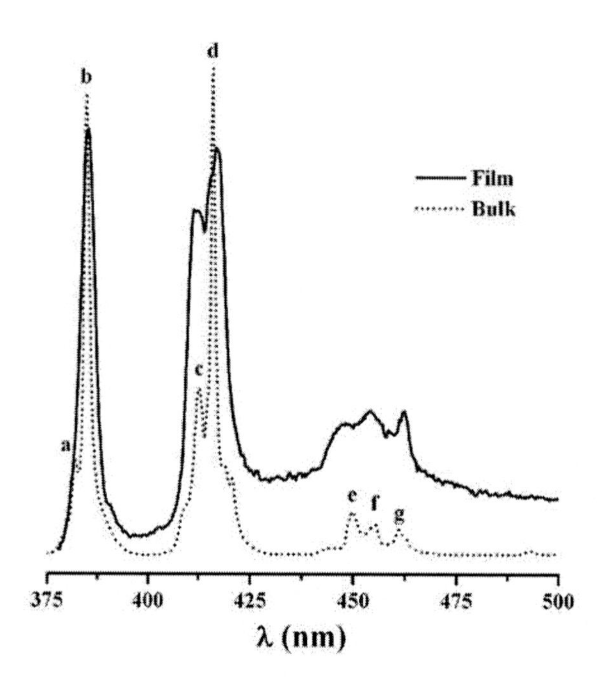

Figure 12. LIF spectrum following 355 nm excitation recorded in the Nd:YLF bulk crystal (dotted line) and in the deposited film (solid line) in the 375 – 500 nm range. The identified lines in the bulk spectrum are: a) $^4D_{3/2} \rightarrow {}^4I_{11/2}$; b) $^2P_{3/2} \rightarrow {}^4I_{9/2}$; c) $^4D_{3/2} \rightarrow {}^4I_{13/2}$; d) $^2P_{3/2} \rightarrow {}^4I_{11/2}$; e) $^4D_{3/2} \rightarrow {}^4I_{15/2}$; f) $^2P_{3/2} \rightarrow {}^4I_{13/2}$; g) $^2D_{5/2} \rightarrow {}^4I_{11/2}$.

fig. 12 (black line) in the range 375-500 nm. All the observed lines could be ascribed to the Nd^{3+} ions presence from the comparison with the LIF spectrum recorded in the bulk crystal under the same excitation, fig. 12 (dotted line). The lines could be identified according to the Nd^{3+} ion levels in an YLF matrix [40].

It is important to underline that the LIF spectrum induced by the 355 nm excitation is not suitable to investigate which kind of film has been deposited. The laser photons are so energetic that very high Nd^{3+} levels are excited. For this reason any presence of energy shifts due to the Nd^{3+} ions being trapped in different matrixes can not be resolved with this check. The shape of the spectrum simply testifies of the presence of the Nd^{3+} ions. The determination of the kind of crystal grown in the film and of its quality is demanded to the analysis performed *ex situ* by exciting the Nd^{3+} fundamental absorption band $^4I_{9/2} \rightarrow {}^4F_{5/2} + {}^2H_{9/2}$ in the 800 nm range. For all the recorded spectra shown in the following, both the monochromator resolution and the intensity scales for the bulk and the film spectra were maintained, respectively, the same in the two different analyzed regions so that a direct comparison among the corresponding spectra is possible.

We deposited two kinds of crystal films: Nd:YF and Nd:YLF.

4.a. Nd:YF Film

The rotating bulk Nd:YLF crystal was ablated with a laser fluency of 10 J/cm^2 in vacuum (1x 10^{-4} Pa). It was cleaned by 1500 laser shots prior to deposition. The YLF substrate was heated to 750°C in 13 hours. At this temperature the film was deposited by using 12000 laser shots. After the deposition the substrate was cooled down to room temperature again in 13 hours. A picture of the film surface taken with the SEM microscope is shown in fig. 13. The deposition appears quite rough, with islands of maximum ~ 1 μm size scattered over an irregular surface in a way much similar to what shown in ref. [35], even if in our case the grain size appears bigger.

Figure 13. SEM image of the Nd:YF film deposited on the YLF substrate heated to 750°C in vacuum with 10 J/cm^2 laser fluency.

4.a.1. Optical Characterization

Figure 14 (a, b) reproduces the unpolarized fluorescence spectrum recorded from the film after excitation with the diode laser output at 806.6 nm. The laser power at the film surface is 250 mW. Figure14 a) shows the spectrum in the 900 nm region, while that in the 1050 nm range is traced in fig. 14 b). In both cases the unpolarized fluorescence spectrum recorded from the bulk crystal under the same excitation is plotted for comparison. In both regions the film spectra show resolved lines which is a first hint of a crystalline nature of the film. Similar spectra were recorded from 3 different locations on the film surface to get always the same spectral profile. On the other hand the matching between the bulk and the film spectrum is poor. There are lines in the film emission that can not be attributed to the Nd^{3+} ions in an YLF matrix. The typical mark of this mismatch is the appearance of two lines in the short wavelength side of the emission: one at 858 nm for the $^4F_{3/2} \rightarrow {}^4I_{9/2}$ manifolds transition and one at 1037 nm for the $^4F_{3/2} \rightarrow {}^4I_{11/2}$ one. This can be explained by assuming that during the ablation and deposition process a fluoride host other than YLF was formed. Since Nd^{3+} can only substitute the Y^{3+} sites, the obvious candidate was YF. This assumption was supported by the comparison of our measured spectra with the calculation of the wavelengths for the analyzed Nd^{3+} ions transitions in YLF and YF hosts [40, 48]. The results are shown in fig. 15 and listed in table 1. As can be seen the agreement between our measurements and the calculated values for Nd:YF is, for the majority of lines, within 1 nm which is quite satisfactory when comparing data taken with different experimental arrangements and in different conditions. The most striking coincidences are the presence both in the film and in the Nd:YF spectra of two lines below 860 nm and 1040 nm and of three peaks around 1060 nm which can not be ascribed to any Nd:YLF transition. This comparison strongly suggests the Nd^{3+} ions remained embedded in an YF matrix during the film growth. A check of the fluorescence intensity of the $^4F_{3/2} \rightarrow {}^4I_{11/2}$ transition versus the exciting wavelength confirmed this hypothesis. In fact, while the maximum intensity in the Nd:YLF bulk was obtained for

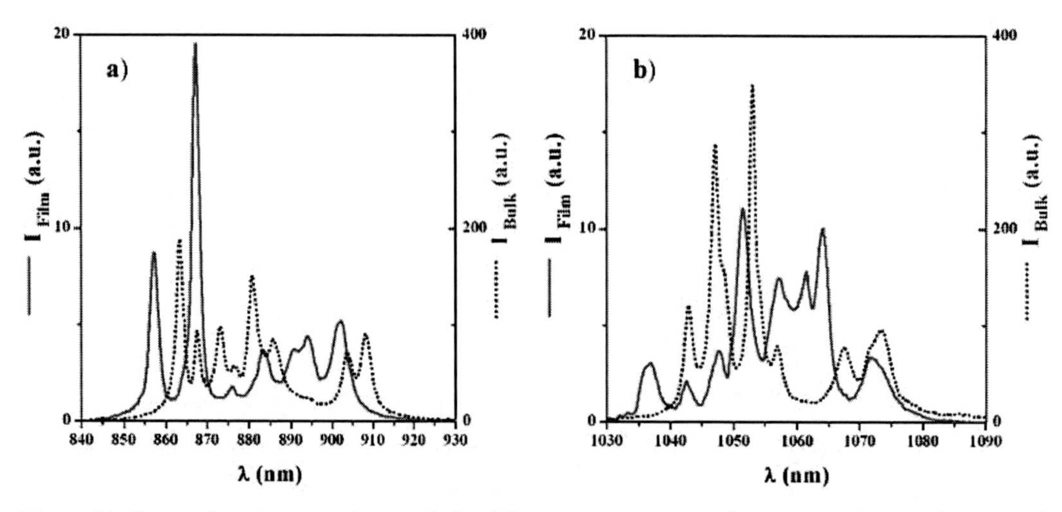

Figure 14. Comparison between the unpolarized fluorescence spectra of the Nd:YF film (solid line, left scale) and that of the bulk crystal (dotted line, right scale) in the regions around 900 nm (a) and 1050 nm (b).

λ_{exc}= 806.6 nm, we realized that in the film it occurred for λ_{exc}= 803 nm which was the minimum wavelength achievable with our excitation system. A shifting of the maximum excitation towards the lower wavelengths is in agreement with the $^4I_{9/2} \rightarrow {}^4F_{5/2} + {}^2H_{9/2}$ transition peaks calculated for Nd:YF according to ref. [40, 48]. Unfortunately we have no Nd:YF crystal to be used as comparison and the available data in the literature for this crystal are limited to the Nd^{3+} ions energy levels and no emission spectrum is reported or analyzed.

Nevertheless we checked also which was the behaviour of the LIF film fluorescence when the latter was polarized with E || or E⊥ to the c- axis of the YLF substrate. Fig. 16 (a, b) shows the two emission spectra recorded according to the two polarizations for both the investigated ranges. It is evident that no fluorescence variation takes place in both the spectral ranges by shifting from one polarization to the other. From these observations we can only infer that the film does not show any preferential orientation somehow related to the substrate one.

Table 1. Values of the calculated $^4F_{3/2} \rightarrow {}^4I_{9/2}$ and $^4F_{3/2} \rightarrow {}^4I_{11/2}$ manifolds transitions wavelengths when the Nd^{3+} ions are embedded either in an YLF matrix or in an YF one, together with the experimental values measured in the Nd:YLF bulk crystal and in the film.

Transition	λ *Nd:YLF(Calc)* [a] (nm)	λ *Nd:YLF Bulk* [b] (nm)	λ *Nd:YF(Calc)* [c] (nm)	λ *Film* [b] (nm)
$^4F_{3/2} \rightarrow {}^4I_{9/2}$	--	--	859.6	857.7
	862.2	863.1	865.3	--
	866.5	867.2	867.3	867.4
	872.1	872.8	--	--
	876.5	876.2	876.2	876.1
	877.0	--	879.0	--
	881.1	880.5	883.1	883.9
	881.4	--	890.3	--
	885.6	885.5	891.4	890.8
	--	--	894.5	894.2
	902.9	903.5	902.1	902.2
	907.6	907.9	--	--
$^4F_{3/2} \rightarrow {}^4I_{11/2}$	--	--	1036.1	1036.6
	1041.7	1042.7	1041.9	1042.6
	1046.5	1047.2	1045.9	--
	1047.9	--	1049.4	1047.3
	1050.5	--	1051.8	1051.4
	1052.7	1053.0	1056.8	--
	1056.9	1056.9	1057.8	1057.2
	--	--	1059.3	--
	1067.2	1067.4	1061.9	1061.5
	1071.4	--	1065.5	1064.3
	1073.8	1073.4	1072.1	1072.8
	1077.9	--	1075.7	--

a) From ref. [40]
b) Data from our spectra
c) From ref. [48]

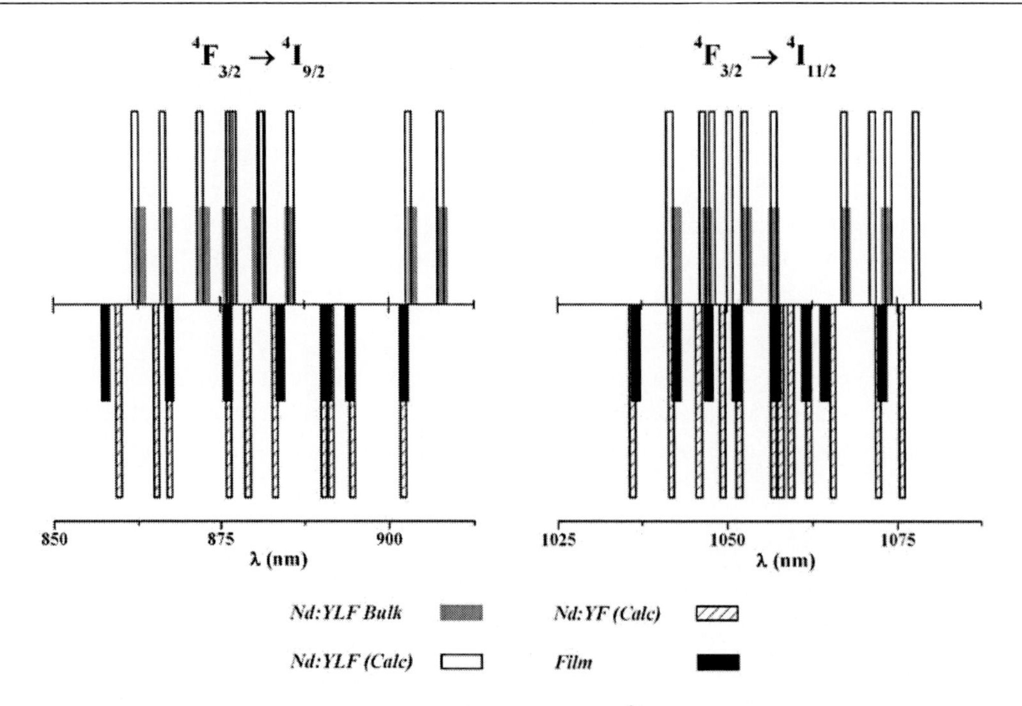

Figure 15. Diagram of the transition wavelengths of the Nd^{3+} ion in different conditions: *Nd:YLF (Calc)*: peaks position of Nd:YLF calculated from the data reported in ref. [40]. *Nd:YLF Bulk*: experimental peaks position corresponding to the dotted line of fig. 14. *Film*: position of the peaks of the film (solid line of fig. 14). *Nd:YF (Calc)*: peaks position of Nd:YF calculated from the data reported in ref. [48].

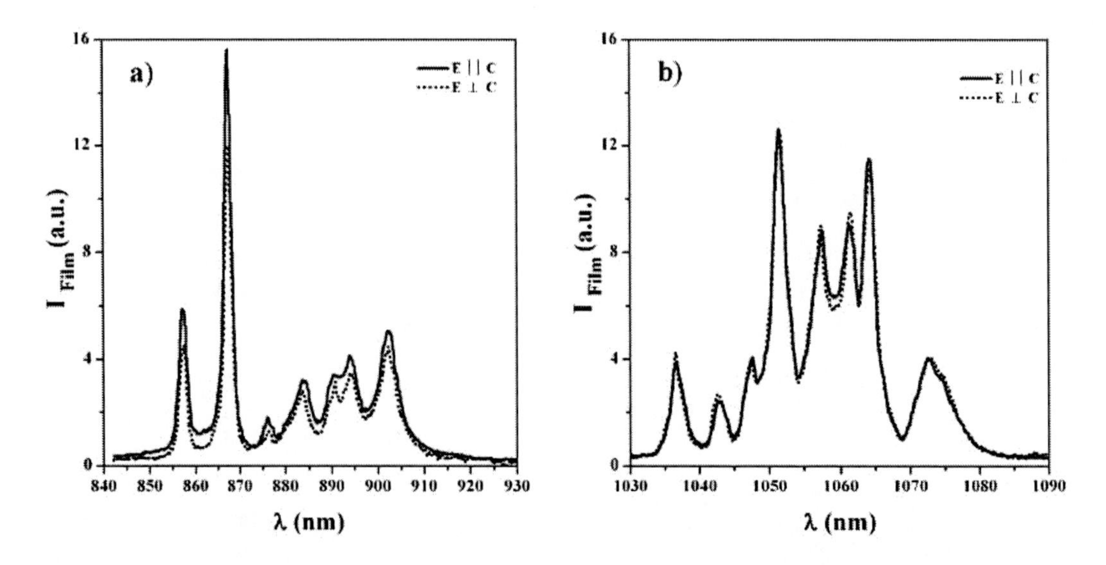

Figure 16. Comparison between the Nd:YF film fluorescence spectra polarized || (solid line) and ⊥ (dotted line) to the substrate c-axis for the $^4F_{3/2} \rightarrow ^4I_{9/2}$ transition (a) and for the $^4F_{3/2} \rightarrow ^4I_{11/2}$ transition (b).

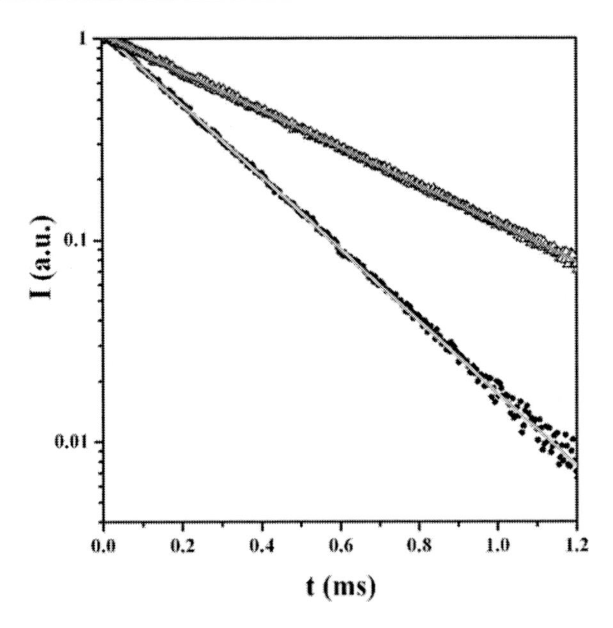

Figure 17. Fluorescence decay signals of the $^4F_{3/2}$ Nd^{3+} manifold in the Nd:YLF (1.5 % at.) crystal (Δ) and in the Nd: YF film (•). The single exponential decay fits (grey lines) are also shown.

We also measured the $^4F_{3/2}$ manifold lifetime in the film by fitting the decay of the time resolved signal of the $^4F_{3/2}$ Nd^{3+} fluorescence and we compared the result with the measurement done in the bulk Nd:YLF (1.5 % at.) crystal. The results are plotted in fig. 17. In both cases the shape is single exponential, but the lifetime values differ considerably. The $^4F_{3/2}$ Nd^{3+} manifold lifetime measured in the film is 242 ± 5 μs, much shorter than the value measured in the Nd:YLF (1.5 % at.) crystal, 462 ± 2 μs. The same measure done in the two other analyzed film surface points gave values within ± 8 %. The value measured in the film can be compared with the only value present in the literature for Nd^{3+} ions embedded in YF matrix [48]. The authors report a value of 240 μs, with the only additional information of the dopant ions concentration in the melt being 1%.

4.b. Nd:YLF Film

A second film was grown in presence of 1Pa He pressure inside the UHV chamber with the same ablation parameters as above and a different substrate temperature. The latter was raised in 12 hours to 650 °C at which temperature the growth took place. At the end the substrate was cooled down to room temperature again in 12 hours. The lack of Li in the precedent film could be related to a too high kinetic energy of the Li species in vacuum. As a possible consequence the latter could bump off from the substrate after the collision. To test this hypothesis, helium was introduced inside the chamber. Helium was chosen to slow down the Li velocity as its mass is the closest to the Li one. Such use of He was already reported in the literature for PLD produced films of LiF [49]. A picture of the film surface taken with the SEM microscope is shown in fig. 18. The presence of a uniform core deposit with islands of up to a few μm dimensions scattered over it well spaced one from the other appears evident. From the comparison with fig. 13 it emerges clearly that the He presence reduced the roughness, in a way similar to what reported in ref. [49], even if in the present film the islands appear as an average bigger and more spaced.

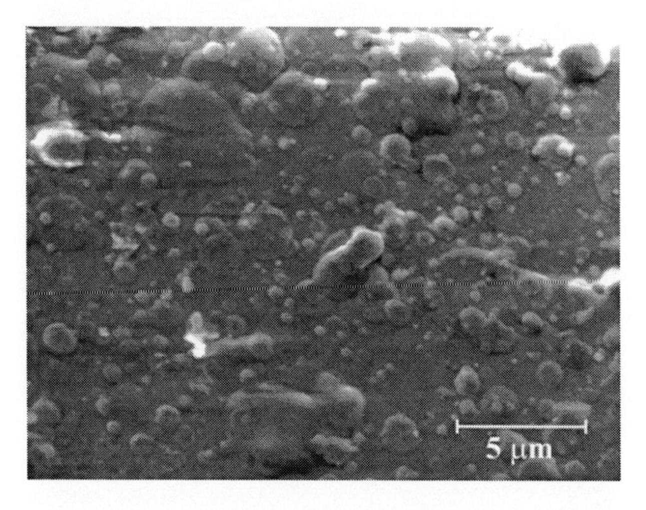

Figure 18. SEM image of the Nd:YLF thin film deposited on the YLF substrate heated to 650°C in presence of 1 Pa He pressure with 10 J/cm^2 laser fluency.

4.b.1. Optical Characterization

This film also was excited by 250 mW of laser diode radiation at 806.6 nm. The unpolarized fluorescence emission in the 900 and 1050 nm regions were compared with the corresponding ones recorded from the target bulk sample under the same excitation. The two spectra

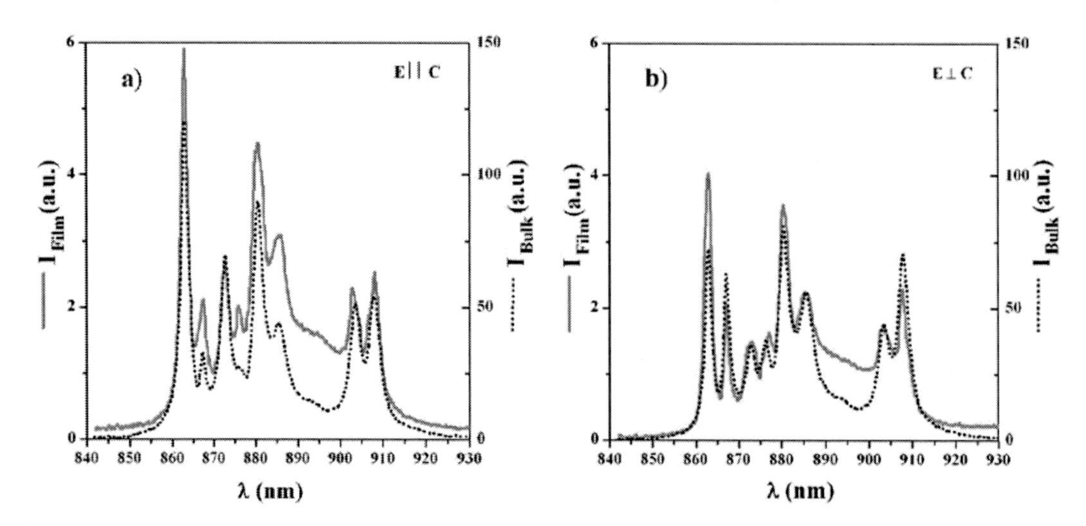

Figure 19. (a) Comparison between the Nd:YLF film fluorescence spectra polarized || to the YLF substrate c-axis (solid line, left scale) and that of the Nd:YLF bulk crystal polarized || to its c-axis (dotted line, right scale). (b) Comparison between the Nd:YLF film fluorescence spectra polarized ⊥ to the YLF substrate c-axis (solid line, left scale) and that of the Nd:YLF bulk crystal polarized ⊥ to its c-axis (dotted line, right scale). The comparison is made for the $^4F_{3/2} \rightarrow {}^4I_{9/2}$ transition.

matched quite well in both ranges which assured that the deposited film was Nd:YLF. Following this first check, the polarized fluorescence spectra of the film with E || or E⊥ to the c- axis of the substrate were recorded and are shown, together with the corresponding E || or E⊥ to the c- axis bulk fluorescence spectra, in fig. 19 for the 900 nm range and in fig. 20 for the 1050 nm range. In both cases the film spectra reproduce all the well resolved lines of the bulk spectra. This behaviour is a strong mark in favour of a crystalline nature of the deposition. In addition the ratio between the intensity of any among the resolved lines and the average intensity of the pedestal underneath them is almost the same in the bulk and in the film spectra. This too is an indication of a large degree of order in the film. Some other important considerations emerge from the comparison between the film and the bulk spectra.

The intensity of the emission in the 900 nm region ($^4F_{3/2} \rightarrow {}^4I_{9/2}$ manifolds transitions, fig. 19) reduces, as an average, both in the film and in the bulk when the fluorescence polarization is shifted from E || to E ⊥ to the c- axis. The profile of the film and of the bulk spectra changes in the same way when the polarization is changed with only a difference in the intensity ratio between the 863 and 867 nm lines which in shifting from E || to E ⊥ to the c- axis in the bulk is reduced twice as much as in the film. For the $^4F_{3/2} \rightarrow {}^4I_{11/2}$ manifolds transitions, fig. 20, both the film and the bulk fluorescence spectra remain the same by changing the polarization and overlap quite well. The only difference here is that for both polarizations the emission at 1047 nm in the film is ~ 20% stronger than that at 1053 nm while in the bulk it is the emission at 1053 nm to be ~ 10% stronger than that at 1047 nm. However, independently on the polarization, the fluorescence in the 1050 nm range is, as an average, both in the film and in the bulk twice stronger than that in the 900 nm range. Also for this film the spectra were recorded from 3 different points on the layer surface, always displaying the same features.

To better characterize the film the $^4F_{3/2}$ manifold lifetime was measured by fitting the $^4F_{3/2}$ fluorescence decay following the pulsed excitation at 806.6 nm of the Nd^{3+} ions, both in the film and in the bulk. The results are plotted in fig. 21. Both signals show a single exponential decay whose fit gives the value of 462 ± 2 μs for the bulk and 468 ± 5 μs for the film. This is an indication of a Nd^{3+} ions concentration in the film very close to that of the bulk, as the $^4F_{3/2}$ manifold lifetime value depends on this parameter. Lower concentrated samples manifest higher lifetimes and vice versa [50]. The Nd^{3+} $^4F_{3/2}$ manifold lifetime measured in the film is position dependent with a maximum 6% difference close to the edge of the layer where the clips can perturb the deposition process.

From all these checks it emerges that the grown film has a large degree of order and manifests several features pointing towards a consistent, if not complete, single crystalline nature. In addition it seems that the substrate orientation is transferred to a certain amount to the film, even if the few mismatches between the film and the bulk spectra underlined above prevent a 100% identification of the film orientation based only on spectroscopic analysis. A transfer of the Nd^{3+} ions concentration from the bulk to the film seems also to be occurred during the ablation/deposition process.

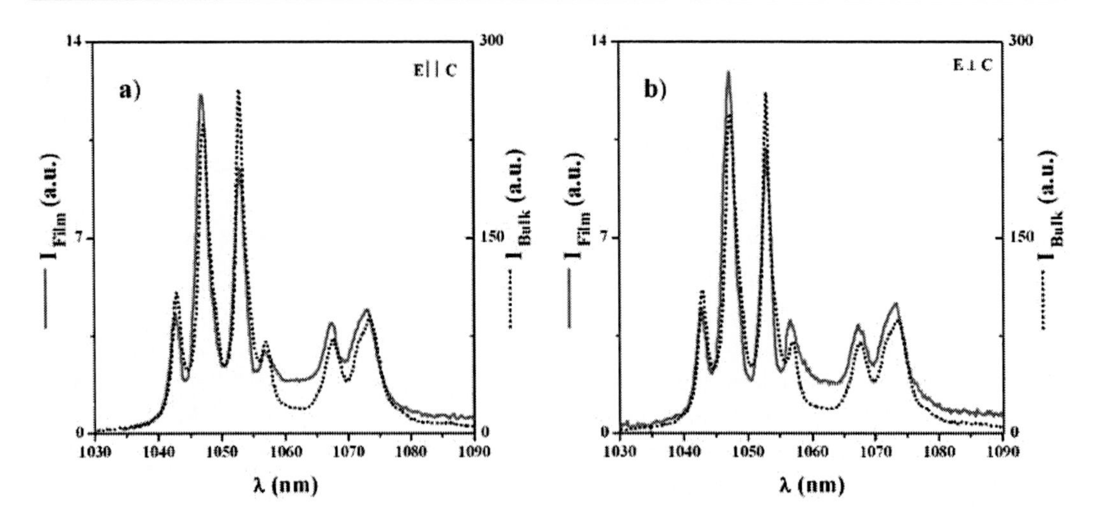

Figure 20. (a) Comparison between the Nd:YLF film fluorescence spectra polarized || to the YLF substrate c-axis (solid line, left scale) and that of the Nd:YLF bulk polarized || to its c-axis (dotted line, right scale). (b) Comparison between the Nd:YLF film fluorescence spectra polarized ⊥ to the YLF substrate c-axis (solid line, left scale) and that of the Nd:YLF bulk polarized ⊥ to its c-axis (dotted line, right scale). The comparison is made for the $^4F_{3/2} \rightarrow {}^4I_{11/2}$ transition. The arbitrary units for the bulk and the film fluorescence intensities are, respectively, the same as in fig. 19.

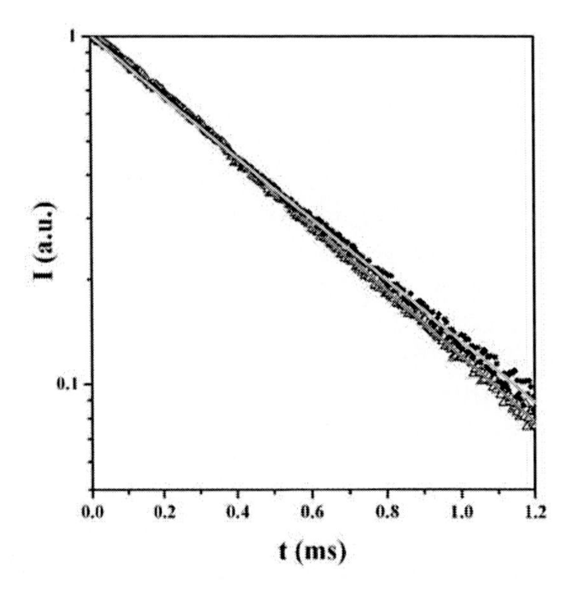

Figure 21. Fluorescence decay signals of the $^4F_{3/2}$ Nd^{3+} manifold in the Nd:YLF (1.5 % at.) crystal (Δ) and in the Nd: YLF film (•). The single exponential decay fits (grey lines) are also shown.

5. Conclusion

This chapter describes the first realization of crystalline nanometric films grown by pulsed laser deposition on a pure monocrystalline YLF heated substrate starting from the laser ablation of a Nd:YLF (1.5% at.) monocrystalline bulk crystal. The films have been optically characterized and a temporal analysis of their IR fluorescence has also been performed to measure the Nd^{3+} $^4F_{3/2}$ manifold lifetime. When PLD takes place in vacuum and the substrate temperature is 750°C, the optical characterization indicates the film grown is Nd:YF and pictures taken with a SEM microscope show a rather irregular surface with grains of maximum ~ 1μm dimensions scattered all over.

A much more promising result is obtained when PLD takes place in presence of a 1Pa He pressure and the substrate temperature is 650°C. In this case the optical characterization indicates the film grown is Nd:YLF. The comparison between the polarized IR film LIF spectra with the corresponding bulk ones shows a considerable degree of orientation in the film associated to a crystalline nature of the deposition which appears, from its optical behaviour, to some extent already monocrystalline. Moreover the Nd^{3+} $^4F_{3/2}$ manifold lifetime measurement indicates that the dopant concentration was transferred from the bulk to the film in the ablation/deposition process. The morphological analysis done with the SEM presents a considerably less rough surface with a core uniform background deposit with islands of up to a few μm dimensions scattered over it well spaced one from the other.

In spite of the islands structure of the deposition, this film shows a reasonable degree of order and of orientation and is very encouraging for the feasibility of monocrystalline rare earth doped fluoride sub-micrometric layers with promising optical qualities. Experiments are in progress to produce other depositions with the modification of some fundamental parameters such as the ablation with lower laser fluency and a larger number of shots. There are reports in the literature showing that closer is the laser fluency to the ablation threshold, less rough is the grown surface [51]. Other parameters that can be changed are the ablation wavelength, the concentration of the dopant ions in the bulk crystal, the laser pulse duration and so on. Other host fluorides can also be used as well as other lanthanides dopant ions to reach the goal of the realization of monocrystalline nanofilms with the optical quality necessary to be used as active optical devices in the 1 - 2 μm range.

Acknowledgments

Thanks are due to Prof. R. Dallai of the Department of Evolutive Biology, University of Siena for letting us use the SEM and to Dr. Eugenio Paccagnini and Mr. Alessandro Gradi who helped us in taking the SEM pictures.

Both the target (Nd:YLF) and the substrate (YLF) crystals were grown and provided by the crystal growth facility set up at the Department of Physics, University of Pisa.

References

[1] Bonner, C. L.; Anderson, A. A.; Eason, R. W.; Shepherd, D. P.; Gill, D. S.; Grivas, C.; Vainos, N. *Opt. Lett.* 1997, 22, 988-990.

[2] Gill, D. S.; Anderson, A. A.; Eason, R. W.; Warburton, T. J.; Shepherd, D. P. *Appl. Phys. Lett.* 1996, 69, 10-12.

[3] Burkhalter, R.; Dohnke, I.; Hulliger, J. *Prog. Cryst. Growth Char. Mater.* 2001, 42, 1-64.

[4] Bunshah, R. F. *Handbook of Deposition Technologies for Films and Coatings*; 2nd Edition, William Andrew Publishing: Norwich, (NY), 1994.
[5] Torrison, L.; Tolle, J.; Tsong, I. S. T.; Kouvetakis, J. *Thin Solid Films* 2003, 434, 106-111.
[6] Chartier, I.; Ferrand, B.; Pelenc, D.; Field, S. J.; Hanna, D. C.; Large, A. C.; Shepherd, D. P.; Tropper, A. C. *Opt. Lett.* 1992, 17, 810-812.
[7] Pierson, H. O. *Hand Book of Chemical Vapor Deposition, Principles, Technology and Applications*; 2nd Edition, William Andrew Publishing: Norwich, (NY), 1999.
[8] Ji, L.; Su, Y.; Chang, S.; Wu, L.; Fang, T.; Xue, Q.; Lai, W.; Chiou, Y. *Mater. Lett.* 2003, 57, 4218-4221.
[9] Lugstein, A.; Weil, M.; Basnar, B.; Tomastik, C.; Bertagnolli, E. *Nucl. Instr. and Meth.* B 2004, 222, 91-95.
[10] Maurizio, C. *Nucl. Instr. and Meth.* B 2004, 218, 396-404.
[11] Chrisey, D. B.; Hubler, G. K., *Pulsed Laser Deposition of Thin Films*; Wiley: New York, (NY), 1994.
[12] Ashfold, M. N. R.; Claeyssens, F.; Fuge, G. M.; Henley, S. J. *Chem. Soc. Rev.* 2004, 33, 23-31.
[13] Willmott, P. R.; Huber, J. R. *Rev. Mod. Phys.* 2000, 72, 315-328.
[14] Afonso, C. N. In *Insulating Materials for Optoelectronic*; Agullo-Lopez, F.; Ed.; World Scientific: Singapore, 1995.
[15] Shen, J.; Gai, Z., Kirschner, J. *Surf. Sci. Rep.* 2004, 52, 163-218.
[16] Pollnau, M.; Romanyuk, Y. E. *C. R. Physique* 2007, 8, 123-137.
[17] Fan, H.; Reid, S. A. *Chem. Mater.* 2003, 15, 564-567.
[18] Fu, Z. W.; Zhou, M. F.; Qin, Q. Z. *Appl. Phys.* A 1997, 65, 445-449.
[19] Wolf, P. J. *Appl. Phys.* A 1996, 62, 553-558.
[20] Calì, C.; Macaluso, R.; Mosca, M. *Spectrochim. Acta* B 2001, 56, 743-751.
[21] Calì, C.; Macaluso, R.; Mosca, M. *Opt. Commun.* 2001, 197, 341-354.
[22] Giardini, A.; Marotta, V.; Morone, A.; Orlando, S.; Parisi, G. P. *Appl. Surf. Sci.* 2002, 197-198, 338-342.
[23] Saenger, K. L. In *Pulsed Laser Deposition of Thin Films*; Chrisey, D. B.; Hubler, G. K.; Wiley: New York, (NY), 1994; pp 581-604.
[24] Chattopadhyay, S.; Ayyub, P.; Multani, M; Pinto, R. *J. Appl. Phys.* 1998, 83, 3911-3913.
[25] Khakani, M. A. E.; Dolbec, R.; Serventi, A. M.; Horrillo, M. C.; Trudeau, M.; Saint-Jacques, R. G.; Rickerby, D. G.; Sayago, I. *Sens. Actuators* B 2001, 77, 383-388.
[26] Martino, M.; Caricato, A. P.; Fernández, M.; Leggieri, G.; Jha, A.; Ferrari, M.; Mattarelli, M. *Thin Solid Films* 2003, 433, 39-44.
[27] Anderson, A. A.; Eason, R. W.; Jelinek, M.; Grivas, C.; Lane, D.; Rogers, K.; Hickey, L. M. B.; Fotakis, C. *Thin Solid Films* 1997, 300, 68-71.
[28] Sonsky, J.; Lancok, J.; Jelinek, M.; Oswald, J.; Studnikca, V. *Appl. Phys. A* 1998, 66, 583-586.
[29] Yi, S.; Bae, J. S.; Choi, B. C.; Shim, K. S.; Yang, H. K.; Moon, B. K.; Jeong, J. H.; Kim, J. H. *Opt. Mater.* 2006, 28, 703-708.
[30] Bar, S.; Scheife, H.; Huber, G. *Opt. Mater.* 2006, 28, 681-684.
[31] Kuzminykh, Y.; Kahn, A.; Huber, G. *Opt. Mater.* 2006, 28, 883-887.
[32] Kaminskii, A. A. *Laser Crystals*; Springer-Verlag: New York, (NY), 1981.

[33] Ferrand, B.; Chambaz, B.; Couchaud, M. *Opt. Mater.* 1999, 11, 101-114.

[34] Renard, S.; Camy, P.; Doualan, J. L.; Moncorge, R.; Couchaud, M.; Ferrand, B. *Opt. Mater.* 2006, 28, 1289-1291.

[35] Hong, B.; Kawano, K. *J. Alloys. Compd.* 2006, 408-412, 838-841.

[36] Qin, G.; Qin, W.; Wu, C.; Huang, S.; Zhang, J.; Lu, S.; Zhao, D.; Liu, H. *J. Appl. Phys.* 2003, 93, 4328-4330.

[37] Camposeo, A.; Fuso, F.; Arimondo, E.; Toncelli, A.; Tonelli, M. *Surf. Coat. Technol.* 2004, 180-181, 607-610.

[38] Xiao, S.; Yang, X.; Yang, L.; Yang, Y.; Liu, Z. *J. Mater. Sci. Lett.* 2002, 21, 1139-1141.

[39] Douysset-Bloch, L.; Ferrand, B.; Couchaud, M.; Fulbert, L.; Joubert, M. F.; Chadeyron, G.; Jacquier, B. *J. Alloy. Compd.* 1998, 275-277, 67-71.

[40] Kaminski, A. A. *Crystalline lasers: physical processes and operating schemes*; CRC Press, Boca Raton, 1996.

[41] Ryan, J. R.; Beach, R. *J. Opt. Soc. Am.* B 1992, 9, 1883-1887.

[42] Cornacchia, F.; Di Lieto, A.; Maroni, P.; Minguzzi, P.; Toncelli, A.; Tonelli, M.; Sorokin, E.; Sorokina, I. *Appl. Phys.* B 2001, 73, 191-194.

[43] Barsanti, S.; Cornacchia, F.; Lieto, A. D.; Favilla, E.; Toncelli, A.; Bicchi, P. In *Atoms & Molecules in Laser & Other External Fields*; Mohan M.; Ed.; Narosa Publishing House Pvt. Ltd: New Delhi, in Press (March 2008).

[44] Barsanti, S.; Cornacchia, F.; Lieto, A. D.; Toncelli, A.; Tonelli, M.; Bicchi, P. *Thin Solid Films* 2007, doi: 10.1016/j.tsf.2007.06.144.

[45] Barsanti, S.; Favilla, E.; Bicchi, P. *Rad. Phys. Chem.* 2007, 76, 512-515.

[46] Geohegan, D. In *Pulsed Laser Deposition of Thin Films*; Chrisey, D. B.; Hubler, G. K.; Wiley: New York, (NY), 1994; pp 129-137.

[47] Dang, H.; Zhou, M.; Qin, Q. *Appl. Surf. Sci.* 1999, 140, 118-125.

[48] Martin, I. R.; Guyot, Y.; Joubert, M. F.; Abdulsabirov, R. Y.; Korableva, S. L.; Semashko, V. V. *J. Alloy. Compd.* 2001, 323-324, 763-767.

[49] Perea, A.; Gonzalo, J.; Afonso, C. N.; Martelli, S.; Montereali, R. M. *Appl. Surf. Sci.* 1999, 138-139, 533-537.

[50] Pinto, A. A.; Fan, T. Y. O*SA Proceedings on Advanced Solid-State Lasers;* Optical Society of America: Washington, DC, 1993; vol. 15, pg. 91.

[51] Henley, S. J.; Ashfold, M. N. R.; Pearce, S. R. J. *Appl. Surf. Sci.* 2003, 217, 68-77.

In: Radiation Physics Research Progress
Editor: Aidan N. Camilleri, pp. 219-246

ISBN: 978-1-60021-988-7

Chapter 5

RADIATION-INDUCED POLLUTION CONTROL

Marilena Radoiu*

BOC Edwards, Exhaust Management Systems, Kenn Business Park,
Kenn Road, Clevedon BS21 6TH, United Kingdom

Abstract

The effect of radiation induced chemistry as a potential alternative to conventional chemistry for air and water pollution control is studied in this paper. The precipitation of sulphur dioxide (SO_2) and nitrogen oxides (NO_X) by irradiation with accelerated electron beams and/or 2.45 GHz microwave was studied in the presence of ammonia and water as reagents. The same technology was studied for the preparation of polymers used for the treatment of wastewaters from the food industry. An original experimental unit easily adaptable for both separate and simultaneous irradiation with accelerated electron beams and microwaves was built for this research. It was shown that the application of synergetic methods such as combined accelerated electron beam (EB) and microwave (MW) results to a significant reduction of electrical energy consumption for the cleaning of flue gases as well as to the obtaining of polymers with improved properties such as higher molecular weight and good water solubility.

Key words: electron beam; microwaves; pollution control; polyelectrolytes; flue gas.

Introduction

There is increased worldwide concern that human activities are endangering – perhaps permanently – the quality of the environment. The public is increasingly aware of the environmental damage caused by air and water pollutants as pesticides, toxic wastes, chlorofluorocarbons, nuclear radiation, oil spills etc.

In order to achieve compliance with future legislative limits intended to reduce pollution, the industry must have a wide selections of abatement technologies from which to choose, given the different designs, ages and capacities of existing plants and procedures. Thus, the

* E-mail address: Marilena.Radoiu@bocedwards.com; Tel. +44 1275 337100; Fax +44 1275 337200;

development of cost effective and clean processes through pollution prevention and waste treatment with enhanced efficiencies is an important chemical engineering task.

1. Air Pollution Control

It is well known that exhaust gases from the manufacturing, transportation and utility industries can contribute to ozone-induced smog and acid rain, as well as to the climate change by their potential global warming and atmospheric ozone depletion mechanisms. Increasingly stringent environmental regulation imposed on both large and small industrial companies has created a growing demand for alternative abatement methods for a variety of hazardous air pollutants.

1.1. Sources of Air Pollution

Most pollutant gases have both natural and human-made emissions; there are significant natural mechanisms (land-based or ocean-based sinks) for removing these gases from the atmosphere. However, increased levels of human-made emissions have pushed the total level of pollutant emissions (both natural and human-made) above the natural absorption rate for these pollutants. This positive imbalance between emissions and absorption has resulted in the continuing growth in atmospheric concentration of these pollutants.

Acidic emissions of sulphur dioxide (SO_2) and nitrogen oxides (NO_X) arise from many industrial sources as a result of combustion processes. Emissions of SO_2 from industry result from power stations burning fossil fuels (i.e. coal and oil) that contain sulphur, some refinery processes and the production of sulphuric acid. Industries also emit NO_X (i.e. NO and NO_2) that can also cause rainfall to become more acidic; agricultural activities (nitrogen fertilization of agricultural soil), road transport and industrial production of adipic and nitric acids are the major sources of NO_x emissions.

1.1.1. Methods for SO_2 and NO_X Removal

Conventional techniques for reducing NO_X and SO_2 emissions to the atmosphere can be classified into two fundamentally different categories: combustion controls and post-combustion controls.

Combustion controls reduce NO_X and SO_2 formation during the combustion process, while post-combustion controls reduce NO_X and SO_2 after they have been formed [1, 2].

NO_x Control Technologies

Combustion controls for NO_X include low-NO_X burners (LNBs), re-burning, over-fire air (OFA), flue gas recirculation (FGR), and operational modifications. Post-combustion controls include selective catalytic reduction (SCR) and selective non-catalytic reduction (SNCR).

SO_2 Control Technologies

Most SO_2 control technologies involve the addition of a calcium- or sodium-based sorbent to the system. Under the proper conditions, these materials react with SO_2 and sulphur trioxide (SO_3) to form sulphite and sulphate salts. Sometimes the sorbent is injected directly into the furnace or flue gas duct, where the dry particles react with SO_2 and are subsequently removed by the boiler's particulate control device. This is known as Dry Sorbent Injection (DSI). In other cases, the sorbent is dissolved in or slurried with water, and the flue gas contacts the solution or slurry in a scrubber. This approach is referred to as wet flue gas desulphurization (FGD). Because of their low cost, limestone and lime are the most frequently used sorbents. Another approach, less frequently used, is to oxidize the SO_2 to SO_3 over a catalyst and absorb the SO_3 in water to form sulphuric acid.

As one of these techniques used to remove SO_2 and NO_X from exhaust gases, electron beams have demonstrated their technical and economical feasibility compared to commercially available techniques [3-11]. In this method, the energy of the electron beam is used directly to dissociate and ionize the background gas. During the ionization by the beam, a shower of secondary electrons is produced, which further produce a cascade of ionization and dissociation processes. This cascading effect produces a large volume of active species that can be used to initiate the removal of NO_X and SO_2. Other important features of this method consist of:

- Both SO_2 and NO_X can be simultaneously removed when the flue gas is irradiated with an electron-beam accelerator;
- Continuous dry treatment of the flue gas is possible by irradiating the gas for a short time;
- When ammonia (NH_3) is used as a reagent, valuable end by-products, such as ammonium-nitrates NH_4NO_3 and ammoniumsulphates $(NH_4)_2SO_4$ are obtained;
- Integration with an electrostatic precipitator (ESP) installed downstream of the electron beam reactor to remove solid by-products.

To achieve simultaneous NO_X and SO_2 removal, minimum energy consumption usually depends on the NO_X removal rate and its initial concentration. Various kinds of additives, such as NH_3, H_2O_2, hydrocarbons, N_2H_4, natural gas and hydrated lime have been used to improve the energy efficiency and to control final by-products. Although treatment with accelerated electron beams is considered one of the most effective methods for simultaneous abatement of SO_2 and NO_X, the electrical energy consumption for effective purification of flue gases is very high; it requires 2-4% of the capacity of the power station. Thus, reduction of electrical energy consumption for the cleaning of flue gases is important and could be solved by the application of synergetic methods such as combined accelerated electron beam (EB) and microwave (MW) induced non-thermal plasma irradiation.

The initial electrons from an electron beam accelerator are monoenergetic and the order of a few keV to MeV. These electrons react with the background gas *via* the following reactions:

$$
\begin{aligned}
&e + AB \rightarrow AB^{+} + 2e && \text{Direct ionization} \\
&e + AB \rightarrow A^{+} + B + 2e && \text{Dissociative ionization} \\
&e + AB \rightarrow A + B + e && \text{Dissociation} \\
&e + AB \rightarrow AB^{-} + h\nu && \text{Radiative attachment} \\
&e + AB \rightarrow A^{-} + B && \text{Dissociative attachment} \\
&e + AB \rightarrow AB^{*} + e && \text{Excitation}
\end{aligned}
\tag{1}
$$

These reactions (inelastic) as well as elastic collision with thermal electrons will generate the plasma. An electron beam accelerator can generate large volume plasmas since most collision cross sections have a maximum in the range of 10 to 100 eV, and hence electron beams can penetrate relatively deep in a flow of gas to generate ions and active species.

Physically, non-thermal equilibrium in atmospheric pressure discharges is a relatively marginal situation. These electrical discharges achieve non-thermal conditions through the production of microdischarges called streamers. Streamers are plasma filaments produced by highly localized space-charge waves. These space-charge waves enhance the applied field in front of the wave and propagate because of electron avalanching in this high field. To date, according to Rea [12] and Veldhuizen [13], environmental applications have mainly involved corona and dielectric barrier discharge (DBD) systems. However, electronic densities in these types of plasmas ($10^9 - 10^{11}$) are not sufficient to achieve appropriate conversion rates of the toxic gases.

Microwave excited plasmas belong to a fundamentally different category. Non-equilibrium results from the fact that at a sufficiently high frequency, electrons respond solely to the applied electromagnetic field. At atmospheric pressure, microwave plasmas exhibit homogeneous densities of $10^{12} - 10^{15}$ cm^{-3} and therefore appear much more attractive for the abatement of toxic gases. Microwave plasmas can be excited inside resonant cavities, within waveguide microwave circuits or by means of surface wave field applicators [14]. The use of microwave energy in addition to electron beam energy, or even microwave energy only, to sustain the energy of the free electrons at optimum levels, depends on several parameters such as electric field amplitude, field distribution, energy distribution, and reaction vessel (applicator) geometry [15].

The aim of this study was to demonstrate the ability of simultaneous irradiation with electron beam and microwaves to abate NO_X and SO_2 and to explore the mechanisms involved in the process in order to optimise it from the point of view of both the removal efficiency and the input energy at a temperature below 70 ^{0}C.

1.2. Experimental

1.2.1. Experimental System

A schematic diagram of the experimental system is shown in Fig. 1. Cylinders with oxygen *1*, nitrogen *2*, sulphur dioxide *3* (SO_2, >99%), nitric oxide *4* (NO, >99%) and carbon dioxide *5* (CO_2, >99%) were purchased from L'Air Liquide, Romania. Individual mass flow controllers *1', 2', 3', 4', 5'* were used to achieve the desired concentration of SO_2, NO and CO_2 in air. A specially made glass chamber *6* packed with glass spheres *7* was used to mix all gases. The resulted flow was passed via a three-way valve *8* into the pre-heating chamber *9* thermostatically controlled, then through a second mixing reactor *11* where a solution of ammonia *10* was added in stoichiometric amount calculated as follows:

$$C_{NH_3} = 2C_{SO_{2\,initial}} + C_{NO_{initial}} \tag{2}$$

where $C_{initial}$ is the initial molar concentration of SO_2 and NO, respectively.

The conditioned gas was treated with microwave and/or electron beams in the irradiation reactor *12*, filtered *13*, neutralized with a solution of sodium hydrogen carbonate *14*, and evacuated into the acid exhaust. The gas effluent was monitored using an on-line quadrupole mass spectrometer *QMS* HAL 200 Gas Analyser, Hidden Analytical, U.K.

Typical composition of the synthetic gas mixture was: oxygen (10%), water (12 – 18%), CO_2 (10%), SO_2 (to 2000 ppm), and NO (to 1000 ppm) and nitrogen (balance). In all experiments, the simulated flue gas was circulated continuously at a flow rate of 10 l/min.

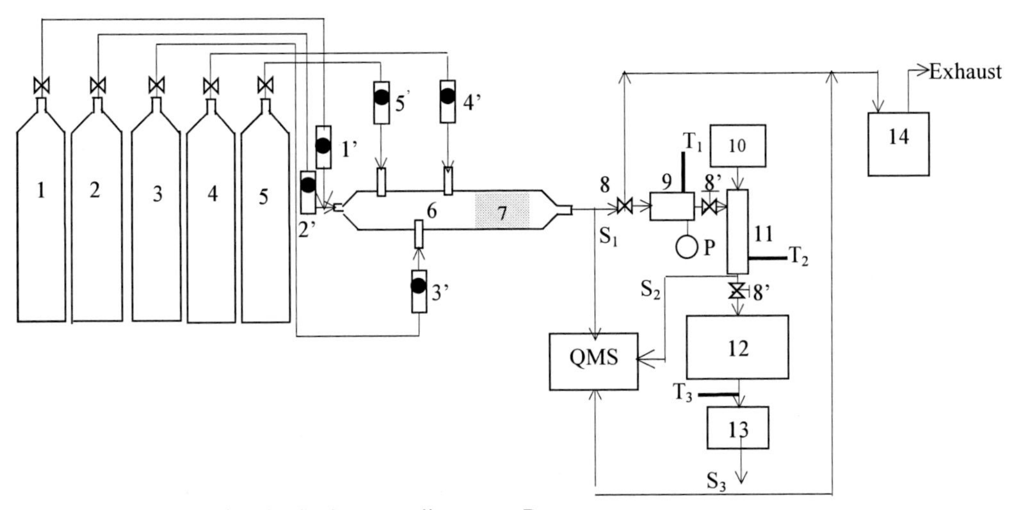

T_1, T_2, T_3 – thermocouples; S_1, S_2, S_3 – sampling ports; P – pressure gauge

1, 2, 3, 4, 5 – Gas cylinders (O_2, N_2, SO_2, NO and CO_2)
1', 2', 3', 4', 5' – Mass flow controllers
6 – Glass chamber
7 – Glass spheres
8 – Three way valve
8' – Isolation valves
9 – Pre-heating chamber
10 – Ammonia solution reservoir
11 – Glass mixing reactor
12 – Irradiation chamber (plasma reactor)
13 – Filter
14 – Neutralization vessel (Dreschel bottle with 10% sol. $NaHCO_3$)

Figure 1. Schematic diagram of the experimental system.

1.2.2. Irradiation Reactor

Laboratory experimental methods and procedures in microwave and accelerated electron beam treatments have been continuously evolving [16]. Early experiments to demonstrate the viability of the technique and to identify and test some of the basic ideas governing the use of microwave power in gaseous systems containing SO_2 were illustrated using commercial microwave ovens modified to contain a quartz reactor and connected to a vacuum line for injection of gaseous reagents. At that time, it was found that the interaction of the microwave energy with the reagents used in experiments (SO_2, H_2O and/or NH_3) was so efficient that it was necessary to reduce the energy in the reactor in order to control the temperature and minimize the formation of NO_X. Therefore a cylindrical microwave cavity (CMC) with controlled variable power 0 – 850 W was built [17]. Fig. 2a shows the CMC, a brass tube 90 mm i.d. and 1000 mm long containing a concentric quartz tube. Various diameters – 20 to 50 mm - of quartz tube were tested. The CMC was provided with properly sized end microwave chokes and a very efficient water-cooling system. The cavity was excited with a rectangular waveguide propagating the microwave electric field parallel to its axis. The microwave injection system consisted of a controlled microwave generator (2.45 GHz and 850 W maximum output power) and a waveguide launcher. Destruction efficiencies at ~80% for SO_2 and 30% for NO_X were obtained in the experimental conditions described above [18]. However, the stability of the system decreased while the SO_2 and NO_X were converted to ammoniumsulfates and nitrates. The salts formed in the chemical reaction built solid deposits on the reactor walls which affected the intensity of the discharge and required the quartz tube to be washed with water. Fig. 2b shows the quartz tube with a periodical deposit of white powdery material on the inner surface.

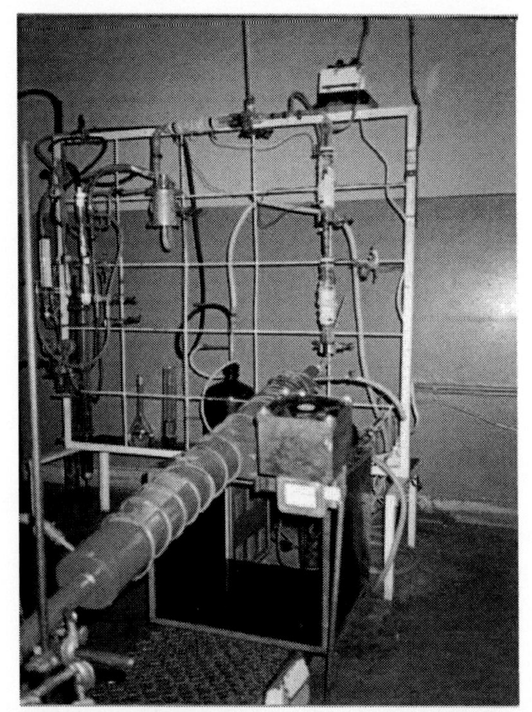

Figure 2a. Cylindrical microwave cavity (CMC).

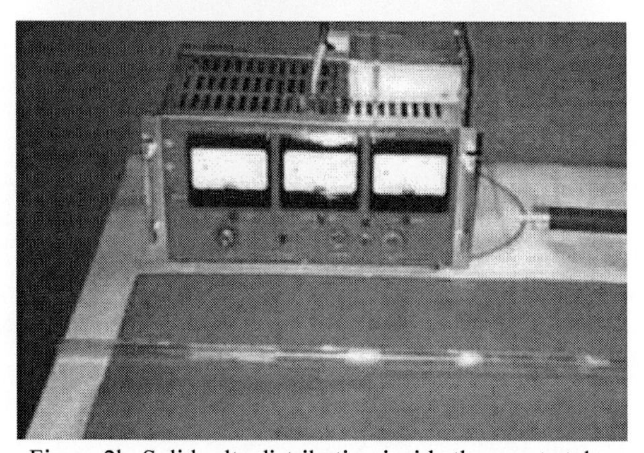

Figure 2b. Solid salts distribution inside the quartz tube.

Figure 2. Cylindrical microwave cavity.

Promising results obtained led us to develop a more suitable test unit (irradiation reactor), designed to facilitate the use of both separate and simultaneous electron beam and microwave energies to produce free radicals.

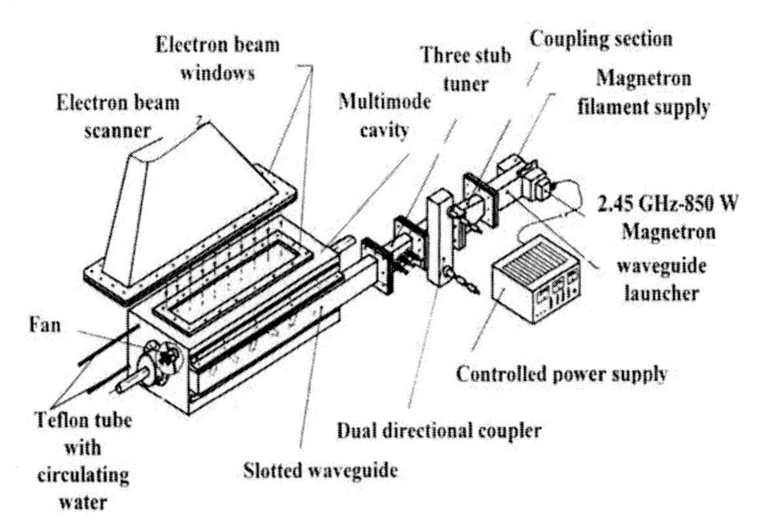

3a. Schematic.

3b. Photograph.

Figure 3. Electron beam and/or microwave irradiation reactor.

The source of electron beams was an industrial linear accelerator, ALID-7, built in the Electron Accelerator Laboratory of the Institute of Atomic Physics, Bucharest [19]. The calibration was made by several chemical systems such as Ceric and Chlorobenzene Dosimeters [20]. The values of the maximum beam power (P_B) and the optimum values for the peak beam current (I_B) and electron energy (E_B) to produce maximum output beam power are as follows: P_B = 670 W (f_r = 250 Hz and t_B = 3.75 μs), I_B = 130 mA, E_B = 5.5 MeV. The microwave injection system consists of a microwave power controlled generator with 2.45 GHz magnetron of 1200 W maximum power output, a rectangular waveguide launcher fitted to a WR430 waveguide, a dual directional coupler for forward and reflected power monitoring, a three stub tuner for impedance matching, and a slotted waveguide (inclined slots cut in the broad wall of a WR430 waveguide) used as a radiating antenna.

In order to minimize the dielectric losses in the internal walls, the rectangular multimode reaction cavity of 240 mm x 240 mm x 450 mm (inner dimensions) was made out off aluminium. The scanned electron beams are introduced perpendicular to the upper-end plate through a 100 μm thick titanium window. The method of coupling the microwave power to the cavity via the slotted waveguide provided good microwave energy transfer and uniformity over a large area. The gas mixture entrance and exit are branched through two sieve metallic end plates. A schematic drawing and a photograph of the experimental unit are shown in Figs. 3a and b.

1.3. Results and Discussion

The reactions taking place in the gas phase within and subsequent to the irradiation can be divided into two categories. The first category contains all those reactions resulting from the direct action of the high energy electrons with the gas molecules resulting in ionization and dissociation.

$$\begin{aligned} &N_2 \rightarrow N_2^+ + e^- \\ &O_2 \rightarrow O_2^+ + e^- \\ &H_2O \rightarrow H_2O^+ + e^- \\ &N_2 \rightarrow 2N \\ &O_2 \rightarrow 2O \\ &H_2O \rightarrow OH + H \end{aligned} \tag{3}$$

Subsequent reactions take place with NO_X and SO_2 as follows:

For NO_X ($NO + NO_2$)

$$\begin{aligned} &NO \xrightarrow{OH,O,HO_2} HNO_2 \\ &NO + O \rightarrow NO_2 \\ &NO + N \rightarrow N_2 + O \\ &NO_2 \xrightarrow{OH,O,H_2O} HNO_3 \\ &NO_2 + N \rightarrow N_2O + O \end{aligned} \tag{4}$$

For SO_2

$$\begin{aligned} &SO_2 \xrightarrow{OH,O,HO_2} HSO_3, HSO_4 \\ &SO_3 \xrightarrow{OH,HO_2,H_2O} H_2SO_4 \end{aligned} \tag{5}$$

Some of these products combine with ammonia to form salts.

$$HNO_3 + NH_3 \longrightarrow NH_4NO_3$$
$$H_2SO_4 + 2NH_3 \longrightarrow (NH_4)_2SO_4 \qquad (6)$$

Tests were carried out to estimate the efficiencies of NO_X and SO_2 removal from the flue gases in different irradiation conditions, and to estimate the reduction of the energy consumption required to obtain the same purification effect both by irradiation with electron beam only and simultaneous by irradiation with electron beam and microwave power.

The NO_X and SO_2 destruction and removal efficiencies (DRE_{NO_X} and DRE_{SO_2}) were calculated with the following equations:

$$DRE_{NO_X}(\%) = \left(1 - \frac{C_{NO_{X\,exit}}}{C_{NO_{X\,initial}}}\right) \times 100 \text{ and } DRE_{SO_2}(\%) = \left(1 - \frac{C_{SO_{2exit}}}{C_{SO_{2initial}}}\right) \times 100 \qquad (7)$$

where $C_{initial}$ and C_{exit} are the initial and final molar concentrations of NO_X ($NO + NO_2$) and SO_2, respectively.

In all experiments, the irradiation was carried out until the gas composition reached a steady state.

1.3.1. Spontaneous Reaction of SO_2 with Ammonia

In the removal process of SO_2 from flue gases by precipitation with electron beams, ammonia is added into the flue gas just before irradiation. It is well known that SO_2 reacts spontaneously with ammonia without additional energy. To determine the amount of SO_2 removed via

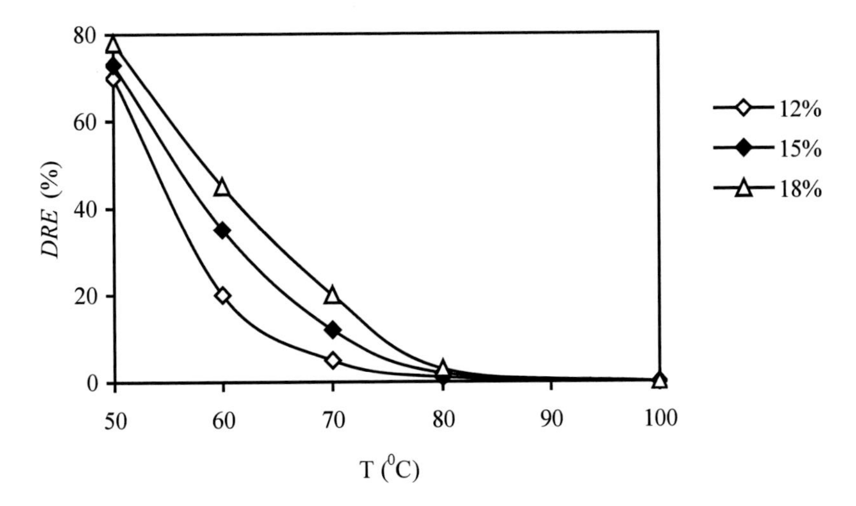

Figure 4. Removal efficiency of SO_2 by spontaneous reaction with NH_3 vs. reaction temperature and water content $C_{SO_2initial}$ = 2000 ppm, $C_{NO_Xinitial}$ = 0, T = 70 ^{0}C.

spontaneous reaction with ammonia, a series of experiments were carried out using a gaseous mixture of simulated gas. In these experiments, mixtures of 2000 ppm SO_2, 10% O_2, 10% CO_2, 12-18% H_2O, 4000 ppm NH_3 and N_2 (balance) were charged into the temperature-controlled chamber *9* with the two-way valve *8'* closed. When the pressure gauge *P* indicated ~ 1 atm, the temperature of the reaction mixture was brought up to 50, 60, 70, 80, 90 and 100 ^{0}C and the pressure of the reactor was carefully controlled not to exceed 1 atm. Heated gas was then passed into the mixing chamber 11 where ammonia from solution was added. As shown in Fig. 4, the reaction was found to have a strong dependence on reaction temperature and on water concentration. In these conditions, it can be observed that the reaction rate decreases with temperature and increases with water concentration. The white solid deposited uniformly on the walls of the mixing chamber 11 were removed easily by spraying with water.

1.3.2. Microwave Irradiation (MW)

The dependence of *DRE* on the irradiation time and power was investigated for gas mixtures containing only SO_2 and only NO_X. The results are shown in Figs. 5 and 6.

The results in Fig. 5 indicate that the microwave system can be applied with good destruction efficiency for SO_2, *DRE* 50 – 90%. A microwave discharge appears and remains stable while especially low values of the reflected microwave power rate (P_r <10%) are realized at forward power (P_f) greater than 800 W. Once the positions of the movable waveguide short circuit and the three screws tuner were adjusted for minimal P_r/P_f, there is hardly any change in the power meter readout.

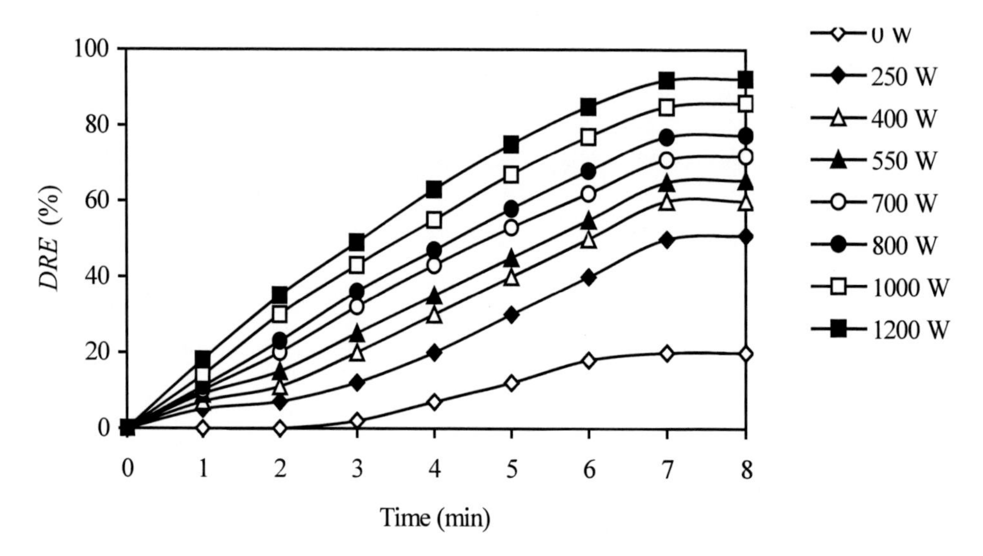

Figure 5. Removal efficiency of SO_2 vs. microwave irradiation time and power $C_{SO_2initial}$ = 2000 ppm, $C_{NO_Xinitial}$ = 0, T = 70 ^{0}C.

Fig. 6 shows that the microwave process is less effective for the removal of NO_X. At low levels of forward microwave power, P_f < 400 W, the ratio of NO_X formation from air is greater than its destruction. At power levels in the range of 400 to 1200 W the destruction of NO_X

exceeds recombination. However, *DRE* remains very low, between 6 – 10%, indicating that microwave irradiation, in the described conditions, cannot be applied as an efficient method for the destruction of NO_X.

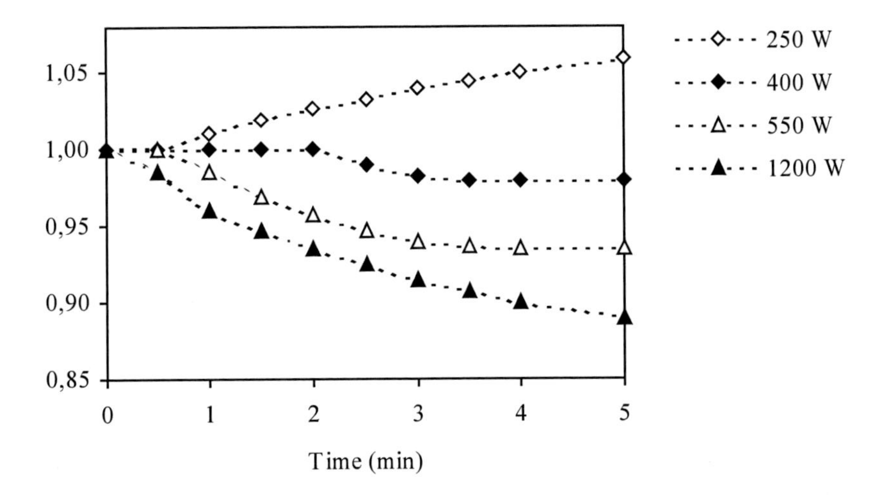

Figure 6. Removal efficiency of NO_X vs. microwave irradiation time and power $C_{SO_2 initial}$ = 0, $C_{NO_X initial}$ = 730 ppm, T = 70 ^{0}C.

1.3.3. Irradiation with Accelerated Electron Beam (EB)

The basic principle of removing SO_2 and NO_X from flue gases by electron beams is the irradiation of the gas mixture with high energetic electrons. In this method, NH_3 is injected into the gas mixture just before the irradiation section and the solid by-products (ammonium salts) are separated by filtration after irradiation.

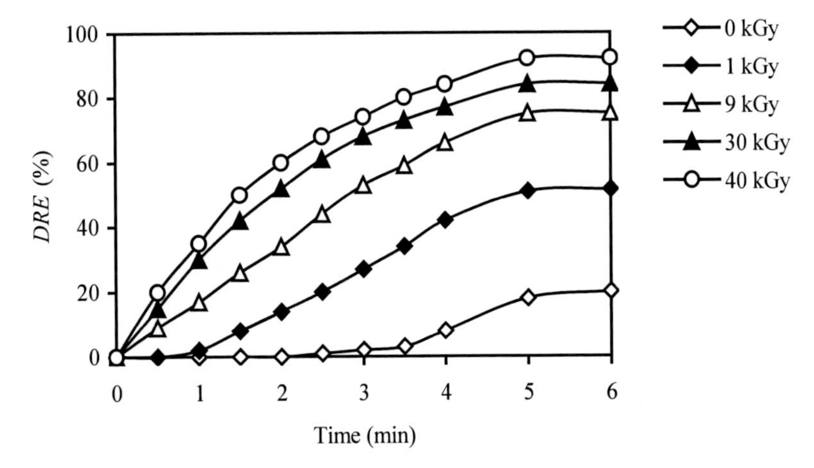

Figure 7. Removal efficiency of SO_2 vs. electron beams irradiation time and total absorbed dose $C_{SO_2 initial}$ = 2000 ppm, $C_{NO_X initial}$ = 0, T = 70 ^{0}C.

Figs. 7 and 8 show the effect of electron beams on the removal rate of SO_2 and NO_X.

SO_2 is easily removed even at low total absorbed dose levels; at 1 kGy, *DRE* was 50% and increased to 90% for 40 kGy.

For SO_2 removal, the initial reaction rates were found to be linear in the studied range of total absorbed doses; for NO_X, the initial reaction rate seemed to reach a plateau, indicating that, especially at high absorbed dose levels, the formation reactions of NO_X (recombination from ions and radicals and from N_2 oxidation) are in equilibrium with those involving its removal.

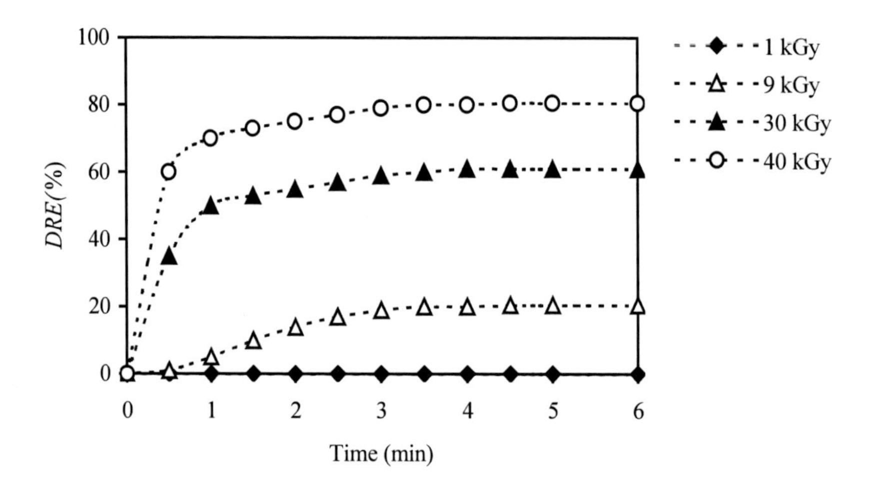

Figure 8. Removal efficiency of NO_X vs. electron beams irradiation time and total absorbed dose $C_{SO_2 initial} = 0, C_{NO_X initial} = 730$ ppm, T = 70 ^{0}C

1.3.4. Simultaneous Irradiation with Electron Beam and 2.45 GHz Microwave (EB+MW)

It has been proved that both external accelerated electron beam and microwave can be regarded as means of generating free electrons and active species. In the case of external electron accelerators, the active species are concentrated along the accelerated electrons path. At low and medium dose rates, the chemical reactions consume the majority of the radicals. At high dose rates, the radical concentration is much higher and radical recombination reactions may no longer be neglected, leading to low SO_2 and NO_X removal efficiencies. With microwave energy applied by properly designed applicators, promising results are expected, as microwaves penetrate large volumes and lead to the formation of a volume of active species, thus avoiding fast radical recombination reactions.

The essential feature of combined microwave and electron beam irradiation is the additional use of microwave energy for increasing the number of free electrons and sustaining their energy at optimum levels. This leads to decreased levels of electron beam total dose by maintaining the same removal efficiency.

Fig. 9 summarizes the SO_2 destruction removal efficiency over time for the simultaneous irradiation experiments. Although the process conditions were slightly varied throughout the many tests performed, the input concentration of SO_2 remained close to 2000 ppm and the temperature maintained at 70 ^{0}C. The air dilution flow rate was 10 l/min.

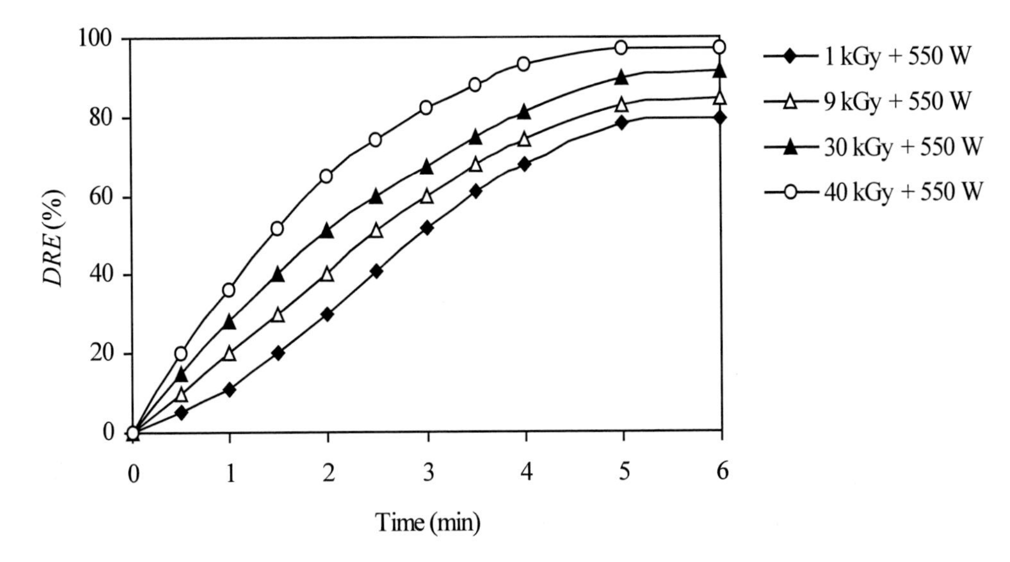

Figure 9. Removal efficiency of SO_2 vs. electron beam total absorbed dose and microwave power and irradiation time $C_{SO_2 initial}$ = 2000 ppm, $C_{NO_X initial}$ = 0, T = 70 ^{0}C.

The additional use of MW energy to the EB energy increases the SO_2 removal speed and efficiency. SO_2 removal efficiency for simultaneous EB and MW irradiation of 1 kGy + 550 W has the same value as for EB irradiation at 9 kGy; thus, for the same removal efficiency, the required dose in simultaneous EB+MW is approximately 10 times smaller than used in EB irradiation. The process carried out at 9 kGy + 550 W gives the same *DRE* as obtained at 30 kGy; the required total absorbed dose level being 3 times lower. This effect decreases with the increase of the absorbed dose level because, at high dose, the EB irradiation becomes very effective by itself – Table 1.

Table 1. The reduction of total absorbed dose in EB+MW method vs. EB in destruction of SO_2; $C_{SO_2 initial}$ = 2000 ppm, $C_{NO_X initial}$ = 0, T = 70 ^{0}C; P_f = 550 W

Total absorbed dose level (kGy)	DRE_{SO2} in EB process (%)	DRE_{SO2} in EB+MW process (%)
1	51	79
9	75	84
30	84	91
40	92	97

A set of experiments was performed to evaluate the efficiency of the process when SO_2 and NO_X are added together in the gaseous mixture. Fig. 10 shows the destruction efficiency of SO_2 and NO_X related to EB and EB+MW methods. The presence of both SO_2 and NO_X in the gas mixture proved to increase the removal efficiency of the toxic gases in both EB and EB+MW processes, especially when low doses were applied.

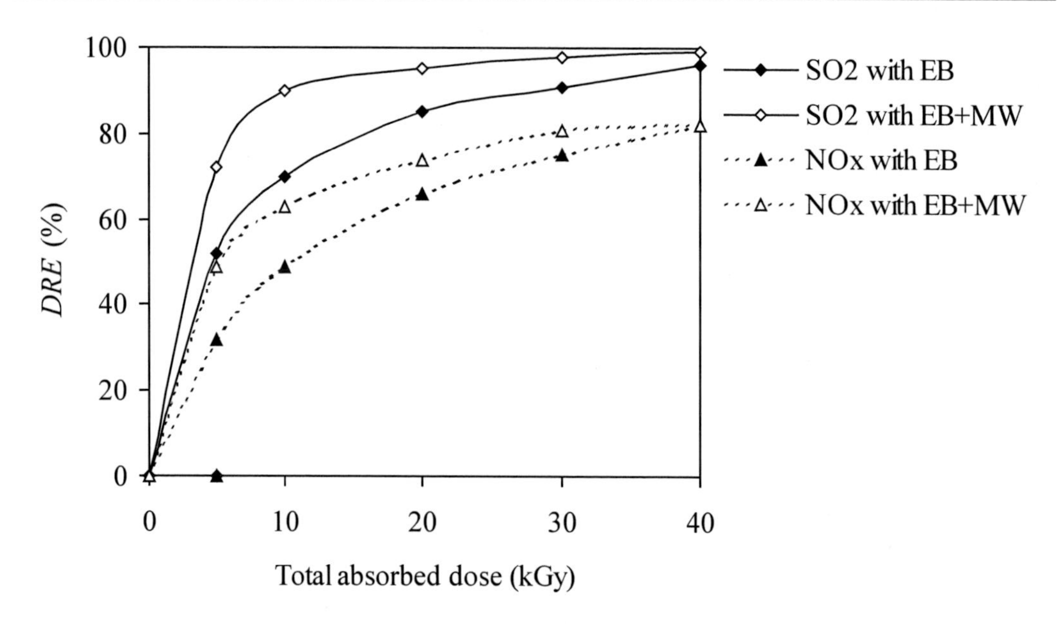

Figure 10. Removal efficiency of SO_2 and NO_X vs. electron beam total absorbed dose and microwave power (P_{MW} = 550 W) $C_{SO_2 initial}$ = 1350 ppm, $C_{NO_X initial}$ = 900 ppm, T = 70 ^{0}C.

The enhancement of removal efficiency of NO_X and SO_2 from flue gases is an important subject to reduce the cost of the flue gas treatment process. Table 2 shows the dependence of DRE_{SO2} and DRE_{NOx} on the irradiation method: destruction removal efficiency (*DRE*) of SO_2 in MW processing is greater than 90% at 1.2 kW (2.45 GHz); the irradiation with EB at 40 kGy total absorbed dose removes ~90% SO_2. The simultaneous treatment EB+MW also provides *DRE* ~90% but the required energy introduced into the system by electron beams is 1.4 times lower.

It was documented that the microwave treatment of gaseous mixtures containing NO_X cannot be applied as an efficient method for the destruction of NO_X. At power levels below 400 W, the reaction rate of recombination is greater than destruction. In the EB and EB+MW processes, NO_X is removed with 80% efficiency.

Table 2. Influence of the irradiation method on the destruction removal efficiency (*DRE*) of SO_2 and NO_X

Irradiation method	DRE_{SO2} (%)	DRE_{NOx} (%)
MW 550 W (2.45 GHz)	65	6.5
EB 40 kGy	89	80
EB+MW 30 kGy + 550 W	91	81

Based on this research and the results obtained by a pilot project built in collaboration with Electrostatica Bucharest, a project for an installation involving simultaneous SO_2 and NO_X removal by irradiation with accelerated electron beams and microwaves has been proposed. The main parameters and the process flow diagram are shown in Table 3 and Fig. 11.

Table 3. Installation for SO_2 and NO_X removal by combined electron beam and microwave irradiation; Proposed for a boiler of 525 t/h at Thermo-Power Plant CET-West-Bucharest of 550 MW

Parameter	
The source of flue gas	Oil + gas burning
Flow rate	345 100 Nm^3/h
Conditioned gas composition	CO_2 = 140.5 g/Nm^3; CO = 0.0035 g/Nm^3 O_2 = 165.2 g/Nm^3; N_2 = 779.51 g/Nm^3 SO_2 = 3.46 g/Nm^3; NO_X = 0.458 g/Nm^3 H_2O = 100.45 g/Nm^3; NH_3 = 3.464 g/Nm^3 Dust = 0.05 g/Nm^3
Conditioned gas temperature	70^0C
EB power per 1 Nm^3	4.40 W
MW power per 1 Nm^3	4.00 W
Total EA + MW power	1520 kW + 1380 kW
Estimated removal efficiency	SO_2 = 95-99% NO_X = 75-80%
The main items of the process	Number of process lines: 2 Process vessel: 1.1 x 1.96 x 0.245 m^3 Number of accelerators/line: 2 Number of MW sources/line: 4 EB energy: 0.8-1 MeV EB power/accelerator unit = 400kW MW power/ source unit = 200 kW

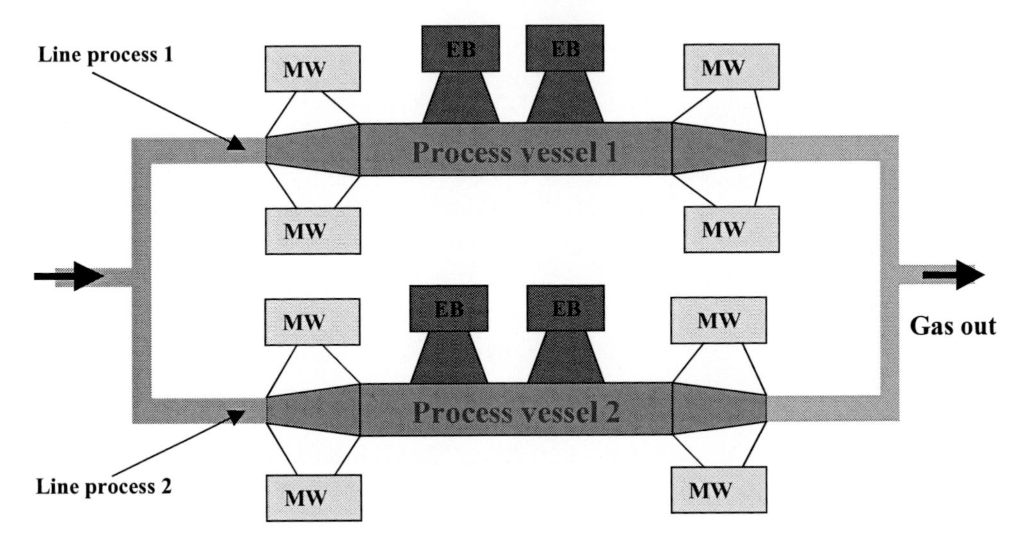

Figure 11. Thermo-Power Plant CET-West Bucharest. Process for simultaneous SO_2 and NO_X removal by combined electron beam and microwave irradiation.

1.3.5. By-Product

X-ray diffraction of the solid by-product separated in the filter installed downstream the irradiation reactor showed characteristic peaks for $(NH_4)_2SO_4$, $(NH_4)_2SO_4x2NH_4NO_3$ and $(NH_4)_2SO_4x3NH_4NO_3$.

Quantitative and qualitative chemical methods described by Mendham et al. [20] were also used to identify and analyse the solid by-product. The results of these analyses are presented in Table 4. However, in order to demonstrate and clarify how the irradiation method and concentration of reactants influence the reaction, we shall further extend these results with more studies focussed on the by-product formation and changes as the reaction proceeds.

Table 4. Results of quantitative and qualitative analyses of the by-product separated after irradiation

Compound	Analytical method	By-product composition % (wt.)
Sulphate (SO_4^{2-})	Barium sulphate	30.50
Nitrate (NO_3^-)	Devarda's alloy and Kjeldahl	41.70
Ammonia (NH_4^+)	Kjeldhal	23.60
Water	Gravimetric	3.10
	TOTAL	98.90

2. Waste-Water Pollution Control

The pursuit for a clean environment creates the need to develop industrial wastewater treatment methods with better performance efficiencies than the conventional methods of stabilisation ponds and activated sludge. The choice of either biological or chemical methods for treatment of industrial wastewaters depends on the composition of the wastes. Industrial wastes from food industries typically contain a significant amount of biodegradable compounds. Especially in the last five to ten years, because of the pressure to comply with wastewater discharge permits and reduction of the governmental grants for constructing new and upgrading existing treatment works, municipal and regional sewer authorities are applying more pressure on industries to reduce their organic and suspended and dissolved solids loading to the sewers.

This is why the food processing industry needs cost-effective reduction and recycling technologies for its wastewaters. These technologies include both source reduction options (technologies to reduce the amount of water used) and treatment options (technologies to reduce the amount of contamination level of wastewater requiring discharge).

Wastewater derived from food production has attributes that are very distinct from other industrial activities. In particular, food-processing wastewater can be characterized as "friendly" in that it generally does not contain conventional toxic chemicals. However, food-processing wastewaters can be subject to bacterial contamination, which represents a special issue for wastewater reuse.

More generally, food processing wastewaters are distinguished by their generally high organic compound (COD) concentrations, high level of dissolved and/or suspended solids including fats, oils and grease (FOG), nutrients such ammonia and minerals (e.g. salts).

The characteristics and generation rates of food wastewater are highly variable, depending on the specific types of food processing operations. One important attribute is the general scale of the operations, since food processing extends from small, local operations to large-scale national or international producers. This difference in scale is relevant not only in identifying sources of wastewaters, but also in determining the appropriate reduction or recycling options.

In addition to the variability in internal operating conditions, external constrains on wastewater management also vary widely. Wastewater disposal costs, which are a key driver for reduction/recycling technologies, will vary based on the location and regulatory requirements which vary by region/city. As a result of these variations in context and applications, the determination of the cost-effectiveness of current and emerging reduction technologies must be generally made on a case-by-case base.

2.1. Principles of Wastewater Treatment

Wastewater treatment consists of applying known technology to improve or upgrade the quality of a wastewater. Physico-chemical methods, which play a considerable role in treating process wastewaters, are used both on their own and in combination with mechanical, chemical and biological methods. Physico-chemical methods are being increasingly used for the preliminary treatment of wastewater before its biochemical purification. This is due to stricter requirements in regard to the degree of purification of wastewater and the need to remove all organic admixtures before it is discharged. The coarsely dispersed particles are easily removed by mechanical treatment although finely dispersed particles and colloid particles remain. To remove them, coagulation methods are needed. The particles in a colloid system agglomerate to form larger particle aggregates which are settled out and removed mechanically from the wastewater. One of the forms of coagulation is flocculation, in which the small suspended particles from loose accumulations of flocs which settle out easily under the influence of specially added substances (flocculants).

The typical and most widely used coagulants include hydrolysing salts of aluminium (Al^{3+}) and iron (Fe^{3+}). When these substances are used as coagulants, the hydrolysis reaction leads to the formation of water-insoluble iron and aluminium hydroxides, which absorb on the developed floc surface in suspension. The reaction also forms finely dispersed colloid substances that sink to the bottom of the tank. Hydrolysis reactions (8, 9) generate mineral acids which may need to be neutralised by the addition of an alkali to produce the required pH.

$$Al_2(SO_4)_3 + 6H_2O \rightarrow 2Al(OH)_3 \downarrow + 3H_2SO_4 \tag{8}$$

$$FeCl_3 + 3H_2O \rightarrow Fe(OH)_3 \downarrow + 3HCl \tag{9}$$

The amount of coagulant needed to carry out the coagulation process depends on the type of coagulant, the throughput and the composition of the wastewater, and the degree of purification required; coagulants are added in amounts of 30-300 mg/l (in most cases 100-150 mg/l). To ensure the decolourisation of highly concentrated and intensively coloured wastewaters, the rate of coagulant addition rises to 1-4 kg/m^3. The volume of sediment obtained as a result of coagulation reaches 10-20% of the volume of the wastewater processed. The coagulation method of purification is mainly used for small flows when coagulants are available.

The use of flocculants that increase the density and the solidity of the flocs formed reduces the consumption of coagulants and increases the reliability of the work and the throughput capacity of the treatment plant.

When dissolved in wastewater, flocculants may be in a non-ionised or ionised state. If they are ionised, they are called soluble polyelectrolytes. Depending on the composition of the polar groups, flocculants can be:

Non-ionogenic polymers containing non-ionogenic groups: -OH etc. (starch, oxyethyl and Na-carboxymethylcellulose, polyvinyl alcohol, polyacrylonitrile, polyethyleneoxide etc.);

Anionic polymers containing anionic groups – COOH, -SO_3H, -OSO_3H (active silicic acid, sodium polyacrylate, sodium alginate obtained from seaweeds, lignosulphonates, polystyrene-sulphonic acid etc.);

Cationic polymers containing cationic groups –NH_2 =N (polyethyleneimine, vinyl-pyridine copolimers, polydimethylaminoethylmetacrylate etc.);

Amphoterico polymers containing simultaneously anionic and cationic groups, polyacrylamide, proteins etc.

The rate and effectiveness of the flocculation process depends on the composition of the wastewater, its temperature, the rate of mixing, and the order in which coagulants and flocculants are introduced into the wastewater. The doses of flocculants applied are 0.1-10 g/m^3 (on average 0.5-1 g/m^3).

The obtaining of tough polymeric flocculants has been the subject of many studies in the past decades. An alternative and innovative way to conventionally initiated polymerisation reactions (heat sensitive peroxide, redox systems etc.) can be the use of ionising radiation, like gamma photons or accelerated electrons. Some significant advantages can be ascribed to this polymerisation route both in terms of purity of the final products and polymerisation temperature and initiation rate range of operation [22].

In this work the polymerisation of acrylamide and of acrylic acid using ionising (electron beams) and non-ionising (microwave) irradiation has been investigated. The aim is to investigate the possibility to favour the formation of polyelectrolytes suitable for wastewaters treatment.

Previous work carried out in our laboratory has shown that electron beams and microwave energies can be applied as the energy source for the polymerisation of acrylamide, acrylic acid, and vinyl acetate in aqueous solutions [23]. Since then we have focused our attention on finding improved irradiation conditions and a better chemical composition of the solutions to be irradiated for increasing the final parameters of the copolymer with regards to water solubility, average molecular weight, residual monomer concentration and efficiency on wastewater cleaning processes as well as to the electrical energy reduction and so, the technological cost.

2.2. Experimental

2.2.1. Materials

The synthesis of the acrylamide – acrylic acid copolymer (denoted throughout the text as polymer PA) was carried out by irradiating aqueous solutions of appropriate acrylamide and acrylic acid quantities. Complexing agents (CX) for the impurities inhibition, chain transfer agents (CT) for the crossed-link structure inhibition and initiators (INI) for the monomer conversion optimisation have been also added to the initial solution of monomers. A typical chemical composition of the aqueous stock solution used for polymerisation is given in Table 5.

Table 5. Typical recipe for irradiation-induced copolymerisation of acrylamide and acrylic acid

Material		Weight ratio (%)
Monomer	Acrylamide	36
	Acrylic acid	4
Complexing agents (CX)	Ethylenediaminetetraacetic acid (EDTA)	0.005 – 0.1
Chain transfer agent (CT)	Sodium formate	0.005 – 0.1
Initiator (INI)	Sodium persulfate	0.005 – 0.8
Dispersion media	Distilled water	~ 60

2.2.2. Polymer Characterization

The copolymer properties were studied through the following parameters:

- *Conversion coefficient* (CC) *of the monomers to copolymer* by brome addition to the double olefin bond as described by Mendham et al. [24]. The reactant, bromine, is generated *in situ* through reaction between 0.1N $KBr/KBrO_3$ and 18% HCl. Unreacted Br_2 is measured by titration with 0.1N $Na_2S_2O_3$ in the presence of 1% starch solution
- *Average molecular weight* (M_W) or *intrinsic viscosity* (η);
- *Chain linearity coefficient* given by Huggin's constant (k_H).

The values of η, M_W, and k_H can be calculated from Mark-Houwink-Sakurade equations for a polymer in aqueous solution of 1N $NaNO_3$ at 30^0C and pH = 7 [24]:

$$\begin{aligned} \eta_{red} &= \eta + C \times tg\alpha \\ \eta_{red} &= \eta_{sp}/C = (\eta_{rel} - 1)/C \\ tg\alpha &= \Delta\eta_{red}/\Delta C \\ k_H &= tg\alpha/\eta^2 \\ \eta &= K \times M_W^a \end{aligned} \tag{10}$$

Where η_{rel} is established from measurements with a Hoppler viscosimeter BH-2.

2.2.3. Irradiation System

The irradiation system consists of an electron beam accelerator (ALID-7 of 5.5 MeV and 670 W output power) and a microwave multimode cavity operated at 2.45 GHz and variable power, up to 850 W – Fig. 3a.

The irradiation of the samples was carried out in Pyrex flasks containing ca. 75 ml monomer solution.

2.2.4. Wastewater Samples

Food wastewater samples were collected from the Slaughter House (abattoir) INCAF, Ploiesti and the Vegetable Oil Plant, Solaris, Bucharest.

For hygienic reasons abattoirs use large amount of water in animal processing operations. This produces large amounts of wastewater that must be treated. Effective primary treatment before secondary treatment will increase the overall effectiveness and efficiency of wastewater treatment systems, as it is cheaper to remove physically the fat and solids than to treat later in secondary and tertiary treatment facilities. Wastewater produced in animal slaughter areas typically has a high BOD. It is also very saline and has high levels of nutrients, suspended solids and bacterial contamination. Appropriate design criteria for coagulation/precipitation are determined by what is to be removed. Different coagulants are needed for different pollutants.

Although most coagulants are inexpensive, the cost can be high for an ongoing supply of them, particularly in some parts of the water control measures. Another disadvantage is the volume of sludge generated, which includes the solids removed from the waste stream as well as the coagulants that are added. If any metals or toxics are coagulated or precipitated, then the sludge must be disposed carefully and cannot be reused. The amount of sludge depends on the amount of coagulant added, the amount of precipitate formed and the amount of solids removed.

2.2.5. Quality Indicators of Wastewaters

Quality parameters such as total suspended matters (TSM), chemical oxygen demand (COD), biological oxygen demand (BOD) and fat, oils and grease (FOG) were measured to investigate the effect of the polymer addition on the degree of purification of wastewater.

The analysis of the impure water was carried out in parallel with a blank determination on pure, double-distilled water [21, 24]. The total chemical oxygen demand (COD) is a measure of the total quantity of oxygen required to oxidize all organic material into carbon dioxide and water; it was measured by the dichromate COD test. A 2-h reflux with concentrated sulphuric acid and potassium dichromate with silver sulphate and mercuric sulphate catalysts is adequate for complete oxidation of all but a few aromatic organic compounds. The 5-day, 20 ^{0}C, biochemical oxygen demand (BOD_5) has been also used for evaluating BOD in wastewaters. BOD_5 is a measure of the amount of oxygen that bacteria will consume while decomposing organic matter under aerobic conditions. The total suspended solids (TSS) content was determined by dry ashing at 105^{0}C and UV-VIS spectrometry of suspended solid extinction (the amount of light stopped or scattered by a suspension). Fat, oil and grease (FOG) content was determined by extraction with petroleum ether.

The degree of purification *RE* was calculated:

$$RE(\%) = \frac{C_i - C_f}{C_i} \times 100 \tag{11}$$

Where C_i and C_f are the initial (before treatment) and final (after flocculation) concentrations of the load, respectively.

2.3. Results and Discussion

Laboratory experimental methods and procedures in ionising (gamma radiation with Co^{60} and accelerated electron beams) and non-ionising (2.45 GHz microwaves) treatments for polymeric materials obtaining have been continuously evolving [26, 27]. Early experiments to demonstrate the viability of the technique and to identify and test some of the basic ideas governing the use of irradiation in monomer aqueous systems have shown that the irradiation induced polymerisation gives, in the proper irradiation conditions and for an adequate chemical composition of the monomeric solution, higher conversion coefficients (>99.9%) and lower residual monomer concentration (< 0.05%) than classical polymerisation. It has also been proven that the characteristics of the PA copolymer are influenced by the chemical composition of the solution to be irradiated, absorbed dose level D, absorbed dose rate level D^*, and the nature of the energy which induces the polymerisation process.

2.3.1. Polymerisation with Accelerated Electron Beams (EB)

In EB polymerisation the radical reaction mechanism depends on the total monomer concentration as well as on the presence of water in the system; however, the radicals originated from water radiolysis have a predominant role on the radicals produced by monomer irradiation. Fig. 12 shows the effect of initial monomer concentration on the conversion of monomer to polymer. It can be easily observed that a total monomer concentration greater than 40% leads to a decrease in reaction efficiency indicating that the water radicals facilitate the polymerisation process.

Although increasing the total absorbed dose (D) gives better reaction efficiency (conversion of monomer to polymer) and high intrinsic viscosity values, the final polymer has very poor water solubility indicating a cross-linked structure. Polymers with good chain linearity (given by k_H) and high molecular weight (given by η) are desirable because they are water-soluble and have good flocculation properties, i.e. fast gravitational sedimentation and compact sediment.

Fig. 13 shows the polymerisation conversion and molecular weight against total absorbed dose level. The dose level dependence of the conversion is represented by S-shaped curves. After a short initial period, the polymerisation rate increases until the higher conversion. This feature can be explained by the gel effect. At the early polymerisation stage, the process mainly occurs in the continuous phase, and the growing oligomers form in the continuous phase. As conversion increases, the viscosity with each particle builds up, decreasing the rate of diffusion-controlled termination, and in turn accelerating the rate of propagation, which results in the presence of the gel effect [28, 29]. The polymer molecular weight (given by the intrinsic viscosity) increases with irradiation time in the early stages and then reaches a steady value, which is also indicative of the existence of the gel effect.

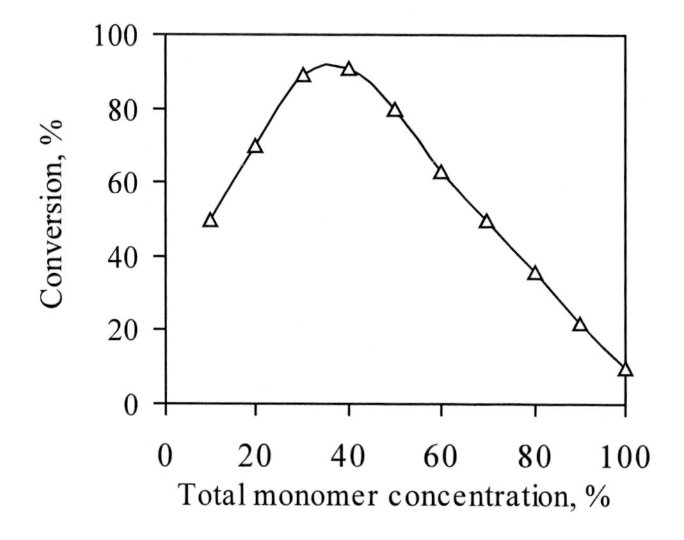

Figure 12. Polymerisation conversion vs. total monomer concentration EB irradiation, D = 1.9 kGy, CX = 0.025%, CT = 0.2%

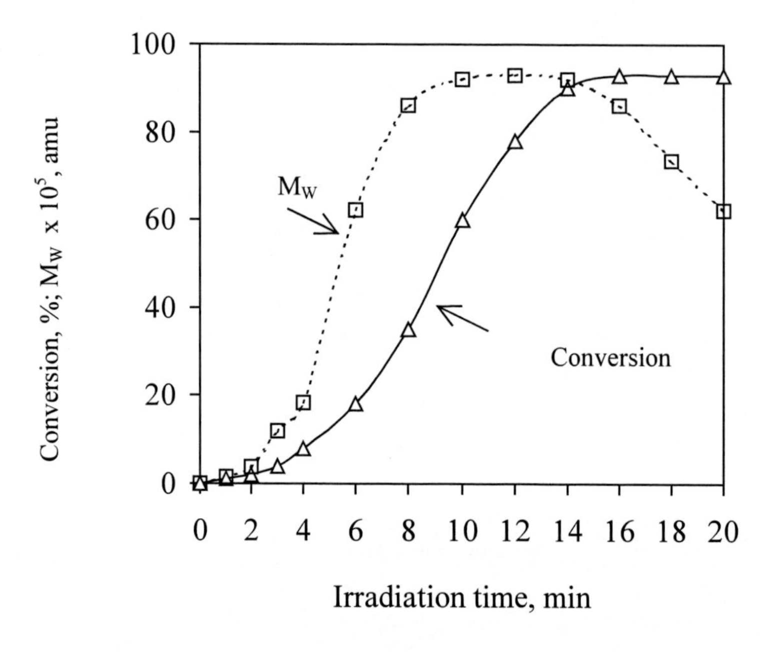

Figure 13. Polymerisation conversion vs. irradiation time and polymer molecular weight (M_W) vs. irradiation time Dose rate 150 Gy/min; CX = 0.025%, CT = 0.1%, INI = 0.5%

Table 6. The effects of EB irradiation mode and chemical composition on the PA parameters. Dose level 1.9 kGy, INI 0.5%, CX 0.025%

CT %	M_w x 10^{-6} amu	k_H	Solubility in water
0.05	15.90	3.12	Insoluble
0.01	9.31	1.52	Insoluble
0.15	4.67	0.88	Partial soluble
0.2	3.38	0.25	Soluble
0.5	2.83	0.15	Soluble
1.0	2.41	0.08	Soluble
2.0	1.25	0.05	Soluble

Fig. 13 shows a decrease in the polymer molecular weight with the irradiation time, from 93×10^5 after 10 minutes of irradiation to about 63×10^5 amu after 20 minutes. This result can be explained considering that during irradiation several processes can occur. They involve monomer polymerisation, chain branching and cross-linking and degradation of the PA polymer already formed; also, interactions among all the components are caused. A competition between all these phenomena can be suggested. PA degradation determines a general decrease of its performance in wastewater cleaning, clearly attributable to the different effects of ionising radiation on the monomer, which affect the final structure of the polymer.

It is noteworthy that polymers with high molecular weight are insoluble in water. The cross-linking effect can be controlled by CT. High quantities of CT prevent the cross-linking effect but lead to lower M_W and polymer conversion values – Table 6.

2.3.2. Microwave Polymerisation (MW)

Alongside ceramic processing, polymer chemistry forms probably the largest single discipline in microwave chemistry, and the methods used are certainly among the most developed. Polar starting materials and very often products, allow rapid and controllable syntheses, the dielectric properties themselves being an excellent indicator of reaction progress. The ability to control syntheses with high accuracy and with direct heating of the reactants has the advantage of large potential savings in energy [30].

The origin of microwave heating lies in the ability of the electric field to polarise the charges in the irradiated material and the inability of this polarisation to follow extremely rapid oscillations (2.45 GHz frequency) of the electric field. Therefore, the polarisation vector lags the applied electric field ensuring that the resulting current has a component in phase with the applied electric field, which results in the dissipation of power within the material. Coupled with these polarisation effects, a dielectric can be heated through direct conduction due to, for example, the redistribution of charge particles under the influence of the externally applied electric field forming conducting paths, particularly in mixtures of heterogeneous materials [31]. From the above consideration, it is clear that polar molecules can absorb microwaves resulting in a greater heating rate. Consequently, a solvent able to couple efficiently with microwaves will increase the reaction rate of the dissolved solutes, even poorly polar ones. Thus, in the microwave-induced polymerisation of acrylamide – acrylic acid solutions, due to the presence of water with strong microwave absorbance, it was expected that the polymerisation reaction rate would be much enhanced. Indeed, the use of microwave heating for acrylamide – acrylic acid aqueous solution polymerisation requires a significantly reduced time than conventional heating, about 50-100 times lower.

Due to rapid, volumetric and selective microwave transfer, the molecular weight dispersion of the polymeric material is low. Median M_W values (3 – 5×10^6 amu) but good water solubility were always obtained in microwave fields – Table 7.

Table 7. Microwave polymerisation of acrylamide – acrylic acid aqueous solutions. Monomer total concentration 40% (w/w), P_{MW} = 250 W; INI 0.5%; CT 0.2%, CX 0.025%

Irradiation time S	$M_W \times 10^{-6}$ amu	k_H	Solubility in water
115	1.05	1.50	Partial soluble
120	2.92	0.96	Soluble
125	4.01	0.61	Soluble
130	4.50	0.52	Soluble
135	4.96	0.36	Soluble
140	3.51	3.26	Insoluble

2.3.3. Simultaneous Polymerisation with Accelerated Electron Beams and Microwaves (EB+MW)

The irradiation with electron beams has been already recognized as a very effective method for material processing because it can induce chemical reactions at any temperature in solid, gas or liquid phase without using catalysts. The main disadvantage of the method is that the required doses are generally very high. In addition, the equipment involved in ionising

Table 8. Polymerisation of acrylamide-acrylic acid solutions under simultaneous irradiation with EB and MW. Monomer concentration 40%; INI 0.5%; CT 0.2%; CX 0.025%; P_{MW} = 250 W; irradiation time 30 s.

Total absorbed dose kGy	$M_W \times 10^{-6}$ amu	k_H	Solubility in water
0.2	2.2	0.30	Soluble
0.3	7.1	0.15	Soluble
0.4	8.9	0.09	Soluble
0.5	9.0	0.05	Soluble
0.6	8.5	0.10	Soluble
0.7	7.5	0.15	Soluble
0.8	6.7	0.19	Soluble
1.0	4.1	0.57	Soluble

radiation chemistry requires expensive shielding and high maintenance costs. Thus, for industrial scale processing, the problem of minimizing the electricity consumption as well as the electron beam cost is especially important.

The results of simultaneous irradiation with electron beams and microwaves (EB+MW) of 40% acrylamide – acrylic aqueous solutions are given in Table 8. The additional use of MW energy to the EB energy results in excellent properties of the polymer, very high M_W simultaneously with low k_H and a reduction in the electron beam dose of 2 – 4 times.

2.4. Wastewater Treatment

Methods of coagulation and flocculation are widely used for treating wastewater in many branches of the industry [32-35]. The effectiveness of the coagulation treatment depends on many factors: the type of colloid particles, their concentration and degree of dispersion, the presence in wastewater of electrolytes and other substances. The main process for the coagulation treatment of the process wastewater is heterocoagulation – the interaction of colloid and finely dispersed particles in the wastewater with the aggregates formed when coagulants are introduced into the wastewater. Suspended particles acquire an electrostatic charge, which, in water treatment is usually negative. The charges produce repulsion between particles, which tends to stabilise the suspension. In colloidal suspensions, which have a maximum particle size of less than 2 microns, this repulsion effectively prevents settling. Coagulating agents are selected to have an opposite charge to that of the suspended solids and effectively neutralise that charge. This destabilises the suspension and allows the particle to come together. Flocculating agents form bridges between particles and lead to the formation of large agglomerates, which can be removed by settlement or flotation.

Table 9. Wastewater from the Slaughter House INCAF, Ploiesti, Romania

Quality indicator	Level of admixture in the water (mg/l)			*RE* (%)	
	Before treatment	After treatment			
		C[a]	C+P[b]	C	C+P
TSS	766	496	112	33.9	85.4
BOD	1500	1050	205	30.0	86.3
COD	480	321	100	33.1	79.1
FOG	332	275	32	17.1	90.0

[a] Classical treatment: 1 l wastewater + 4 ml $Al_2(SO_4)_3$ 10%;

[b] Combined treatment: 1 l wastewater + 2 ml $Al_2(SO_4)_3$ 10% + 4 ml PA 0.1%

Preliminary experiments comprising investigation of the flocculation capacity of PA polyelectrolyte were carried out in order to quantify parameters such as total suspended matters (TSM), chemical oxygen demand (COD), biological oxygen demand (BOD) and fat, oils and grease (FOG). High average molecular weight (9×10^6 amu) and low k_H (0.05) PA obtained by EB+MW treatment was employed. Two wastewaters, one from a slaughterhouse and the other from a vegetable oil plant, were tested. The coagulant (aluminium sulphate from 10% solution) and the flocculant (PA from 0.1% solution) were gradually poured into the water with 20-30 rpm stirring. The pH was fixed at 7 with $Ca(OH)_2$ and the temperature of the treated water was 20 ^{0}C. The results are presented in Tables 9 and 10.

The experimental data shown in Tables 9 and 10 indicate that the use of PA polyelectrolyte in combination with a classical coagulant $Al_2(SO_4)_3$ much increases the degree of purification of both wastewaters.

Table 10. Wastewater from the Vegetable Oil Plant SOLARIS, Bucharest, Romania

Quality indicator	Level of admixture in the water (mg/l)			*RE* (%)	
	Before treatment	After treatment			
		C[a]	C+P[b]	C	C+P
TSS	1020	971	50	4.8	95.1
BOD	2300	1400	298	39.1	87.0
COD	4664	2889	395	38.1	91.5
FOG	3497	82	8	97.7	99.8

[a] Classical treatment: 1 l wastewater + 4 ml $Al_2(SO_4)_3$ 10%;

[b] Combined treatment: 1 l wastewater + 2 ml $Al_2(SO_4)_3$ 10% + 4 ml PA 0.1%

Conclusion

The simultaneous irradiation with electron beams and microwave (EB+MW) has been proved to have advantages over the conventional processes due to the ability of the combined irradiation process to induce a variety of physical and chemical phenomena, which can play a significant role in the development of new and existing technologies in toxic gases abatement.

To make EB+MW removal process more economically attractive, further investigation of the mechanism of the chemical effects from microwaves and electron beams, on reactor technology and process technology is needed. The general conclusion of this study is that simultaneous electron beam and microwave irradiation is a viable and promising method for flue gas cleaning showing the following main advantages over conventional methods:

- Simultaneous removal of SO_2 and NO_X at high efficiency levels;
- Dry process which can be easily controlled via three main parameters: ammonia (and water) dosage, electron beam dose and microwave power;
- The by-product obtained is a mixture of ammoniumsulphates and nitrates which can be used as a fertilizer;
- No wastewater is formed in the process;
- The energy reduction due to simultaneous irradiation with microwaves is 10-25%.

Regarding the irradiation of acrylamide – acrylic acid solutions, both external accelerated electron beams and microwaves can be also considered as sources for the polymerisation. The examination of the results suggested that the combination between the ionising effects of external accelerated electron beams and the dielectric heating feature of microwaves could give new and promising results in the polymer chemistry. Expected improvements include a reduction of the total required absorbed dose of electron beam irradiation and also, an improvement of the polymer properties such as higher molecular weight and good water solubility.

References

[1] Breihofer, D., Mielenz, A. & Rentz, O. *Emission Control of SO_2, NO_X and VOC at Stationary Sources in the Federal Republic of Germany.* Karlsruhe: IIP, 1991.

[2] White, D.M. & Maibodi, M. *Assessment of control technologies for reducing emissions of SO_2 and NO_X from existing coal-fired utility boilers: final report.* Washington, D.C., U.S.: Environmental Protection Agency, Office of Research and Development, 1990.

[3] Penetrante, B.M. & Schultheis, S.E. (1993). *Non-Thermal Plasmas Techniques for Pollution Control. Part A: Overview, Fundamentals and Supporting Technologies* and *Part B: Electron Beam and Electrical Discharge Processin,*. Berlin: Springer.

[4] Kawamura, K.; Hirasawa, A.; Aoki, S.; Kimura, H.; Fuji, T.; Mizutani, S.; Higo, T.; Ishiwara, R.; Adachi, K. *Rad. Phys. Chem.*, 1979, 13, 12-17.

[5] Tokunaga, O.; Nishimura, K.; Suzuki, N.; Washino, M. *Intern. J. Appl. Rad. Isot.*, 1979, 30, 19-26.

[6] Kawamura, K.; Aoki, S.; Kimura, H.; Adachi, K., *Rad. Phys. Chem.*, 1980, *16*, 133-138.

[7] Willibald, U.; Paltzer, K.H.; Witting, S., *Rad. Phys.Chem.*, 1990, 35, 422-426.

[8] Platzer, K.H.; Willibald, U.; Gottstein, I.; Tremmel, A.; Angeli, H.I.; Zellner, K., *Rad. Phys. Chem.*, 1990, 35, 427-430.

[9] Zimek, Z.; Chimieleswsky, A.; Artiuch, I.; Frank, N. *A Process for Removal of SO_2 and NO_x from Combustion Flue Gases and an Apparatus Used therefore*, International Patent, WO 92/20433 (B01D 53/34), Nov. 26, 1992.

[10] Hoshi, Y.; Mizusawa, K.; Kashiwagi, M. *Electron Beam Processing Systems for Environmental Applications*, International Symposium on Radiation Technology for Conservation of Environment, Zakopane, Poland, September 8-12, 1997, IAEA-Sm-350/36.

[11] Niessen, W.R. (Ed), *Combustion and Incineration Processes*, Marcel Dekker, New York, 2002.

[12] Rea, M. In *Combustion flue gas treatment;* Chang, J.S.; Kelly, A.J.; Crowley, J.M. Handbook of Electrostatic Processes, Marcel Dekker, Inc. 1995, pp. 607.

[13] Van Veldhuizen, E.M. *Electrical Discharges for Environmental Purposes: Fundamentals and Applications*, Nova Sci. Publishers, Inc., N.Y., USA, 2000.

[14] Moisan, M.; Pelletier, J.; *Microwave Excited Plasmas in Plasma Technology, 4,* Elsevier Science B.V. 1992.

[15] LePrince, P.; Marec, J.; *Microwave Excitation Technology*, Plasma Technology, Plenum Press, New York, 1992, pp. 153.

[16] Martin, D.; Niculae, D.; Radoiu, M.; Indreias, I.; Cramariuc, R.; Margaritescu, A.; Mihailescu, A., *J. Microwave Power and Electromagnetic Energy*, 1997, 32, 215-219.

[17] Martin, D.; Jianu, A.; Ighigeanu, D., *IEEE Transaction on Microwave Theory and Techniques*, 2001, 49, 542-549.

[18] Martin, D. ; Dragusin, M. ; Radu, A. ; Oproiu, C. ; Radoiu, M. ; Cojocaru, G. ; Marghitu, S., *Nucl. Instrum. Meth. B,* 1996, 113, 106-109.

[19] Radoiu, M.; Martin, D.; Calinescu, I., *J. Haz. Mat. B*, 2003, 97, 145-158.

[20] Dvornik, I.; In *Manual on Radiation Dosimetry*, Holm, N.W.; Berry, R.J.; Marcel Dekker, New York, 1970.

[21] Mendham, J.; Denney, R.C.; Barnes, J.D.; Thomas, M., *Vogel's Textbook of Quantitative Chemical Analysis*, 6th Edition, Prentice Hall, 2000, pp.356-455.

[22] Chapiro, A., *J. Appl. Polym. Sci. Pol. Symp.*, 1975, 50, 181-188.

[23] Martin, D.; Mateescu, E.; Ighigeanu, D.; Jianu, A.; *J. Microwave Power and Electromagnetic Energy*, 2000, 35, 216-220.

[24] Marr, I.L.; Cresser, M.S.; *Environmental Chemical Analysis*, International Textbook Company, Glasgow, 1983, pp. 116-121.

[25] Flory, J.P.; *Principles of Polymer Chemistry*, Cornell University Press, Ithaca and London, 1978, pp.267-314.

[26] Dragusin, M.; Martin, D.; Radoiu, M.; Moraru, R.; Oproiu, C.; Marghitu, S.; *Progress in Colloid and Polymer Science*, 1996, 102, 123-126.

[27] Martin, D.; Dragusin, M.; Radoiu, M.; Moraru, R.; Radu, A.; Oproiu, C.; Cojocaru, G.; *Progress in Colloid and Polymer Science*, 1996, 102, 147-150.

[28] Liu, J.; Chew, C.; Wong, S.; Gan, L., *Polymer*, 1998, 39, 283-288.

[29] Gibanel, S.; Heroguez, V.; Focarda, J., *Macromolecules*, 2001, 35, 2467-2472.

[30] Kingston, H.M.; Haswell, S.J.; *Microwave-Enhanced Chemistry*, American Chemical Society, Washington, DC, 1997, pp. 4-14.

[31] Metaxas, A.C.; Meredith, R.J.; *Industrial Microwave Heating*, Peter Peregrinus Ltd., 1993, pp. 5-102.

[32] Edzwald, J.K.; *Water Science &Technology*, 1993, 27, 21-30.

[33] Fetting, J.; Ratnaweera, H.; Odegaard, H.; *Water Supply*, 1991, 9, 19-27.

[34] McCormick, C.L.; Hester, R.D.; Morgan, S.E.; Safieddine, A.M., *Macromolecules*, 1990, 23, 2132-2139.

[35] McCormick, C.L.; Hester, R.D.; Morgan, S.E.; Safieddine, A.M., *Macromolecules*, 1990, 23, 2124-2131.

In: Radiation Physics Research Progress
Editor: Aidan N. Camilleri, pp. 247-286
ISBN: 978-1-60021-988-7

Chapter 6

MODELING STUDY OF ELECTRON-BEAM VOCS DECOMPOSITION IN INDUSTRIAL FLUE GASES

Gennady Gerasimov
Institute of Mechanics, Moscow State University,
Michurinsky avenue 1, 119192 Moscow, Russia

Abstract

The generalized mathematical model of the electron-beam volatile organic compounds (VOCs) treatment in industrial flue gases is proposed. The model gives theoretical description of physical and chemical processes induced by the electron-beam irradiation of gaseous mixture and resulting in VOCs decomposition: the formation of active species (ions, atoms and radicals) by the action of accelerated electrons on gas macro-components; the gas-phase oxidation of main admixtures, SO_2 and NO, in reactions with active species yielding sulfuric and nitric acids; the formation of the aerosol droplets upon the binary volume condensation of the sulfuric acid and water vapors; the heterogeneous oxidation of SO_2 in the aerosol droplets; VOCs decomposition and their formation from precursors under influence of active species. The most toxic representatives of VOCs such as polycyclic aromatic hydrocarbons, their chlorinated and nitrated derivatives, polychlorinated dibenzo-*p*-dioxins and dibenzofurans, and polychlorinated biphenyls are included in consideration. Results of calculations are compared with available experimental data.

Introduction

The present level of organic fuels utilization in various spheres of human activities causes considerable environment pollution by combustion products, which contain a variety of toxic admixtures such as sulfur dioxide (SO_2), nitrogen oxides (NO_x), volatile organic compounds (VOCs), and others. These impurities, in spite of the fact that their concentrations in combustion products are low, have adverse influence on the biosphere and bring appreciable material damage. The organic fraction of combustion products includes volatile aliphatic and aromatic hydrocarbons, their chlorinated derivatives, and heterocyclic compounds of nitrogen, oxygen and sulfur. The content of these compounds in the flue gases has a particular interest in environmental protection due to carcinogenic and mutagenic properties some of them.

The electron-beam dry scrubbing (EBDS) process for the removal of sulfur and nitrogen oxides from industrial flue gases is the most intensely developed area of radiation technology application to solve environmental problems [1]. This cleaning method has a number of advantages over traditional ones: the simultaneous removal of SO_2 and NO_x, the simplicity and reliability in operation, low capital and operation costs, the possibility of by-products utilization, etc. [2]. Originally developed in pilot-scale conditions [3-5], it has been implemented at full-scale in an industrial plants [6,7].

The electron-beam VOCs decomposition in industrial flue gases is a new direction of radiation technology application. The most dangerous representatives of VOCs are polycyclic hydrocarbons (PAHs), their chlorinated (CPAHs) and nitrated (NPAHs) representatives. Polychlorinated dibenzo-*p*-dioxins and dibenzofurans (PCDD/Fs) are the most toxic VOCs according to EPA. The problem of environmental pollution by these substances arose in connection with the fact that noticeable amounts of dioxins were found in the products of municipal solid waster incineration [8]. However, these compounds are formed in almost all processes associated with the combustion of organic fuels in the presence of various chlorine compounds: industrial and domestic coal combustion, medical waster incineration, sinter plants, cement industry, ferrous and non-ferrous metallurgy, motor transport, etc. [9]. There are also increasing concerns over the exposure of human population and ecosystems by polychlorinated biphenyls (PCBs) and in particular those PCB congeners that are considered to have “dioxin-like” toxicity [10]. These compounds are recalcitrant and prone to decomposition in the environment via only a limited number of pathways.

The decomposition of VOCs under electron-beam irradiation has been investigated in numerous laboratory experiments that represent in the main the radiation-induced transformation of chlorinated aliphatic hydrocarbons (see, for example, [11]). Concerning the most toxic VOCs there are a quantity of tests that demonstrate the electron-beam decomposition efficiency for aromatic hydrocarbons [12] and chlorinated ones [13,14], PAHs [15], NPAHs [16], and PCDD/Fs [17]. The received experimental data are very important to strategize further technology applications of EBDS method to removal of VOCs from industrial flue gases.

The mathematical modeling of radiation-induced VOCs decomposition plays an important role in understanding of process mechanisms, and allows to estimate the efficiency of the EBDS purification technology as applied to various kinds of VOCs at the various process conditions (initial concentrations of main, minor and trace species, temperature, humidity, absorbed dose, etc.). The detailed numerical calculations are the base of process optimisation that is hardly realized or impossible at the experimental study.

The previous modeling investigations of the electron-beam induced chemical reactions in gaseous streams as applied to EBDS-process were directed on the construction of kinetics models and computer codes for description of SO_2 and NO conversion under irradiation conditions. In this direction the detailed and reduced kinetic schemes of the processes in gas-phase were proposed [18-21]. These schemes include such steps as energy absorption of electron-beam with active species (ions, atoms and radicals) generation, and gas-phase chemical reactions leading to conversion of SO_2 and NO into corresponding acids.

Electron-beam induced chemical processes in flue gases take place at the typical temperatures near 60 ^{o}C. At these temperatures the formation of aerosol droplets upon binary volume condensation of sulfuric acid and water vapours occurs. This process gives rise to dissolution of other gaseous species (SO_2, OH, NH_3, etc.) in droplets, and to liquid-phase chain oxidation of SO_2. Therefore, the further development of modeling study was connected with inclusion in consideration of liquid-phase processes [22].

Modeling studies of electron-beam VOCs treatment in flue gases were based in the main on kinetic models of gas-phase decomposition of organic molecules in their reactions with active species produced by the gas irradiation [17,23,24]. The recent kinetic model [25] proposed for description of PAHs and NPAHs behaviour under irradiation includes aromatic molecules decomposition in gas-phase reactions, their liquid-phase conversion in the aerosol droplets, and the formation of PAHs and NPAHs under interaction of active species with aliphatic and aromatic hydrocarbons.

In the present work, the generalized kinetic model of the electron-beam VOCs decomposition in industrial flue gases is proposed. The most toxic representatives of VOCs such as PAHs, CPAHs, NPAHs, PCDD/Fs, and CBP are included in consideration. The model contains sub-models of gas-phase and liquid-phase reactions leading to VOCs decomposition, and their formation from precursors under influence of active species.

Kinetic Model of Radiation-Induced Processes

The electron-beam dry scrubbing process was proposed as the most effective method for the simultaneous removal of various harmful admixtures from industrial flue gases. It is based on the transformation of the main pollution species, SO_2 and NO, into corresponding acids. The acids are converted into salts by the addition of neutralizing species (e.g. ammonia, NH_3), and removed from the flue gases by dry electrostatic precipitators.

The presence of VOCs in the radiation zone has negligible influence on the kinetics of SO_2 and NO oxidation since their mole fractions in gas is much lower. In these conditions, the concentration of active species, which are responsible for the conversion of VOCs molecules, depends on the set gas-phase and liquid-phase chemical processes related to the oxidation of SO_2 and NO.

The fundamental physical and chemical processes induced by electron-beam irradiation of the industrial flue gases include: (1) the formation of active species (ions, atoms and radicals) by the action of accelerated electrons on gas macro-components (N_2, O_2, CO_2, and H_2O); (2) the gas-phase oxidation of SO_2 and NO in reactions with active species yielding sulfuric and nitric acids, respectively; (3) the formation of the aerosol droplets upon the binary volume condensation of the sulfuric acid and water vapors; (4) the heterogeneous oxidation of SO_2 in the aerosol droplets; (5) thermal SO_2-NH_3 interaction, and (6) transformation of trace VOCs impurities in reactions with the active species.

Formation of Active Species under Irradiation

The interaction of accelerated electrons with gaseous mixture depends both on the electron energy and properties of gas macro-components. In the EBDS-process, the energy of accelerated electrons is in the range 0.1 – 1 MeV typically. This energy is sufficient to initiate chemical reactions leading to the harmful admixtures transformation, but is too low to permit close electron-nuclei interactions. A full description of the energy release by accelerated electrons is fairly complicated. Therefore, integral values such as total path-length of accelerated electrons and primary radiation yields of active species have gained practical importance. The first one is changed from approximately 0.2 m for 0.1 MeV electrons up to 6 m for 1 MeV electrons in air at the normal conditions [26].

The formation of active species (excited molecules, positive ions, thermalized electrons, atoms and radicals) in irradiated gas in assumption that radiation energy absorption occurs uniformly in space is described by means of the primary radiation yields G_{ij}, the number of intermediates X_j generated from pure species X_i per each 100 eV of absorbed energy:

$$-G_{ii}X_i \rightarrow \sum_j G_{ij}X_j. \quad (1)$$

In gaseous mixture the radiation yield G_j of reactant X_j is defined by the expression:

$$G_j = \sum_i G_{ij}x_i z_i / \sum_i x_i z_i, \quad (2)$$

where x_i is the molar fraction of macro-component i, and z_i is a weight factor. The primary radiation yields and the weight factors for various pure gases entered into Eq. (2) are determined in radiation-chemical experiments. For the macro-components of industrial flue gases the weight factors z_i are equal: 7.02, 6.24, 12.0, and 10.9 for N_2, O_2, H_2O, and CO_2 accordingly [27]. The numerical values of G_{ij} in accepted kinetic model were adopted on the base of analysis of available experimental data [28], and active species generation mechanism (1) may be written as follows:

$$\begin{aligned}
&5.32\ N_2 \rightarrow 2.27\ N_2^+ + 0.69\ N^+ + 2.96\ e^- + 3.05\ N + 2.36\ N(^2D),\\
&5.30\ O_2 \rightarrow 2.07\ O_2^+ + 1.23\ O^+ + 3.30\ e^- + 2.80\ O + 2.43\ O(^1D),\\
&5.86\ H_2O \rightarrow 1.99\ H_2O^+ + 1.99\ e^- + 0.29\ O + 3.58\ OH + 4.16\ H,\\
&7.47\ CO_2 \rightarrow 2.24\ CO_2^+ + 0.21\ O^+ + 2.45\ e^- + 5.02\ O + 5.02\ CO.
\end{aligned} \quad (3)$$

Gas-Phase Chemical Reactions Leading to SO_2 and NO Oxidation

Many kinds of computer codes as applied to gas-phase chemistry of SO_2 and NO oxidation in EBDS-process are in existence [18-21]. On the base of these kinetic models the optimal kinetic mechanism of the process was proposed [22]. It was constructed with help of special numerical procedure, which generates kinetic models with minimum sets of gas species and processes when given sensitivity level of appropriate goal function (NO concentration in this case) is achieved. This mechanism consists of 34 chemical reactions and 31 gas species, and can predict the behavior of main species concentrations when various process parameters (dose, temperature, initial composition etc.) change in wide range. It was chosen as a base for present modeling study of electron-beam VOCs decomposition in the industrial flue gases.

When accelerated electrons are absorbed in irradiated gas, they cause ionization and excitation of gas molecules followed by formation of primary active species (ions, atoms and radicals) and thermalization of secondary electrons that occurs within 1 ns at normal gas conditions. The subsequent charge transfer reactions lead to formation of comparable stable ions:

$$(N_2^+, O_2^+, O^+, CO_2^+) \xrightarrow{H_2O} H_2O^+ \xrightarrow{H_2O} H_3O^+ + OH. \tag{4}$$

The corresponding reverse reactions are negligible. These sequences lead to production of OH-radicals that are the main SO_2, NO and VOCs oxidants. The second group of charge species reactions represents the formation and transformation of negative ions:

$$e^- \xrightarrow{O_2} O_2^- \xrightarrow{NO_2} NO_2^- \xrightarrow{HNO_3} NO_3^-. \tag{5}$$

Ionic recombination is the final stage of charge species chemistry, and in accepted kinetic model it is described by the reaction:

$$H_3O^+ + (NO_2^-, NO_3^-) \rightarrow H + H_2O + (NO_2, NO_3). \tag{6}$$

Reaction group describing SO_2 oxidation in adopted kinetic scheme is sufficiently easy, and represented by the reaction chain leading to sulfuric acid vapor formation:

$$SO_2 \xrightarrow{OH} HSO_3 \xrightarrow{O_2} SO_3 \xrightarrow{H_2O} H_2SO_4. \tag{7}$$

The results of experimental study [29] show that the rate constant of the first reaction in this chain is not depend only on temperature and pressure but also on the water vapor concentration. Therefore, in the kinetic scheme [22] the tree-molecular reaction of SO_2 and OH interaction with $M = H_2O$ as a third body is picked out as a separate reaction.

Reaction group describing formation and transformation of nitrogen containing compounds has more ramified structure. The main chain of reactions resulting in nitric acid vapor formation may be represented as:

$$NO \xrightarrow{OH} HNO_2 \xrightarrow{OH} NO_2 \xrightarrow{OH} HNO_3. \tag{8}$$

The presence of ammonia that is added to the chemical system for conversion of acids into salts leads on the one hand to N_2O generation, and on the other hand to noticeable decrease of NO concentration:

$$NH_3 \xrightarrow{OH} NH_2 \xrightarrow{NO_2} N_2O, NO + NH_2 \rightarrow N_2 + H_2O. \tag{9}$$

As it is evident from (7) and (8), OH radicals generated in the flue gases both in radiolytic reactions (3) and in charge transfer reactions (4) play an important role in the radiation-induced SO_2 and NO oxidation. Reactions of these oxides with other active species, in particular, with thermalized electrons e^- are not so significant, and were omitted at the construction of optimal kinetic mechanism.

Kinetic Model of Liquid-Phase SO_2 Oxidation

At the typical temperatures of the EBDS-process (T = 60 – 80 °C), there are realized the conditions of bulk binary condensation of water vapor and sulfuric acid one, which is formed in gas-phase chemical processes. This leads to spontaneous aerosol droplets formation and to vapors condensation on the existing ash particles [30]. As the concentration of sulfuric acid in the gas is much lower than that of water vapor and Henry's Law constant for H_2SO_4 is very high, it may be assumed that the whole amount of sulfuric acid formed transfers into solution and a change in the partial vapor pressure of water can be neglected.

The volume condensation in the H_2O-H_2SO_4 system as applied to EBDS-process conditions was considered in [22]. It includes such stages as binary nucleation of H_2O and H_2SO_4 vapors, heterogeneous condensation and coagulation. Increase in droplets size can also be due to the liquid-phase SO_2 oxidation, which leads to H_2SO_4 formation. Numerical investigation of the condensation process showed that formation of liquid phase is fast in comparison with gas-phase H_2SO_4 generation. In this case, liquid-phase content in gas L is determined by H_2SO_4 concentration in chemical system, and the mass concentration of the acid in droplets y is a function of the temperature and partial pressure of the water vapor:

$$L = (\mu_{H_2SO_4} / \rho_l) c_{H_2SO_4} (1 - y) / y, \qquad y = y(T, p_{H_2O}) \tag{10}$$

where $\mu_{H_2SO_4}$ is H_2SO_4 molecular weight, ρ_l is liquid density, $c_{H_2SO_4}$ is mole-mass concentration of sulfuric acid in the reacting system.

The liquid-phase formation is accompanied with dissolution of other gaseous species, such as SO_2, NH_3, OH etc. in the aerosol droplets. This process gives rise to liquid-phase SO_2 oxidation which includes the following steps: the diffusion of gas-phase species to the surface of aerosol droplet, the dissolution of these species in the surface layer of the droplet, the diffusion of the dissolved species to the interior of the droplet, the dissociation of dissolved sulfur dioxide, and its reactions. The estimation of the characteristic times of these steps shows that the last one limits the rate of the process. Therefore the process of dissolution is sufficiently fast, and one can assume the existence of quasi-equilibrium among gas and liquid for these species. In this approximation the mole-mass concentration c_i for dissoluble component i is the concentration of species aggregate consisted of one gas phase and one or several liquid phase components that are connected with each other by the use of equilibrium conditions.

The liquid-phase SO_2 oxidation in aerosol droplets plays a significant role in EBDS-process [31]. The hydroxyl radicals keep a central place in this oxidation. These radicals are formed in the gas-phase via radiation chemical reactions and then dissolved in aerosol droplets. The radiolysis of liquid water gives insignificant contribution to OH-concentration owing to small liquid-phase content in the gas [22] and big primary radiation yields (G-values) for water vapour compared to those of liquid water [32]. The reaction of OH-radicals with dissolved SO_2 initiates a chain liquid-phase oxidation of SO_2 in droplets. The accepted kinetic scheme of the liquid-phase reactions was developed in [22] on the base of available kinetic schemes, and can be written as:

$$HSO_3^- \xrightarrow{OH} SO_5^- \xrightarrow{HSO_3^-} SO_5^- + HSO_5^- \xrightarrow{HSO_3^-} 2\,SO_4^{2-}$$

$$SO_5^- \swarrow S_2O_8^{2-} \qquad SO_5^- \searrow SO_4^- \xrightarrow{HSO_3^-} SO_5^- + SO_4^{2-} \qquad (11)$$

Initiation of the chain SO_2 oxidation in this scheme proceeds via the reaction of liquid SO_2 component, HSO_3^-, with hydroxyl radical followed by chain SO_5^- transformation to SO_4^{2-}. The termination reaction leads to peroxydisulfate $S_2O_8^{2-}$ production, which is converted to SO_4^{2-}.

Thermal SO_2-NH_3 Interaction

Simultaneously with relatively fast gas- and liquid-phase chemical reactions there is slower, direct interaction of gaseous ammonia and sulfur dioxide in the absence of electron-beam irradiation – "thermal" SO_2-NH_3 reaction [33]. This reaction leads to additional reduction of SO_2 concentration and formation of solid ammonia sulfate. The kinetic mechanism of this process is not studied completely. At room temperature the reaction is irreversible, and has first order with respect to both SO_2 and NH_3 [34]. The following kinetic scheme of the process was proposed:

$$SO_2(g) + NH_3(g) \rightarrow NH_3 \cdot SO_2(g) \xrightarrow{NH_3(g)} (NH_3)_2 \cdot SO_2(s) \xrightarrow{O_2, H_2O} (NH_4)_2 \cdot SO_4(s). \quad (12)$$

The first, gas-phase reaction in (12) is the limiting stage with rate constant $2.74 \times 10^{-5} \exp(9000/T)$ $cm^3 mol^{-1} s^{-1}$, which was restored in [22] from experimental data [35] at temperatures 20 - 120 °C.

Kinetic Model of PAHs Molecules Transformation

The OH-radicals interaction with PAHs molecules plays a dominant initiating role in the aromatic compounds transformation at atmospheric conditions [36] as well as at high temperature hydrocarbons combustion [37]. This interaction is characterized by the reaction rate constant k_{OH} as a non-monotonic function of temperature [38]. Two temperature ranges can be distinguished, which differ in reaction pathways and, correspondingly, in behaviour of $k_{OH} = k_{OH}(T)$ curves. At low temperatures, the reaction proceeds in accordance with OH-addition mechanism. For single-ringed compounds, in particular, benzene this range is lower 100 °C. Higher-ringed aromatics satisfy the OH-addition conditions at elevated temperatures; namely, upper temperature limit for anthracene is near 400 °C [39]. The dominant mechanism of reaction at high temperatures is H atom abstraction from aromatic molecule.

Low-Temperature Destruction of PAHs Compounds

In low temperature range, the interaction of the OH radical with the aromatic molecule A_i, containing i aromatic rings, can be represented as the reaction sequence of OH radical addition to A_i, the formation of a bi-cyclic radical, and the subsequent decomposition of the aromatic ring in the interaction of the bi-cyclic radical with molecular oxygen and nitrogen oxide [36]:

OH + → H OH $\xrightarrow{O_2}$ H OH O_2 →
(OH-adduct) (peroxyl radical)
→ H OH O O $\xrightarrow[NO_2]{O_2,\ NO}$ H OH O O H O → H OH O O H O = C etc.

(13)

The conversion of aromatic molecules occurs in direction of decrease in the number of aromatic rings. The rate of this process is practically independent of temperature and governed by the rate of first step with rate constant k_{OH}, which depends on the type of compound.

It is necessary to note, that as a result of the low temperature decomposition process new PAHs are formed [15]. These compounds have smaller number of the aromatic rings, and occasionally can be even more toxic, than the original ones. The formation of such reaction products can be included to proposed kinetic model, but requires special consideration for every individual compound.

The dissolution of gaseous PAHs in the aerosol droplets leads to appearance of the liquid-phase channel for the decomposition of these compounds under conditions of EBDS-process. The OH radicals play a principal role in aqueous organics oxidation [40]. The kinetic scheme of aromatic molecules decomposition can be represented as OH radical addition to PAH molecule followed by the decay of formed adduct under its interaction with molecular oxygen [41]. The rate constant k_{OH} for the first, rate-determining step of the considering process is approximately equal: $k_{OH} \approx 10^{10}$ l mol^{-1}s^{-1} for various PAHs [42,43]. The Henry's Law constants H that describe PAHs solubility in aerosol droplets change in a wide range for the different kinds of aromatic compounds, and are strong temperature dependent [44,45].

High-Temperature PAHs Transformation

At high temperatures ($T >$ 120 °C for single ring aromatics), the reaction of aromatic molecule with OH radical leads to hydrogen atom abstraction from A_i and the formation of $A_i \bullet$ radical. The subsequent transformation of this radical leading to growth of aromatic structure can be described by the following polymerization type sequence with participation of acetylene molecules [46]:

$$\begin{array}{ccccccc} \longrightarrow A_i & \xrightarrow{\mathrm{OH,\,H}} & A_i \bullet & \xrightarrow{\mathrm{C_2H_2}} & A_i\mathrm{C_2H_2} \bullet & \xrightarrow{\mathrm{C_2H_2}} & A_{i+1} \longrightarrow \\ \downarrow \mathrm{O} & & \downarrow \mathrm{O_2} & & & & \downarrow \mathrm{O} \\ A_{i\text{-}1} & & A_{i-1} & & & & A_i \end{array} \qquad (14)$$

where $A_i\mathrm{C_2H_2} \bullet$ is aromatic radical formed by the addition of acetylene molecule to $A_i \bullet$. On the one hand, this reaction sequence is a simplified kinetic mechanism of PAHs growth. On the other hand, it adequately represents the main steps in the PAHs growth: hydrogen abstraction in the interaction of the aromatic molecule with hydrogen atoms and OH radicals, and the subsequent addition of C_2H_2. The decomposition of the aromatic molecules occurs due to their interaction with O atoms and O_2 molecules.

The formation of a first aromatic ring in conditions of the EBDS-process is described by interaction of the active species with aliphatic hydrocarbons. Under conditions favourable for incomplete fuel combustion in combustion chamber, the residual concentrations of these hydrocarbons in combustion products are sufficiently high. One of the possible pathways of aliphatic hydrocarbons transformation into single-aromatic compounds is the formation of the divinyl radical n-C_4H_5 followed by its cyclization into benzene in the reaction with acetylene [47]:

$$\begin{array}{ccccccc} \mathrm{C_2H_4} & \xrightarrow{\mathrm{OH}} & \mathrm{C_2H_3} \bullet & \xrightarrow{\mathrm{C_2H_2}} & n-\mathrm{C_4H_5} \bullet & \xrightarrow{\mathrm{C_2H_2}} & \mathrm{C_6H_6}, \\ \downarrow \mathrm{O} & & \downarrow \mathrm{O_2} & & \downarrow \mathrm{O_2} & & \\ \mathrm{CO, CH_2O} & & \mathrm{CO, CH_2O} & & \mathrm{CO, C_2H_2} & & \end{array} \qquad (15)$$

which was used in [25] for the estimation of conversion degree in given temperature range. This mechanism was called in question by Miller and Melius [48], who suggested that n-C_4H_5 radical could not be presented in gas in sufficiently high concentration to form benzene because it transforms rapidly to resonantly stabilized isomer i-C_4H_5. In this case the recombination of propargyl radicals (C_3H_3) is the most important source of benzene production. The corresponding reaction chain can be written as:

$$\begin{array}{ccccccc} \mathrm{C_2H_2} & \xrightarrow{\mathrm{O}} & {}^1\mathrm{CH_2} \bullet & \xrightarrow{\mathrm{C_2H_2}} & \mathrm{C_3H_3} \bullet & \xrightarrow{\mathrm{C_3H_3\bullet}} & \mathrm{C_6H_6}. \\ & & \downarrow \mathrm{O_2} & & \downarrow \mathrm{O_2} & & \\ & & \mathrm{CO} & & \mathrm{CO, CH_2CO} & & \end{array} \qquad (16)$$

Table 1. Kinetic Scheme of the First Aromatic Ring Formation

Reaction[a]	*A*	*n*	*E*/*R*, K
$C_2H_2 + O \rightarrow {}^1CH_2 + CO$	1.0E+07	2.0	956
${}^1CH_2 + O_2 \rightarrow H + OH + CO$	2.8E+13	-	-
$C_2H_2 + {}^1CH_2 \rightarrow C_3H_3 + H$	4.0E+13	-	-
$C_3H_3 + C_3H_3 \rightarrow C_6H_6$	5.0E+13	-	-
$C_3H_3 + O_2 \rightarrow CH_2CO + HCO$	3.0E+10	-	1450

[a] Rate constant: $k = AT^n \exp(-E/RT)$, $cm^3 mol^{-1} s^{-1}$.

Rate constants for the single reactions in simplified kinetic scheme of benzene formation (16) were taken from [37,49] (see Table 1). The oxidation of aliphatic hydrocarbons under irradiation results in formation of aldehydes, alcohols, acids, peroxides, etc. In proposed simplified kinetic model, carbon oxides (CO and CO_2), formaldehyde (CH_2O), and ketene (CH_2CO) act as the products of radiation-induced oxidation of C_2H_4 and C_2H_2. It is necessary to point out that the kinetic data connected with transformation of aliphatic hydrocarbons are usually originated from combustion chemistry. Extrapolation of the high temperature rate constants to relatively low-temperature conditions peculiar to the EBDS-process needs care.

Chlorination of Aromatic Compounds

Organic fuels contain trace amounts of chlorine. For example, Cl-content in coal varies from a few mg kg^{-1} up to 1 g kg^{-1}, depending on the origin. During combustion, chlorine is released in the form of gaseous hydrogen chloride. HCl can react with organic matter in conditions of fuel combustion forming chlorinated aliphatic and aromatic compounds [50]. At smaller temperatures typical for the EBDS-process, HCl molecules interact with active species (O, OH, H), which leads to generation of Cl-atom [51]. The latter plays significant role in chlorination of the aromatic compounds.

Similar to the OH-radical, Cl-atom can react with benzene via the following two reaction channels [52]:

$$Cl + C_6H_6 \rightarrow HCl + C_6H_5\bullet, \quad (17)$$

$$Cl + C_6H_6 = [C_6H_6Cl] \rightarrow \text{products.} \quad (18)$$

Reaction (17) is the H-abstraction one leading to phenyl formation. Reaction (18) is the Cl-atom addition to the aromatic ring, where $[C_6H_6Cl]$ is the unstable intermediate that can either decompose back to the original reactants or further react to form reaction products. The relative importance of these two channels depends on the stability of the $[C_6H_6Cl]$ intermediate and the energy barrier on the potential energy surface leading to the formation of this intermediate. For the OH-radical reaction with benzene, the OH-addition is predominant channel in the low temperature range leading to decomposition of formed intermediate. As product analyses shows, the Cl-atom addition reaction with benzene leads to chlorobenzene formation [52]:

$$Cl + C_6H_6 = [C_6H_6Cl] \xrightarrow{O_2} C_6H_5Cl + HO_2. \tag{19}$$

The estimation of the reaction rates for the channels (17) and (19) shows that the last one is predominant [53]. Corresponding rate constants for the direct reactions in (19) are 6.0 × 10^{12} $cm^3mol^{-1}s^{-1}$, and 5.0 × 10^7 $cm^3mol^{-1}s^{-1}$. Rate constant for the reverse reaction of intermediate decomposition to the original reactants is 10^7 s^{-1}.

Kinetic Model of Radiation-Induced PCDD/F Transformation

Problems of PCDD/Fs destruction under electron beam irradiation were experimentally investigated at various process conditions [17,54]. Theoretical process description is based on the PCDD/Fs decomposition in gas-phase reactions with OH radicals produced by the irradiation of flue gases [17]. The presence of active species (atoms and radicals) in a radiation-chemical zone and their reactions with unburned gaseous and condensed aromatic compounds can result in an opposite effect: the formation of PCDD/Fs in considerable amounts [55]. Under certain process conditions, the concentration of these compounds can exceed their initial concentration. Modeling study of such processes allows solving the problem of optimization of electron-beam process parameters with respect to minimization of PCDD/Fs concentrations in the flue gases.

Destruction of PCDD/F under Irradiation

As it follows from previous consideration, OH-radicals play an important role in radiation-induced oxidation of SO_2, NO and PAHs molecules. These radicals are also the main active species responsible for decomposition of PCDD/Fs molecules [56]. The simplified scheme of the process in gas phase can be represented by analogy with decomposition of aromatic compounds as the reactions sequence of OH radical addition to PCDD/Fs molecule and formation of OH-adduct followed by the decomposition of this adduct in the reaction with O_2 and NO. The rate of this process is determined by the rate of the first step with the rate constant k_{OH}, which depends on the nature of the compound [57]. To evaluate the total PCDD/Fs removal efficiency, the mole fractions x_i must be multiplied by toxicity equivalence factors TEF_i, which relate the toxicity of a given PCDD/F to the toxicity of 2,3,7,8-TCDD [58]. This procedure gives the toxicity equivalence (TEQ) concentration of PCDD/Fs. Table 2 summarizes the mole fractions x_i and corresponding values of k_{OH} and TEF_i for various PCDD/Fs. It should be noted that the reaction of OH radical addition to the aromatic molecule dominates only at low temperatures ($T \leq 100$ °C for single ring aromatics), where the rate constant k_{OH} practically does not depend on temperature. For PCDD/Fs molecules there is no experimental evidence for change in reaction mechanism from OH addition to H abstraction for temperatures up to 600 °C [59].

Table 2. Physical and Chemical Properties of PCDD/F.

Compounds	x_i (ppt)	k_{OH}, $cm^3 mol^{-1} s^{-1}$	TEF_i	K_{298}, $mol\ l^{-1} Pa^{-1}$
ΣCl_4DD	0.245	7.24E+12	1.0	4.67E-04
ΣCl_5DD	0.738	6.03E+12	0.5	3.76E-03
ΣCl_6DD	1.15	3.24E+12	0.1	9.23E-04
ΣCl_7DD	1.75	2.09E+12	0.01	7.86E-04
Cl_8DD	2.40	9.55E+11	0.001	1.46E-03
PCDF	11.3	3.63E+12	0.1	6.84E-04

The phase state of PCDD/Fs is an important question in the context of the applicability of radiation technology to the removal of these compounds. Dioxins undergo partial absorption in the pores of soot and ash particles, which are present in combustion products, and thus don't more participate in gas-phase chemical reactions. The distribution of a given component i between phases is described by the partition coefficient $\varphi_i = c_G /(c_G + c_P)$, where c_G and c_P are its content (g/m^3 of gas) in the gas and solid particles, respectively. The radiation-induced reduction of the PCDD/Fs concentrations in the gas-phase causes the re-establishment of the gas-particulate partitioning for these compounds, resulting in the reduction of the PCDD/Fs concentrations in the particulate phase [17].

Dioxins are readily soluble in the liquid phase, thus resulting in the appearance of the liquid-phase channel for the decomposition of these compounds at temperatures $T \leq 85$ °C, where the formation of aerosol droplets under bulk binary condensation of water and sulfuric acid vapors occurs. The OH and NO_3 radicals play a principal role in aqueous organics oxidation [40]. These radicals are formed in the gas-phase via radiation chemical reactions and then dissolved in aerosol droplets. By analogy with the liquid-phase oxidation of aromatic compounds [41], the kinetic scheme of PCDD/Fs decomposition can be represented as OH radical addition to PCDD/Fs molecule followed by the decay of formed OH-adduct under its interaction with molecular oxygen. The rate constant for the first, rate-determining step of the process is approximately equal: $k_{OH} \approx 10^{10}$ $l\ mol^{-1}s^{-1}$ [42]. The rate constant of PCDD/Fs reaction with NO_3 radicals may be taken about three orders of magnitude smaller than the above one [60]. Table 2 summarizes the Henry's Law constants K_{298} for various PCDD/Fs at 298 K [61]. The temperature dependences $K = K_{298}\exp[-\Delta H / R(1/T - 1/298)]$, for these values may be obtained on the base of solubility data [62] and saturated vapor pressures [63]. Standard enthalpy change ΔH for TCDD used in following calculations is equal: $\Delta H / R = -5.05 \times 10^3$ K.

Radiation Induced Heterogeneous and Homogeneous PCDD/Fs Formation

The results of experimental studies indicate that thermal PCDD/Fs formation at high temperatures ($T \geq 300$ °C) is constituent of the process of aromatic structures burning down in the combustion products of organic fuels [8,64]. At lower temperatures, when the concentration of the active species in the chemically reacting system sharply decrease, the reactions of PCDD/Fs formation occur at a very low rate or are frozen at all. Nevertheless, they are greatly accelerated in the presence of an additional source of active species, such as irradiation in the case of the EBDS-process.

Two mechanisms of the heterogeneous process were proposed. One of them involves the formation of PCDD/Fs molecules from structural elements of the unburned organic mass in solid particles (mechanism *de novo*). The other considers the formation of PCDD/Fs from small aromatic molecules-precursors, which undergo adsorption on the surface of ash particles and then react with each other (mechanism of *precursors*). Homogeneous, gas-phase PCDD/Fs formation from precursors also plays a significant role in overall process [65].

The available kinetic models of the thermal heterogeneous PCDD/Fs formation in *de novo* process are attached to particular test conditions and don't describe all features of the process [66,67]. These models are as a rule empirical and global ones, and don't consider the chemical processes on elementary level. This prevents from their application to conditions that are typical for the EBDS-process. The most appropriate kinetic model for description of the heterogeneous oxidation of residual carbon C_s [68] contains main chemical stages leading to the dioxins formation, and allows the consideration of low temperature radiation induced processes in flue gases. The activation of C_s is considered as the formation of free valence in condensed aromatic structure under interaction of a constituent $\geq C - H$ group with O atoms and OH radicals [69]:

$$C_s + O \rightarrow (CO)_s\bullet + H,\ C_s + OH \rightarrow C_s\bullet + H_2O, \qquad (20)$$

where $(CO)_s\bullet$ is the condensed aromatic structure with one free valence, which contains the group $\geq C - O\bullet$. Activated carbon $C_s\bullet$ is converted into $(CO)_s\bullet$ in the reaction with O_2 molecules followed by the destruction of the corresponding benzene ring of the aromatic structure and the release of CO molecules into the gas phase:

$$C_s\bullet + O_2 \rightarrow (CO)_s\bullet \rightarrow CO + \text{products}. \qquad (21)$$

The reaction of dioxins (DDs) formation can be represented in a simplified manner as a reaction of two activated carbon atoms in the form of $(CO)_s\bullet$ followed by the liberation of the dioxin molecule from the new aromatic structure formed during this reaction:

$$(CO)_s\bullet + (CO)_s\bullet \rightarrow DD + \text{products}. \qquad (22)$$

The presence of various chlorine compounds results in the chlorination of both reactants and products. The destruction of the DDs molecules occurs in reactions with OH radicals (see above). In contrast to the high temperature range ($T > 300\ ^oC$), where O atoms and OH radicals are formed via the thermal decomposition of main gas species, the concentrations of these active species under conditions of the EBDS-process are determined as a result of competition between their radiation-induced formation, and consumption in the course of SO_2 and NO oxidation.

The kinetic parameters of the model are given in Table 3. As the rate of C_s oxidation at temperatures $T \leq 300^oC$ is low, the heterogeneous reaction proceeds in the kinetic regime. The molar-volume concentrations of both gas-phase and condensed species were chosen as the main variables in calculations. This allows writing the set of kinetic equations in a unified form. The reaction rate constants were taken in accordance with data [37] for aromatic compounds. The chlorination of dioxins is not considered in this kinetic scheme.

Table 3. Mechanisms of Dioxins Formation under Electron-beam Irradiation

Reaction[a,b]	*A*	*n*	*E*/*R*, K
Heterogeneous *de novo* mechanism			
$C_s + OH \rightarrow C_s\bullet + H_2O$	1.6E+08	1.4	7.30E+02
$C_s\bullet + O_2 \rightarrow (CO)_s\bullet + O$	2.1E+12	-	3.76E+03
$C_s + O \rightarrow (CO)_s\bullet + H$	2.2E+13	-	2.28E+03
$(CO)_s\bullet \rightarrow CO$ + products	2.5E+11	-	2.21E+04
$(CO)_s\bullet + (CO)_s\bullet$ = DD + products	1.1E+23	-2.9	8.00E+03
DD + OH $\rightarrow$ products	1.0E+12	-	-7.13E+02
Gas-phase *precursors* mechanism			
$P + OH \rightarrow P\bullet + H_2O$	1.0E+12	-	-
$P\bullet + OH \rightarrow$ products	1.0E+12	-	-
$P\bullet + O_2 \rightarrow$ products	1.0E+07	-	-
$P\bullet + P\bullet \rightarrow PD$	1.0E+11	-	-
$PD + OH \rightarrow DD + H_2O$	1.0E+12	-	-

[a] Rate constant: $k = AT^n \exp(-E/RT)$, $cm^3mol^{-1}s^{-1}$.

[b] Names of compounds see in text.

The gas-phase *precursors* mechanism of the DDs formation from polychlorinated phenols is one of the main process channels in addition to the usually assumed heterogeneous reactions [65]. The available kinetic models of this mechanism [70,71] are originated from Shaub and Tsang [72] model, and may be used in the low temperature range for estimation of radiation induced DDs formation. The simplified kinetic scheme of the process can be represented as [70]:

$$P \xrightarrow{OH} P\bullet \xrightarrow{P\bullet} PD \xrightarrow{OH} DD, \quad \downarrow^{O_2,\,OH} \text{products} \tag{23}$$

where P – polychlorinated phenol, P• - polychlorinated phenoxi radical, and PD – polychlorinated 2-phenoxyphenol.

The model [70] was applied in the present study to description of the gas-phase polychlorinated phenols transformation to the DDs molecules under electron beam irradiation. The main initiators of this transformation are OH radicals. The rate constants of separate reactions in (23) are given in Table 3. Although HCl can be an important product of combustion process, the presence of Cl atoms in the low temperature range is negligible, and reactions with participation of these atoms are omitted.

It should be noted that the gas-phase *precursors* mechanism of DDs formation from polychlorinated phenols is valid only at temperatures $T \geq 120$ °C where initiation step (interaction of phenol molecule with OH radical) proceeds on the way of H atom abstraction from P. At lower temperatures the reaction channel changes on H atom addition to P followed by the destruction of formed adduct (see above).

Radiation-Induced PCBs Transformation

As was pointed out below, over recent years there has been increasing interest in the exposure of the human population to polychlorinated biphenyls (PCBs) and in particular those PCBs congeners that are considered to have "dioxin-like" toxicity. Levels of "dioxin-like" PCBs measured in the industrial flue gases show that well-controlled modern combustion plants with comprehensive pollution control give low emissions, typically about 5-10% of the toxic equivalent of the emissions of PCDD/Fs at the same plants and below the widely used standard of 0.1 ng TEQ/Nm^3 [10]. Emission factors for the domestic burning of coal are higher and average 1000 ng/kg fuel for ΣPCBs (near 100 ng/Nm^3 of flue gases) and 100 ng/kg fuel for ΣPCDD/Fs [73].

Table 4. Kinetic Mechanism of Biphenyls Formation

Reaction[a]	*A*	*n*	*E*/*R*, K
$C_6H_5\bullet + C_6H_6 \rightarrow C_6H_5\text{-}C_6H_5 + H$	7.31E+26	-4.42	6.60E+03
$C_6H_5\bullet + C_6H_5Cl \rightarrow C_6H_5\text{-}C_6H_5 + Cl$	1.00E+12	-	2.17E+03
$C_6H_5\bullet + C_6H_5Cl \rightarrow C_6H_5\text{-}C_6H_4Cl + H$	4.50E+49	-10.65	1.61E+04
$C_6H_5\bullet + C_6H_5\bullet \rightarrow C_6H_5\text{-}C_6H_5$	1.40E+13	-	5.60E+01
$C_6H_4Cl\bullet + C_6H_4Cl\bullet \rightarrow C_6H_4Cl\text{-}C_6H_4Cl$	1.80E+13	-	-
$C_6H_4Cl\bullet + C_6H_5Cl \rightarrow C_6H_4Cl\text{-}C_6H_5 + Cl$	2.60E+12	-0.13	2.23E+03
$C_6H_4Cl\bullet + C_6H_5Cl \rightarrow C_6H_4Cl\text{-}C_6H_4Cl + H$	5.43E+15	-1.39	1.94E+03
$C_6H_6 + OH \rightarrow C_6H_5\bullet + H_2O$	1.58E+08	1.42	7.30E+02
$C_6H_5Cl + OH \rightarrow C_6H_4Cl\bullet + H_2O$	1.58E+08	1.42	7.30E+02
$C_6H_5\bullet + O_2 \rightarrow$ products	2.10E+12	-	3.76E+03
$C_6H_4Cl\bullet + O_2 \rightarrow$ products	2.10E+12	-	3.76E+03
$C_6H_6 + O \rightarrow$ products	2.20E+13	-	2.28E+03
$C_6H_5Cl + O \rightarrow$ products	2.20E+13	-	2.28E+03
PCBs + OH $\rightarrow$ products	1.58E+08	1.42	7.30E+02
PCBs + O $\rightarrow$ products	2.20E+13	-	2.28E+03

a Rate constant: $k = AT^n \exp(-E/RT)$, cm3mol-1s-1.Formation of Organic Nitrates and Nitro-Aromatic Compounds

The major removal pathway for PCBs from the industrial flue gases just as for other considered pollutants is their reaction with OH radicals [74]. Measured second-order rate constants k_{OH} for these species are ranged from 3.0×10^{12} $cm^3mol^{-1}s^{-1}$ for 3-chlorobiphenyl to 2.4×10^{11} $cm^3mol^{-1}s^{-1}$ for 2,2',3,5',6-pentachlorobiphenyl [75] confirming theoretical estimations that PCBs contained smaller number of chlorine atoms react with the OH radical at faster rates than the more highly chlorinated congeners [76]. The mechanism of reaction as in the case of other aromatic compounds consists of reactions sequence of OH radical addition to PCBs molecule and formation of OH-adduct followed by the decomposition of this adduct in the reaction with O_2.

The formation of PCBs takes place in high temperature range of the EBDS-process (see *High-Temperature PAHs Transformation* above) via phenyl radical addition to either chlorobenzene or benzene, or direct recombination of phenyl and chlorinated ones radicals. The first reaction goes through a substituted cyclohexadienyl intermediate formation:

$$C_6H_5\bullet + C_6H_5Cl = [C_6H_5\text{-}Cl\text{-}C_6H_5\bullet] \rightarrow C_6H_5C_6H_5 \text{ (or } C_6H_5C_6H_4Cl) + Cl \text{ (or H)} \quad (24)$$

Calculations show that more than 90% of the collisions result in biphenyl + Cl for phenyl addition to the chlorobenzene ipso position [77]. This appears to be a primary route to biphenyl production in this reaction. Phenyl addition at C-H position in benzene or chlorobenzene results in nearly equal amounts of stabilized intermediate and the biphenyl or chlorobiphenyl + H products. Recombination of phenyl radicals leads directly to biphenyl molecules formation [78]. The simplified kinetic scheme of the PCBs formation process has been adopted on the base of available kinetic data (see Table 4). Rate constants for the separate reactions are taken from [37,77,78].

The OH radicals are the main active component that initiates the reaction sequences leaded to formation of nitrated organic compounds. For example, this sequence for alkanes can be written generally as [79]:

$$\begin{array}{c} \text{ONIT} \\ \uparrow \scriptstyle NO \\ RH \xrightarrow{OH} R\bullet \xrightarrow{O_2} RO_2\bullet \longrightarrow RO\bullet \xrightarrow{O_2} ALD + HO_2\bullet \\ \downarrow \scriptstyle HO2 \\ \text{ROOH} \end{array} \quad (25)$$

where $R\bullet = C_nH_{2n+1}$, $ALD = C_{n-1}H_{2n-1}CHO$ ($n = 2,3,...$). Aldehydes (ALD) and organic nitrates (ONIT) are the main products of reaction. Hydroperoxyde, ROOH, is a by-product of reaction from standpoint of subsequent peroxyacetyl nitrates (PAN) formation. In case of olefins, the set of chemical reactions leading to aldehydes formation have about the same form. For example, for propene (PRO) one can write:

$$PRO + OH \xrightarrow{O_2} CH_3C(OO\bullet)HCH_2OH \xrightarrow{NO+O_2} ALD + HCHO + HO_2 + NO_2. \quad (26)$$

Organic peroxy radical $CH_3C(OO\bullet)HCH_2OH$ (OPR) appears in the role of intermediate. Peroxyacetyl nitrates are generated in the sequence of transformations [80]:

$$\begin{array}{l} ALD \xrightarrow{OH} R'CO\bullet \xrightarrow{O_2} R'C(O)O_2\bullet \xrightarrow{NO_2} PAN \\ \qquad\qquad\qquad\qquad {\scriptstyle NO}\downarrow \qquad \downarrow {\scriptstyle HO_2} \\ \qquad\qquad\qquad\quad R'O_2\bullet \quad R'C(O)O2H \end{array} \quad (27)$$

where R'• = $C_{n-1}H_{2n-1}$. The rate constants of corresponding reactions are given in [25]. In the process of ethane oxidation, the rate constant of the first step is about one order smaller then the corresponding one for higher alkanes. Therefore, in the accepted kinetic scheme the process of ethane interaction with OH radicals is presented as the separate reaction.

As we see above, the interaction of the OH radical with the aromatic molecule A_i in low temperature range ($T \leq$ 80 °C) leads to formation of aromatic radical A_i-OH•. The subsequent decomposition of aromatic ring under interaction of A_i-OH• with molecular oxygen is in competition with formation of nitro-aromatic molecules. The appropriate reaction sequence can be schematically represented as [81]:

$$\begin{array}{l} \text{PAH} \xrightarrow{\text{OH}} \text{PAH-OH} \bullet \xrightarrow{\text{O}_2,\text{NO}} 2\text{RO}_2 \bullet + \text{products} \\ \qquad\qquad\quad \text{NO}_2 \downarrow \\ \qquad\qquad\quad \text{NPAH} \xrightarrow{\text{OH}} \text{NPAH} - \text{OH} \bullet \xrightarrow{\text{O}_2,\text{NO}} \text{products} \\ \qquad\qquad\qquad\qquad\qquad \text{NO}_2 \downarrow \\ \qquad\qquad\qquad\qquad\qquad \text{di-NPAH} \xrightarrow{\text{OH}} \text{products} \end{array} \tag{28}$$

where di-NPAH are dinitro-PAH compounds. This sequence describes main pathway of aromatic nitro-compounds formation. The process rate is governed by the rate of first step with rate constant k_{OH}, which depends on the type of compound. For benzene and toluene (slow reacting aromatics), k_{OH} is equal about to 10^{12} cm^3 mol^{-1} s^{-1}; for phenols, xylenes and cresols (fast reacting aromatics), k_{OH} is equal about to 10^{13} cm^3 mol^{-1} s^{-1}. Rate constant for reaction of OH radical with NPAH molecule is taken equal to corresponding one for nitrophenol [38].

Mathematical Model of Radiation-Induced Processes

Mathematical models commonly used in chemical kinetics are directed to solution of the *direct kinetic problem*: the calculation of time dependencies of species concentrations (kinetic curves) and thermochemical parameters of a chemical system with a given mechanism of complex chemical reaction, known reaction rate constants, and defined initial conditions. Direct kinetic problem in formal representation is a system of ordinary differential equations of first order with specified initial conditions (Cauchy mathematical problem):

$$\mathrm{d}c_i / \mathrm{d}t = f_i(c_1,...,c_n), \qquad c_i(0) = c_{i0}, \qquad i = 1,...,n, \tag{29}$$

where c_i are the species concentrations, time t is change in the interval $0 < t < t_{max}$, c_{i0} are the initial conditions of species concentrations, and n is the number of chemical species. Solution of direct chemical problem in this work is carried out with numerical methods.

Sub-model of Radiation Induced Gas-Phase Processes

The active species generation during electron beam irradiation of multi-component gas system followed by gas phase chemical reactions can be represented by kinetic equation:

$$\mathrm{d}c_i / \mathrm{d}t = \sum_{j=1}^{n} k_{ij} c_j + R_i(\mathrm{gas}), \tag{30}$$

where k_{ij} is the rate constant of component j formation owing to radiation impact on component i, and $R_i(\mathrm{gas})$ is rate of change of i component concentration because of secondary gas phase reactions. The rate constant k_{ij} in (30) is defined by the expression:

$$k_{ij} = 1.036 \times 10^{-7} I G_{ij} \sum_{k=1}^{n} x_k \mu_k, \tag{31}$$

where I is the dose rate (kGy s^{-1}), G_{ij} is the primary radiation yield (particles per 100 eV of absorbed energy), x_i and μ_i are the mole fraction and molecular weight of component i. In the case of gas phase chemical reactions the rate of change of the ith component concentration can be expressed in the form:

$$R_i(\mathrm{gas}) = \rho^{-1} \sum_{j=1}^{m} (\nu_{ij}^{-} - \nu_{ij}^{+})[k_{fj} \prod_{k=1}^{n} (\rho c_k)^{\nu_{kj}^{+}} - k_{bj} \prod_{k=1}^{n} (\rho c_k)^{\nu_{kj}^{-}}], \tag{32}$$

where ν_{ij}^{+} and ν_{ij}^{-} are the stoichiometric coefficients for reactants and products of jth reaction, k_{fj} and k_{bj} are the reaction rate constants in forward and backward directions, ρ is the gas density, c_k is mole-mass species concentrations.

Sub-model of Liquid-Phase Chemical Processes in Droplets

The H_2SO_4 molecules produced in gas-phase oxidation process rapidly acquire water vapour and nucleate to form small H_2SO_4/H_2O droplets or condense on existing solid particles. As stated above, the formation of liquid phase is fast in comparison with gas-phase H_2SO_4 generation. In this case, liquid-phase content in gas L is determined by H_2SO_4 concentration in chemical system and can be calculated from (10). The liquid-phase formation in flue gases is accompanied with dissolution of gaseous species in the aerosol droplets, which gives rise to liquid-phase SO_2 oxidation. The process of dissolution is sufficiently fast, and one can assume the existence of quasi-equilibrium among gas and liquid for deliquescent species. When liquid-phase reactions proceed, the kinetic equation (30) contains additional term R_i(liquid), the rate of change of i th component concentration owing to liquid-phase reactions:

$$R_i(\text{liquid}) = L\sum_{j=1}^{m}(\nu_{ij}^- - \nu_{ij}^+)[k_{fj}\prod_{k=1}^{n}(c_k/L)^{\nu_{kj}^+} - k_{bj}\prod_{k=1}^{n}(c_k/L)^{\nu_{kj}^-}], \quad (33)$$

It is important to note that some species in Eq. (30) are included in some components aggregates consisted of one gas phase and several liquid phase components, which are connected with each other by the use of equilibrium conditions. For example, the equilibrium correlations for SO_2-containing components are:

$$c_{SO_2} = p_{SO_2}(\rho RT)^{-1} + L([SO_2 \cdot H_2O] + [HSO_3^-] + [SO_3^{2-}]),$$

$$[SO2 \cdot H2O] = p_{SO_2} H_{SO_2}, \quad (34)$$

$$[HSO_3^-] = p_{SO_2} H_{SO_2} H_{1,SO_2} [H^+]^{-1},$$

$$[SO_3^{2-}] = p_{SO_2} H_{SO_2} H_{1,SO_2} H_{2,SO_2} [H^+]^{-2},$$

where p_{SO_2} is the partial pressure of SO_2, R is the gas constant, $[X_i] = c_i L^{-1}$, H_{SO_2} is the Henry's Law constant for SO_2, H_{1,SO_2} and H_{2,SO_2} are the dissociation constants for SO_2 in solution. Similar equilibrium correlations take place for soluble aromatic compounds. Ion concentrations in liquid phase must satisfy to the electro-neutrality equation:

$$[H^+] + [NH_4^+] = [OH^-] + [HSO_3^-] + 2[SO_3^{2-}] + [HSO_4^-] + [SO_4^{2-}] + \ldots \quad (35)$$

Substitution of equilibrium correlations (34) in (35), where lumped concentrations c_i are defined from Eq. (30), gives nonlinear algebraic equation for pH determination.

Results and Discussion

The content of main harmful admixtures, SO_2 and NO, in the industrial flue gases is much higher than VOCs one. Therefore, the concentration of active species, which are responsible for the conversion of VOCs molecules in the EBDS-prosess, depends on the set gas-phase and liquid-phase chemical reactions related to the oxidation of SO_2 and NO. Testing of the elaborated mathematical model, which describes this oxidation, is the first and important step in the numerical investigation of VOCs decomposition in the EBDS-process.

SO_2 and NO Oxidation under Irradiation

The testing of SO_2 and NO oxidation kinetic model was performed on the base of comparison of calculated results with available experimental data. The role of various process parameters such as initial SO_2 and NO concentration, temperature, humidity, absorbed dose, dose rate etc. was investigated. Some results of such comparison of model and measured data are given below.

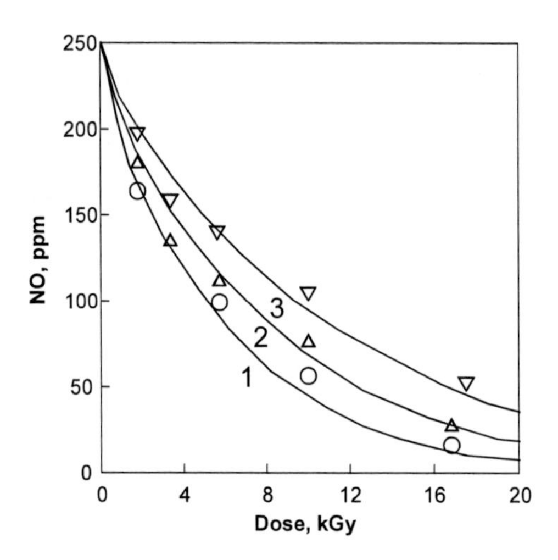

Figure 1. Calculated (lines) and measured (points) concentrations of NO as a function of dose in mixture of 12% O_2, 8% H_2O in N_2 at various temperatures: T = 100 (1), 150 (2), and 200 °C (3). Experimental data are of Tokunaga et al. [33].

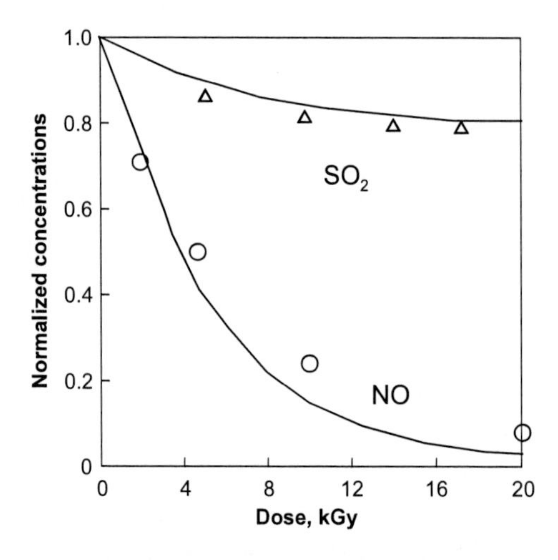

Figure 2. Calculated (lines) and measured (points) concentrations of SO_2 and NO as a function of dose in mixture of 12% O_2, 8% H_2O in N_2 at T = 100 °C. Experimental data are of Tokunaga and Suzuki [82].

Fig. 1 demonstrates the influence of temperature variation on NO conversion in the high temperature range. Testing gas contains only NO and NH_3 with initial concentrations 250 and 1500 ppm, accordingly [33]. One can see that the removal efficiency increase with temperature lowering but this increase is not impressive. In all temperature interval the degree of NO oxidation amount to 90% at absorbed radiation dose D = 20 kGy, which is typical for the EBDS-process.

The overall effect of SO_2 and NO oxidation during the process of gas-phase chemical reactions is given in Fig. 2. Initial concentrations of SO_2, NO, and NH_3 are equal 340, 250, and 750 ppm, accordingly [82]. It is observed a good agreement of calculated and measured results. One can see that the gas-phase SO_2 oxidation don't lead to the considerable removal efficiency for this gas component. At absorbed radiation dose D = 20 kGy, the degree of SO_2 oxidation hardly reach 30%.

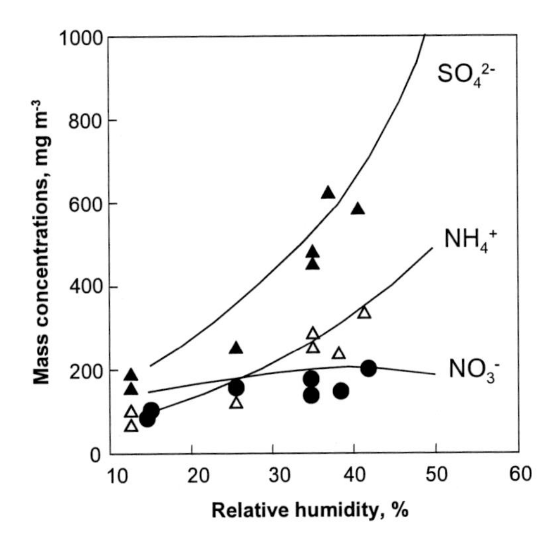

Figure 3. Chemical composition of the aerosol droplets as a function of the relative humidity in mixture of 8% O_2, 16% H_2O in N_2 at dose D = 10 kGy. Solid lines are model predictions; points are experimental data of Paur and Jordan [30].

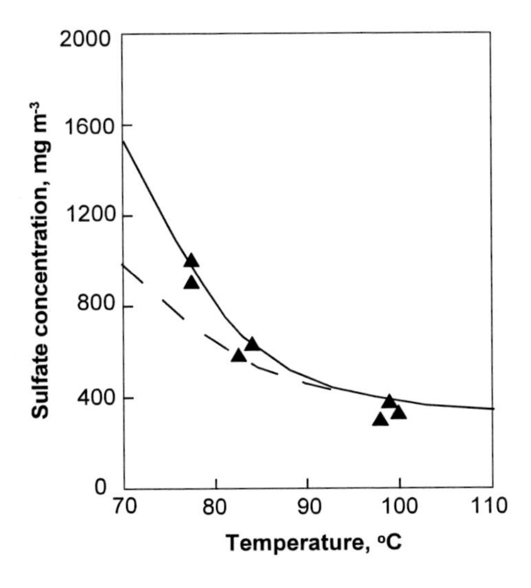

Figure 4. Sulfate concentration in the aerosol droplets as a function of temperature in previous mixture at dose D = 15 kGy. Solid and dotted line is model predictions, accordingly, with and without liquid-phase chemical reactions; points are experimental data of Paur and Jordan [30].

Figures 3 and 4 demonstrate the calculation results of SO_2 oxidation, which includes gas-phase and liquid-phase chemical reactions, and their comparison with experimental data of Paur and Jordan [30]. The initial concentrations of SO_2 and NO are equal, accordingly, 500 and 300 ppm. Water vapor content is equal to 16%. Ammonia content is stoichiometric. The calculations are carried out at the constant dose rate, I = 10 kGy s^{-1}. As one can see from Fig. 3, it is observed a good agreement of calculated and measures results for chemical composition of the aerosol droplets as a function of the relative gas humidity (ratio of partial steam pressure to saturated one at the given temperature). The increase in relative humidity up to 80 % at the given water vapor content (correspondingly, decrease of the gas temperature down to 60 °C) leads to approaching of the sulfate concentration in gas to its limit value 1760 mg m^{-3}. It is clear from Fig. 4 where sulfate concentration as a function of temperature is given. The comparison of calculated and measured data shows that the full kinetic model, which includes gas-phase and liquid-phase chemical reactions, gives better agreement. The contribution of the liquid-phase SO_2 oxidation to total SO_2 transformation grows at the temperature lowering and reaches 40% at 70 °C.

The overall effect of sulfur dioxide removal, which includes gas-phase chemical reactions (7), the chain liquid-phase oxidation of SO_2 in the aerosol droplets (11), and direct thermal SO_2 transformation into salts (12), is shown in Fig. 5. Calculations were carried out as applied to experimental conditions [83]. The initial mole concentrations of admixtures are equal: SO_2 = 1000, NO = 300, NH_3 = 2300 ppm. Temperatures on the outlet of EB-reactor are equal 70 and 80 °C. The process time is equal 0.5 s. It is assumed that the increase of absorbed dose on 1 kGy leads to rise of the process temperature on 1 °C. As one can see from Fig. 5, the results of numerical simulation are in good agreement with reported plant test results.

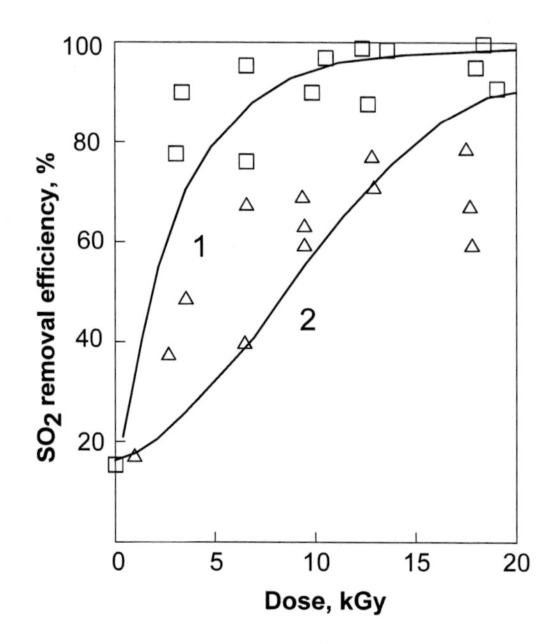

Figure 5. Calculated (lines) overall SO_2 removal efficiency as a function of dose and its comparison with plant test results [83] (points) for the industrial flue gases, containing 6% O_2, 12% H_2O, and 12 % CO_2. Temperature on the outlet of reactor is equal 70 (1) and 80 (2) °C.

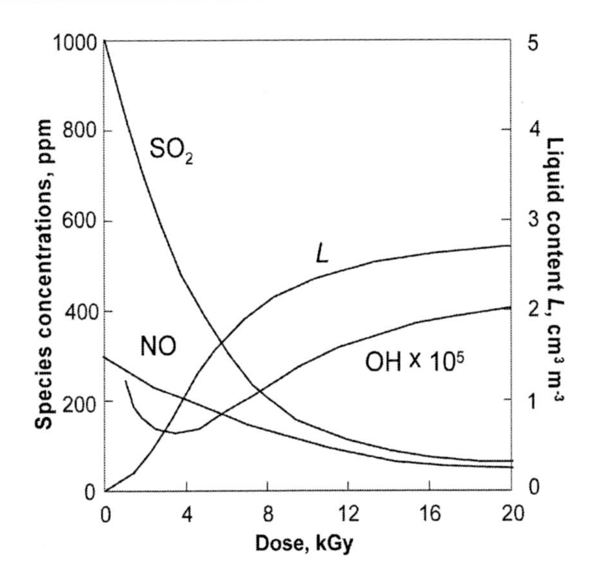

Figure 6. Behavior of the main parameters of the EBDS-process influenced on VOCs transformation under irradiation. The gas composition is the same as that for Fig. 5. Temperature of the process is equal 70 °C.

The main parameters of the EBDS-process, which influences on VOCs transformation, are presented in Fig. 6. Calculations were carried out at the same conditions as in Fig. 5. It is observed quick decrease of SO_2 and NO concentrations with dose growth, leading to corresponding increase of the liquid-phase content L in gas, which is determined in accordance with Eq. (10). The concentration of OH-radical behaves as non-monotonous function of the dose lowering down to 1.6 ppb on the interval $D = 1 - 4$ kGy, and then increasing up to 4 ppb at $D = 20$ kGy. Acidity of the aerosol droplets slowly decreases from pH = 5.3 at $D = 1$ kGy to pH = 4.5 at $D = 20$ kGy

Transformation of Aromatic Hydrocarbons

The kinetics of aromatic hydrocarbons behavior under irradiation was investigated separately for the above mentioned two temperature ranges, namely, low temperature ($T \leq 80$ °C) and high temperature ($T > 120$ °C) ones. At low temperatures, the gas composition was taken typical for the electron-beam treatment of organic-fuel combustion products in EBDS-process: $N_2 = 70$, $O_2 = 6$, $H_2O = 12$, $CO_2 = 12$ vol.%, $SO_2 = 1000$, NO = 300, $NH_3 = 2300$ ppm [83]. The initial concentration of aromatic compounds in the flue gases depends on combustion conditions and changes on a large scale from 10 ppb [50] up to 10 ppm [73]. The calculations were performed at pressure $p = 1$ atm, and radiation dose rate $I = 10$ kGy s^{-1}.

Removal efficiencies of single ring aromatic compounds in moist air at room temperature are given in Fig. 7 in comparison with experimental data [84]. Rate constants k_{OH} for calculations were taken equal to 8.1×10^{12} and 4.5×10^{13} cm^3 mol^{-1} s^{-1} for *o*-xylene and ethylbenzene, respectively [38]. The reported value of k_{OH} for chlorobenzene (5.6×10^{11} cm^3 mol^{-1} s^{-1}) gives understated removal efficiency for this compound. Therefore, in case of chlorobenzene calculations were carried out with rate constant enlarged by a factor of five. Thus, the good agreement of calculations with results of measurements demonstrates that the initial OH attack to aromatic molecules is the main pathway for the removal of these compounds under irradiation conditions.

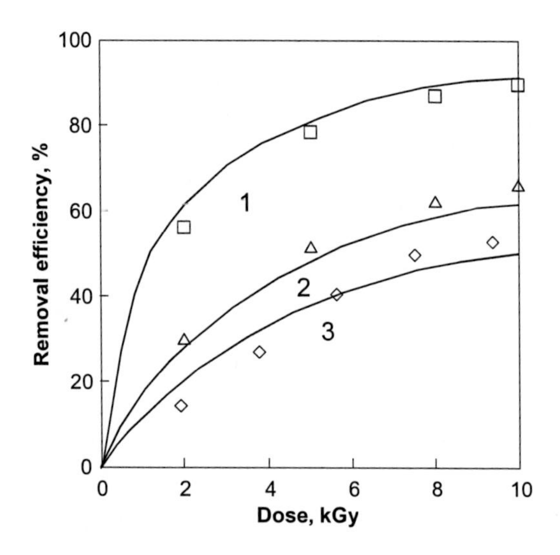

Figure 7. Comparison of model predictions (lines) with experimental data [84] (points) for removal efficiency of *o*-xylene (1), ethylbenzene (2), and chlorobenzene (3) in electron-beam irradiated moist air: H_2O = 1%, T = 25 °C.

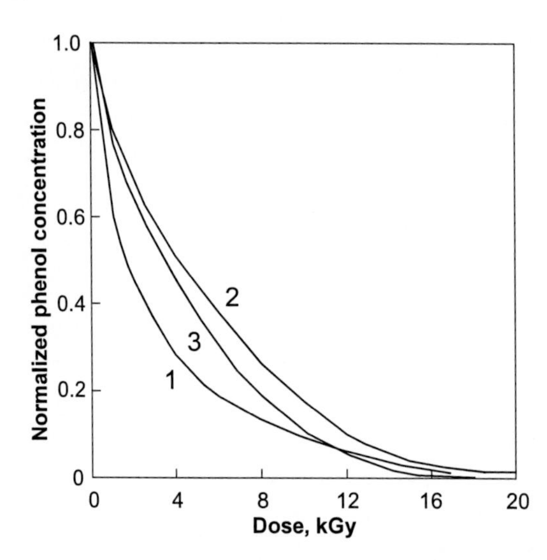

Figure 8. Influence of the liquid-phase processes on phenol oxidation: (1) – only gas-phase reactions, (2) – total kinetic scheme without liquid-phase phenol oxidation, (3) – total kinetic scheme with liquid-phase phenol oxidation.

The results of numerical study of liquid-phase reactions influence on the oxidation of aromatic molecules in EBDS-process are presented in Fig. 8 with an example of phenol at temperature T = 70 °C. The rate constant k_{OH} for gas-phase interaction of phenol molecules with OH radical is taken equal to 1.7×10^{13} $cm^3\ mol^{-1}\ s^{-1}$ [38]. The Henry's Law constant H for this compound is sufficiently high and equal to $3.17 \times 10^3 \exp[5850(1/T - 1/298)]$ mol l^{-1} atm^{-1} [44]. Liquid-phase oxidation of phenol molecules in aerosol droplets is described by rate constant $k_{OH} = 6.6 \times 10^9$ l $mol^{-1}s^{-1}$ [85]. It can be seen that, under test conditions, taking into account only gas-phase chemical reactions gives maximum decrease of phenol concentration. This is explained by low efficiency of SO_2 removal in gas-phase reactions, which leads to more high concentration of the OH radicals in the chemical system. The influence of liquid-phase phenol oxidation is small, and reaches 10% from total at D = 8 kGy. It must be pointed that in most cases the Henry's Law constant for aromatic compounds is much less [45], and influence of liquid-phase processes on these aromatic molecules oxidation under EBDS-irradiation conditions is negligible.

There are a moderate number of data in respect to the rate constants k_{OH} for PAHs molecules up to four-ringed [39,86-89]. Rate constants k_{OH} for five-ringed aromatic molecules were received in the present work by means of the fit of their calculated concentrations to measured ones in assumption that the initial concentrations of six-ringed compounds, precursors of five-ringed compounds formation, are negligible (see experimental data [15]). In this way the rate constants k_{OH} were determined for benzo(e)pyrene (BeP), benzo(a)pyrene (BaP), perylene (Pe), and dibenzo(a,h)anthracene (dBahA), which are cited in Table 5.

Fig. 9 presents the results of calculations for the dependence of some aromatic compounds normalized concentrations on absorbed dose D, and the comparison of these results with experimental data [15] obtained at D = 8 kGy, and T = 70 °C. Rate constants k_{OH} for calculations were taken equal to 3.7×10^{13}, 1.4×10^{13}, and 7.5×10^{12} $cm^3\ mol^{-1}\ s^{-1}$ for acenaphthene, naphthalene, and benzo(a)pyrene, respectively [88]. It can be seen that under the test conditions the concentrations x_i are decreased by more than an order of magnitude at the radiation dose D = 20 kGy for compounds with k_{OH} value of the order of 10^{13} $cm^3\ mol^{-1}\ s^{-1}$. The calculations show also that the formation of aromatic molecules in the reaction sequence (16) has not noticeable influence on the dynamics of the process in this temperature range.

Table 5. Estimated rate constants k_{OH} and concentrations for anthracene forming compounds

Compounds	k_{OH}, $cm^3mol^{-1}s^{-1}$	$(x_{i0})_{exp}$	$(x_i/x_{i0})_{exp}$	$(x_i/x_{i0})_{calc}$
Benzo(e)pyrene	1.2E+13	1.0E-08	0.40	0.40
Benzo(a)pyrene	7.5E+12	4.4E-08	0.56	0.56
Perylene	7.1E+12	1.4E-08	0.58	0.58
Dibenzo(a,h)anthracene	1.4E+13	2.9E-09	0.35	0.35
Anthracene	4.3E+13	3.4E-09	0.36	0.35
Benzo(a)anthracene	1.0E+14	6.4E-09	0.26	0.28

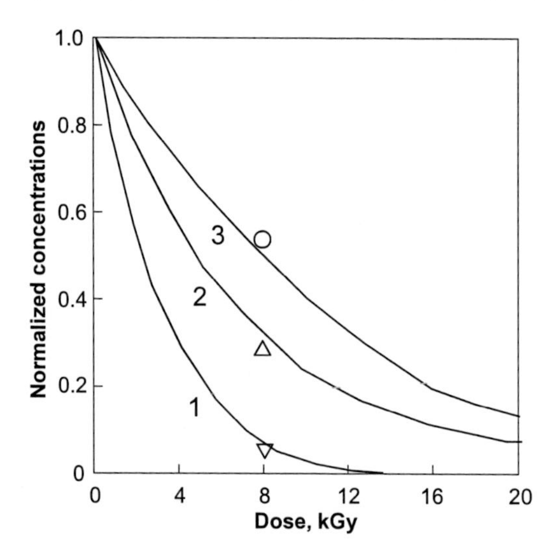

Figure 9. Calculated (lines) and measured (points) concentrations of acenaphthene (1), naphthalene (2), and benzo(a)pyrene (3) as a function of dose at T = 70 °C. Experimental data are of Chmielewski et al. [15].

Real flue gases contain various PAHs kinds differed in number of rings and in their interdependence. The interaction of OH-radicals with higher-ringed PAHs leads to their decomposition in the reaction sequence (13) and to the formation of smaller-ringed compounds as the radiation process products. Therefore, some of these compounds may have higher concentration after irradiation than the initial one. In the present work, the attempt was made to include such PAHs transformation in the computation scheme, which requires special consideration for every individual compound. The numerical study was carried out on the base the pilot plant tests of Chmielewski et al. [15] examining the influence of electron beam on the chosen PAHs during simultaneous SO_2 and NO removal.

The analysis of aromatic compounds, which are presented in the experimental data [15], allows supposing that the formation of anthracene (A) takes place in the reaction sequence:

$$\text{BaP, BeP, dBahA, Pe} \xrightarrow{\text{OH}} \text{BaA} \xrightarrow{\text{OH}} \text{A}, \tag{36}$$

followed by destruction of anthracene molecules in the reaction with OH-radicals. First step of this sequence, which leads to benzo(a)anthracene (BaA) formation, proceeds with rate constants: k_{OH} for perylene and 0.4 k_{OH} for benzo(a)pyrene, benzo(e)pyrene, and dibenzo(a,h)anthracene in accordance with ring structure of these aromatic molecules. Second step proceeds with rate constant 0.25 k_{OH}, where k_{OH} is rate constant of benzo(a)anthracene interaction with OH-radical. It is supposed that the other reaction products are formed in this kinetic scheme, which don't lead to anthracene formation.

The results of calculations for anthracene and benzo(a)anthracene concentrations received at the dose D = 8 kGy are shown in Table 5. Rate constant k_{OH} for anthracene was taken from the experimental data [39], where its value measured using a refined pulsed laser photolysis/pulsed-induced fluorescence (PLP/PLIF) technique in temperature range between 373 and 923 K is approximated by the following modified Arrhenius equation:

$k_{OH} = 4.92 \times 10^{38} T^{-8.3} \exp(-3170/T)$ cm^3 mol^{-1} s^{-1}. The corresponding value for benzo(a)anthracene was restored as the result of anthracene concentration fit to its experimental quantity. It can be seen from received data that the applied procedure gives wholly satisfactory description of PAHs compounds transformation from higher-ringed compounds to smaller-ringed ones.

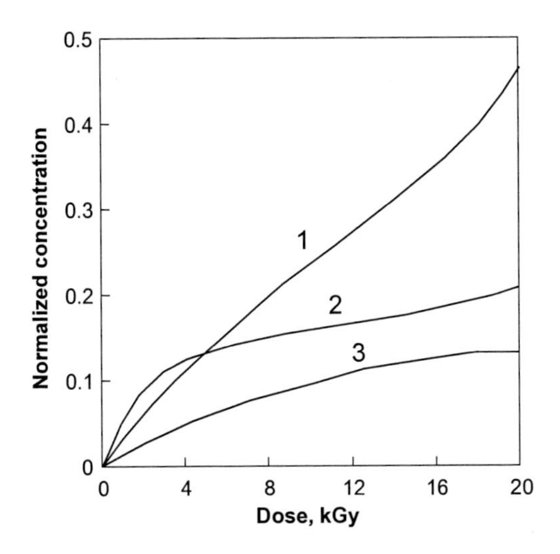

Figure 10. The fraction of benzene: (1) destructed on exposure to oxygen, (2) formed from acetylene (× 10^4), and (3) converted into naphthalene (× 10), as a function of absorbed radiation dose. The process temperature: T = 227 °C.

It will be noted that the phase state of PAHs may be important in the context of the description of these compounds behavior under electron-beam irradiation. The low-ringed PAHs compounds with number of rings less than four exist almost exclusively in the vapor phase. The high-ringed aromatic hydrocarbons undergo partial absorption in the pores of soot and ash particles, which are present in combustion products. The four-ringed PAHs are distributed already uniformly between gas and particles. The percentage of five-ringed PAHs in particulate phase increase up to 70% [90]. As was demonstrated with the example of polychlorinated dioxins [17], the electron-beam decomposition of the aromatic hydrocarbons in the particulate phase was also observed but with lower efficiency as compared with that in the gas phase. On the other hand the reduction of the PAHs concentrations in the gas phase caused the re-establishment of the gas-particle partitioning of PAHs resulting in the reduction of the PAHs concentrations in the particulate phase. In accordance with this remark the received kinetic data can be regarded as average values describing the electron-beam PAHs decomposition both in gas and particulate phases.

In high temperature range (T > 120 °C), the reaction of aromatic molecule with OH-radical leads to competition of two processes: growth of the aromatic structure under interaction with acetylene and other components presented in flue gases as well as its destruction on exposure to oxygen. Taking benzene as an example, the influence of these processes on the dependence of aromatic molecules concentration from the absorbed radiation dose is shown in Fig. 10. The following gas composition was taken for the calculations: N_2 = 65, O_2 = 2, H_2O = 19, CO_2 = 14 vol.%. These values correspond to the combustion products of lignite with air excess $\alpha = 1.1$. The initial concentrations of admixtures were chosen to be equal to the typical values at the exit from combustion zone: $(SO_2)_0 = 1000$, $(NO)_0 = 300$, $(C_2H_4)_0 = (C_2H_2)_0 = 100$ ppm, $(C_6H_6)_0 = 1$ ppm. A stoichiometric amount of ammonia $(NH_3)_0 = 2300$ ppm was added to the system in order to neutralize the resulting acids. One can see from Fig. 10 that the decomposition of benzene in reaction sequence (14) plays the main role in comparison with its formation from aliphatic hydrocarbons and conversion to naphthalene.

Fig. 11 demonstrates the temperature dependence of benzene decomposition process. The replacement of reaction channel in OH-radical interaction with aromatic molecule from OH-radical addition at low temperatures ($T \leq 80$ °C) to H-atom abstraction at high temperatures ($T > 120$ °C) leads to abrupt lowering of the rate constant and accordingly to decrease of benzene decomposition (see curve 1 in Fig. 11). The further increase of temperature gives quick growth of the rate constant k_{OH}, and the EBDS-process efficiency improves. The behaviour of naphthalene concentration as a function of absorbed radiation dose and temperature is represented in Fig. 12. The fraction of benzene converted into naphthalene is comparable with the decomposed one, only at temperatures close to 125 °C. At temperatures higher than 300 °C, the process of naphthalene formation from benzene is in competition with its destruction. This leads to decrease of naphthalene concentration on the end stage of irradiation process (see curve 3 in Fig. 12. The concentration of three-ringed aromatic molecules in calculations is approximately two orders of magnitude lower than that of naphthalene.

As mentioned above, during combustion chlorine reacts with organic matter forming chlorinated aliphatic and aromatic compounds [50]. At conditions of the EBDS-process, hydrogen chloride, the main chlorine component in the combustion products, interacts with active species resulting in Cl-atoms generation, which are responsible for chlorination of aromatic compounds in reaction sequence (19). Calculations show that this process doesn't give noticeable increase of chlorinated aromatic compounds concentrations because on the one hand the concentration of Cl-atoms doesn't exceed 1 ppt at initial HCl concentration in

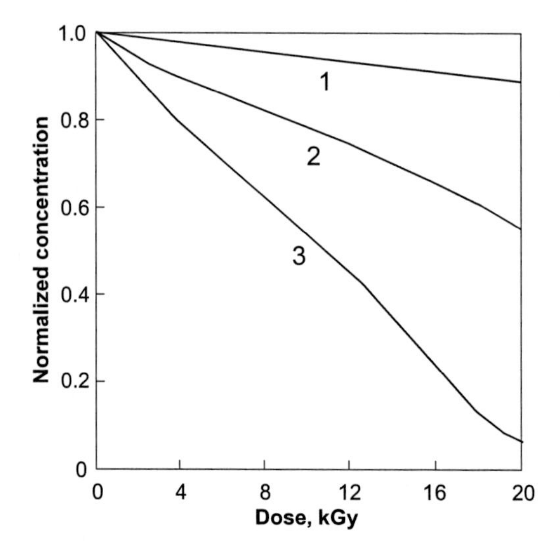

Figure 11. The behavior of benzene concentration as a function of absorbed dose. The process temperature: $T = 127$ (1), 227 (2), and 327 °C (3).

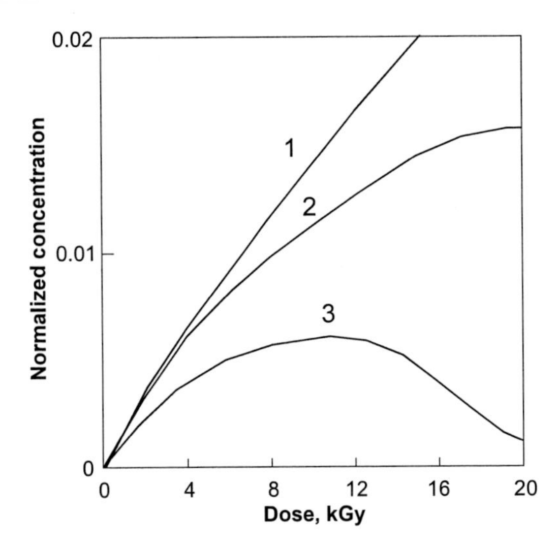

Figure 12. The behavior of naphthalene concentration as a function of absorbed dose. The process temperature: T = 127 (1), 227 (2), and 327 °C (3).

gas of the order of 100 ppm. On the other hand the rate constant of the reaction intermediate [C_6H_6Cl] transformation to C_6H_5Cl under interaction with O_2 in (19) is very small in comparison with the corresponding value for the reverse reaction of intermediate decomposition to the original reactants, C_6H_6 and Cl. Therefore, the chlorination of aromatic compounds in conditions of the EBDS-process is of no importance in contrast to combustion zone.

Radiation Induced PCDD/F Decomposition and Formation

The kinetics of the dioxins decomposition and formation under irradiation was studied by the example of the municipal solid waster incinerator (MSWI) flue gases. The initial gas composition was chosen as follows: N_2 = 57, O_2 = 10, CO_2 = 8, H_2O = 25 vol%; SO_2 = NO = 100, HCl = 400 ppm [17]. The mole fractions of the condensed carbon C_s and polychlorinated phenols changed within the ranges of 1-100 ppm, and 1-100 ppb, respectively. The initial concentration of the PCDD/Fs was taken equal 1 ppt, which corresponds approximately to10 ng-TEQ/m^3_N.

Fig. 13 shows the calculated decomposition efficiencies of PCDD/Fs as a function of dose in comparison with experimental data of Paur et al. [54] measured at T = 85 °C. The concentrations of solid carbon (C_s) and polychlorinated phenols (P) are equal zero. Calculations were carried out as applied to reduced PCDD concentration: $x_{\mathrm{PCDD}} = \sum (x_{\mathrm{PCCD}})_i (\mathrm{TEF}_{\mathrm{PCDD}})_i$. The corresponding values of mole fractions x_i, rate constants k_{OH}, and toxicity equivalence factors TEF_i were taken from Table 2. At the selection of the initial information for evaluation of PCDD/Fs decomposition efficiency in EBDS-process, it should be taken into account that PCDD/Fs concentrations for various MSWI differ greatly both in amount and composition of PCDD/Fs in flue gases. It is also important to have the information on the isomer content for a given homologue, because the TEF-values for various isomers may differ by two orders of magnitude [58]. It can be seen from Fig. 13 that a dose of approximately 8 kGy causes a 90% decomposition efficiency of PCDDs. The appropriate dose for PCDFs is equal roughly 20 kGy.

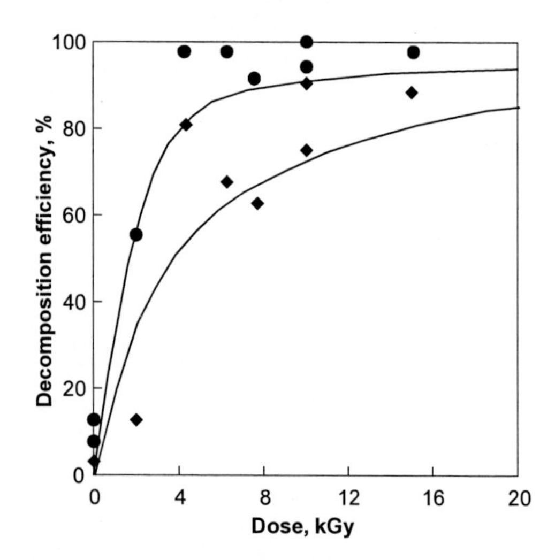

Figure 13. Comparison of model predictions (lines) with experimental data [50] (points) for decomposition efficiency of PCDDs (•) and PCDFs (♦) in MSWI flue gases under electron beam irradiation at T = 85 °C.

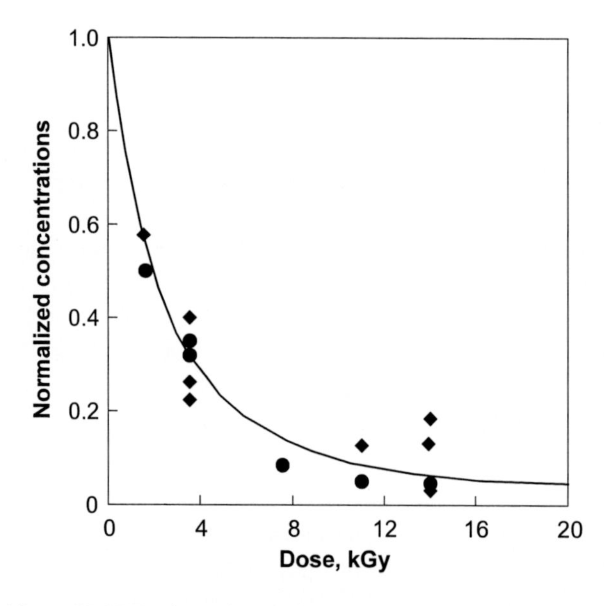

Figure 14. Decomposition of PCDDs homologues in MSWI flue gases under electron beam irradiation at T = 200 °C. Solid line – model predictions, points – experimental data of Hirota et al. [17]: (•) - TCDD; (♦) - OCDD.

The decomposition of PCDDs homologues in MSWI flue gases under electron beam irradiation at temperature T = 200 °C is shown in Fig. 14. The comparison of experimental data [17] with calculated results is carried out for TCDD and OCDD because the high temperature rate constants k_{OH} are accessible only for these PCDDs homologues:

$k_{OH} = 1.0 \times 10^{12} \exp(713/T)$ cm^3 mol^{-1} s^{-1} for 1,2,3,4-TCDD and $k_{OH} = 1.9 \times 10^{13} \exp(-667/T)$ cm^3 mol^{-1} s^{-1} for OCDD [59]. These rate constants at room temperatures practically coincide with theoretical estimations of Atkinson [57]. It can be seen from Fig. 14 that numerical estimations are in scattering limits of experimental data. The calculated curves for TCDD and OCDD are fused as their rate constants, k_{OH}, are equal with each other at T = 200 °C.

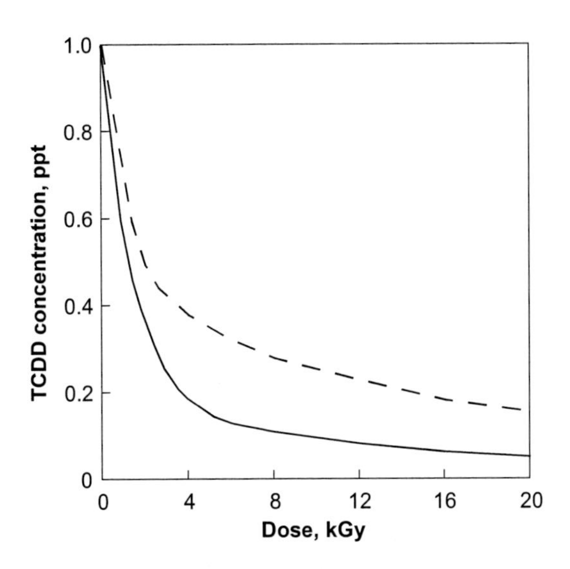

Figure 15. Calculated concentration of TCDD as a function of dose at T = 70 °C: solid line – total EBDS-process mechanism, dash line – gas-phase mechanism.

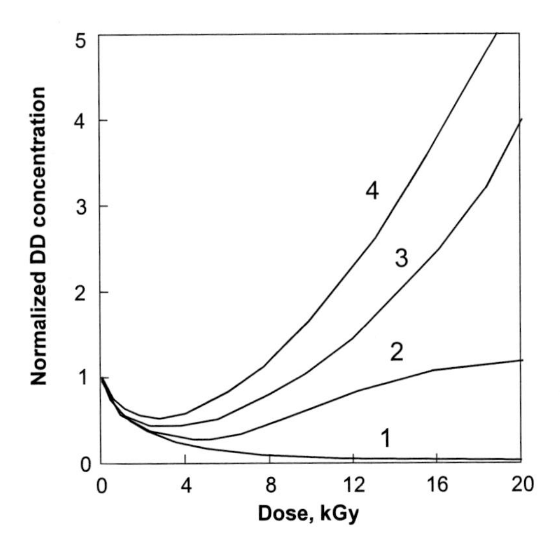

Figure 16. Influence of solid carbon (C_s) and polychlorinated phenols (P) initial concentrations on dioxins behavior under electron beam irradiation at T = 200°C: (1) $(C_s)_0 = 0$, $(P)_0 = 0$; (2) $(C_s)_0 = 0$, $(P)_0$ = 50 ppb; (3) $(C_s)_0$ = 1 ppm, $(P)_0 = 0$; (4) $(C_s)_0$ = 1 ppm, $(P)_0$ = 50 ppb. Initial DD concentration: $(DD)_0$ = 1 ppt.

The sufficiently good agreement of calculated and measured data suggests that the adopted kinetic model of EBDS-process [22] and kinetic mechanism of PCDD/Fs decomposition based on the primary attack of PCDD/Fs molecules by OH radicals give adequate description of these compounds behavior under electron beam irradiation. The principal role in consumption of OH radicals during EBDS-process belongs to main pollutants of the flue gases such as sulfur and nitrogen oxides. The inclusion of the liquid-phase processes in the kinetic model leads to the chain liquid-phase oxidation of SO_2 and to decrease of OH radicals consumption rate [31]. According to this the OH concentration rises with simultaneous increase of PCDD/Fs decomposition efficiency. This is illustrated in Fig. 15 where considerable decrease of calculated TCDD concentration is achieved owing to inclusion of liquid-phase processes into consideration. This decrease is resulted exclusively from the change of OH concentration in gas phase. The liquid-phase TCDD oxidation plays appreciable role only at high SO_2 concentrations (~1000 ppm) when a large amount of the liquid phase is formed during irradiation process.

Thermal formation of PCDD/Fs proceeds only in the presence of various chlorine compounds. Independently of the character of these compounds in the initial fuel, chlorine in the flue gases is predominantly represented as HCl. In this connection, the effect of HCl additives on the dynamics of PCDD/Fs oxidation was examined. The kinetic scheme of reactions involving chlorine compounds was taken from [51]. The results of the calculations demonstrate that the presence of HCl has almost no effect on the PCDD/Fs decomposition efficiency at HCl concentration of the order of 400 ppm that are typical for the MSW incinerators flue gases.

The dependence of the dioxins concentration from absorbed radiation dose at considerable contents of solid carbon (C_s) and polychlorinated phenols (P) in the flue gases is determined by the competition of two processes, namely, the DDs molecules formation from C_s and P, and their decomposition in reaction with radicals OH. Fig. 16 demonstrates the behavior of DDs at various C_s and P initial concentrations. It can be seen that the concentration of dioxins quickly decreases at the initial stage of the irradiation process. This decrease can be explained by a low rate of activated carbon C_s• and phenoxi radicals P• formation, which makes the process of DDs decomposition to be predominant. An increase of D changes the behavior of the concentration curves depending on the initial concentration of C_s and P. At high values of $(C_s)_0$ and $(P)_0$ the formation of DDs begins to dominate over their decomposition, and the $(DD)/(DD)_0$ ratio becomes greater than unit. At low values of $(C_s)_0$ and $(P)_0$ the production of dioxins is insignificant and an increase of the radiation dose leads to consumption of these compounds. Thus, the results of calculations demonstrate that electron beam treatment of the flue gases can't ensure the efficient removal of PCDD/Fs at high initial concentrations of the solid carbon and polychlorinated phenols ($C_s \geq 1$ ppm, $P \geq$ 50 ppb) as precursors of dioxins formation in the radiation zone.

PCBs Formation under Irradiation

Radiation induced behavior of biphenyl compounds in conditions of the EBDS-process was investigated as applied to the combustion products of lignite burning with air excess $\alpha = 1.1$: $N_2 = 65$, $O_2 = 2$, $H_2O = 19$, $CO_2 = 14$ vol.%. The initial concentrations of admixtures were chosen to be equal to the typical values at the exit from combustion zone: $(SO_2)_0 = 1000$,

$(NO)_0$ = 300, $(C_6H_6)_0$ = $(C_6H_5Cl)_0$ = 1 ppm. The content of ammonia added to system is stoichiometric. The initial concentrations of the biphenyl congeners are equal 10 ppt [73]. The calculations were performed at the constant dose rate, I = 10 kGy s^{-1}.

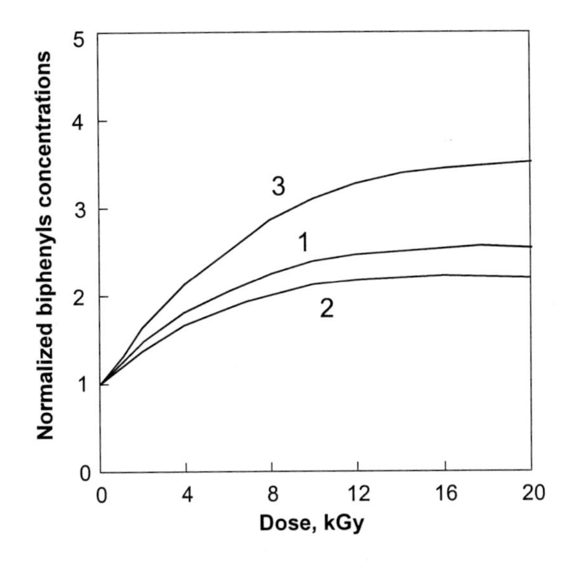

Figure 17. Normalized concentrations of biphenyl (1), chlorobiphenyl (2), and dichlorobiphenyl (3) as a function of absorbed dose. The initial concentrations of biphenyl compounds are equal 10 ppt. The process temperature T = 227 °C.

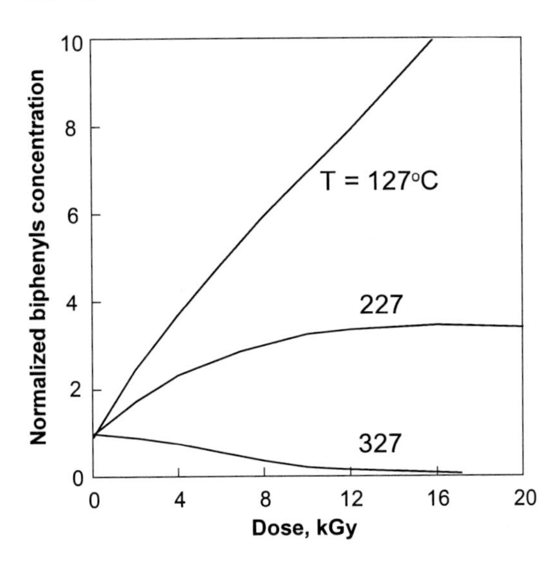

Figure 18. Normalized biphenyls concentration as a function of absorbed dose and process temperature. The initial concentrations of biphenyl compounds are equal 10 ppt.

Fig. 17 gives comparative behavior of the normalized biphenyls concentrations, (x_i / x_{i0}), at the process temperature T = 227 °C (high temperature range of the EBDS-process), which is determined by the competition of two chemical channels: the formation of these compounds from benzene and chlorobenzene as well as their destruction on exposure to oxygen (see Table 4). At the initial stage of the process, moderate rise of concentrations is to be observed. Further increase of absorbed dose leads to the lowering of benzene and chlorobenzene concentrations due to their destruction under the action of electron-beam irradiation that results in reduction of the biphenyl formation rate. As one can see from Fig. 17, the formation of of biphenyl, chlorobiphenyl, and dichlorobiphenyl takes place at about equal rates.

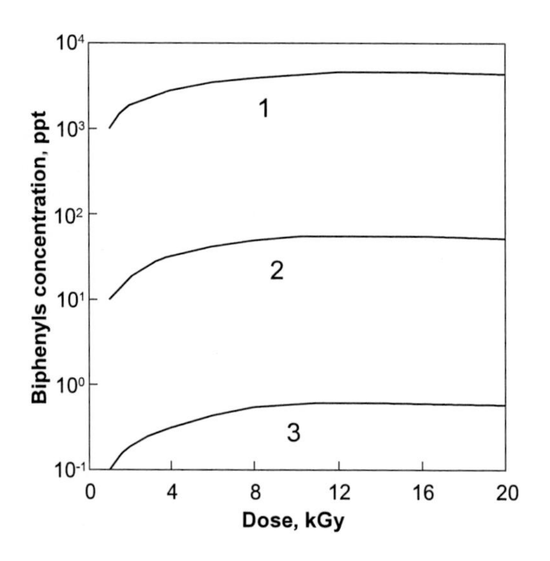

Figure 19. Biphenyls concentration as a function of dose and initial precursors concentration: $(C_6H_6)_0$ = $(C_6H_5Cl)_0$ = 10 (1); 1 (2); and 0.1 ppm (3). The initial concentrations of biphenyl compounds are zero. The process temperature T = 227 °C.

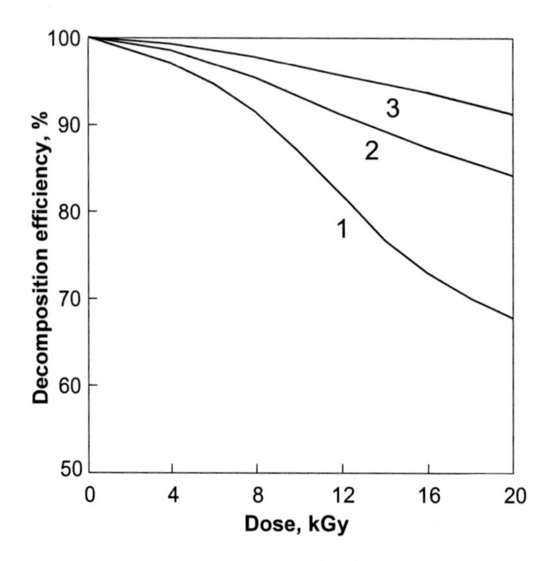

Figure 20. Calculated decomposition efficiency of 2,4-dichlorobiphenyl (1), 2,4',5-trichlorobiphenyl (2), and 2,3,3',4',6-pentachlorobiphenyl (3) as a function of dose at T = 70 °C.

The temperature dependence of normalised biphenyls concentration is shown in Fig. 18. At low temperature, T = 127 °C, the rate of biphenyls formation exceeds the destruction one resulting in growth of total biphenyls concentration by more than an order of magnitude at the radiation dose D = 20 kGy. The increase of process temperature leads to replacement of main reaction channel from biphenyls formation to their decomposition. At temperatures near 300°C, the decomposition process is predominant, and total biphenyls removal efficiency at the radiation dose D = 20 kGy exceeds 90%.

The dependence of biphenyls concentration from initial content of benzene and chlorobenzene in flue gases is shown in Fig. 19. The initial concentrations of biphenyl compounds in calculations are taken to be zero. Therefore, the calculated curves $x_{PCB} = x_{PCB}(D)$ give the absolute values of biphenyls concentration. It can be seen from Fig. 19 that the increase of precursors concentration by the order of magnitude gives corresponding growth of biphenyls concentration by two orders.

The simplified kinetic mechanism of biphenyls behaviour under irradiation given in Table 4 includes formation of biphenyl congeners up to dichlorobiphenyl in high temperature range of the EBDS-process. The kinetic information for more chlorinated compounds, which are more toxic, contains only rate constants of their decomposition in reaction with OH-radicals at room temperature [74,75]. This information allows estimating of PCBs removal efficiency in low temperature range of the EBDS-process ($T \le 80$ °C) where the reaction of PCBs molecules proceeds in accordance with OH-addition mechanism, and rate constants k_{OH} are faintly dependent on temperature. Fig. 20 demonstrates the dependence of the removal efficiency from absorbed radiation dose for some chlorinated compounds. Rate constants k_{OH} for calculations are taken from [75]. One can see from Fig. 20 that the radiation induced decomposition of PCBs decreases as number of chlorine atoms in biphenyl molecule grows. At typical EBDS-process doses, $D \le 20$ kGy, total removal efficiency of PCBs is small and don't exceed 30%.

Nitro-Compounds Formation

The numerical investigation of peroxyacetyl nitrates and aromatic nitro-compounds formation was carried out under the same process conditions. The initial concentrations of aliphatic and aromatic hydrocarbons in calculations were taken as: ETH = PRO = OLE = 10, RH = 50, PAH = 50 ppm. Fig. 21 demonstrates the dependence of ratio $x_{PAN}/(x_{HC})_0$ from absorbed dose D and process temperature T, where x_{PAN} is mole fraction of PAN in gas, and $(x_{HC})_0$ is the initial mole fraction of aliphatic hydrocarbons. The calculations show that temperature increase leads to abrupt decrease of PAN formation rate and to lowering of x_{PAN}. This behavior of concentration curves may be explained by strong temperature dependence of PAN monomolecular destruction rate. The destruction of PAN molecules at temperatures about 80 °C and below is insignificant, and the curve $x_{PAN} = x_{PAN}(D)$ is practically independent from T. The absolute value of x_{PAN} at dose D = 20 кГр and T = 80 °C is equal to 20 ppm. The background concentration of PAN in atmosphere is about 1 ppb [80].

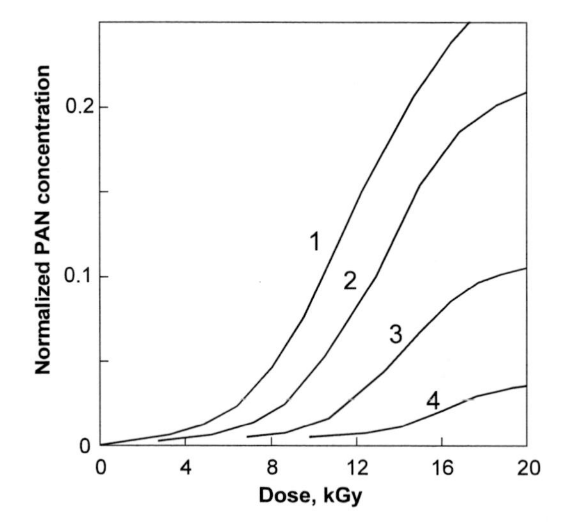

Figure 21. Formation of PAN as a function of absorbed radiation dose under electron-beam irradiation. Process temperature $T = 100$ (1), 120 (2), 140 (3), and 160 °C (4).

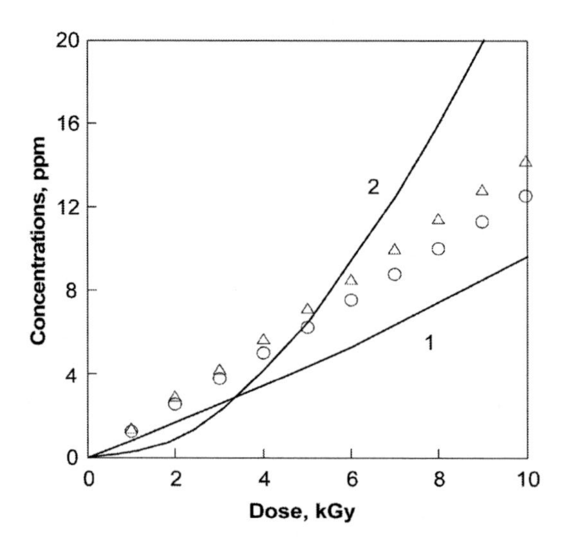

Figure 22. Formation of organic nitrates (1, Δ) and nitro-aromatic compounds (2, O) in γ-radiolysis of benzene/air mixture: $[H_2O]_0 = 1.3\%$, $[C_6H_6]_0 = 1\%$, $T = 40$ °C. Solid lines are model predictions; points are experimental data fit [16].

The model predictions of nitro-compounds formation and their comparison with experimental data are given in Fig. 22. The experimental points were restored from *G*-values received under the conditions of the γ-radiolysis of humidified air containing 4×10^{-4} mol dm^{-3} benzene [16]. The corresponding experimental values for electron-beam irradiation are more than 1 order of magnitude smaller. This difference is explained by higher gas dilution in electron-beam experiments and pulse nature of the electron-beam with its high dose rate, which signifies a transiently high free-radical concentration. Both of these facts favor the self-termination of the primary radicals relative to their reaction with the organic material, as compared with γ-radiolysis [16]. It can be seen from Fig. 22 that the model predictions give satisfactory agreement with experimental data.

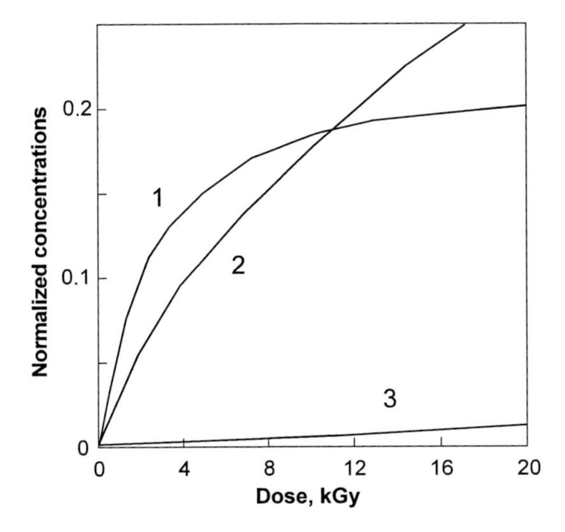

Figure 23. Mole fractions of aromatic hydrocarbons Δx_{PAH}, which are destroyed in reaction with O_2 (1) and turned into mono-NPAH (2) and di-NPAH (3), as a function of absorbed radiation dose. Initial PAHs concentration $(x_{PAH})_0 = 70$ ppm, gas temperature $T = 80$ °C.

Fig. 23 demonstrates the normalized mole fraction of aromatic hydrocarbons, $\Delta x_{PAH}/(x_{PAH})_0$, which are destroyed in reaction with molecular oxygen under the influence of radiation and turned into NPAHs, as a function of absorbed radiation dose D at typical conditions of EBDS-process ($NO_0 = 300$ ppm, $T = 80$ °C). Results of calculations show that these two competitive processes of PAHs transformation proceed approximately with equal rates. Concentration of dinitro-PAH compounds is more than 1 order of magnitude smaller than corresponding value for mononitro-PAH compounds.

Conclusion

The results of modeling study show that the EBDS-process is acceptable for simultaneous removal of SO_2, NO and VOCs from the industrial flue gases. At relatively low temperatures, $T \leq 80$ °C, typical for the EBDS-process the removal efficiency of SO_2 and NO is more than 90% at the radiation dose $D = 20$ kGy. The significant role in SO_2 radiation induced transformation plays liquid-phase oxidation of sulfur dioxide in aerosol droplets formed upon binary volume condensation of sulfuric acid and water vapours. The concentrations of aromatic representatives of VOCs such as PAHs and PCDD/Fs are decreased by more than an order of magnitude for compounds with value of the rate constant of their interaction with OH-radicals, playing a dominant initiating role in the aromatic compounds transformation, exceeded 10^{12} cm^3 mol^{-1} s^{-1}. Total removal efficiency of PCBs is smaller and don't exceed 30%.

At elevated temperatures, the reaction of OH-radical with aromatic molecule leads to the formation of aromatic radical followed by its transformation in competition of two processes: formation of more complicated aromatic structures with participation of the other aliphatic and aromatic compounds as well as its destruction on exposure to oxygen. In particular, the results of calculations demonstrate that electron beam treatment of the industrial flue gases can't ensure the efficient removal of PCDD/Fs at high initial concentrations of the solid carbon C_s and polychlorinated phenols P ($C_s \geq 1$ ppm, $P \geq 50$ ppb) as precursors of dioxins formation in the radiation zone.

The numerical investigation of aromatic nitro-compounds formation in low temperature range show that at typical NO concentrations in the combustion products (near 250 ppm) the destruction of PAHs molecules under irradiation and their transformation to NPAHs proceed with about equal rates. Consequently, the electron beam PAHs treatment is accompanied by their intensive nitration, which cans results in increase of the gas toxicity.

Acknowledgments

This work was supported by the International Atomic Energy Agency, Contract No. F23024-13139 in the framework of the IAEA Coordinated Research Project "Electron-beam treatment of organic pollutants containing in gaseous streams", and this financial support is gratefully acknowledged.

References

[1] Chmielewski, A.G.; Haji-Saeid, M. *Radiat. Phys. Chem.* 2004, 71, 17-21.
[2] Frank, N. *Radiat. Phys. Chem.* 1995, 45, 989-1002, 1017-1019.
[3] Kawamura, K.; Katayama, T.; Kawamura, K. *Radiat. Phys. Chem.* 1981, 18, 389-398.
[4] Frank, N.; Kawamura, K.; Miller, G. *Radiat. Phys. Chem.* 1985, 25, 35-45.
[5] Jordan, S. *Radiat. Phys. Chem.* 1988, 31, 21-29.
[6] Benijing, M.; Bonan, D. *Environ. Protect.* 2004, 9, 15-18.
[7] Chmielewski, A. G.; Licki, G.; Pawelec, A. et al. *Radiat. Phys. Chem.* 2004, 71, 441-444.
[8] Tuppurainen, K.; Halonen, I.; Ruokojärvi, P. et al. *Chemosphere* 1998, 36, 1493-1511.
[9] Douben, P. E. T. *Chemosphere* 1997, 34, 1181-1189.
[10] Dyke, P. H.; Foan, C.; Fiedler, H. *Chemosphere* 2003, 50, 469-480.
[11] Koch, M.; Cohn, D.R.; Patrick, R.H. et al. *Environ. Sci. Technol.* 1995, 29, 2946-2952.
[12] Hirota, K.; Mätzing, H.; Paur, H.-R.; Woletz, K. *Radiat. Phys. Chem.* 1995, 45,649-655.
[13] Prager, L.; Langguth, H.; Rummel, S.; Mehnert, R. *Radiat. Phys. Chem.* 1995, 46, 1137-1142.
[14] Han, D.-H.; Stuchinskaya, T.; Won, Y.-S. et al. *Radiat. Phys. Chem.* 2003, 67, 51-60.
[15] Chmielewski, A. G.; Ostapczuk, A.; Zimek, Z. et al. *Radiat. Phys. Chem.* 2002, 63, 653-655.
[16] Prager, L.; Mark, G.; Mätzing, H. et al. *Environ. Sci. Technol.* 2003, 37, 379-385.
[17] Hirota, K.; Hakoda, T.; Taguchi, M. et al. *Environ. Sci. Technol.* 2003, 37, 3164-3170.
[18] Nishimura, K.; Suzuki, N. *J. Nuclear Sci. Technol.* 1981, 18, 878-886.
[19] Busi, F.; D'Angelantonio, M.; Mulazzani, Q.G. et al. *Radiat. Phys. Chem.* 1985, 25, 47-55.
[20] Person, J. C.; Ham, D. O. *Radiat. Phys. Chem.* 1988, 31, 1-8.
[21] Mätzing, H. *Adv. Chem. Phys.* 1991, 80, 315-402.

[22] Gerasimov, G. Y.; Gerasimova, T. S.; Makarov, V. N.; Fadeev, S. A. *Radiat. Phys. Chem.* 1996, 48, 763-769.

[23] Sun, Y.-X.; Hakoda, T.; Chmielewski, A.G. et al. *Radiat. Phys. Chem.* 2001, 62, 353-360.

[24] Nichipor, H.; Dashouk, E.; Yacko, S. et al. *Radiat. Phys. Chem.* 2002, 65, 423-427.

[25] Gerasimov, G. *Radiat. Phys. Chem.* 2007, 76, 27-36.

[26] Henly, E.; Johnson, E. *The Chemistry and Physics of High Energy Reactions*; University Press: Oxford, UK, 1969.

[27] Klots, C.E. In *Fundamental Processes in Radiation Chemistry*; Ausloos, P.; Ed.; Interscience: New York, US, 1968; pp 1-57.

[28] Willis, C.; Boyd, A.W. *Radiat. Phys. Chem.* 1976, 8, 71-111.

[29] Mätzing, H.; Namba, H.; Tokunaga, O. *Radiat. Phys. Chem.* 1993, 42, 673-677.

[30] Paur, S.N.; Jordan, S. *Radiat. Phys. Chem.* 1988, 31, 9-13.

[31] Yermakov, A.N.; Zhitomorsky, B.M.; Poskrebyshev, G.A. *Radiat. Phys. Chem.* 1992, 39, 455-461.

[32] Getoff, N. *Radiat. Phys. Chem.* 1996, 47, 581-593.

[33] Tokunaga, O.; Nishimura, K.; Suzuki, N. et al. *Radiat. Phys. Chem.* 1978, 11, 299-303.

[34] Harley, E.M.; Matteson, M.J. *Ind. Eng. Chem. Fundam.* 1975, 14, 67-72.

[35] Fuchs, P.; Roth, B.; Schwing, U. et al. *Radiat. Phys. Chem.* 1988, 31, 45-56.

[36] Atkinson, R. *Atmos. Environ.* 2000, 34, 2063-2101.

[37] Wang, H.; Frenklach, M. *Combust. Flame* 1997, 110, 173-221.

[38] Atkinson, R. *Chem. Rev.* 1986, 86, 69-201.

[39] Ananthula, R.; Yamada, T.; Taylor, P.H. *J. Phys. Chem. A* 2006, 110, 3559-3566.

[40] Dutot, A.-L.; Rude, J.; Aumont, B. *Atmos. Environ.* 2003, 37, 269-273.

[41] Popov, P.; Getoff, N. *Radiat. Phys. Chem.* 2005, 72, 19-24.

[42] Buxton, G.V.; Greenstock, C.L.; Helman, W.P., Ross, A.B. *J. Phys. Chem. Ref. Data* 1988, 17, 513-886.

[43] Schuler, R.; Albarran, G. *Radiat. Phys. Chem.* 2002, 64, 189-195.

[44] Harrison, M. A. J.; Cape, J. N.; Heal, M. R. *Atmos. Environ.* 2002, 36, 1843-1851.

[45] Görgényi, M., Dewulf J., Van Langenhove, H. *Chemosphere* 2002, 48, 757-762.

[46] Frenklach, M. *Phys. Chem. Chem. Phys.* 2002, 4, 2028-2037.

[47] Frenklach, M.; Wang, H. In *Soot Formation in Combustion. Mechanisms and Models;* Bockhorn, H.; Ed..; Springer Series in Chemical Physics; Springer-Verlag: Berlin, 1994; Vol. 59, pp 165-192.

[48] Miller, J.A.; Melius, C.F. *Combust. Flame* 1992, 91, 2-39.

[49] Appel, J.; Bockhorn, H.; Frenklach, M. *Combust. Flame* 2000, 121, 122-136.

[50] Fernandez-Martinez, G.; Lopez-Mahia, P.; Muniategui-Lorenzo, S. et al. *Atmos. Environ.* 2001, 35, 5823-5831.

[51] Lee, K.Y.; Puri, I.K. *Combust. Flame* 1993, 94, 191-204.

[52] Shi, J.; Bernhard, M.J. *Int. J. Chem. Kinet.* 1997, 29, 349-358.

[53] Sokolov, O.; Hurley, M.D.; Wallington, T.J. et al. *J. Phys. Chem. A* 1998, 102, 10671-10681.

[54] Paur, H.-R.; Baumann, W.; Mätzing, H.; Jay, K. *Radiat. Phys. Chem.* 1998, 52, 355-359.

[55] Khachatryan, L.; Asatryan, R.; Dellinger, B. *J. Phys. Chem. A* 2004, 108, 9567-9572.

[56] Kwok, E.S.C.; Arey, J.; Atkinson, R. *Environ. Sci. Technol.* 1994, 28, 528-533.

[57] Atkinson, R. *Sci. Total Environ.* 1991, 104, 17-33.

[58] Barnes, D.G. *Sci. Total Environ.* 1991, 104, 73-86.
[59] Taylor, P.H.; Takahiro, Y.; Neuforth, A. *Chemosphere* 2005, 58, 243-252.
[60] Herrmann, H.; Ervens, B.; Jacobi, H.-W. et al. *J. Atmosph. Chem.* 2000, 36, 231-284.
[61] McLachlan, M.S.; Welsch-Pausch, K.; Tolls, J. *Environ. Sci. Technol.* 1995, 29, 1998-2004.
[62] Doucette, W.J.; Andren, A.W. *Chemosphere* 1988, 17, 243-252.
[63] Eizer, B.D.; Hites, R.A. *Sci. Total Environ.* 1991, 104, 9-15.
[64] Addink, R.; Olie, K. *Environ. Sci. Technol.* 1995, 29, 1425-1434.
[65] Sidhu, S.S.; Maqsud, L.; Dellinger, B. *Combust. Flame* 1995, 100, 11-20.
[66] Milligan, M.S.; Altwicker, E.R. *Environ. Sci. Technol.* 1996, 30, 230-236.
[67] Stanmore, B.R. *Chemosphere* 2002, 47, 565-573.
[68] Gerasimov, G.Y. *Combust. Explos. Shock Waves* 2001, 37, 148-152.
[69] Chelliah, H.K. *Combust. Flame* 1996, 104, 81-94.
[70] Babushok, V.I.; Tsang, W. *Chemosphere* 2003, 51, 1023-1029.
[71] Khachatryan, L., Asatryan, R.; Dellinger, B. *Chemosphere* 2003, 52, 695-708.
[72] Shaub, W.M., Tsang, W. *Environ. Sci. Technol.* 1983, 17, 721-730.
[73] Lee, R.G.M.; Coleman, P.; Jones, J.L. et al. *Environ. Sci. Technol.* 2005, 39, 1436-1447.
[74] Totten, L.A.; Eisenreich, S.J.; Brunciak, P.A. *Chemosphere* 2002, 47, 735-746.
[75] Anderson, P.N.; Hites, R.A. *Environ. Sci. Technol.* 1996, 30, 1756-1763.
[76] Kwok, E.S.C.; Atkinson, R.; Arey, J. *Environ. Sci. Technol.* 1995, 29, 1591-1598.
[77] Ritter, E.R.; Bozzelli, J.W. *J. Phys. Chem. A* 1990, 94, 2493-2504.
[78] Park, J.; Lin, M.C. *J. Phys. Chem. A* 1997, 101, 14-18.
[79] Stockwell, W.R. *Atmos. Environ.* 1986, 20, 1615-1632.
[80] Wunderli, S.; Gehric, R. *Atmos. Environ.* 1991, 25A, 1599-1608.
[81] Leone, J.A.; Seinfeld, J.H. *Atmos. Environ.* 1985, 19, 437-464.
[82] Tokunaga, O.; Suzuki, N. *Radiat. Phys. Chem.* 1984, 24, 145-165.
[83] Frank, N.; Hirano, S., Kawamura, K. *Radiat. Phys. Chem.* 1988, 31, 57-82.
[84] Hirota, K.; Hakoda, T.; Arai, H.; Hashimoto, S. *Radiat. Phys. Chem.* 2002, 65, 415-421.
[85] Bonin, J.; Janik, I.; Janik, D.; Bartels, D.M. *J. Phys. Chem. A* 2007, 111, 1869-1878.
[86] Biermann, H.W.; MacLeod, H.; Atkinson, R. et al. *Environ. Sci. Technol.* 1985, 19, 244-248.
[87] Atkinson, R.; Arey, J.; Zelinska, B.; Aschmann, S.M. *Int. J. Chem. Kinet.* 1990, 22, 999-1014.
[88] Brubaker, W.W.,Jr.; Hites, R.A. *J. Phys. Chem. A* 1998, 102, 915-921.
[89] Kwok, E.S.C.; Harger, W.P.; Arey, J.; Atkinson, R. *Environ. Sci. Technol.* 1994, 28, 521-527.
[90] Harner, T.; Bidleman, T.F. *Environ. Sci. Technol.* 1998, 32, 1494-1502.

In: Radiation Physics Research Progress
Editor: Aidan N. Camilleri, pp. 287-300
ISBN: 978-1-60021-988-7

Chapter 7

PACKAGING DETERMINES COLOR AND ODOR OF IRRADIATED GROUND BEEF

***K.C. Nam*[1], *B.R. Min*[2], *K.Y. Ko*[2], *E.J. Lee*[2], *J. Cordray*[2] *and D.U. Ahn*[2,*]**

[1] Exam. Div. of Food & Biological Resources,
Korean Intellectual Property Office, Daejeon 302-701, Korea
[2] Department of Animal Science, Iowa State University, Ames, IA 50011-3150

Abstract

Irradiation is an excellent method to ensure the microbial safety, but causes color changes, lipid oxidation and off-odor production in irradiated meat. Irradiation of ground beef under aerobic conditions oxidized myoglobin and drastically reduced color a*-values. Under vacuum or non-oxygen conditions, however, irradiation did not influence the redness of ground beef. Also, the red color of ground beef was maintained even after the irradiated beef was exposed to aerobic conditions. Vacuum-packaged irradiated ground beef had lower met-myoglobin content and lower oxidation-reduction potential than the aerobically packaged ones. Irradiating ground beef under vacuum-packaging conditions was also advantageous in preventing lipid oxidation and aldehydes production. Vacuum-packaged irradiated beef, however, produced high levels of sulfur volatiles during irradiation and maintained their levels during storage, which resulted in the production of characteristic irradiation off-odor. Double-packaging (V3/A3: vacuum-packaging during irradiation and the first 3 days of storage and then aerobic-packaging for the remaining 3 days) was an effective alternative in maintaining original beef color (red), and minimizing lipid oxidation and irradiation off-odor. The levels of off-odor volatiles in double-packaged irradiated ground beef were comparable to that of aerobically packaged ones, and the degree of lipid oxidation and color changes were close to those of vacuum-packaged ones. Ascorbic acid at 200 ppm level was not effective in preventing color changes and lipid oxidation in irradiated ground beef under aerobic conditions, but was helpful in minimizing quality changes in double-packaged irradiated ground beef. This suggested that preventing oxygen contact from meat during irradiation and early storage period (V3/A3 double-packaging) and double-packaging+ascorbic acid combination are excellent strategies to prevent quality changes in irradiated ground beef.

---

* E-mail address: duahn@iastate.edu; Tel: 515-294-6595; Fax: 515-294-9143; (Corresponding author)

Keywords: irradiation, oxygen, color, lipid oxidation, volatiles, ground beef

Introduction

Irradiation is among the most effective technologies in preventing pathogenic microorganisms, but negatively influences quality of ground beef. Beef color is a prime quality parameter that determines consumer acceptance. Thus, maintaining the color of irradiated beef within a normal color range is one of the most important factors if irradiation can be used as a pathogen-reduction tool in beef (Seideman et al., 1984).

Color changes induced by irradiation are different depending on animal species, muscle type, irradiation dose, and packaging type (Satterlee et al., 1971; Luchsinger et al., 1996; Ahn et al., 1998): light meat such as pork loin and poultry breast meat produce pink color while dark meat such as beef becomes brown or gray after irradiation (Millar et al., 1995; Nanke et al., 1998; Ahn et al., 1998; Kim et al., 2002). Studies on the mechanisms and pigments involved in the color changes of irradiated meats indicated that the pink color in irradiated light meats was mainly caused by carbon monoxide-heme pigment complexes (Nam and Ahn, 2002). Lee and Ahn (2004) reported that the production of CO by irradiation was via the radiolytic degradation of meat components such as glycine, asparagine, glutamine glyceraldehydes, pyruvate, α-ketoglutarate and phospholipids, and the production of CO via radiolytic degradation was closely related to the structure of molecules.

Both light and dark meats produced carbon monoxide by irradiation and the amounts of carbon monoxide produced were irradiation dose-dependent (Nam and Ahn, 2002; Lee and Ahn, 2004). The formation of carbon monoxide-heme complex, however, could not explain the color changes in irradiated dark meat such as ground beef. While the amount of carbon monoxide produced in beef by irradiation was similar to those of light meats (Kim et al., 2002), the pigment content in beef is about 10 times higher than that in light meats. The amount of carbon monoxide produced from meat by 2-3 kGy of irradiation was 1-3 ppm CO/g meat, which was just enough to react with 1-2 mg Mb/g meat (Nam and Ahn, 2002). Therefore, the contribution of carbon monoxide-heme pigment to the color of ground beef was much smaller than that of light meats. However, the mechanisms and causes of color changes in irradiated beef are not fully understood yet.

Ascorbic acid is a reducing agent, which inhibits myoglobin oxidation and prevents brown color development in nonirradiated beef (Wheeler et al., 1996; Lee et al., 1999; Sanchez-Escalante et al., 2001). Packaging, which determines oxygen availability to the meat in a packaging bag, is also a critical factor that influences color of meat because the chemical conditions of heme pigments are dependent upon the partial pressure of oxygen. Therefore, use of ascorbic acid and/or modification of packaging conditions can be a good strategy to control irradiation-dependent color changes in ground beef.

Irradiation produces off-odor volatiles: the major volatile compounds responsible for off-odor in irradiated meats are sulfur compounds. Volatile sulfur compounds were produced from the radiolytic degradation of the side chains of sulfur-containing amino acids such as methionine and cysteine (Jo and Ahn, 2000; Ahn, 2002; Ahn and Lee, 2002). The sulfur volatiles produced by irradiation had characteristic odor, and the intensity of irradiation off-odor diminished over storage period as the sulfur volatiles disappeared during storage under aerobic conditions (Nam and Ahn, 2003). Using the high volatility of sulfur compounds under aerobic conditions, Nam and Ahn (2002) devised an efficient packaging method that can remove irradiation off-odor in irradiated meat during storage. The method is called "double-packaging", which combines both aerobic and vacuum conditions. Double-packaging was also effective in minimizing color changes and lipid oxidation in irradiated turkey and pork (Nam and Ahn, 2003; Nam et al., 2004).

The objective of this study was to determine the effect of ascorbic acid and double-packaging on color, oxidation-reduction potential, lipid oxidation, met-myoglobin formation, and volatiles production in irradiated ground beef during storage. Although maintaining attractive color in irradiated ground beef is very important, little work has been done to improve the color of irradiated beef. This study can be of help in understanding the causes and mechanisms of color changes and developing an efficient method to control quality changes in irradiated ground beef.

Materials and Methods

Sample Preparation

Beef loins (*Longissimus dorsi*) from 4 different carcasses were obtained from Meat Laboratory at Iowa State University (Ames, IA, USA). Muscles taken from each carcass were treated as a replication. Each replication was ground separately through a 3-mm plate. Seven different treatments were prepared: 1) aerobically packaged, non-irradiated, 2) aerobically packaged, irradiated, 3) double-packaged I (A3/V3: aerobically packaged during irradiation and the first 3 days of storage and then vacuum-packaging for the next 3 days), irradiated, 4) vacuum-packaged, irradiated, 5) double-packaged II (V3/A3: vacuum-packaging during the irradiation and the first 3 days of storage and then aerobic packaging for the next 3 days), irradiated, 6) ascorbic acid-added, aerobically packaged, and irradiated, and 7) ascorbic acid-added, double-packaged II (V3/A3), and irradiated.

For ascorbic acid-added treatments, ascorbic acid (Sigma, St Louis, MO, USA) was dissolved in minimal amount of distilled water and added at 200 ppm to the ground beef (final concentration) and mixed for 1 min in a bowl mixer (Model KSM 90; Kitchen Aid Inc., St. Joseph, MI, USA). The mixed meat samples were ground again through a 3-mm plate to ensure even distribution of ascorbic acid. For rest of the treatments, water with no ascorbate was added, mixed, and ground at the same conditions as in ascorbic acid-added treatments.

Ground beef (approximately 75 g) was individually packaged in an oxygen-permeable bag (polyethylene, 4 x 6, 2 MIL, Associated Bag Company, Milwaukee, WI, USA), an oxygen-impermeable bag (nylon/polyethylene, 9.3 mL $O_2/m^2/24$ h at 0°C; Koch, Kansas City, MO), or double-packaged. For double-packaging, ground beef was individually packaged in an oxygen permeable bag and then vacuum-packaged in a larger oxygen-impermeable bag.

The packaged ground beef were irradiated at 4.5 kGy using a Linear Accelerator Facility (LAF, Circe IIIR; Thomson CSF Linac, St. Aubin, France) with 10 MeV energy and 10.2 kW power level. The average dose rate was at 83.5 kGy/min. Alanine dosimeters were placed on the top and bottom surfaces of a sample and were read using a 104 Electron Paramagnetic Resonance Instrument (Bruker Instruments Inc., Billerica, MA, USA) to check the absorbed dose. The dose range absorbed at meat samples was 4.449 to 4.734 kGy (max/min ratio was 1.06). Aerobically packaged non-irradiated ground beef was used as a control. The samples were stored at 4 °C for 6 days. Color, met-myoglobin, oxidation-reduction potential, lipid oxidation, and volatiles of the samples were determined during the storage.

Color Measurement

CIE color values were measured on the surface of meat samples using a LabScan colorimeter (Hunter Associated Labs. Inc., Reston, VA, USA) with a 1.225-cm aperture. The colorimeter was calibrated against a black and a white reference tiles covered with the same packaging bags used for samples. The color L* (lightness)-, a* (redness)-, and b* (yellowness)-values were obtained (AMSA, 1991). An average value from 2 random locations on each sample surface was used for statistical analysis.

Met-Myoglobin Contents

Met-myoglobin content was determined by the modified method of Krzywicki (1982). Meat sample (1g) was homogenized with 9 mL of 0.04 M phosphate buffer (pH 6.8, 4 °C) using a Brinkman Polytron (Type PT 10/35; Brinkman Instrument Inc., Westbury, NY) for 10 s at high speed. Meat homogenate (1 mL) was centrifuged at 8,000 × *g* for 1 min and the absorbances of the supernatant were immediately measured at 525, 572, and 700 nm using a spectrophotometer (Beckman DU 640, Beckman Instruments, Inc., Fullerton, CA). Total metmyoglobin percentage was calculated using the following equation:

$$\text{Metmyoglobin (\%)} = \{1.395 - [(A_{572} - A_{700})/(A_{525} - A_{700})]\} \times 100$$

Oxidation-Reduction Potential (ORP)

To reduce the deviation of ORP values depending on the location of a meat patty, meat homogenate was used to determine ORP. Meat sample (5 g), 15 mL deionized distilled water, and 50 μL butylated hydroxytoluene (7.2% in ethanol) were placed in a 50-mL test tube and homogenized using a Brinkman Polytron (Type PT 10/35, Brinkman Instrument Inc., Westbury, NY, USA) for 15 s at high speed. The ORP values of homogenates were determined using a pH/ion meter (Accumet 25; Fisher Scientific, Fair Lawn, NJ, USA) equipped with a platinum electrode filled with an electrolyte solution (4 M KCl saturated with AgCl).

2-Thiobarbituric Acid-Reactive Substances (TBARS)

Lipid oxidation was determined using a TBARS method (Ahn et al., 1999). Minced sample (5 g) was placed in a 50-mL test tube and homogenized with 15 mL of deionized distilled water (DDW) using a Brinkman Polytron (Type PT 10/35) for 15 s at high speed. The meat homogenate (1 mL) was transferred to a disposable test tube (13 x 100 mm), and butylated hydroxytoluene (7.2%, 50 μL) and thiobarbituric acid/trichloroacetic acid [20 mM TBA and 15% (w/v) TCA] solution (2 mL) was added. The sample was mixed using a vortex mixer, and then incubated in a 90 °C water bath for 15 min to develop color. After cooling for 10 min in cold water, the samples were vortex mixed and centrifuged at 3,000 x *g* for 15 min at 5 °C. The absorbance of the resulting upper layer was read at 531 nm against a blank prepared with 1 mL DDW and 2 mL TBA/TCA solution. The amounts of TBARS were expressed as mg of malonedialdehyde (MDA) per kg of meat.

Volatile Compounds

Volatiles of samples were analyzed using a Solatek 72 Multimatrix Vial Autosampler/Sample Concentrator 3100 (Tekmar-Dohrmann, Cincinnati, OH, USA) connected to a GC/MS (Model 6890/5973, Hewlett-Packard Co., Wilmington, DE, USA) according to the method of Ahn et al. (2001). Sample (3 g) was placed in a 40-mL sample vial, flushed with helium gas (40 psi) for 3 s, and then capped airtight with a Teflon*fluorocarbon resin/silicone septum (I-Chem Co., New Castle, DE, USA). The maximum waiting time for a sample in a loading tray (4 °C) was less than 2 h to minimize oxidative changes before analysis. The meat sample was purged with He (40 mL/min) for 14 min at 40 °C. Volatiles were trapped using a Tenax/charcoal/silica column (Tekmar-Dohrmann) and desorbed for 2 min at 225 °C, focused in a cryofocusing module (-80 °C), and then thermally desorbed into a column for 60 s at 225 °C. An HP-624 column (7.5 m, 0.25 mm i.d., 1.4 µm nominal), an HP-1 column (52.5 m, 0.25 mm i.d., 0.25µm nominal), and an HP-Wax column (7.5 m, 0.250 mm i.d., 0.25 µm nominal) were connected using zero dead-volume column connectors (J &W Scientific, Folsom, CA, USA). Ramped oven temperature was used to improve volatile separation. The initial oven temperature of 0 °C was held for 1.5 min. After that, the oven temperature was increased to 15 °C at 2.5 °C per min, increased to 45 °C at 5 °C per min, increased to 110 °C at 20 °C per min, and then increased to 210 °C at 10 °C per min and held for 2.25 min at that temperature. Constant column pressure at 22.5 psi was maintained. The ionization potential of MS was 70 eV, and the scan range was 19.1 to 350 m/z. The identification of volatiles was achieved by the Wiley library (Hewlett-Packard Co.). The area of each peak was integrated using ChemStation™ software (Hewlett-Packard Co.) and the total peak area (total ion counts x 10^4) was reported as an indicator of volatiles generated from the samples.

Statistical Analysis

A completely randomized design with 7 treatments and 4 replications was used. Data were analyzed using the generalized linear model procedure of SAS software (SAS Institute 1995). Student-Newman-Keul's multiple range test was used to determine significant differences between the mean values of treatments. Mean values and standard error of the means (SEM) were reported. Significance was defined at $P< 0.05$.

Results and Discussion

Color Values of Irradiated Ground Beef

The color of aerobically packaged ground beef was significantly influenced by irradiation (Table 1). Electron beam-irradiation at 4.5 kGy reduced the CIE L*-, a*-, and b*-values of ground beef at Days 0 and 1, but the extent of decrease in a*- and b*-values were greater than that of L*-value. L-values in all irradiated meat gradually increased with storage except for A3/V3 treatment. The decrease of a*-values in ascorbic acid-added, aerobically packaged irradiated ground beef was the greatest at Day 0. After 1 day of storage, however, the a*-values in irradiated ground beef under aerobic and A3/V3 conditions were the lowest. After 6 days of storage, color a*-value of irradiated meat with V3/A3 double-packaging+ascorbate treatment was the highest, and irradiated meat with vacuum-packaging, V3/A3 double-packaging, or ascorbic acid treatment showed higher a*-values than nonirradiated control. The decrease of b*-values in irradiated ground beef was greater under vacuum than aerobic conditions.

The visual color of aerobically packaged ground beef changed from a bright red to a greenish brown immediately after irradiation, which should be unattractive to consumers. The color of ground beef was more stable when irradiated and stored under vacuum conditions. One day after irradiation, the color defects of irradiated ground beef was getting worse under aerobic conditions and the a*-values of aerobically packaged irradiated ground beef decreased by approximately 3 units from Day 0. However, the a*-values of irradiated ground beef under vacuum conditions did not change much and were significantly higher than those under aerobic conditions at Day 1. At Day 6, aerobically packaged non-irradiated control discolored to dark brown due to pigment oxidation. The degree of discoloration at Day 6 was the most severe in aerobically packaged or A3/V3 double-packaged irradiated ground beef. When irradiated ground beef was vacuum-packaged, V3/A3 double packaged, or ascorbic acid-added, the a*-value of irradiated ground beef were greater than that of non-irradiated control at Day 6. Addition of ascorbic acid to ground beef at 200 ppm (w/w) was also effective in reducing color changes by irradiation in aerobically conditions. This suggested that elimination of ascorbic acid as well as oxygen during irradiation is very important to minimize color changes in irradiated ground beef during storage.

Nam and Ahn (2003) reported that ascorbic acid incorporated to beef at the level of 0.1% was highly effective in maintaining redness (a*-values) of irradiated ground beef but sesamol + tocopherol had no effect. When the irradiated ground beef was added with ascorbic acid and V3/A3 double-packaged, the color of ground beef was the best among the treatments after 6 days of storage (Table 1). The a*-value of ground beef with ascorbic acid + V3/A3 double-packaging treatment was greater than that of the non-irradiated aerobically packaged beef due to the synergistic effect of ascorbic acid and vacuum conditions during the first 3 days of storage period, which helped maintaining heme pigments in reduced state.

Table 1. Color values of ground beef affected by irradiation, oxygen, and ascorbic acid during the refrigerated storage

Irradiation	Package	Ascorbic	0 day	1 day	6 day	SEM
			---------------- *L*-value*	--------------		
0 kGy	Aerobic	0 ppm	52.2ax	49.2a	50.6ab	1.0
4.5 kGy	Aerobic	0 ppm	43.2bz	46.9by	51.7abx	0.9
4.5 kGy	A3/V3	0 ppm	45.1by	45.8b	46.0c	1.2
4.5 kGy	Vacuum	0 ppm	40.6bz	45.1bcy	48.0bcx	0.8
4.5 kGy	V3/A3	0 ppm	41.0by	42.6cxy	49.5abcx	0.8
4.5 kGy	Aerobic	200 ppm	44.8by	46.4bxy	49.4abcx	1.0
4.5 kGy	V3/A3	200 ppm	41.6by	43.0cy	52.1ax	0.9
SEM			1.1	0.8	1.0	
			---------------- *a*-value*	--------------		
0 kGy	Aerobic	0 ppm	33.0ax	28.8ay	16.6cz	0.9
4.5 kGy	Aerobic	0 ppm	18.4cx	15.0dy	12.7dz	0.4
4.5 kGy	A3/V3	0 ppm	18.1cx	15.4dy	12.3dz	0.6
4.5 kGy	Vacuum	0 ppm	21.0b	20.1bc	19.6b	0.4
4.5 kGy	V3/A3	0 ppm	20.5bxy	21.4bx	19.7by	0.4
4.5 kGy	Aerobic	200 ppm	16.8cy	18.8cx	19.1bx	0.5
4.5 kGy	V3/A3	200 ppm	18.8cz	20.3bcy	23.7ax	0.4
SEM			0.6	0.5	0.5	
			---------------- *b*-value*	--------------		
0 kGy	Aerobic	0 ppm	30.7ax	28.6ay	21.7az	0.7
4.5 kGy	Aerobic	0 ppm	20.4b	19.9c	18.8c	0.6
4.5 kGy	A3/V3	0 ppm	20.3b	19.6c	19.3c	0.7
4.5 kGy	Vacuum	0 ppm	13.8cy	14.9dx	15.6dx	0.3
4.5 kGy	V3/A3	0 ppm	15.4cy	16.4dy	21.2abx	0.4
4.5 kGy	Aerobic	200 ppm	19.0by	21.7bx	19.9bcy	0.4
4.5 kGy	V3/A3	200 ppm	15.4cy	15.7dy	22.3ax	0.5
SEM			0.6	0.5	0.5	

a-d) Mean values with different letters within a column are significantly different ($P < 0.05$).

x-z) Mean values with different letters within a row are significantly different ($P < 0.05$). n = 4.

V3/A3: Aerobically packaged during irradiation and the first 3 days of storage and then vacuum packaging for the next 3 days.

V3/A3: Vacuum-packaged during the irradiation and the first 3 days of storage and then aerobic packaging for the next 3 days.

Forms of Heme Pigments in Irradiated Ground Beef

Ground beef irradiated under aerobic conditions produced higher amounts of met-myoglobin than those irradiated under vacuum conditions at Day 1 (Table 2), and the proportion of met-myoglobin in meat and color a*-value was negatively correlated. Aerobic conditions during storage further increased the percentage of met-myoglobin, but addition of ascorbic acid helped slowing down met-myoglobin formation. Vacuum conditions during storage reduced the level of met-myoglobin in irradiated ground beef. This suggested that the brown discoloration of ground beef upon irradiation under aerobic conditions was caused by the met-myoglobin formation, and the color was totally different from that of irradiated pork or poultry breast, which was pink or red (Nam et al., 2002; Nam and Ahn, 2003). Nam and Ahn (2002) reported that the complex of heme pigments with carbon monoxide produced by irradiation was responsible for the increased redness in irradiated turkey breast. The formation of CO-heme complex, however, could not explain the color changes in beef after irradiation. Kim et al. (2002) reported that the production of carbon monoxide in beef by irradiation was similar to that of light meats, but the pigment content in beef is much higher (> 10-fold) than that in light meats. Therefore, the contribution of CO-heme pigment to the color of ground beef was much smaller than that of the light meats.

Because the color intensity of ferrous heme pigment is stronger than that of ferric form, oxidation-reduction potential (ORP) plays a very important role on the color changes of meat. Both vacuum conditions and ascorbic acid were effective in maintaining low ORP values in irradiated ground beef (Table 3). Giroux et al. (2001) speculated that free binding sites of myoglobin can react with free radicals such as hydroxyl or sulfuryl radicals produced by irradiation and forms met-myoglobin and sulf-myoglobin responsible for brown and green color, respectively. However, the lowered ORP values by vacuum conditions and/or ascorbic acid maintained heme pigments in ferrous state and stabilized the color of irradiated ground beef. At Day 6, vacuum-packaged irradiated ground beef had lower met-myoglobin content and ORP values than the V3/A3 double-packaged ones, indicating that the color of vacuum-packaged irradiated ground beef could be bloomed when it is exposed to aerobic conditions even after 6 days of storage. Although the chemical forms of heme pigments in irradiated beef have not been clearly identified, the color changes in irradiated ground beef could be reverted to the desirable red color if the reducing power of meat is maintained by vacuum conditions and/or ascorbic acid.

Table 2. Met-myoglobin percentage in ground beef affected by irradiation, oxygen, and ascorbic acid during the refrigerated storage

Irradiation	Package	Ascorbic	1 day	6 day	SEM
			(%)		
0 kGy	Aerobic	0 ppm	33.3by	60.4ax	2.5
4.5 kGy	Aerobic	0 ppm	48.9ay	58.2ax	1.3
4.5 kGy	A3/V3	0 ppm	46.4ax	29.4cy	2.3
4.5 kGy	Vacuum	0 ppm	12.8c	11.9d	0.9
4.5 kGy	V3/A3	0 ppm	10.5cy	39.3bx	1.0
4.5 kGy	Aerobic	200 ppm	35.6b	41.6b	3.2
4.5 kGy	V3/A3	200 ppm	10.6cy	28.5cx	0.3
SEM			1.9	1.9	

a-d) Mean values with different letters within a column are significantly different (P < 0.05).

x, y) Mean values with different letters within a row are significantly different (P < 0.05). n = 4.

V3/A3: Aerobically packaged during irradiation and the first 3 days of storage and then vacuum packaging for the next 3 days.

V3/A3: Vacuum-packaged during the irradiation and the first 3 days of storage and then aerobic packaging for the next 3 days.

Table 3. ORP values of ground beef affected by irradiation, oxygen, and ascorbic acid during the refrigerated storage

Irradiation	Package	Ascorbic	1 day	6 day	SEM
			---------------- (*mV*)	--------------	
0 kGy	Aerobic	0 ppm	181.9ay	244.9ax	8.6
4.5 kGy	Aerobic	0 ppm	185.6ay	224.0bx	1.0
4.5 kGy	A3/V3	0 ppm	185.0ay	208.6cx	0.8
4.5 kGy	Vacuum	0 ppm	154.9b	159.1d	2.7
4.5 kGy	V3/A3	0 ppm	148.9by	197.1cx	2.4
4.5 kGy	Aerobic	200 ppm	136.9by	199.1cx	8.7
4.5 kGy	V3/A3	200 ppm	112.0cy	131.8ex	2.6
SEM			6.2	2.7	

a-e) Mean values with different letters within a column are significantly different ($P < 0.05$).

x, y) Mean values with different letters within a row are significantly different ($P < 0.05$). n = 4.

V3/A3: Aerobically packaged during irradiation and the first 3 days of storage and then vacuum packaging for the next 3 days.

V3/A3: Vacuum-packaged during the irradiation and the first 3 days of storage and then aerobic packaging for the next 3 days.

Lipid Oxidation of Irradiated Ground Beef

The TBARS values of ground beef irradiated and stored under aerobic conditions were not different from that of control at Day 1, but the rate of lipid oxidation was faster in irradiated than non-irradiated control as the storage time increased (Table 4). Vacuum-packaged irradiated beef was resistant to lipid oxidation during the 6 days of storage. Double-packaged irradiated ground beef showed significant differences in TBARS values depending on whether they were aerobic/vacuum or vacuum/aerobic packaged. V3/A3 double-packaged irradiated ground beef had lower TBARS values than the A3/V3, indicating that hydroxyl radicals produced by irradiation can initiate lipid oxidation but cannot propagate without the presence of oxygen. Addition of ascorbic acid at 200 ppm was not effective in preventing lipid oxidation when the irradiated beef was aerobically packaged.

When irradiated ground beef was stored for 6 days under aerobic conditions, significant amounts of volatile aldehydes (propanal, pentanal, and hexanal) were produced in irradiated ground beef (Table 5). Hexanal, a major indicator of lipid oxidation (Shahidi et al., 1987), was the most prominent aldehyde, and large amounts of hexanal were detected in aerobically packaged ground beef at Day 6. Addition of ascorbic acid was not effective in inhibiting production of volatile aldehydes in irradiated beef at Day 6 as shown in TBARS values. However, vacuum packaging or double-packaging+ascorbic acid combination was effective in minimizing the production of volatile aldehydes in irradiated ground beef. Therefore, lipid oxidation of irradiated ground beef was highly dependent upon the availability of oxygen to meat during storage. Addition of 200 ppm ascorbate to double-packaged ground beef was helpful in slowing down the development of lipid oxidation in irradiated ground beef.

Table 4. TBARS values of ground beef affected by irradiation, oxygen, and ascorbic acid during the refrigerated storage

Irradiation	Package	Ascorbic	1 day	6 day	SEM
			----------- (*mg MDA/kg meat*)	----------	
0 kGy	Aerobic	0 ppm	2.26ay	3.92bx	0.21
4.5 kGy	Aerobic	0 ppm	2.94ay	5.61ax	0.18
4.5 kGy	A3/V3	0 ppm	2.98ay	4.31bx	0.26
4.5 kGy	Vacuum	0 ppm	0.73b	0.84e	0.05
4.5 kGy	V3/A3	0 ppm	0.95by	2.45cx	0.13
4.5 kGy	Aerobic	200 ppm	2.75ay	5.46ax	0.16
4.5 kGy	V3/A3	200 ppm	0.97by	1.67dx	0.10
SEM				0.20	0.16

a-e) Mean values with different letters within a column are significantly different ($P < 0.05$).

x, y) Mean values with different letters within a row are significantly different ($P < 0.05$). $n = 4$.

V3/A3: Aerobically packaged during irradiation and the first 3 days of storage and then vacuum packaging for the next 3 days.

V3/A3: Vacuum-packaged during the irradiation and the first 3 days of storage and then aerobic packaging for the next 3 days.

Table 5. Production of aldehydes in ground beef affected by irradiation, oxygen, and ascorbic acid during the refrigerated storage

Irradiation	Package	Ascorbic	1 day	6 day	SEM
Propanal			----------- (*10^4 ion counts*)	----------	
0 kGy	Aerobic	0 ppm	513aby	2973cx	452
4.5 kGy	Aerobic	0 ppm	1184ay	12073bx	383
4.5 kGy	A3/V3	0 ppm	1085ay	4508cx	583
4.5 kGy	Vacuum	0 ppm	313b	1575c	388
4.5 kGy	V3/A3	0 ppm	626ab	2426c	677
4.5 kGy	Aerobic	200 ppm	764aby	15970ax	1819
4.5 kGy	V3/A3	200 ppm	252by	1603cx	24
SEM			166	1135	
Pentanal			----------- (*10^4 ion counts*)	----------	
0 kGy	Aerobic	0 ppm	613bcy	1163bx	132
4.5 kGy	Aerobic	0 ppm	1696a	3528a	478
4.5 kGy	A3/V3	0 ppm	949b	1457b	131
4.5 kGy	Vacuum	0 ppm	469bc	1034b	183
4.5 kGy	V3/A3	0 ppm	98cy	1677bx	123
4.5 kGy	Aerobic	200 ppm	670bcy	3470ax	376
4.5 kGy	V3/A3	200 ppm	107cy	643bx	27
SEM			166	319	

Table 5. Continued.

Irradiation	Package	Ascorbic	1 day	6 day	SEM
Hexanal			----------- *(10^4 ion counts)* ----------		
0 kGy	Aerobic	0 ppm	5997bc	16121b	3079
4.5 kGy	Aerobic	0 ppm	13505ay	49217ax	2171
4.5 kGy	A3/V3	0 ppm	12934a	14504b	2181
4.5 kGy	Vacuum	0 ppm	3123c	8820bc	2074
4.5 kGy	V3/A3	0 ppm	2791cy	18041bx	2046
4.5 kGy	Aerobic	200 ppm	10595aby	49732ax	2125
4.5 kGy	V3/A3	200 ppm	2134c	2689c	154
SEM			1643	2543	

a-d) Mean values with different letters within a column are significantly different ($P < 0.05$).

x, y) Mean values with different letters within a row are significantly different ($P < 0.05$). n = 4.

V3/A3: Aerobically packaged during irradiation and the first 3 days of storage and then vacuum packaging for the next 3 days.

V3/A3: Vacuum-packaged during the irradiation and the first 3 days of storage and then aerobic packaging for the next 3 days.

Sulfur Volatiles of Irradiated Ground Beef

Irradiated ground beef produced significant amounts of sulfur-methyl ester ethanethioic acid and dimethyl disulfide (Table 6). The amounts of sulfur volatiles in aerobically packaged irradiated ground beef, however, were much smaller than those in vacuum-packaged ones at Day 1. Aerobically packaged and double-packaged irradiated ground beef did not have sulfur-methyl ester ethanethioic acid and only a small amount of dimethyl disulfide was detected at Day 6. Most of the sulfur volatiles in irradiated ground beef disappeared after 6 days of storage under aerobic or double-packaged conditions because they were highly volatile. Nam et al. (2004) showed that irradiation increased the amounts of total volatiles and a few sulfur volatiles. The sulfur volatiles newly generated by irradiation were sulfur-methyl ester ethanethioic acid, dimethyl sulfide, dimethyl disulfide, and dimethyl trisulfide. The production of sulfur-containing volatiles were highly dependent upon irradiation dose (Ahn et al., 2000), and sulfur volatiles were responsible for the characteristic off-odor in irradiated meat although the amounts and compositions may be different depending on meat species and muscle types (Ahn et al., 2001; Ahn, 2002; Ahn and Lee, 2002).

Table 6. Production of sulfur volatiles in ground beef affected by irradiation, oxygen, and ascorbic acid during refrigerated storage

Irradiation	Package	Ascorbic	1 day	6 day	SEM
S-methyl ester ethanethioic acid			----------- *(10^4 ion counts)* ----------		
0 kGy	Aerobic	0 ppm	0c	0b	0
4.5 kGy	Aerobic	0 ppm	47c	0b	33
4.5 kGy	A3/V3	0 ppm	209bcx	0by	47
4.5 kGy	Vacuum	0 ppm	375bx	95ay	32
4.5 kGy	V3/A3	0 ppm	442bx	0by	34

Table 6. Continued.

Irradiation	Package	Ascorbic	1 day	6 day	SEM
4.5 kGy	Aerobic	200 ppm	330bx	0by	60
4.5 kGy	V3/A3	200 ppm	803ax	0by	82
SEM			68	21	
Dimethyl disulfide			----------- *(10^4 ion counts)* ----------		
0 kGy	Aerobic	0 ppm	0d	0c	0
4.5 kGy	Aerobic	0 ppm	294cx	106bcy	37
4.5 kGy	A3/V3	0 ppm	740bx	195by	99
4.5 kGy	Vacuum	0 ppm	1137ax	1605ay	69
4.5 kGy	V3/A3	0 ppm	1178ax	242by	25
4.5 kGy	Aerobic	200 ppm	374c	144bc	99
4.5 kGy	V3/A3	200 ppm	1079ax	189by	107
SEM			95	42	

a-d) Mean values with different letters within a column are significantly different ($P < 0.05$).

x, y) Mean values with different letters within a row are significantly different ($P < 0.05$). n = 4.

V3/A3: Aerobically packaged during irradiation and the first 3 days of storage and then vacuum packaging for the next 3 days.

V3/A3: Vacuum-packaged during the irradiation and the first 3 days of storage and then aerobic packaging for the next 3 days.

Conclusion

Because color changes in irradiated ground beef is a major defects (from bright red to a greenish brown, especially under aerobic conditions), it would be very difficult to utilize irradiation technology in ground beef without controlling the discoloration problems. Although addition of ascorbic acid at 200 ppm (w/w) to ground beef prior to irradiation stabilized color and slowed down the development of lipid oxidation in ground beef during the aerobic storage, vacuum and V3/A3 double-packaging conditions during irradiation were also effective in producing the desirable beef color when it was exposed to aerobic conditions later. Absence of air (oxygen) during irradiation and early storage reduced met-myoglobin formation in beef. Aerobically packaged irradiated ground beef produced higher amounts of volatile aldehydes than the vacuum-packaged ones. Double-packaged irradiated ground beef produced more aldehydes than vacuum-packaged one after 6 days of storage, but the amounts were significantly lower than those of aerobically packaged beef, especially when ascorbate was added. Considering all these facts, double-packaging concept (vacuum packaging during irradiation and early storage and then aerobic packaging for the remaining storage period) or double-packaging in combination with ascorbate can be a good strategy to prevent overall quality changes in irradiated ground beef.

References

[1] Ahn, DU; Lee, EJ. 2002. Production of off-odor volatiles from liposome-containing amino acid homopolymers by irradiation. *J. Food Sci.* **67**(7), 2659-2665.

[2] Ahn, DU. 2002. Production of volatiles from amino acid homopolymers by irradiation. *J. Food Sci.*, **67**(7), 2565-2570.

[3] Ahn, DU; Jo, C; Olson, DG. 1999. Headspace oxygen in sample vials affects volatiles production of meat during the automated purge-and-trap/GC analyses. *J. Agric. Food Chem.* **47**, 2776-2781.

[4] Ahn, DU; Jo, C; Olson, DG; Nam KC. 2000. Quality characteristics of pork patties irradiated and stored in different packaging and storage conditions. *Meat Sci.* **56**, 203-209.

[5] Ahn, DU; Nam, KC; Du, M; Jo, C. 2001. Volatile production in irradiated normal, pale soft exudative (PSE) and dark firm dry (DFD) pork under different packaging and storage conditions. *Meat Sci.* **57**, 419- 426.

[6] Ahn, DU; Olson, DG; Jo, C; Chen, X; Wu, C; Lee, JI. 1998. Effect of muscle type, packaging, and irradiation on lipid oxidation, volatile production, and color in raw pork patties. *Meat Sci.* **47**, 27-39.

[7] AMSA (American Meat Science Association), 1991. Guidelines for meat color evaluation. In Proceedings of the 44th reciprocal meat conference. Chicago, IL: Nat. Livestock Meat Board.

[8] Giroux, M; Ouattara, B; Yefsah, R;, Smoragiewicz, W; Saucier, L; Lacroix, M. 2001. Combined effect of ascorbic acid and gamma irradiation on microbial and sensorial characteristics of beef patties during refrigerated storage. *J. Agric. Food Chem.* **49**, 919-925.

[9] Jo, C; Ahn, DU. 2000. Production volatile compounds from irradiated oil emulsions containing amino acids or proteins. *J. Food Sci.* **65**, 612-616.

[10] Kim, YH; Nam, KC; Ahn, DU. 2002. Color, oxidation-reduction potential, and gas production of irradiated meat from different animal species. *J. Food Sci.* **67**, 1692-1695.

[11] Krzyiwicki, K. 1982. The determination of haem pigments in meat. *Meat Science* **7**, 29-33.

[12] Lee, BJ; Hendricks, DG; Cornforth, DP. 1999. A comparison of carnosine and ascorbic acid on color and lipid stability in a ground beef patties model system. *Meat Sci.* **51**, 245-253.

[13] Lee, EJ; Ahn, DU. 2004. Sources and mechanisms of carbon monoxide production by irradiation. *J. Food Sci.* **69**(6), C485-90.

[14] Luchsinger, SE; Kropf, DH; Garcia-Zepeda, Hunt, MC; Marsden, JL; Rubio-Canas, EJ; Kastner, CL; Kuecker, WG; Mata, T. 1996. Color and oxidative rancidity of gamma and electron beam irradiated boneless pork chops. *J. Food Sci.* **61**, 1000-1005, 1093.

[15] Millar, SJ; Moss, BW; MacDougall, DB., Stevenson, MH. 1995. The effect of ionizing radiation on the CIELAB color co-ordinates of chicken breast meat as measured by different instruments. *Int. J. Food Sci. Technol.* **30**, 663-674.

[16] Nam, KC., Ahn, DU. 2002. Carbon monoxide-heme pigment complexes are responsible for the pink color in irradiated raw turkey breast meat. *Meat Sci.* **61**(1), 25-33.

[17] Nam, KC; Ahn, DU. 2003. Effects of ascorbic acid and antioxidants on the color of irradiated beef patties. *J. Food Sci.*, **68**(5), 1686-1690.
[18] Nam, KC; Du, M; Jo, C; Ahn, DU. 2002. Effect of ionizing radiation on quality characteristics of vacuum-packaged normal, pale-soft-exudative, and dark-firm-dry pork. *Innov. Food Sci. Emerg. Technol.* **3**, 73-79.
[19] Nam, KC; Ahn, DU. 2003. Combination of aerobic and vacuum packaging to control color, lipid oxidation and off-odor volatiles of irradiated raw turkey breast. *Meat Sci.* **63**(3),389-395.
[20] Nam, KC; Min, BR; Lee, SC; Cordray, J; Ahn, DU. 2004. Prevention of pinking, off-odor, and lipid oxidation in irradiated pork loin using double packaging. *J. Food Sci.* **69**(3): FTC 214-219.
[21] Nanke, KE; Sebranek, JG; Olson, DG. 1998. Color characteristics of irradiated vacuum-packaged pork, beef, and turkey. *J. Food Sci.* **63**, 1001-1006.
[22] Sanchez-Escalante, A; Djenane, D; Torrescano, G; Beltran, JA; Roncales, P. 2001. The effects of ascorbic acid, taurine, carnosine and rosemary powder on colour and lipid stability of beef patties packaged in modified atmosphere. *Meat Sci.* **58**, 421-429.
[23] SAS Institute Inc. 1995. SAS/STAT User's Guide. Cary, NC: SAS Institute.
[24] Satterlee, LD; Wilhelm, MS; Barnhart, HM. 1971. Low dose gamma irradiation of bovine metmyoglobin. *J. Food Sci.* **36**, 549-551.
[25] Seideman, SC; Cross, HR; Smith, GC; Durland, PR. 1984. Factors associated with fresh meat color: A review. *J. Food Qual.* **6**, 211-237
[26] Shahidi, F; Yun, J; Rubin, LJ; Wood, DF. 1987. The hexanal content as an indicator of oxidative stability and flavor acceptability in cooked ground pork. *Can. Inst. Food Sci. Technol.* **20**, 104-106.
[27] Wheeler, T.; Koohmaraie, M; Shackelford, SD. 1996. Effect of vitamin C concentration and co-injection with calcium chloride on beef retail display color. *J. Anim. Sci.* **74**, 1846-1853.

In: Radiation Physics Research Progress
Editor: Aidan N. Camilleri, pp. 301-328

ISBN: 978-1-60021-988-7

Chapter 8

X-RAY RADIATION ARISING IN FREE ELECTRON/SUBSTANCE INTERACTION

G.V. Pavlinsky, Dukhanin A. Yu and M.S. Gorbunov
Irkutsk State Univ., Institute of Applayed Physics

Abstract

The processes of arising of characteristic and bremsstrahlung x-ray, conditioned by flux of photo- , Auger and Compton electrons in a irradiated sample are considered. The theoretical dependencies are obtained which allow carrying out calculations of intensity of arising radiation.

It is shown that in long-wave range of x-ray fluorescent spectrum at x-ray tube primary radiation these processes can be insignificant.

The contribution of ionizing action of photo- , Auger electrons of different shells of atoms of containing matrix is estimated in forming of x-ray fluorescence. It has appeared that for fluorine, oxygen and nitrogen this contribution is essential, and for carbon, boron and beryllium it becomes to be dominant.

The role of photo- and Auger electrons increases at the decreasing concentration of elements with low atomic number and is defined by composition of containing matrix. As rival process of cascade transition is considered which transform energy of primary photons to long-wave L-radiation of matrix elements. This radiation is able to effectively excite atoms of elements with low atomic number. For Fe, Zn, Ni matrixes the contribution of processes of cascade transitions to excitation of x-ray fluorescence of these elements was estimated. The conditions are studied allowing to obtain maximum intensity of analytical signal. The results of theoretical calculations are confirmed by comparison to experimental values of relative intensities of carbon measured for different carbonaceous bonds.

The bremsstrahlung of photo- , Auger and Compton electrons arising at its interaction with nuclei of atoms of matrix elements is studied. The realized calculations are allowed to define dependence of separate components of this radiation from composition of irradiated matrix. It is shown that bremsstrahlung of Compton electrons becomes to be significant only at high energies of primary photons. At usual voltages of x-ray tubes (up to 60 kV) this component can be neglected. The comparison with intensity of scattered polychromatic radiation of x-ray tube is allowed to define wave-length range where bremsstrahlung of considered electrons becomes to be significant. The bremsstrahlung of these electrons is considered in two approximations. The first approximation supposes that electrons brake in point of primary photon absorption. The second approximation takes into account that track-length of considered electrons can appear to be comparable with depth of penetration of long-wave bremsstrahlung. The bremsstrahlung of photo- , Auger and Compton electrons can appear to be dominant in forming of x-ray background of analytical lines of elements with low atomic number. The background genetically concerned with origination of x-ray fluorescence therefore it is not eliminated.

1. Introduction

Irradiation of matter by x-ray photons flux leads to arising photo- , Auger and Compton electrons. Coulomb interaction of these electrons with nuclei, which are called – *free,* makes condition to arising bremsstrahlung of these electrons; their ionization of atoms is accompanied with characteristic x-ray. The contribution of *free* electrons into excitation of characteristic x-rays radiation of sample as rule is neglected in comparison with processes of direct ionization by primary photons, and their bremsstrahlung radiation of x-ray is neglected in comparison with scattered primary X-ray tube bremsstrahlung radiation. However, there is situation when neglect of *free* electrons can lead to significant errors in defining of X-ray radiation intensity of sample. In case of excitation of elements with low atomic number (F, O, N, C, B, Be) the long-wave primary radiation of x-ray tube, effectively exciting atoms of these elements, is sufficiently absorbed by beryllium window x-ray tube. In this case the role of *free* electrons is essential. Besides the filtering of primary x-ray by beryllium window leads to bremsstrahlung radiation of considered electrons is determinative in long-wave length of fluorescent spectrum.

Since processes of excitation of characteristic radiation and bremsstrahlung are different, logically it is considered separately.

2. Characteristic X-ray Radiation of Elements with Low Atomic Numbers

Current theory of excitation of X-ray fluoresce is based on considering processes of primary radiation photoelectric absorption. Other possible variants of ionization are not taken into account. In particularly, ionize action of appeared in a sample the photo-, Auger- electrons is neglected. The first estimation of photoelectrons contribution in x-ray fluorescence was implemented in 1971 by *Ebel et al.* [1]. It was shown that in case of bremsstrahlung of x-ray tube (voltage 30-40kV) the contribution of K – photo electrons into intensity of fluorescent radiation of copper sample is tenth parts of percent.

Increasing of this contribution was to be expected by essential exceeding energy of photo-, Auger electrons over energy of binding of electrons in atoms of excited element. Implemented in 1974 estimation[2] showed that in case of monochromatic primary radiation with 30 keV energy contribution of these electrons in excitation of K-shell of Aluminum atoms (10% Al, 90% Fe) is approximately 30%, but at excitation of low contents of Na by ^{109}Cd radiation source with 22.1 keV energy this contribution reached to 60%[3]. The obtained results prejudiced accuracy of theoretical modeling of processes of florescence excitation of silicate group elements without taking in into account ionization atoms by photo-, Auger electrons. However considered condition of fluorescence excitation is far from reality. At polychromatic primary radiation of X-Ray tube with Rh anode (voltage 25 kV, thickness of beryllium window 300 μm) and 1% content of Na in binary mixture with other elements (Z<40) the considered contribution was less than 10%, according to works [4,5]. The calculations implemented by Borkhodoev [6] in the same conditions for real rock alloyed with lithium tetra borate , showed that ionize action of photo-, Auger electrons is neglected. This conclusion is fairly for modern x-ray tubes with 75 μm thickness of beryllium window.

The problem of ionize action of photo- , Auger electrons appeared again when multilayer pseudo crystal was created[7] and serial x-ray apparatus was equipped detectors for low atomic number elements radiation (B, C, N, O, F). The place of K-edge of absorption of carbon about spectral allocation of primary radiation of X-Ray tube with 75 μm thickness of window schematically presented on figure 1. It follows from figure 1 that in range of wave length ~ *10 Å - 43.6 Å* doesn't content primary radiation which is able to effectively excite carbon atoms whereas atoms with high atomic number is completely excited by primary radiation (look at the place of some element j absorption *K*-edge).

In case of conditions which are showed on figure 1 is to be expected that contribution of these electrons in forming X-Ray fluorescence of carbon atoms will be important. The process of cascade transitions in elements matrix is not fairly researched, which transform short wavelength primary radiation to essentially longer wavelength. The latest is able to effectively excite atoms with low atomic number Z. This process is variety of selective excitation.

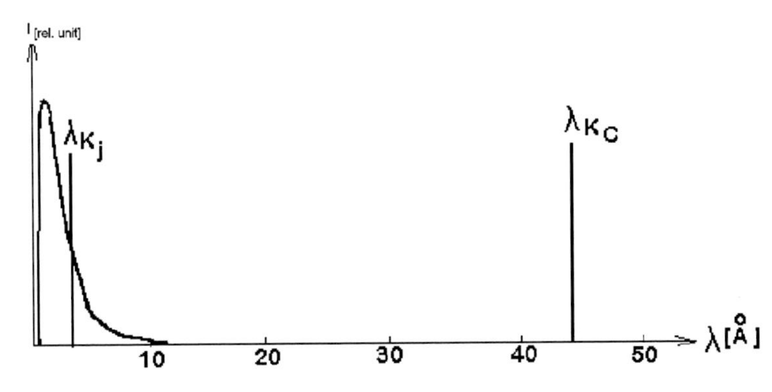

Figure 1. Schema to basis of photo-, Auger electrons role in excitation of X-Ray fluorescence by low atomic number elements. On scale of wave length absorption K-edge of carbon and absorption K_j-edge of some elements *j*, which is source of photo-, Auger electrons, is indicated.

2.1. Theory

The intensity of X-Ray florescence N_i (expressed by counts of photons per time unit, per perpendicular to beam direction square unit) for some *i* elements in massive sample at polychromatic primary radiation can be presented as [5, 9]:

$$N_i = K(\lambda_i) \int_{\lambda_0}^{\lambda_q} \frac{N(\lambda)\theta_i(\lambda)d\lambda}{\frac{\mu(\lambda)}{Sin\varphi} + \frac{\mu(\lambda_i)}{Sin\psi}} \tag{1}$$

where $K(\lambda_i)$ coefficient depending on the irradiated area of fluorescence sample, geometry of spectrometer and detecting condition of *i* element,

$N(\lambda)$ - spectral intensity of primary radiation expressing by numbers of photons per time unit,

$\mu(\lambda)$ and $\mu(\lambda_i)$ - mass attenuation coefficient of the primary and fluorescence radiation in a sample, accordingly,

φ and ψ - angles of primary x-ray incidence and takeoff of fluorescent x-ray, respectively,

$\theta_i(\lambda)$ - probability of transformation of primary x-ray photons with λ wavelength into photon of fluorescent radiation with λ_i wavelength .

This probability is defined by sum of probabilities of all processes leading to ionization of atom *i*:

$$\theta_i(\lambda) = \theta_{i1}(\lambda) + \theta_{i2}(\lambda) + \theta_{if}(\lambda) + \theta_{iAu}(\lambda) \qquad (2),$$

where $\theta_{i1}(\lambda)$- probability of direct ionization, $\theta_{i2}(\lambda)$- probability of selective ionization, $\theta_{if}(\lambda)$- probability of ionization by photoelectrons of sample, $\theta_{iAu}(\lambda)$- probability of ionization by Auger electrons of sample. Consider each process separately.

For direct ionization.

$$\theta_{i1}(\lambda) = \tau_{q_i}(\lambda)\omega_{qi}p_{qk}C_i \qquad (3),$$

where $\tau_{q_i}(\lambda)$ - partial coefficient photoelectric absorption primary radiation for *i* element for q- shell, ω_{qi} and p_{qk} are yield of fluorescence of element *i* and probability of emission of its *k*- line of q- shell, respectively, $\tilde{N}_i$ -concentration of element in sample.

For selective ionization.

$$\theta_{i2}(\lambda) = P_1 L P_2 \qquad (4)$$

where $P_1 = \tau_{qj}(\lambda)C_j\omega_{qj}p_{qjm}$ - probability of transformation primary photon to photon of selective excited element *j*. $P_2 = \tau_{qi}(\lambda_j)C_i\omega_{qi}p_{qin}$- probability of transformation of photon of elements *j* to registrable by detector photon of fluorescent radiation of element *i* , and L – same transfer function , which is defined by expression:

$$L = \frac{1}{2}\sum_j \left\{ \theta_{j1}(\lambda)\theta_{i1}(\lambda_j)\left[\frac{\sin\varphi}{\mu(\lambda)}\ln\left(1+\frac{\mu(\lambda)}{\mu(\lambda_j)\sin\varphi}\right) + \frac{\sin\psi}{\mu(\lambda_i)}\ln\left(1+\frac{\mu(\lambda_i)}{\mu(\lambda_j)\sin\psi}\right)\right]\right\}$$

Summation is conducted along all lines of elements *j*, which radiation is able to excite atoms of element *i*.

For photoelectrons ionization.

If point of atom ionization occurs in the point of origination of element j electrons, we can write following expression:

$$\theta_{if}(\lambda) = \omega_{qi} p_{qk} \sum_j C_j \sum_m [\tau_{mj}(\lambda) n_{mj}(C_i, E)] \qquad (5)$$

where $n_{mj}(C_i, E)$ - probability of ionization atoms of i elements by m-shell electrons of j elements. The quantity of $n_{mj}(C_i, E)$ is presented, for example, in paper [10]. Summation is conducted along all shells m of elements j, which radiation is able to excite q-shell of element i.

For Auger electrons ionization.

The probability of Auger transition in some m- shell of j elements is defined as $\tau_{mj}(\lambda)(1-\omega_{mj})$.

The absence of reliable information about relation of probability of individual group of Auger –transition leads to necessity to use assumption that all Auger electrons of j elements from the same series have equal energy. This assumption doesn't carry in significant error in results of calculations for individual groups of L- and M- series Auger electrons, because the energies of these are not fairly distinguished. Hence,

$$\theta_{iAu}(\lambda) = \omega_{qi} p_{qik} \sum_j C_j \sum_m \tau_{mj}(\lambda)(1-\omega_{mj}) n(\tilde{N}_i, E_{mj}) \qquad (6).$$

Exception is only K-series of Auger electrons, for which are accepted[11] that probability of K-LL transitions is approximately 70% and others are 30%. With taking into account this fact, $n(C_i, E_{mj})$ is defined by sum $n(C_i, E_{mj}) = 0.7n(C_i, E_{KL_2L_3}) + 0.3n(C_i, E_{KL_2M_3})$.

Expressions (3)-(6) for probabilities of transformation of primary photons into fluorescent photons use for implementation of X-Ray florescence intensity calculations by formula (1). All calculations suppose that the cathode of X-ray tube is grounded and return back to anode of scattered electrons of cathode beam increases intensity of primary radiation [12, 13].

2.3. The Ratio of Long Wavelength of X-ray Florescence Components

Estimate contribution of selective excitation to intensity of fluorescence of elements with low atomic number. This contribution is defined by expression (4) and can appear in three forms: ionization atoms by K- radiation of matrix, by L- radiation of matrix and by L - radiation of matrix originated as result of cascade transitions. The two first considered effects is well known [15], but the third is only just mentioned in work [16, 17].

The cascade transition transforms primary photons of matrix elements to low energy photons, which are more effectively to excite elements with low atomic number. The energy of L – radiation of Fe (~0.7 keV) is much nearer to energy of bond of K – electrons of C, N, O, F, than energy of K-radiation of Fe matrix (~ 6.4 keV). The coefficient of K- L-

transformation of cascade transitions, which is defined by ω_L - yield of X-Ray *L*-fluorescence, proves to be sufficiently low (according to data by work [6] , ω_L(Fe) ~ 0.6%). That's why it hard to estimate the meaning of these variant of selective excitation without prior calculations. Also it is clearly, that with increasing atomic number of excited element the *K* – edge absorption energy approaches to energy of *L* – line radiation of Fe, and the role of *L* –selective excitation in forming *K* – fluorescence of considered elements has to increase.

If energy of primary photons is more than biding energy of *K*-shell, then cascade transitions will make general contribution in forming of *L*- radiation [18]. This contribution strongly decreases when general part of polychromatic primary radiation gets to range between edges of absorption of *K* – and *L* – shells. The ratio between these considered parts primary radiation fairly depends on material of anode and voltage and window thickness.

In case of cascade transitions the probability P_l from equation 4 is defined as:

$$P_1 = \tau_{Kj}(\lambda)c_j P_{KL}\omega_{Lj}p_{Ljm} \tag{7}$$

Table 1. The intensities (conventional units) and contributions (%) of processes of analytical signal forming of elements with low atomic number. X-ray tube: W-anode, 300 µm thickness of beryllium window, 40 kV voltages.

Matrix	Processes										
	Direct Excitation		Selective Excitation							photo- and Auger excitation	
			K –		L –		K – L –		Σ		
	Conv. unit	%	Conv. unit	%	Conv. unit.	%	Conv. unit.	%	%	Conv. unit	%
Carbon											
Fe	57	21	0.7	0.3	9.8	4	0.8	0.4	4.7	204	74.3
Ni	45.4	20	0.5	0.2	8.8	3.9	0.8	0.4	4.5	173	75.5
Zn	**36.4**	**21**	**0.1**	**0.1**	**5.8**	**3.3**	**0.2**	**0.2**	**3.6**	**135**	**75.4**
Nitrogen											
Fe	340	43	5.4	0.7	68.7	8.7	5.8	0.8	10.2	372	46.8
Ni	373	42	3.3	0.5	61	9.5	5.8	0.9	10.9	303	47.1
Zn	**219**	**47**	**1**	**0.2**	**40**	**8.5**	**1.7**	**0.4**	**9.1**	**206**	**43.9**
Oxygen											
Fe	1475	58	29.9	1.2	353	14	31.3	1.3	16.5	638	25.5
Ni	1187	58	18.1	0.9	314	15	31	1.5	17.4	512	24.6
Zn	**957**	**64**	**5.7**	**0.4**	**205**	**14**	**9.1**	**0.6**	**15**	**328**	**21**
Fluorine											
Fe	4952	65	134	1.8	1406	19	134	1.8	22.6	981	12.4
Ni	3992	64	81	1.3	1257	20	134	2.1	23.4	800	12.6
Zn	3232	70	26	0.6	827	18	40	0.9	19.5	497	10.5

P_{KL} - probability of transformation K –shell vacancy to L- shell vacancy $P_{KL} = 0.85\omega_{Kj} + 0.9(1-\omega_{Kj})$, where the first component define probability of radiation transition without taking into account K_β -line, the second component define non-radiation transition of Auger electrons, which is concerned with ionization only L-shell. This probability, according to work [11], equals approximately 0.9 . In that case, according to [18], expression (4) can be presented as: $P_1 \approx \tau_{Kj}(\lambda)c_j(0{,}9-0{,}05\omega_{Kj})\omega_{Lj}p_{Ljm}$ (8).

The result of calculations of components of process of analytical signal forming of elements with low atomic number for different elements matrix was taken from [19] and presented in table 1(99,9 %– matrix, 0.1 % - element). L- radiation of these matrixes excite K-fluorescence of considered elements. The calculations are implemented with taking into account ionization by photo- and Auger electrons. The spectrum of primary radiation corresponds with work [13, 20]. The M- radiation of tungsten X-ray tube anode is taken into account in these calculations.

It follows from table 1 that with atomic number increasing of considered elements role of direct excitation increases to the prejudice of photo- and Auger electrons excitation. Simultaneously the contribution of selective excitation rapidly increases from units of percent for Carbon to 22.6 for Fluorine. Thus, these effects can be neglected for Carbon but it is necessary to consider for fluorine.

The most significant of these three effects of selective excitation is L- shell radiation selective excitation. It is necessary to notice; that the contribution of selective effects resulted in all tables to total intensity of x-ray fluorescence of elements with small Z is defined with a significant error. This error is conditioned by inaccuracy of existing data of L-shell fluorescence yields of considered elements. The dispersion of this data for Zn, Fe, Ni [6, 22] is characterized by roof-mean-square deviation about 20-40%. The average quantity of ω_L, according to [6], equals 0.6% for Fe, 1.09 – for Ni, 1.25 – for Zn. The small quantity of ω_L is the cause of the contribution of selective excitation is too less than numbers of arising L-vacancy.

The significant contribution of photo- and Auger electrons in x-ray fluorescence excitation of considered elements (look at table 1) necessitate studying more attentive the content of their contribution.

If sample consists of one element with low atomic number then only K-photo electrons influent on intensity of its fluorescence (the probability of L-shell ionization is neglected, and energy of K - Auger electrons is rather low to excite own K-shell). In this case the contribution of photo- and Auger electrons is rather low. The quantity of this contribution for Rh-anode X-ray tube with 40 kV voltage and 75 μm beryllium window is presented in table 2.

It follows from table 2 that considered contribution rapidly decrease with atomic number increasing. It's conditioned by approaching of absorption edges to primary radiation wavelengths and correspondingly by increasing role of photoelectric atom ionization.

Table 2.The contribution of *K*-photo electrons in excitation of x-ray fluorescence of monoelement matrix with low atomic number.

Element	B	C	N	O	F
$\frac{I_i^{ôå}}{I_i + I_i^{ôå}}$	33.3	17.7	9.0	4.8	2.5

With low content of considered elements in sample the contribution of photo- and Auger electrons in fluorescence repeatedly increases because accommodating matrix get a lot of free electrons with high energy. The contribution of photo and Auger electrons in x-ray fluorescence forming of Carbon at primary radiation of Rh x-ray tube with 75 μm thickness of window presented in table 3. It is supposed that samples content 0.1 % of carbon and 99.9% of stuff elements with *Z* atomic number.

From table 3 it follows that total contribution of photo- and Auger electrons in x-ray fluorescence forming of carbon is dominant. With increasing of atomic number of stuff elements in the carbon excitation, photo- and Auger electrons of more remote shells insensibly join in. For $Z_{stuff} < 30$ at calculations of intensity of x-ray fluorescence of carbon it is possible to be limited by taking into accounts only ionization action of *K*-, *L*- photo and *K*- Auger electrons.

On figure 2, which is taken from [9] work, the intensities of carbon lines, excited by photo- and Auger electrons, is related to intensity conditioned of direct ionization by primary x-ray. The conditions of excitation correspond to conditions referred above in the table 2. The selective excitation of carbon is neglected and it was not taking into account in our calculations.

Table 3. The contribution of photo- and Auger electrons in x-ray fluorescence forming of carbon at the primary radiation of Rh x-ray tube with 75 μm thickness of output window (voltage 40 kV). Sample content: 0.1 % of carbon and 99.9% of stuff element.

Z stuff elements	K-photo, %	L-photo, %	M-photo, %	N-photo, %	K-Auger, %	L-Auger, %	M-Auger, %	All, %
7	24.47	0.70	-	-	0.26	-	-	25.42
11	38.59	4.61	-	-	15.01	-	-	58.20
17	12.89	12.11	-	-	50.53	-	-	75.53
23	14.09	39.33	-	-	18.56	0.82	-	72.80
29	6.95	51.63	-	-	8.58	13.09	-	80.25
35	2.23	39.73	15.70	-	2.03	28.97	-	88.66
41	0.77	15.79	22.02	-	1.10	49.80	-	89.48
48	0.19	18.49	44.44	8.72	0.15	14.49	-	86.48
53	-	14.62	49.38	11.60	-	12.55	-	88.16
59	-	12.28	49.81	15.99	-	10.79	-	88.87
65	-	10.32	41.28	22.01	-	9.39	5.49	88.49
71	-	8.40	27.49	27.85	-	7.87	16.27	87.89
77	-	6.09	22.63	25.92	-	5.89	27.18	87.71
82	-	4.41	20.37	22.42	-	3.77	37.84	88.81

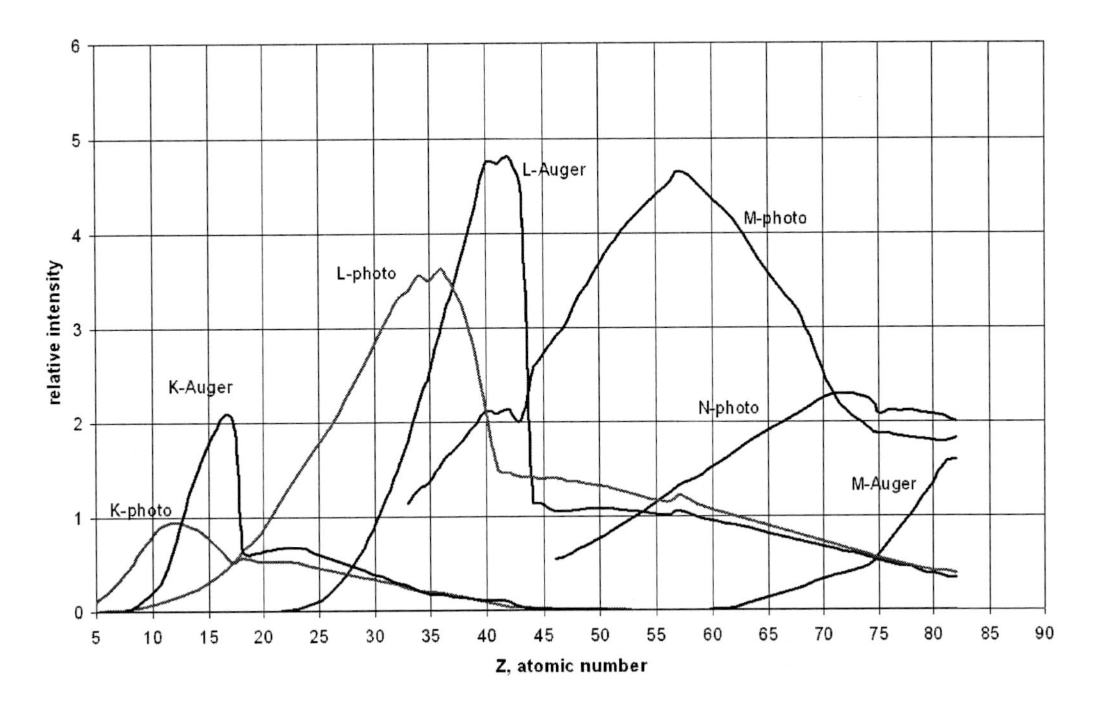

Figure 2. The dependence of relative intensity of carbon excited by all kinds of photo- and Auger electrons to atomic number of stuff elements. Primary radiation of Rh x-ray tube with 75 μm thickness of output window (voltage 40 kV).

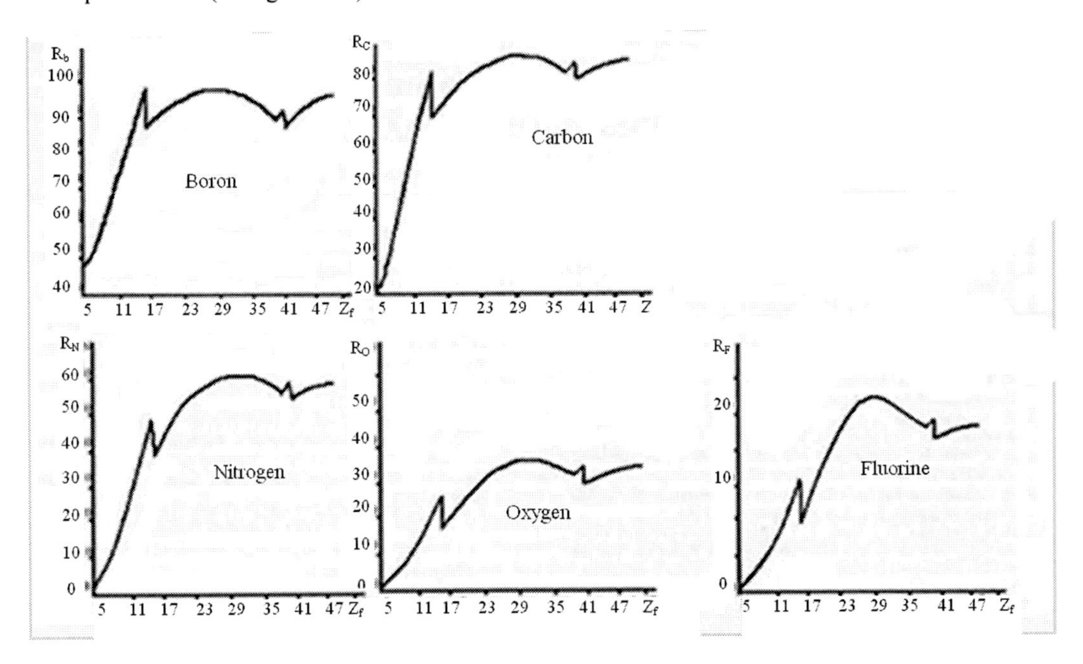

Figure 3. The total contribution of photo- and Auger electrons for different containing matrix with Z_R to intensity of x-ray fluorescence of light elements (mass concentration 0.1%). The excitation conditions: x-ray tube with Mo anode with 40 kV voltage and 300 μm thickness of beryllium window, $\varphi=50^{\circ}$ and $\psi=30^{\circ}$.

From fig. 2 it follows that intensity of carbon excited by photo- and Auger electrons repeatedly exceeds intensity excited directly by primary photons. The each related intensity of x-ray fluorescence of carbon is conditioned by different photo- and Auger electrons excitation has some maximum. It is conditioned by the following fact that with increasing of atomic number of stuff element increase the probability of its atom ionization by primary monochromatic radiation of tube pro rata increasing of coefficient of photoelectrical absorption i.e. increase number of photo- and Auger electrons. In this case energy of photoelectrons decreases and counts of ionizations correspondingly fall per electron. The action of these two contra factors leads to maximum arising for photoelectrons. In case of Auger electrons at monochromatic primary radiation with increasing Z of stuff element its energy does not decrease. Hence, there is moment when the energy of primary photons becomes insufficient to ionize the shell of stuff element, and the contribution of Auger electrons rapidly falls to zero.

The same processes occur with the polychromatic primary radiation, maximums are more degraded because of uncertainty of wavelength of this radiation. These degraded maximums are showed on fig. 2 for all shell of stuff element.

It is evidently that for more remote shells from atom these maximums are shifted to elements with high atomic numbers Z_{stuff}.

The role of photo- and Auger electrons in forming of x-ray fluorescence of boron, carbon, nitrogen, oxygen, fluorine defined in work[8] and is showed on fig. 3.

From fig. 3 it follows that the contribution of photo – and Auger electrons to total x-ray fluorescence at low content of fluorescent element increase over eight times at the decreasing of atomic number Z from Z=9 (fluorine) to Z=5 (boron).

2.3. The Comparison of Experimental and Theoretical Data

The intensity of x-ray fluorescence of carbon was measured on ARL9800 spectrometer. The anode of x-ray tube is Rh , thickness of beryllium output window - 75 μm, voltage - 40 kV, current – 60 mA. As basis of sample is used collection of carbonaceous compounds, which content is selected to provide significant variations of contribution of photo- and Auger electrons to carbon atom excitation. There was graphite among the analyzed materials; for which the contribution of considered electrons is minimal to florescence excitation of carbon. The mixture of carbonaceous compounds with another elements was not considered to reduce to minimum influence of particle size of irradiate sample to measured intensity of analytical fluorescence line of carbon. The samples were prepared by pressing irradiate material to cavity from boric acid. The choice of angle divergence of collimator and regime of amplitude impulse analyzer allows minimizing of superposing of 2-order lines of oxygen on analytical peak of carbon. Residual superposition of this line is taking into account by graphical subtraction method. The confidence interval was defined by double-ply (with one month interval) measuring intensity of five different samples of each compound for three parallel defining.

The result of calculation and experimental data for intensity of carbon related to carbon intensity in graphite, presented in [9] work, shows in table 4.

Table 4. The comparison of experimental and theoretical data of relations of carbon intensity in different compounds to its intensity in graphite. The X-ray tube with Rh anode, thickness of beryllium output window - 75 μm, voltage - 40 kV, current – 60 mA.

Compound	Experimental data	Calculations	
		With taking into account photo- and Auger electrons	Without taking into account photo- and Auger electrons
Li_2CO_3	0.088 ±0.004	0.087	0.080
$MgCO_3$	0.050 ±0.004	0.051	0.035
K_2CO_3	0.074 ±0.006	0.077	0.044
$K_4Fe(CN)_6$	0.121 ±0.006	0.120	0.073
$Y_2(CO_3)_3\ 3H_2O$	0.051 ±0.006	0.054	0.019
$La(CH_2COO)_3$	0.274 ±0.006	0.286	0.093
$Tm_2(CO_3)_3\ 3H_2O$	0.063 ±0.006	0.072	0.015
$Pb(CH_3COO)_2$	0.286 ±0.006	0.298	0.034

It is clear from table 4 that experimental data is fairly conformed to result of calculations taken into account photo- and Auger electrons contribution and is not conformed without taking into account this contribution. It is not hard to mark that accordance of experimental data and calculation results is degraded for matrix with high Z. It can by explained by degrading reliability of *M*- and *N*- shell fundamental parameters information used for calculations.

2.4. The Influence of Excitation Conditions on X-ray Fluorescence of Elements with Low Atomic Number

The essential contribution of photo- and Auger electrons in x-ray fluorescence of elements with low atomic number forming require revision of conditions for which the maximum of its analytical signal can be achieved. The first factor influenced on considered intensity is thickness of x-ray tube beryllium output window. The results of calculation of x-ray fluorescence intensity of considered elements for x-ray tube primary radiation at beryllium output window of x-ray tube absence related to its intensity at *d* output window thickness presented in table 5. The replacement elements of containing matrix changes result presented in this table incidentally.

Table 5. The ration of x-ray fluorescence of elements with low atomic number (0.1% content of element in Si matrix) at presence beryllium window with d thickness and its absence.

thickness d [μm]	Ratio $K = I_0 / I_d$ for:					
	Na	F	O	N	C	B
0	1	1	1	1	1	1
20	1.3	1.6	6.0	18.4	34.2	57.6
50	1.8	2.2	7.6	24.1	42.2	68.0
75	2.1	2.6	8.8	28.2	48.4	76.7

The decreasing of thickness of x-ray tube output window concerned with degrading reliability of x-ray tube work but advantage in intensity of x-ray fluorescence for F, O, N, C, B elements is not essential (for changing thickness from 75 μm to 50 μm the advantage is less 16%). That's why this way to increasing intensity x-ray fluorescence of elements with low atomic number is hardly well-taken. The perfect way is absence of this window, it would allow to increase intensity of N, C and B in dozen of times in compare with modern 75 thickness x-ray tubes. However in this case the significant background appears which is defined by scattered long wavelength bremsstrahlung primary radiation in range of considered elements analytical lines.

Notice, that the contribution of photo- and Auger electrons in excitation of considered elements proves to be neglected at beryllium output window absence. The dependence presented on fig. 4 and taken from [14] work is evidence of this.

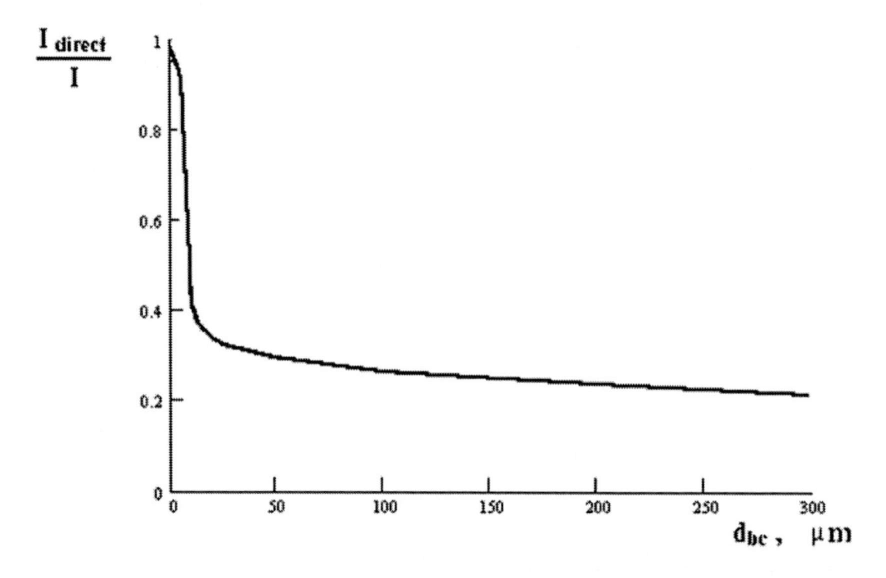

Figure 4. The dependence of contribution of photoelectric excitation of carbon atoms to total intensity from beryllium output x-ray tube window thickness. The carbon content is 0.1% in Si matrix. Rh anode x-ray tube with 40 kV voltage.

Table 6. The variation of intensity of carbon x-ray fluorescence in some matrix at increasing x-ray tube voltage from 20kV to 60 kV.

$(N_i)_{60} / (N_i)_{20}$ for contents			
100%C	0.1%C+99.9%Si	0.1%C+99.9%Fe	0.1%C+99.9%Mo
1.93	2.08	2.42	2.32

The other possibility to increase intensity of x-ray fluorescence of elements with low atomic number is to change voltage on x-ray tube. In short wavelength range the dependence of x-ray fluorescence from voltage can be defined as [23]:

$$N_i = Const(V - V_{qi})^n$$

where V- tube voltage, V_{qi}- ionization potential q – shell of i-element, the n- index changes in the wide range as rule more than 1. That's why increasing x-ray tube voltage is preferred way to increase intensity of x-ray fluorescence instead of increasing anode current.

In the long wavelength range of x-ray fluorescent spectrum there is another situation. For example, the dependence of intensity of x-ray fluorescence of carbon for little contents from voltage in some matrix for 75μm thickness of beryllium window is presented in table 6.

From table 6 it follows that at changing x–ray tube voltage in 3 times the intensity of x-ray fluorescence will changes only in 2-2.5 times. The exponent n in (7) equation appears to be less than 1. Therefore, the more rational way for efficient increasing intensity of x-ray fluorescence is the increasing x-ray tube current rather than voltage. At the three-ply power increasing due to current increasing, the intensity of x-ray fluorescence of carbon increases up to 3 times that are more essential values, that shown in table 6. The third possibility to increasing x-ray fluorescence is selection conditions of optimal primary radiation. If the basic process is atomic photo ionization then characteristic primary radiation is rationally to maximally approach to edge of absorption of fluorescent element. It is right for high concentration of elements with low atomic number; and is confirmed by calculation results presented in table 7 for mono elemental samples. The calculations were implemented at the following conditions: x-ray tube with ground anode, voltage – 40 kV, thickness of output window – 75 μm.

Table 7. The relative intensities of mono elemental samples of carbon, nitrogen, oxygen, fluorine at different x-ray tube primary radiation.

The x-ray tube anode material	Elements			
	C	N	O	F
Ti	0.53	0.54	0.57	0.61
Cu	0.25	0.24	0.24	0.25
Rh	1	1	1	1
W	1.41	1.27	1.13	1.07

From table 7 it follows that the optimal anode material of x-ray tube is W (tungsten) among the considered materials, the M-line characteristic primary radiation($\sim 7\mathring{A}$) appears to be more long wavelength and to be able to pass through beryllium output window of 75 μm thickness. The second by efficiency material is Rh (rhodium), its *L*-line has *~4.5Á* wavelength. The characteristic primary radiation of Ti and Cu is *~2.8Á* and *~1.5Á* accordingly, it is defined its efficiency for excitation of considered elements.

At the low content of considered elements the primary factor of fluorescence excitation, as mentioned above, becomes ionization by photo- and Auger electrons. That's why achievement of analytical signal maximum of considered elements should be provide the most yield of that electrons which contribution to fluorescence intensity is more significant. This definition is accord with data presented in table 8 for x-ray fluorescence intensity of carbon at different x-ray tube primary radiation.

From table it is clear that the primary radiation of Ti x-ray tube is optimal to achieving maximal x-ray fluorescent intensity in Ag matrix. It can be explained by the following fact that the *K*- line radiation of titan well excites *L*-shell of argentums and Auger electrons of *L*-series of titan is more efficient to argentums atom ionizations. For Fe-matrix is most preferable is x-ray tube with W (tungsten) anode because its *L*-radiation is efficient to excite Fe *K*-shell, and *M*-radiation is efficient to excite Fe *L*-shell. The flow of photo- and Auger electrons appears to be significant from *K*- and *L*- shells of Fe ionized carbon atoms.

The one of the important parameters is concentration sensitivity of analytical signal in analytical chemistry (it is reaction of analytical signal to changing of element concentration) which defines error of analyze results. This concentration sensitivity of boron and carbon is reflected in table 9 which calculated with and without photo – and Auger electrons excitation of atoms. The source of primary radiation is x-ray tube with Rh –anode and 40 kV voltage.

Table 8. The relative intensities of carbon x-ray fluorescence (low contents in some matrix) at x-ray tube primary radiation with Ti , Rh, Cu anodes. The conditions of excitation are equal to table 7.

The anode material	0.1% C in compound matrix			
	Al	Fe	Ge	Ag
Ti	0.78	0.65	0.73	1.81
Cu	0.37	0.97	0.38	0.88
Rh	1	1	1	1
W	1.28	1.66	1.20	1.67

Table 9. The content dependence of concentration sensitivity ($\Delta N/\Delta c$) of elements with low atomic number in irradiated sample.

Carbon in *Fe* matrix

Carbon concentration (%)		1	5	10	15	20
$\frac{\Delta N}{\Delta c}$ (conv. units)	With photo- and Auger electrons	5.7	5.4	5.0	4.6	4.0
	Without photo- and Auger electrons	0.82	0.88	0.96	1.1	1.2

Boron in *Si* matrix

Boron concentration (%)		1	5	10	15	20
$\frac{\Delta N}{\Delta c}$ (conv. units)	With photo- and Auger electrons	4.9	4.7	4.6	4.5	4.4
	Without photo- and Auger electrons	0.29	0.32	0.35	0.39	0.45

It follows from table 9 that at 1% element concentration the ionization action of photo- and Auger electrons increases approximately concentration sensitivity of carbon in 7 times and in 17 times – for boron. With concentration increasing of considered elements the concentration sensitivity which was obtained with taking into account ionization action of free electrons decreases as expected.

3. The Bremsstrahlung of Free Electrons

The basic component of the x-ray background in a short-wave length range of the x-ray spectrum of fluorescence is x-ray tube primary radiation scattered on a sample [2, 3]. The long-wave length part of primary radiation is absorbed by the tube window. Therefore, the basic components of the background in a long-wave length range of x-ray fluorescence could be the bremsstrahlung of the electrons originating in the material irradiated. The existing research of bremsstrahlung of these electrons [26, 27] is not admitted as exhaustive. In particular, the definition of wave-length range seems to be important, where bremsstrahlung becomes to be significant up-to scattered primary radiation by sample. It is necessary to estimate the influence of condition of excitation of x-ray fluorescence and elemental content of fluorescent material on ration of contribution of photo-, Auger and Compton electrons in total bremsstrahlung.

3.1. Theory

Spectral intensity $\frac{dN}{d\lambda}$ of bremsstrahlung of free electrons originating in the fluorescing sample at polychromatic primary radiation defined [5] by the same expression (1) , where probability $\theta_i(\lambda)$ obtain another meaning:

$$\frac{dN}{d\lambda} = K(\lambda)\int_{\lambda_0}^{\lambda}\frac{N(\lambda_1)\theta(\lambda_1,\lambda)d\lambda_1}{\frac{\mu(\lambda_1)}{Sin\varphi}+\frac{\mu(\lambda)}{Sin\psi}} \tag{8}$$

where $\theta(\lambda_1,\lambda)$ - probability of transformation of primary photon with λ_1-wavelength to photon λ-wavelength of bremsstrahlung of *free* electrons.

$\mu(\lambda)$ and $\mu(\lambda_i)$ - mass attenuation coefficient of the primary and fluorescence radiation in a sample, accordingly,

φ and ψ - angles of primary radiation incidence and takeoff of fluorescent x-ray, respectively,

$N(\lambda_1)$ -counts of photons of primary radiation in interval $d\lambda_1$.

The equation (8) is right in supposition that the track-length of free electrons before moment of interaction with advent of x-ray bremsstrahlung photons is noticeably less than length passed by originated photon in sample. In this case it can be considered, that x-ray bremsstrahlung photons arises from point of origination of *free* electrons and $\theta(\lambda_1, \lambda)$ - probability is product:

$$\theta(\lambda_1, \lambda) = \theta(\lambda_1, E_e) \cdot \theta(E_e, \lambda) \tag{9}$$

where $\theta(\lambda_1, E_e)$ - probability of primary photon transformation with wavelength λ_1 into free electron with energy E_e (this probability is essentially defined by processes of origination of electron); $\theta(E_e, \lambda)$ – probability of transformation of free electron with energy E_e into bremsstrahlung photon with a wavelength (this probability does not depend on processes of origination of electron and is defined by *Kramers* equation). At using *Kramers* equation, according to [20], it should be intend, that *free* electrons arise in depth of sample and absorption of arisen bremsstrahlung is considered in component $\frac{\mu(\lambda)}{Sin\psi}$ in denominator of formula (8). Therefore is not necessity to consider correction on absorption inserted at consideration of bremsstrahlung of cathode beam. Besides it should to neglect correction of back scattering of electrons because the leaving on the surface of sample is improbable.

Hence,

$$\theta(E_e, \lambda) = \frac{dN_{Kramers}}{d\lambda} = 7.52 \cdot 10^{-5} \cdot Z \cdot (\frac{1}{\lambda e} - \frac{1}{\lambda}) \cdot \frac{1}{\lambda} \cdot B(\lambda) \cdot \frac{T}{L} \text{ [photons/s/electron/st/Å]} \tag{10}$$

where $T = \pi / \sqrt{3}$, $B(\lambda) = \left(\frac{\lambda}{2\lambda_0}\right)^{\alpha}$, λ - bremsstrahlung wave-length, λ_e - short-wave boundary of bremsstrahlung spectrum, corresponding to energy *Ee* of free electron, *Z*- atomic number, $L = \ln\left(\frac{1166}{J} \frac{2\frac{hc}{\lambda_e} + \frac{hc}{\lambda}}{3}\right)$, J – average potential of ionization $J = 11{,}5 \cdot Z$,

$E_e = \frac{hc}{\lambda_e}$, $E = \frac{hc}{\lambda}$.

Probabilities $\theta(\lambda_1, E_e)$ and E_e are different for processes of origination of photo- and Auger and Compton electrons. Consider each process separately.

Photoelectrons.

Probability of photoelectron emergence $\theta^i_{ph}(\lambda_1, E_e)$ for q – shell of an element i is defined by a partial coefficient of primary radiation absorption. $\theta^i_{ph}(\lambda_1, E_e) = \tau_{qi}(\lambda_1)$, or taking into account the possibility to ionize some shells - $\theta^i_{ph}(\lambda_1, E_e) = \sum_q \tau_{qi}(\lambda_1)$. For a multi-component sample different probabilities are summed up with all its elements considered:

$$\theta_{ph}(\lambda_1, E_e) = \sum_i C_i \sum_q \tau_{qi}(\lambda_1). \tag{11}$$

where C_i – content of element i in a sample.

Thus, with taking into account (9), (10) and (11) equations for spectral intensity of bremsstrahlung and in accordance with (8) we express it as:

$$\frac{dN_{ph}}{d\lambda} = K(\lambda) \cdot 7.52 \cdot 10^{-5} \cdot T \cdot Z_{eff} \cdot \frac{1}{\lambda} \cdot \sum_i C_i \int_{\lambda_0}^{\lambda} \frac{N(\lambda_1)}{\frac{\mu(\lambda_1)}{Sin\varphi} + \frac{\mu(\lambda)}{Sin\psi}} \cdot \sum_q \tau_{qi}(\lambda_1) \cdot \left(\frac{1}{\lambda_{qi}} - \frac{1}{\lambda} \right) \frac{B(\lambda)}{L} d\lambda_1 \tag{12}$$

where $Z_{eff} = \sum_i C_i Z_i$ - effective atomic number of a sample.

Auger electrons.

Probability of Auger electron emergence $\theta^i_{Auger}(\lambda_1, E_e)$ for q – shell of element i is defined by $\tau_{qi}(\lambda_1)$ multiplied by probability of non-radiative transition $(1 - \omega_{qi})$, where ω_{qi} – yield of x-ray fluorescence of element i, and multiplied by probability Π_{qi} of emergence of a certain group of Auger transitions in their q – series:

$$\theta^i_{Auger}(\lambda_1, E_e) = \tau_{qi}(\lambda_1) \cdot \Pi_{qi} \cdot (1 - \omega_{qi})$$

.We consider only Auger electrons of K – shell, and will accept, that in all elements probability Π_{KLL} of group KLL – transitions makes up about 70%, while the other groups *KLM* – it is only 30%[11]. Hence, after summation of all elements, forming the sample, for *KLL* – transitions we get:

$$\theta^{KLL}_{Auger}(\lambda_1, E_e) = 0.7 \times \sum_i C_i \tau_{Ki}(\lambda_1)(1 - \omega_{Ki}) \tag{13}$$

and for the rest transitions, assumed for *KLM* it will be:

$$\theta_{Auger}^{KLM}(\lambda_1, E_e) = 0.3 \times \sum_i C_i \tau_{Ki}(\lambda_1)(1-\omega_{Ki}) \tag{14}$$

Using equations (9), (13) and (14) for spectral intensity of bremsstrahlung of Auger *K* – electrons in accordance with (8) the expression is written as:

$$\frac{dN_{Auger}}{d\lambda} = K(\lambda)\cdot 7.52\cdot 10^{-5}\cdot T\cdot\frac{1}{\lambda}Z_{eff}\cdot\sum_i C_i\int_{\lambda_0}^{\lambda}\frac{N(\lambda_1)\cdot\tau_{Ki}(\lambda_1)\cdot(1-\omega_{Ki})}{\frac{\mu(\lambda_1)}{Sin\varphi}+\frac{\mu(\lambda)}{Sin\psi}}\times \tag{15}$$

$$\times\left[0.7\cdot\left(\frac{1}{\lambda_i^{KLL}}-\frac{1}{\lambda}\right)+0.3\left(\frac{1}{\lambda_i^{KLM}}\right)\right]\cdot\frac{B(\lambda)}{L}d\lambda_1$$

where λ_i^{KLL} and λ_i^{KLM} - are the wavelengths, corresponding to energy *KLL*- and *KLM* – of Auger electrons of element *i*.

Compton electrons.

The probability of origination of the Compton electron is defined by the probability of the process of incoherent scattering of the primary photon with wavelength λ_1 in direction Ω, at which it acquires energy E_e. This probability represents atomic differential section of incoherent scattering determined by differential electron coefficient of scattering $\frac{d\sigma^{KN}}{d\Omega}$ multiplied on atomic form-factor *S*:

$$\theta_c(\lambda_1, E_e) = \frac{d\sigma^{KN}}{d\Omega}\cdot S\cdot(2\pi\sin\vartheta)\cdot\left(\frac{1}{dE_e/d\Omega}\right), \tag{16}$$

where atomic form-factor can be calculated [28] by formula:

$$S = Z(1-e^{5\upsilon}),$$

with $\upsilon = \frac{2.21}{\lambda_1}\times\sin\frac{\vartheta}{2}\times Z^{-\frac{2}{3}}$; $\sin\frac{\vartheta}{2} = \sqrt{\frac{E_e}{2(E_1-E_e)k}}$, $E_1 = \frac{hc}{\lambda_1}$ and $k = \frac{E_1}{mc^2}$.

The differential electron coefficient of scattering in formula (16) is described by equation of Klein – Nishina[29]:

$$\frac{d\sigma^{KN}}{d\Omega}=\frac{r_0^2}{2}\left(\frac{E_1-E_e}{E_1}\right)^2\left[\left(\frac{E_1}{E_1-E_e}\right)+\left(\frac{E_1-E_e}{E_1}\right)-\sin^2\vartheta\right],$$

where r_0 – classic radius of electron.

Appearance of efficient $(2\pi\sin\vartheta)$ reflects the fact in equation (16), that electrons with the same energy produce a cone with the angle at vertex ϑ.

Introduction of coefficient $\left(\frac{1}{dE_e/d\Omega}\right)$ in equation (16) is necessitated by transformation of 6angular dependence of differential coefficient of scattering into its dependence on energy of Compton electrons:

$$\left(\frac{1}{dE_e/d\Omega}\right)=\frac{1}{E_1\cdot k\cdot\left(\frac{1}{1+k\cdot(1-\cos(\vartheta))}-\frac{k\cdot(1-\cos(\vartheta))}{(1+k\cdot(1-\cos(\vartheta)))^2}\right)}$$

In contrast to the cases of bremsstrahlung of photo- and Auger electrons, the energy of Compton electrons is dissimilar even at monochromatic primary radiation. This is why for receiving a full probability of bremsstrahlung photons emergence within the interval of energies from E to $E+\Delta E$ it is reasonable to integrate the equation using all possible energies of these electrons.

For a multi-component sample we get:

$$\frac{dN_C}{dE}=K(\lambda)\cdot 7.52\cdot 10^{-5}T\cdot Z_{eff}\cdot\sum_i C_i\int_{E_0}^{E}\frac{N(\lambda_1)}{\frac{\mu(E_1)}{\sin\varphi}+\frac{\mu(E)}{\sin\psi}}\int_{E}^{E_{\max i}}\theta(E_1,E_{ei})(E_{ei}-E)\frac{B(E)}{L}dE_e dE_1, \quad (17)$$

where $E_{max\,i}$ – maximum energy of the Compton electrons of element i.

Probability $\theta(E_1,E_e)$ in equation (17) is defined by relationship (16), in which wavelength λ_1 of primary radiation is replaced by relationship $\lambda_1=\frac{hc}{E_1}$.

Transition to a spectral intensity after wavelengths is performed using formula

$$\frac{dN_C}{d\lambda}=-\frac{dN_C}{dE}\cdot\frac{hc}{\lambda^2}.$$

Equations (12), (15) and (17) are used to calculate the spectral intensity of bremsstrahlung of photo-, Auger and Compton electrons.

3.2. The Ratio of Bremsstrahlung of Photo-, Auger and Compton Electrons and Polychromatic Primary Radiation Scattered on Sample

If we take into account that polychromatic primary radiation is generated by x-ray tube then its spectral intensity $\frac{dN_{sc}}{d\lambda}$ after scattering on a sample can be calculated by:

$$\frac{dN_{sc}}{d\lambda} = K(\lambda) \cdot N_1(\lambda) \cdot \frac{\frac{d\sigma(\lambda)}{d\Omega} \exp[-\mu_{Be}(\lambda)\rho_{Be}d_{Be}]}{\left(\frac{1}{\sin\varphi} + \frac{1}{\sin\psi}\right)\mu(\lambda)} \tag{18}$$

where $N_1(\lambda)$ - the spectral intensity of primary radiation before x-ray tub output window. $\frac{d\sigma(\lambda)}{d\Omega}$ - differential coefficient of scattering of primary radiation with λ wave-length into solid angle $d\Omega$. $\mu(\lambda)$ - mass attenuation coefficient of scattered radiation in a sample. φ and ψ - angles of primary radiation incidence and takeoff of fluorescent x-ray.

The exponent in formula (11) characterizes attenuation of primary radiation in the beryllium window of x-ray tube.

In the long-wavelength range the Compton displacement is negligible, and hence the coherent and incoherent scattering may be characterized by the same wavelength.

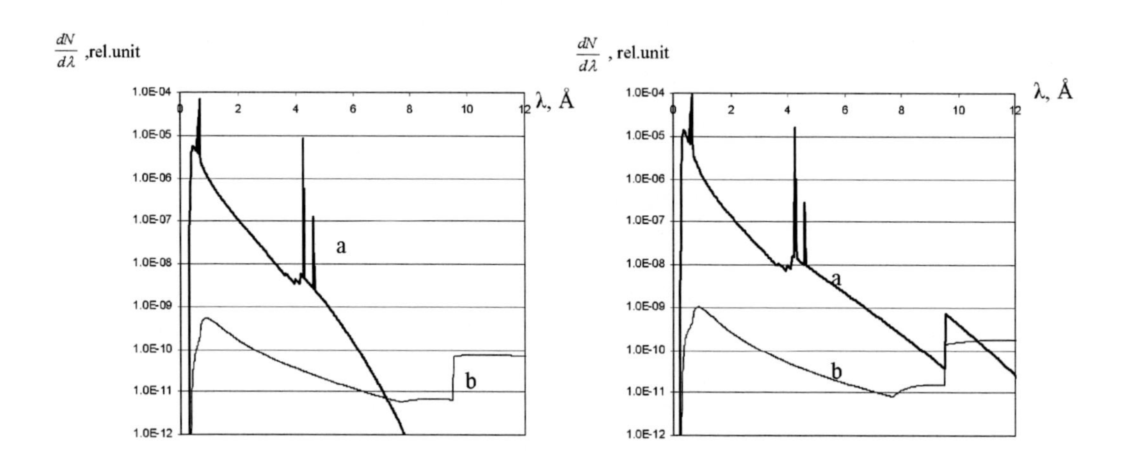

a.– scattered primary radiation of x-ray tube, b.– bremsstrahlung of *free* electrons

Figure 5. Spectrum of x-ray primary radiation of x-ray tube with Rh anode and with beryllium window thickness 300 μm (left) and 75 μm (right) scattered on the magnesium sample, and bremsstrahlung spectrum of free electrons .

On fig.5 the total bremsstrahlung of photo-, Auger and Compton electrons of magnesium sample is compared with polychromatic primary radiation scattered on the sample calculated using equation (18). X-ray tube anode – *Rh,* thickness of beryllium output window is 75 μm and 300 μm.

It follows from Fig.1 that with a certain boundary wavelength the predominance of scattered x-ray primary radiation is replaced by a dominance of free electron bremsstrahlung.

For 300 μm thickness of beryllium output window it occurs at wave-lengths in order 6-7 Å(it is region of K_α –line of phosphorus and silicon). For 75 μm thickness of beryllium output window - 10-11 Å (it is region of K_α–line of magnesium and sodium).

The influence of elemental composition is insignificant on this boundary wave-length location.

3.3. The Comparative Analysis of Spectral Intensity of Bremsstrahlung of Photo-, Auger and Compton Electrons

The dependence of ratio of spectral intensity of bremsstrahlung of photo-, Auger and Compton electrons in the range NaK_α -lines from energy of primary photons is presented on fig.6.

The cases of aluminum and copper irradiated samples at 75 μm thickness of beryllium window are considered.

It is evident from Figure 6, that the ratio of contributions of the components under observation essentially depends on the composition of the fluorescing sample. It remarkable that in the range of primary photon energy exceeds 60 - 80 keV, the bremsstrahlung of Compton electrons becomes dominant both for the aluminum and copper samples. With the energies of primary photons less than 60 keV the contribution of Compton electrons to formation of x-ray background is insignificant. Therefore, with real voltage on the x-ray tube this component of background can be ignored. As shown in Figure 7, this statement is confirmed by the dependences for bremsstrahlung of the electrons in the range NaKα - the lines from voltage on the x-ray tube with Rh anode.

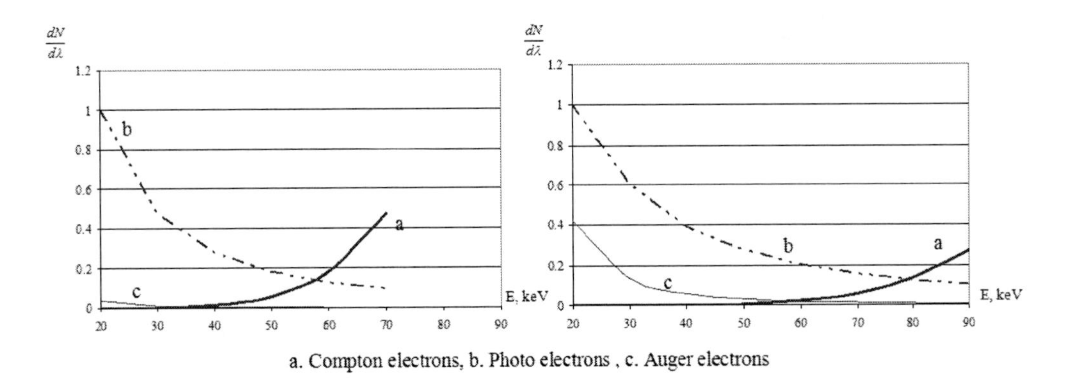

Figure 6. Contribution of photo-, Auger and Compton electrons to formation of the bremsstrahlung intensity in the range NaK_α - lines (λ = 11.9 Å) for aluminum (left) and copper (right) samples as a function of primary photon energy.

In case with aluminum sample, disclosed in Fig.7, it is also characteristic of the other x-ray tube and fluorescing materials: the spectral intensity of photo- and Auger electrons bremsstrahlung in the range NaKα - lines at primary radiation of any x-ray tubes surpass considerably the intensity of bremsstrahlung of Compton electrons.

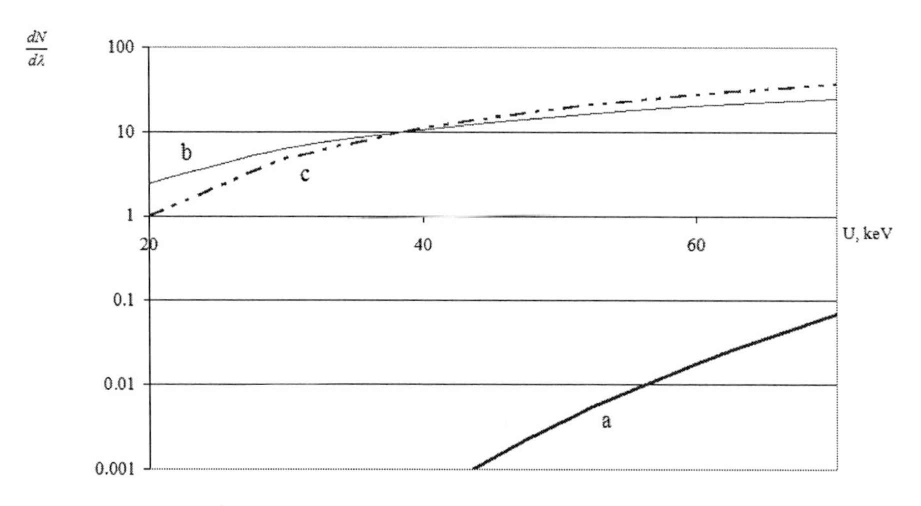

Figure 7. Contribution of photo-, Auger and Compton electrons to formation of bremsstrahlung on the wavelength Na K_{α} - lines (λ = 11.9 Å) for the aluminum sample, as the function of voltage on the x-ray tube anode (x-ray tube with Rh anode, thickness of beryllium window 75 μm).

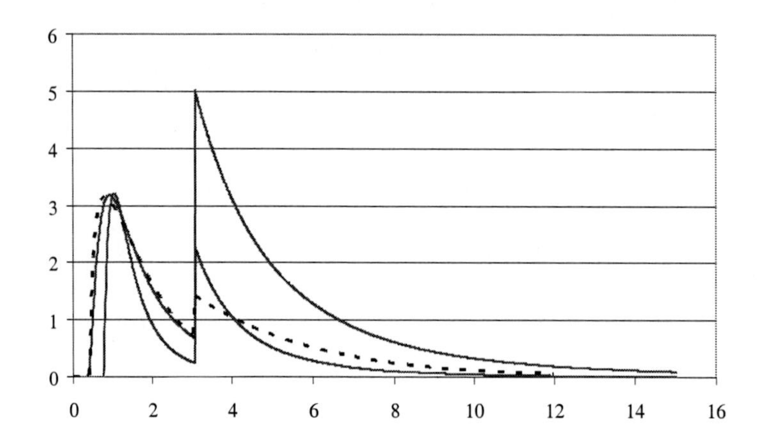

Figure 8. Spectral distribution of bremsstrahlung of free electrons at radiation of Ca sample by monochromatic radiation with energy 20 keV (a); and x-ray tube primary radiation with Rh – anode at voltage 40 kV (K_{α} line Rh 20.2 keV)(b); bremsstrahlung of cathode beam electrons is calculated using theory of *Kramers* (c). The maxima of bremsstrahlung at 1.24 Å are superposed.

It is interested of the comparison of spectral intensity of bremsstrahlung of photo-, Auger and Compton electrons with the classical spectral intensity of bremsstrahlung of electrons of cathode beam calculated by using theory of *Kramers*[31] . Figure 8 provides comparison of a total spectral intensity of bremsstrahlung of photo- and Auger electrons, obtained with the calcium sample for monochromatic primary radiation with the energy 20 keV and x-ray tube with Rh (Z=45) operating at voltage 40 kV. The bremsstrahlung of cathode beam electrons is calculated using theory of *Kramers* for 40 kV. (The maxima of bremsstrahlung at 1.24 Å are superposed.)

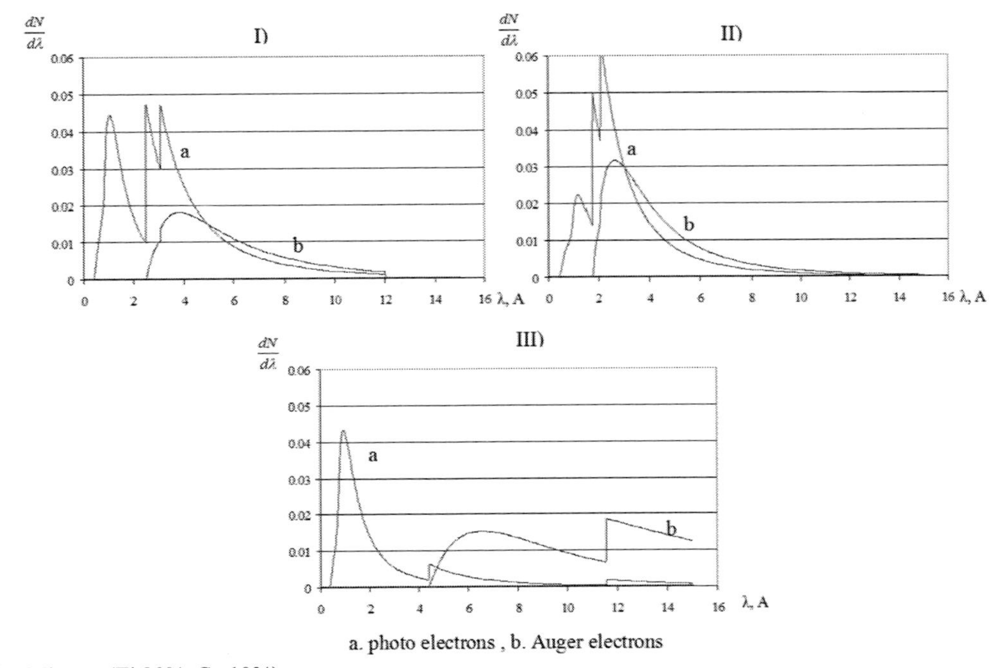

(I) – Mixture (Ti 90%, Ca 10%).
(II) – Alloyed steel (Fe 80%, Cr 15%, Ni 5%).
(III) – Sodium Chloride (Na 40%, Cl 60%).

Figure 9. Spectral distribution of bremsstrahlung of photo- and Auger electrons while radiating some samples by x-ray tube with Rh – anode, operating at voltage 40 kV

The data in Fig.8 indicate that the spectral distribution of bremsstrahlung of photo- and Auger electrons, in general terms corresponds to Kramers theory. The difference consists in significant jumps of absorption, which is explained by a large depth of origination of free electron bremsstrahlung. In addition, with polychromatic spectrum the wavelength component appreciably exceeds the one with monochromatic primary radiation, and the maximum of spectral distribution becomes more indistinct.

In case of multi-component sample, each element contributes to formation of bremsstrahlung spectrum. As a result, the summary bremsstrahlung seems fairly complex. Figure 9 exemplifies the spectral distribution of free electron bremsstrahlung in some samples.

From fig. 9 it follows that the elemental composition of fluorescent sample essentially changes the ratio of bremsstrahlung of photo- and Auger electrons. In the short-wave range of fluorescent spectrum this fact is not important because the primary radiation scattered on a sample is dominate. However, in long-wave range uncontrolled change of intensity of background bremsstrahlung can affect on accuracy of results of x-ray fluorescent analysis for elements with low atomic number.

3.4. The Long-Wave Bremsstrahlung of *Free* Electrons

In long-wave range of fluorescent spectrum the track length of electron appears to be comparable or more than track length of x-ray bremsstrahlung photon. The depth of penetration of x-ray bremsstrahlung photon with 1 keV energy (wavelength range ~ 12Å) equals 0.11 μm for copper sample, and *full* track length of electron with 20 keV energy according to Tomas-Widdington law equals 1.5 μm. However, this estimation is too much great than the depth of origination of bremsstrahlung.

In this case equation (8) becomes to be incorrect, because supposition about braking of *free* electrons in point of origination becomes to be incorrect. For the purpose of specification of process of origination bremsstrahlung we shall consider the schema, presented on fig.10. Let primary radiation penetrate on depth x in a massive flat sample and is absorbed in elementary layer dx. Then number of *free* electrons, originated in elementary volume sdx (s- the area of an elementary layer on depth x):

$$dN_{el} = N(\lambda_1)e^{\left(-\frac{\mu(\lambda_1)}{\sin\varphi}x\right)}\theta(\lambda_1, E_e)sdx,$$

where $N(\lambda_1)$ - number of primary photons with λ_1 wavelength on a sample surface , $\theta(\lambda_1, E_e)$ - the probability of transformation of primary photons to *free* electrons with energy E_e.

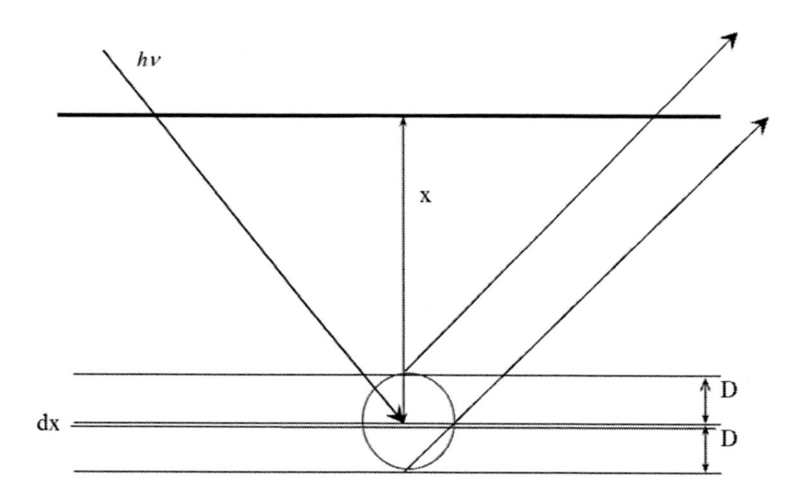

Figure 10. To calculation of intensity of long-wave component of bremsstrahlung of *free* electrons.

The arisen electrons form some "cloud" beyond the elementary volume sdx. The average distance between dx layer and layer in which electrons generate bremsstrahlung, we define as $D(\rho, Z, E_e)$.

Let a half of electrons occurs above layer *sdx,* and other half occurs below. At that case the intensity of bremsstrahlung of upper half-space will equal $\frac{dN^{\uparrow}}{d\lambda}=\frac{dN_{el}}{2}\theta(E_e,\lambda)$, where $\theta(E_e,\lambda)$ is defined by expression (10). The similar expression we get for lower half-space. The total intensity of bremsstrahlung fallen in the detector, occurred under ψ angle to a surface of sample, is defined by expression:

$$\frac{dN}{d\lambda}=K(\lambda)\frac{dN_{el}}{2}\theta(E_e,\lambda)\left\{\int_0^{\infty} e^{-\left(\frac{\mu(\lambda_1)}{\sin\varphi}x+\frac{\mu(\lambda)}{\sin\psi}(x-D)\right)}dx+\int_0^{\infty} e^{-\left(\frac{\mu(\lambda_1)}{\sin\varphi}x+\frac{\mu(\lambda)}{\sin\psi}(x+D)\right)}dx\right\}$$

The part of *free* electrons, originated in surface layer can leave irradiated sample without interaction. Let the considered depth of surface layer equals D_0 . Then the spectral intensity can be written as

$$\frac{dN}{d\lambda}=K(\lambda)\frac{dN_{el}}{2}\theta(E_e,\lambda)\left\{\int_{D_0}^{\infty} e^{-\left(\frac{\mu(\lambda_1)}{\sin\varphi}x+\frac{\mu(\lambda)}{\sin\psi}(x-D)\right)}dx+\int_0^{\infty} e^{-\left(\frac{\mu(\lambda_1)}{\sin\varphi}x+\frac{\mu(\lambda)}{\sin\psi}(x+D)\right)}dx\right\}$$

By integration this expression by x, we obtain:

$$\frac{dN}{d\lambda}=\frac{1}{2}\left(e^{-\left(\frac{\mu(\lambda_1)}{\sin\varphi}D_0+\frac{\mu(\lambda)}{\sin\psi}(D_0-D)\right)}+e^{-\frac{\mu(\lambda)}{\sin\psi}D}\right)K(\lambda)\frac{dN_{el}\theta(E_e,\lambda)}{\frac{\mu(\lambda_1)}{\sin\varphi}+\frac{\mu(\lambda)}{\sin\psi}}$$

From obtained expression it follows that at $\mu(\lambda_1) << \mu(\lambda)$:

$$\frac{dN}{d\lambda}=\frac{1}{2}\left(e^{-\left(\frac{\mu(\lambda)}{\sin\psi}(D_0-D)\right)}+e^{-\frac{\mu(\lambda)}{\sin\psi}D}\right)K(\lambda)\frac{dN_{el}\theta(E_e,\lambda)}{\frac{\mu(\lambda_1)}{\sin\varphi}+\frac{\mu(\lambda)}{\sin\psi}}$$

Accepting that $\frac{1}{2}\left(e^{-\left(\frac{\mu(\lambda)}{\sin\psi}(D_0-D)\right)}+e^{-\frac{\mu(\lambda)}{\sin\psi}D}\right)=P(D,D_0)$ for spectral intensity of bremsstrahlung of photo- electrons we obtain:

$$\frac{dN}{d\lambda} = K(\lambda)\frac{N(\lambda_1)\theta(\lambda_1, E_e)\cdot\theta(E_e, \lambda)}{\frac{\mu(\lambda_1)}{\sin\varphi}+\frac{\mu(\lambda)}{\sin\psi}} \times P(D, D_0) \qquad (19)$$

Or at polychromatic primary radiation

$$\frac{dN}{d\lambda} = K(\lambda)\left(\int_{\lambda_0}^{\lambda}\frac{N(\lambda_1)\cdot\theta(\lambda_1, \lambda)}{\frac{\mu(\lambda_1)}{\sin\varphi}+\frac{\mu(\lambda)}{\sin\psi}}d\lambda_1\right)P(D, D_0) \qquad (20)$$

The obtained expression (20) differ from used equation (8) for more short wave range by presence of correction coefficient $P(D,D_0)$. Both these parameters D and D_0 approximately equally decrease with reduction of energy free electrons.

The value of parameter D is calculated by equation

$$D\rho = a(U-1)^{\alpha} \qquad (21),$$

where $a = 3.2Z^{4,42}\cdot 10^{-11}$ g/cm^2 and $\alpha = 5.85Z^{-0,45}$ for K-series , $a = 2.25Z^{4,51}\cdot 10^{-11}$ g/cm^2 and $\alpha = 3.58Z^{-0,206}$ for L-series. This equation was proposed in [32] for maximum of distribution of electrons of cathode beam by the depth in a sample. The parameter D_0 we define as depth of 'thin' layer where electrons can leave surface without interaction as: $D_0 = \frac{1}{5}D$.

The dependence of $P(D,D_0)$from wave-length of bremsstrahlung of free electrons for copper target at polychromatic primary radiation of x-ray tube with Pd anode is presented on fig. 11.

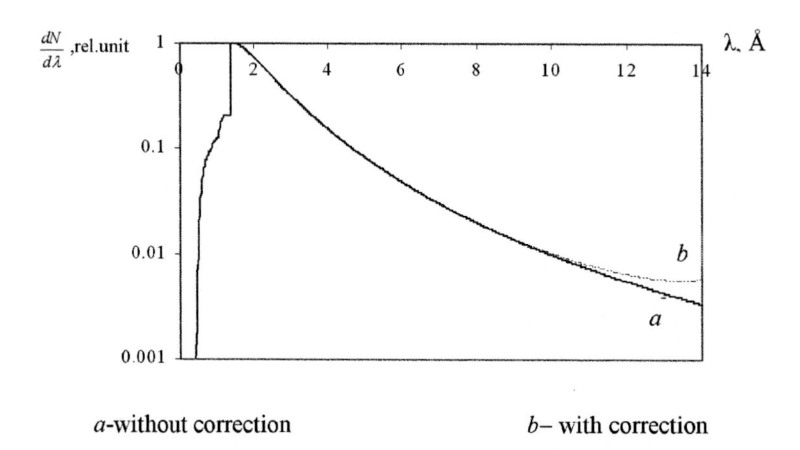

Figure 11. The dependence of $P(D,D_0)$from wave-length of bremsstrahlung of free electrons. (copper sample , source – x-ray tube with Pd anode).

From fig. 11 it follows that this correction fast rise in range of wavelength more than 11-12Å . Correspondingly, bremsstrahlung intensity increases in this range. The energy of Auger electrons as rule is less than energy of photoelectrons, therefore considered correction for Auger electons can be less than represented on fig.11.

Conclusion

The results of carried out researches can be used for x-ray fluorescent analysis at defining of concentration of elements with low atomic number. The realized in our work theoretical and program study allows simulating different analytical situations for the purpose of finding optimal conditions of carrying-out of analysis of these elements. Besides, at defining of its concentration it becomes possible realization of methods of x-ray fluorescent analysis required carrying-out calculation of intensities of analytical signal (methods of fundamental parameters, fundamental coefficients, theoretical corrections, alpha- corrections and others).

It should be mentioned, that the forming of background in long-wave range of x-ray fluorescent spectrum is determined not only bremsstrahlung of *free* electrons, but a partial loss of energy of photons when registered (long-wave “tails” of lines and scattered primary bremsstrahlung). This affect appears in any detection devices. The suppression (removal) or calculation allows using the iteration process realized in methods of fundamental parameters for synchronous taking into account intensity of an analytical signal and background from elemental content of analyzed material.

References

[1] Ebel H., Landler E. and Dirschmid H. *Z. Naturforsch.*, Teil A, 1971, v.25, p.927

[2] Pavlinsky G.V., Gulayeva V.T. In *Issledovaniya v oblasti physiki tverdogo tela.* Irkutsk,1974, т.2, .230 c. (in Russian)

[3] Pavlinsky G.V., Vladimirova L.I. In *Apparatura I metodi rentgenovskogo analiza..* Leningrad, 1988, vol.37, c.65. (in Russian)

[4] Afonin V.P., Piskunova L.F., *Zavoskaya. lab,* 1978, т.44, c.1083. (in Russian)

[5] Pavlinsky G.V., *Apparatura I metodi rentgenovskogo analiza.* Mashinostroenie:S.-Petersburg, 1992, т.41, c.83. (in Russian)

[6] Borhodoev V.Ya. *Rentgenofluorescentniy analiz gornih porod sposobom fundamentalnih parametrov.* Magadan, 1999, 279 c. (in Russian)

[7] Vinogradov A.V., Zeldovich B.Ja. *Appl. Optics*, 1977, v.16, p.89.

[8] Pavlinsky G.V., Dukhanin A.Ju. *X-Ray Spectrometry*, 1994, v.23, p.221.

[9] Dukhanin A.Ju., Pavlinsky G.V., Baranov E.O., Portnoy A.Ju. *X-Ray Spectrometry*, 2006, v.35, No.1, p.34.

[10] Philibert J, Tixier R. *Quantitative Electron Probe Microanalysis*, NBS, Spec. Publ., N. 298 – Washington , 1968, p.13.

[11] Bambinek W., Grasemann B., Fink R.W., Fround H.U., Mark H., Price R.E., Venugopala Rao P., Swift C.D. *Rev. Mod. Phys.*, 1972, v.44, p.716.

[12] Pavlinsky G.V. and Portnoi A.Ju. *Radiation Physics and Chemistry*, 2001, v.62, No.2-3, p.207-213.

[13] Pavlinsky G.V., Portnoy A.Ju. *X-Ray Spectrometry*, 2002, v.31, №3, p.247-251
[14] Pavlinsky G.V., Dukhanin A.Ju, Portnoy A.Ju., Kiun A.V. *Radiation Physics and Chemistry*, 2005, v.72, No.4, p,429-435.
[15] Sherman J., *Spectrochimica Acta*, 1955,v.7, p.286-306.
[16] Weber F.A., Da Silva L.B., Barbee T.W., Ciarlo D., Mantler M. *Adv. X-Ray Analysis*, 1997 (?), v.39, p.821-829
[17] Mantler M. *Adv. in X-Ray Analysis*., 1996, v.40, p. 625..
[18] Pavlinsky G.V , Kitov B.I., Tumencev V.N. *Apparatura I metodi rentgenovskogo analiza*.- Mashinostroenie:S-Petersburg, 1987, vol.36, c. 49-53(in Russian)
[19] Dukhanin A.Yu., Pavlinsky G. V. *X-Ray Spectrometry*, 2006, v.35, No.2, p.237-240.
[20] Finkelshtein A.L,, Pavlova T.O. *X-Ray Spectrometry*, 1999, v.28, No.1, p.27-32.
[21] Pella P.A., Feng L., Small J.A. *X-Ray Spectrometry*, 1991, v.20, p.109 – 110.
[22] Puri S., Mehta D., Chand B. et al. *X-Ray Spectrometry*, 1993, v.22, p.358-361.
[23] Blohin M.A. *Physics of x-ray*, GITLL: Moscow, 1957, 518c.
[24] Losev N.F., Smagunova A.N. *Basics of x-ray fluorescent analysis*. Himiya: Moscow, 1982, 207 c. (in Russian)
[25] Revenco A.G. *X-ray fluorescent analysis of rock*. Nauka: Novosibirsk, 1994, 267 c.
[26] Pavlinsky G.V., Imikshenova N.N. , Ivshev D.V., *Apparatura I metodi rentgenovskogo analiza*. Mashinostroenie:S-Petersburg, 1992, вып.41, c.113-119. (in Russian)
[27] Portnoy A.Yu., Pavlinsky G.V., Duhanin A.Yu., Zuzaan P., Erdemchimeg B. *Analytics and Control*, 2002, vol.6, 4, c. 390-394. (in Russian)
[28] Bahtiarov A.V., Pshenichniy G.A. *Apparatura I metodi rentgenovskogo analiza*. Mashinostroenie:S-Petersburg, 1973, вып.12, c.68-72. (in Russian)
[29] Klein O., Nishina Y. *Physik*, 1929, v.52, p. 853-868.
[30] Bahtiarov A.V., *X-ray fluorescent analysis in geology and geochemistry* . Nedra:Leningrad, 1985, 144 c.
[31] Kramers H.A. *Phil. Mag.*, 1923, v.46, No.275, p.836-871.
[32] L.A.Pavlova ;O.Yu. Belozerova; L.F.Paradina; L.F.Suvorova. *Rentgenospektralniy electronno-zondoviy mikroanaliz prirodnih objectov*. Nauka: Novosibirsk, 2000 (in Russian)

In: Radiation Physics Research Progress
Editor: Aidan N. Camilleri, pp. 329-353

ISBN: 978-1-60021-988-7

Chapter 09

RADIATION HARDNESS OF SCINTILLATION CRYSTALS

Peter Kozma
Institute of Technological Applications,
Vratičova 277/11, 155 21 Prague, Czech Republic

Abstract

Scintillation crystals such NaI:Tl, BGO, CsI, BaF_2, CeF_3, PbF_2 and PWO are widely used in high-energy and nuclear physics for photon detection: electromagnetic calorimeters (EMC) made of scintillation detectors achieve the best energy resolution for photons and electrons. These scintillation detectors have also been considered among the detectors for applications in nuclear medicine, particularly in computerised tomography scanners (CTS), because of their density and large light yield. New scintillation crystals such GSO and LSO are also considered as imaging detectors for single proton emission computed topmography (SPECT) and positron emisson tomographs (PET).

One of the most important characteristics of scintillation crystals is their radiation hardness, i.e. their ability to retain scintillation efficiency and uniformity of light output over the crystal volume after exposure to ionizing radiation. In this paper we report on experimental study of radiation hardness of these crystals at accumulated doses of low-energy gamma-rays up to 10^5 Gy . The radiation sensitivity has been examnined by the measurement of the change in optical transmission spectra before and after irradiations. The resuls are discussed in the framework of the light output degradation and appropriate induced absorption coefficiens. The recovery from radiation damage has been treated as a multistep process.

Keywords: Scintillation crystals, detectors, radiation hardness, gamma-rays, irradiations, optical transmission, light output degradation, induced absorption coefficients, recovery of damaged crystals

1. Introduction

Scintillation crystals are substrates able to convert energy lost by ionizing radiation into pulses of lights. In the most scintillation counting applications, the ionizing radiation is in the form of gamma-rays, X-rays and alpha- or beta-particles ranging in energy between a few thousand to several milion eV (from keV to MeV). Read-out systems of pulses of lights emitted by scintilation crystals are photomultiplier tubes (PMT) and phoptosensitive diodes (PD). The photocatodhe of PMT, situated on the backside of the entrance window, converts the lights (photons) into electrons (photo-electrons), which are accelerated by an electrical field towards the dynodes of the PMT where the multiplications process takes place. The result is that each light scintillation pulse produces a current pulse on the anode of PMT that can be subsequently detected by an appropriate electronics chain. The PMT quantum sensitivity lies in the visible region (300 – 800 nm). Scintilattion light pulses from a crystal can be also detected by silicon photodiode. PD is a silicon device consisting of a thin layer of silicon in which the ligth is absorbed after which free charge carriers (electrons and holes) are created. When these PD´s are optically coupled to a scintillation crystal, a scintillation light pulse will therefore generate a small charge pulse in the diode which can be processed by a charge sensitive preamplifier. The quantum efficiency of silicon PD is typically between 500 and 900 nm. The combination of a scintillation crystal with an appropriate read-out (PMT or PD) is called a scintillation detector.

Intensity of the light pulse emitted by a scintillation crystal is proportional to the energy of the absorbed radiation, therefoer, it is possible te measure this energy: scintillation detectors are therefore widely used for radiation detection and nuclear spectroscopy. To detect nuclear radiation with a given efficiency, dimensions of scintillation crystals should be chosen such that the desired fraction of the radiation is absorbed. For penetrating radiation, such as gamma-rays, a scintillator with high density is required. Moreover, light pulses produced somewhere in the scintillation crystal must pass the material to reach the light detector. These requirements of both, stopping power and optical transparency, have been met by widely used scintillation crystals such NaI:Tl - thalium doped sodium iodide and CsI:Tl - thalium doped cesium iodide. Both these crystals combine a good detection efficnecy with a high light output and, therefore, can be used for high-resolution nuclear gamma-rays spectroscopy. The most important characteristics of inorganic scintillation crystals are:

- high density
- high atomic number
- light output
- decay time (durration of the scintillation pulse)
- mechanical and optical properties .

Of course, today, a large number of scintillation crystals exists for a variaty of applications. For high energy applications the bismuth germanate $Bi_4Ge_3O_{12}$ (BGO) and BaF_2 scintillators have been developped. Fluorides (BaF_2, CeF_3, PbF_2, etc.) with high density and atomic number improve the lateral confinement of light shower, and, along with tungstates ($PbWO_4$ and $CdWO_4$) can also been considered between candidates for detection in high-emergy and particle-physics experiments. To detect fast changes in the intensity of X-rays, as e.g. in computerised tomography, imaging scintillation crystals such GSO (cerium-doped gadolinium orthosilicate) and LSO (cerium-doped luthetium orthosilicate) have been developped particularly for nuclear medicine applications. The most important applications of scintillation crystals are summarised in chapter 2.

2. Application of Scintillation Crystals

Scintillation crystals have been developped as a promissing candiates for dense and fast radiators for gamma-ray detectors which can be used in high-radiation environment. High atomic number crystals, such as bismuth germanate BGO ($Bi_4Ge_3O_{12}$), fluorides (BaF_2, CeF_3, PbF_2) and tungstates ($PbWO_4$, $CdWO_4$) are considered among the favourite scintillators in nuclear and high-energy physics experiments [1] . These crystals are also widely used in nuclear medicine diagnostic systems, particularly positron emission tomography (PET) and computerised tomography scanners (CTS) [2] . Scintillation crystals of GSO (cerium-doped gadolinium orthosilicate: Gd_2SiO_5:Ce) and LSO (cerium-doped luthetium orthosilicate: Lu_2SiO_5:Ce) are new scintillatiors considered as imaging detectors for nuclear medicine applications [3] .

Scintillation crystals of bismuth germanate BGO have been considered as a scintillation material for electromagnetic calorimeters in high-energy physics experiments [1]: about 11000 pcs of these large volume crystals have been grown by the Shanghai Institute of Ceramics (SIC) for the L3 experiment on LEP at CERN [4] . In order to optimize the radiation resistance of these large crystals (2,0 x 2,3 x 3,24 cm^3), systematic experimental radiation damage studies have been performed on a large amount of full-size crystals [5]. After irradiations, neither the fluorescence spectrum nor the efficiency of luminiscences spectrum was modified [6] , but the light yield of BGO was reduced as a result of formation of colour centres. The absorption spectra of these colour centres have been studied for different kinds of irradiation, different doses and during recovery of crystals.

In the framework of the experimental project SPHERE [7,8] realized at the Laboratory of High Energies of the JINR Dubna, the properties of fluorides (BaF_2, CeF_3 and PbF_2) have been investigated. It has been reported on the tests of large volume BaF_2 crystals in beams of relativistic particles [9] , cerium fluoride scintillators working with different photomultipliers [10] as well as experimental study of radiation sensitivity of large volume (3 x 3 x 6 cm^3) CeF_3 crystals [11] . Lead fluoride, PbF_2 was first considered as Cherenkov radiator for electromagnetic calorimetry [12-14]. Two major advantages of PbF_2 over three promising scintillation crystals BGO, BaF_2 and CeF_3 considered for electromagnetic calorimetry (properties of the scintillation crystals as detectors for electromagnetic calorimetry in nuclear and high-physics experiments are listed in Table 2.1), are that it is less expensive and, being a Cherenkov radiator, its performance is not limited by a long decay constant. Results on radiation hardness of a 4 x 4 x 24 cm^3 PbF_2 scintillation crystals are reported in the paper [15]. The results are also compared with radiation resistivity measurements for large volume 4 x 4 x 40 cm^3 Pb-glass SF_5 and SF_6, respectively (relevant properties of PbF_2 crystals and lead glasses SF_5 and SF_6 are summarized in Table 2.2).

Lead tungstate $PbWO_4$ (or PWO), has been intensively investigated as the most promising scintillator for use in the CMS electromagnetic calorimeter in the LHC project at CERN [1, 16] . The relevant properties of this crystal are its high density, short radiation length and small Moliere radius (see Table 1). Because of these properties, as well as relatively large light yield, lead tungstate $PbWO_4$ as well as cadmium tungstate $CdWO_4$ crystals can also be considered among their most prominent scintillation detectors for application in nuclear medicine, particularly in CTS [2] . Radiation damage of these crystals is known to be a crucial parameters . Light yield dependence on radiation dose seems to be the most decisive parameter for the practical use of these scintillators in various applications. It has been found [17] that in some $PbWO_4$ crystals the light yield decreases already at doses of a few Gy only, and that radiation hardness of $PbWO_4$ scintillating crystals could be improved by La-doping [18] . Radiation damage of $PbWO_4$ crystals due to irradiation by gamma-rays have also been studied [19].

Table 2.1. Properties of prominent scintilation crystals

Scintillation crystal	BGO	CeF_3	BaF_2	$PbWO_4$	$CdWO_4$	CsI
Density [g/cm^3]	7.13	6.16	4.87	8.30	7.90	4.51
Melting point [^{o}C]	1050	1285	1280	1123	1325	834
Radiation length [cm]	1.1	1.7	2.1	0.9	1.1	1.9
Moliere radius [cm]	2.7	2.6	4.4	2.0	2.2	1.2
Refractive index [a)]	2.1	1.6	1.9	2.2	2.3	1.8
Wavelength [nm]	480	300	325	480	470	315
Decay time [b)] [ns]	300	25	0.6 / 620	50 / 10	20 / 5	16
Light yield [c)] [%]	15-20	25	20	30-40	25-30	6-10

a) At the wavelength of the maximum
b) Slow / fast component
c) Relative to NaI:Tl for gamma-rays (see Table 2.3)

Table 2.2. Properties of PbF_2 crystals and lead gasses SF_5 and SF_6.

Scintillation crystal	PbF_2	SF_5	SF_6
Density [g/cm^3]	7.77	4.08	5.20
Pb Content [% by weight]	85%	51.5%	64.8%
Radiation length [cm]	0.93	2.54	1.69
Moliere radius [cm]	2.22	3.70	2.70
Refractive index [a)]	1.82	1.67	1.81

a) At the wavelength of the maximum

Scintillation crystals of cesium iodide CsI are also widely used in high-energy physics for photon detection: calorimeters made of CsI crystals achieve the best ebergy resolution for photons and electrons [20,21] . The interest to the radiation hardness of this material increased with the development and construction of B-, Φ-, c-τ factories, i.e. storage rings with high luminosity which implies high positron and electron currents, causing high radiation dose absorbed by the detector components. Recently, this subject became topical again due to the development of new B-factory project s with super high luminosity [22,23] . Radiation hardness of CsI crystals experimantal study have been motivated particularly by the application of this crystals in the BELLE detector [24] in cooperation with KEK colider with luminosity of 10^{34} cm^{-2}s^{-1}. Radiation background at the BELLE calorimeter corresponds to the absorbed dose of up to several hundreds rads per year. This background is expected to be one order of magnitude higher at the planned super B-factories. That is why the interest concerning radiation sensitivity of CsI crystals concentrates particularly on a study of radiation hardness of these crystals in the range up to hundreds kilorads.

Small volume BGO crystals are widely used in nuclear medicine diagnostic systems, parrticularly PE Tomographs (PET) and CT Scanners (CTS). In PET systems the scintillation crystals are used to detect pairs of the 511 keV gamma-rays produced when a positron emitted from a positron emitter (^{11}C, ^{13}N, ^{15}O, etc.) annihilates with an atomic electron. In earlier stage of CTS or gamma camera systems NaI:Tl crystals have been used as the scintillation detectors of gamma-rays. After BGO was invented in the late 1970s, it gradually took the place of NaI:Tl as the scinillation detector in most PET and CTS systems because of its high stopping power, light yield and decay time, as well. The BGO block detectors [25], which was invented in the 1990s, are also used in a multi-slice PET systems with high spatial resolution.

In radioisotope imaging techniques (scintigraphy, single photon emision computed tomography SPECT and positron emission tomography PET) [26], used to obtain the relevant information of energy ranging from 100 to 511 keV (nowadays, nuclides ^{99m}Tc, ^{18}F and ^{15}O are used as main tracers in SPECTs and PETs) have to be detected with the highest possible efficiency and localized with a spatial resolution desirably of a few milimeters or better. Detector systems for radiation imaging have been dominated so far by two scintillation crystals: NaI:Tl and BGO [27]. In recent years, technological progress achieved in detection of elementary particles and nuclei (electromagnetic calorimetry) has also opened new detection possibilities in medical imaging. BGO scintillation crystals, introduced in the late 1970s, gradually took place of NaI:Tl in most medical imaging systems. At present, sensitive imaging systems are essential clinical tools in nuclear medicine, and various kinds of scintillation crystals are used as detectors of radiation from radiation tracers. The demand for high density, large yield and fast scintillators (properties of the most prominent imaging scintillation crystals are summarized in Table 2.3) for this applications favors particularly GSO and LSO crystals as imaging detectors.

GSO crystals have been introduced in 1983 [28] and used as gamma-ray scintillation detectors in 1992 [29] . The crystal is characterized by a good balance of scintillation properties, such as high density (higher than NaI:Tl) , large lifght output and short decay time (shorter than BGO) as shown in Table 2.3. LSO crystals have been discovered and investigated in 1992 [30, 31]. They have high gamma-ray detection efficiency due to their density, and the radiation length, X_o, is very similar to BGO (see Table 2.3). LSO is a fast scintillator and due to unique combination of high emission intensity, speed, high density and atomic number appears to be the most attractive candidate for medical imaging applications.

Table 2.3. Properties of the most prominent imaging scintillation crystals

Scintilation crystal	NaI:Tl	BGO	GSO	LSO
Density [g/cm^3]	3.67	7.13	6.71	7.40
Melting point [^{o}C]	651	1050	1950	1850
Radiation length [cm]	2.6	1.1	1.4	1.1
Moliere radius [cm]	1.3	2.7	2.2	2.6
Refractive index [a)]	1.8	1.8	1.8	1.9
Wavelength [nm]	415	480	430	420
Decay time [ns]	230	300	30	40
Light yield [b)] [%]	100	15-20	20	70

a) At the wavelength of the maximum

b) Relative to NaI:Tl for gamma ray

3. Radiation Damage

Radiation damage is defined as the change in scintillation characteristics caused by prolonged exposure to intense radiation. The damage manifests itself by a decrease of the optical transmissions of a crystal which causes a decrese in pulse height and deterioration of energy resolution of the detector. Radiation damage, other than radio-activation, is usually partially reversible, i.e. the absorption bands can disappear slowly in time. In general, after irradiations, the light transmission of scintillation crystals decreases because of formation of color centres. Therefore, optical spectra before and after irradiations will have to be compared for different kinds of irradiation and different doses, as well. The stability of light output under irradiation and the recovery after irradiation can be estimated.

The radiation hardness of heavy scintillation crystals is known to be a crucial parameter because of degradation of light output and must be studied very carefully. Light yield dependence on the irradiation dose seems to be one of the most decisive parameters for practical use of these scintillation detectors in SPECT´s and PET´s applications. In high-physics experiments and other similar applications, radiation hardness of scintillation detectors caused by low-energy gamma-rays is decisive for choice of these systems.

4. Radiation Resistivity of Heavy Scintillation Crystals

Radiation hardness of scintillation crystals can be examined by the measurement of transmission spectra before and after irradiations. Irradiation conditions and optical transmission spectra measurements are described in details in [32] . The crystal samples are irradiated by a ^{60}Co gamma-ray source placed longitudinally at a given distance, r (cm), from the gamma-ray source. The dose fluence D_{exp} ($Gy.s^{-1}$) can be calculated from

$$D_{\exp} = \frac{A}{4\pi r^2} \sum_i p_i E_i \frac{\mu_i}{\rho} \tag{4.1}$$

where A is the activity of a given gamma-ray source, E_i are appropriate energies ($E_1 = 1.173$ MeV and $E_2 = 1.332$ MeV for ^{60}Co), p_i are appropriate branching ratios of energy lines (two lines $p_1 = p_2 = 1$ were taken for ^{60}Co) and μ_i/ρ are mass-energy coefficients ($\mu_1/\rho = 2.68 \times 10^{-2}$ cm^2g^{-1} and $\mu_2/\rho = 2.59 \times 10^{-2}$ cm^2g^{-1} were taken for ^{60}Co gamma-ray source). For the activity of ^{60}Co gamma-ray source of about 2.05×10^{15} GBq (i.e. 2.05×10^{15} s^{-1}) , appropriate irradiations runs can be calculated (about 1.0 and 10.0 hours for 10^4 and 10^5 Gy accumulated dose, respectively).

The optical transmisson spectra (measured longitudinally through optically polished crystal samples) before and after ^{60}Co gamma-ray irradiations (accumulated doses are attached to each curve) for a large volume CeF_3 sample of dimensions 3x3x6 cm^3 are illustrated in Fig.4.1. These transmission spectra for CeF_3 with other prominent large volume scintillation crystals of BGO, BaF_2 and $PbWO_4$ are compared in Fig.4.2. The absolute degradation of transmission at peak emmission wavelength, per unit radiation length [1], is compared in Table 4.1. The induced absorption coefficient Δk, defined by

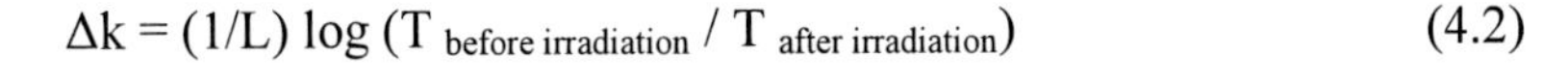

$$\Delta k = (1/L) \log (T_{\text{before irradiation}} / T_{\text{after irradiation}}) \quad (4.2)$$

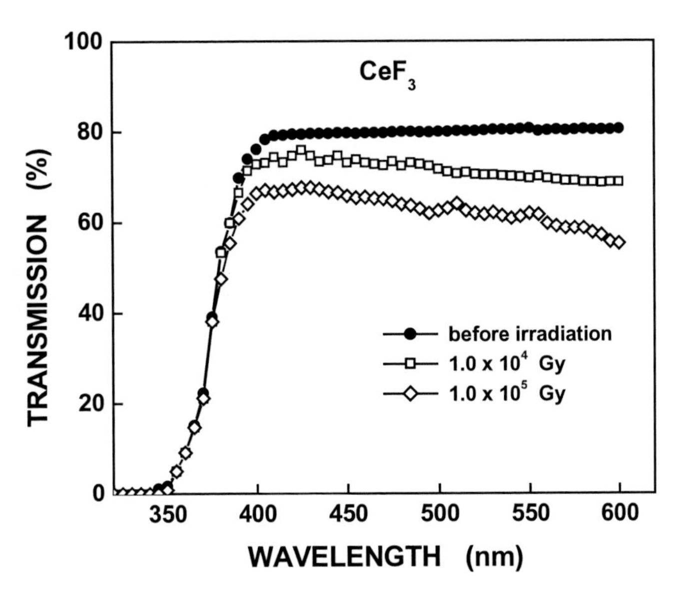

Figure 4.1. Transmission spectra for CeF_3 before and after ^{60}Co gamma-ray irradiations. The accumulated doses are attached to each curve. The optical transmissions were measure longitudinally (through the crystal).

where L is the length of crystal, are also listed in Table 4.1. The recovery time of radiation damage has been estimated from measurement of transmission spectra during 0.5 – 60 days after irradiations at 10^5 Gy. The experimental points at the peak emission wavelength have been least-square fitted by

$$T_{irr}(t) / T(0) = 1 - \exp(-\beta t) , \quad (4.3)$$

Table 4.1. Absolute degradation of transmission, induced absorption coefficients (Δk) and recovery time coefficients (β) for scintillation crystals at peak emission wavelength

Scintillation crystal	Peak emission wavelength [nm]	Gamma-rays irradiations: Degradation 10^4Gy	10^5Gy	Δk [m^{-1}] 10^4Gy	10^5Gy	Recovery time afer 10^5 Gy [day^{-1}]	Fast neutron irradiations: Degradation 10^5Gy	Δk [m^{-1}] 10^5Gy	Recovery time after 10^5 Gy [day^{-1}]
BGO	480	3.4 %	7.5 %	0.62	1.24	2.6 ± 0.4	4.1 %	1.75	14.5 ± 1.2
BaF_2	325	6.7 %	12.0 %	0.49	0.90	1.2 ± 0.2	7.7 %	1.34	10.1 ± 1.0
CeF_3	300	2.8 %	10.1 %	0.44	0.79	1.3 ± 0.2	3.6 %	1.22	9.5 ± 1.0
$PbWO_4$	420	7.5 %	12.5 %	0.48	0.85	3.6 ± 0.6	9.5 %	1.10	15.5 ± 1.4
$CdWO_4$	470	4.8 %	8.0 %	0.21	0.39	3.3 ± 0.5	5.3 %	1.05	15.0 ± 1.3

where T_{irr} (t) is the transition measured at time t (days) , T(0) the transition before irradiation and the fitting parameter β (day^{-1}) describes the recovery. The results are also listed in Table

4.1. Recovery time for optical transmission of two BaF_2 samples of dimensions 3.0x3.0x12.5 cm^3 and 3.0x3.0x24.0 cm^3 after ^{60}Co gamma-ray irradiation of 1.0 x 10^5 Gy are displayed in Fig.4.3. The results obtained for gamma-rays irradiations are compared in Table 4.1 also with those, obtained for fast neutrons (14.7 MeV) irradiations [33]. The neutron generator, based on the neutron output from the reaction $^2H(d,n)^3He$, has been used as the neutron irradiation source in these experiments.

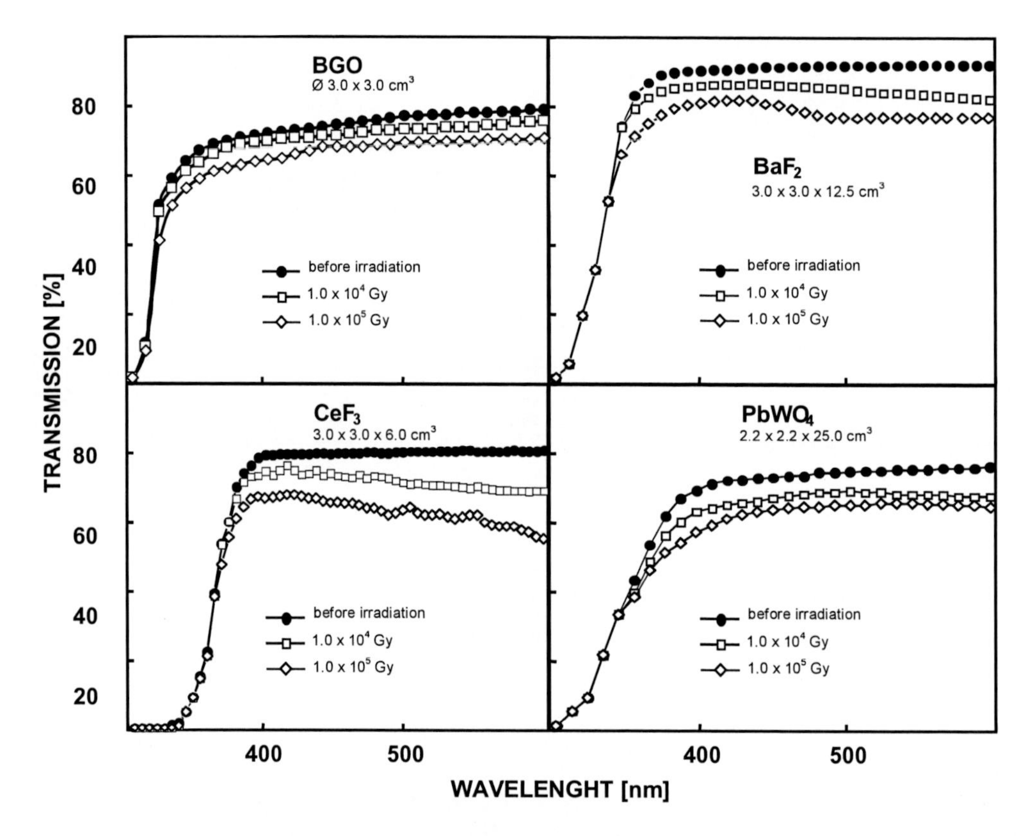

Figure 4.2. Comparison of transmission spectra before and after ^{60}Co gamma-ray irradiations for BGO, $PbWO_4$, BaF_2 and CeF_3 scintillation crystals. The optical transmissions were measured longitudinally (through the crystals).

The experimental results for radiation resistance of large volume heavy scintillation crystals presented in Table 4.1 can be summarized as follows:

(a) Radiation damage of BGO, fluoride and tungstate scintillation crystals, at 10^4 Gy accumulated low energy gamma-ray dose, were found to be very similar. However, radiation damage of these crystals is significant at 10^5 Gy accumulated dose: the absolute degradation of transmission is about 10% .

(b) Induced absorption coefficients substantially increase for a 10^5 Gy gamma-ray accumulated dose.

(c) Induced absorption coefficients for a 10^5 Gy gamma-ray accumulated dosed were found to be substantially lower than those obtained for fast neutron irradiations.

(d) Radiation damage due to low energy gamma-rays is recovered on only a few days, recovery time for fast neutron irradiations is substantially longer.

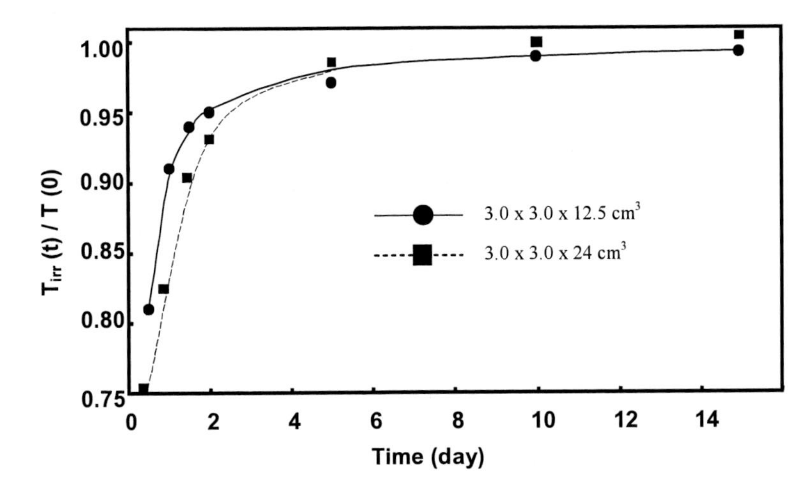

Figure 4.3. Recovery time for optical transmission of BaF_2 after ^{60}Co gamma-ray irradiation 1.0 x 10^5 Gy.

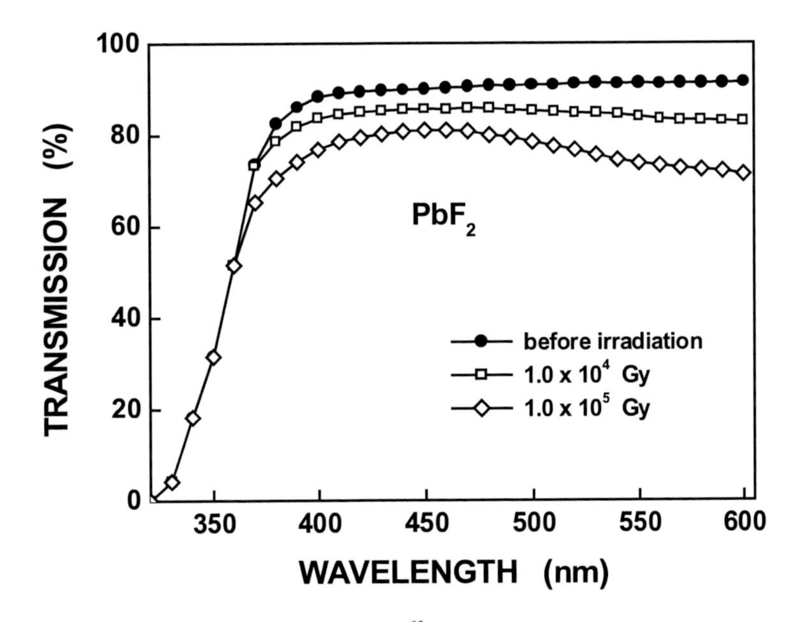

Figure 4.4. Transmission spectra before and after ^{60}Co gamma ray irradiation for PbF_2. The optical transmission was measured longitudinally.

It should be noted that radiation sensitivity of fluoride scintillation crystals, particularly BaF_2 and CeF_3, has been intensively studied in 1990s [1,9-11, 34-37], since these crystals have been listed between most promising candidates for high-energy physics experiments SPHERA [7,8] at JINR Dubna and LHC [1] at CERN Geneva. Even if radiation sensitivity of fluorides to gamma-rays and neutron flux was found to be fair up to 10^5 Gy accumulated doses, the production target costs of these crystals (price of about 8-10 USD/cm^3) was found

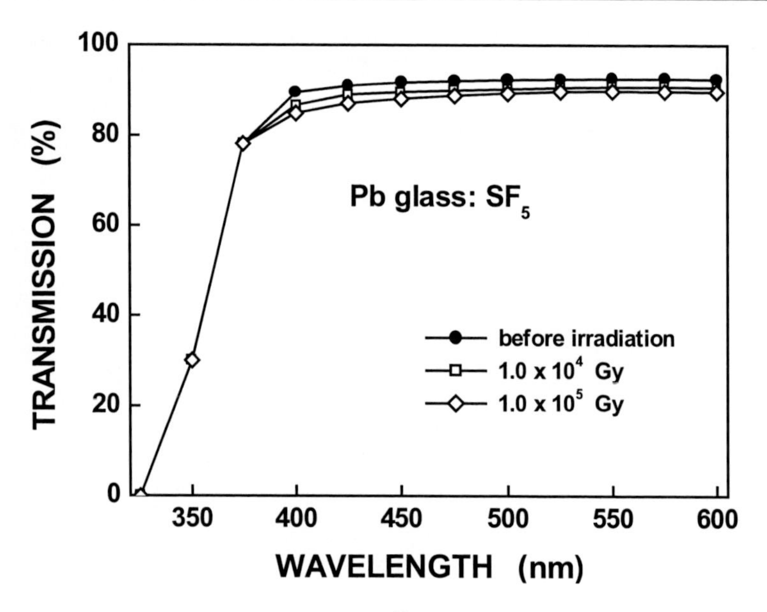

Figure 4.5. Transmission spectra before and after ^{60}Co gamma ray irradiation for SF_5. The optical transmission was measured longitudinally.

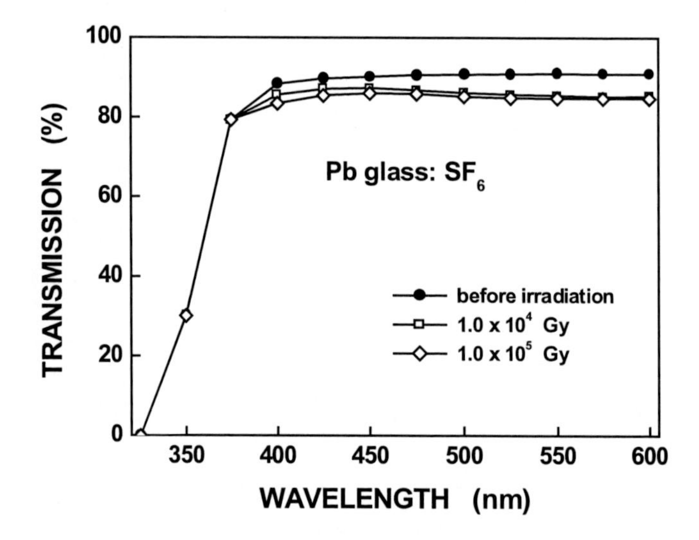

Figure 4.6. Transmission spectra before and after ^{60}Co gamma ray irradiation for SF_6. The optical transmission was measured longitudinally.

to be very high. Among other candidates for electromagnetic calorimeters, a (relatively) cheap crystal of PbF_2 has also been considered. Lead fluoride, PbF_2, was first considered as a Cherenkov radiator for electromagnetic calorimetry [12-14] . The relevant properties of this crystal are its high density, short radiation length and transmission extending to UV. Properties of PbF_2 and the most prominent lead glasses SF_5 and SF_5 are listed in Table 2.2. Two major advantages of PbF_2 over the promising fluorides BaF_2 and CeF_3 as well as BGO scintillation crystals are that it is less expensive, and, being a Cherenkov radiator, its performance is not limited by a long decay constant. However, the radiation hardness of lead fluoride is known to be a crucial parameter and therefore the radiation hardness of many PbF_2

samples has also been intensively investigated . New results [15] on radiation hardness of a 4x4x24 cm^3 PbF_2 crystal samples at accumulated doses of low energy gamma-rays 10^4 and 10^5 Gy have been obtained (see Fig.4.4). The results have also been compared with radiation measurements for two 4x4x40 cm^3 Pb-glass SF_5 (see Fig. 4.5) and SF_6 (see Fig.4.6), respectively. The comparison of transmission spectra for a 3x3x6 cm^3 CeF_3 and 4x4x24 cm^3 PbF_2 crystals for doses 10^4 Gy nd 10^5 Gy displayed in Fig.4.7. The appropriate induced absorption coefficients of PbF_2 at these doses are displayed in Fig.4.8. The recovery time of the same crystal has been estimated from measurements of transmissiion spectra during 14 days (see Fig.4.9) after irradiations at 10^5 Gy. The recovery time as estimated from this dependence is of the order of few days only. The experimental results concerning radiation hardness of large volume lead fluoride scintillation crystal can be summarized as follows:

(a) Radiation damage of lead fluoride crystal and lead fluoride glasses at 10^4 Gy accumulated low energy gamma-rays dose were found to be very similar.
(b) Radiation damane of PbF_2 is significant at 10^5 Gy accumulated dose; both types of lead glasses SF_5 and SF_6 have substantially better radiation hardness than examined lead fluoride crystal.
(c) Radiation damage of large volume crystals PbF_2 and CeF_3 is comparable at 10^4 Gy and 10^5 Gy accumulated doses, respectively.
(d) Radiation damage due to low energy gamma-rays is recovered by a few days only.

Lead tungstate scintillation crystals, $PbWO_4$ (also designated as PWO), have recently been chosen for the design and construction of electromagnetic calorimeter at CMS detector at LHC = Large Hadron Collider, CERN project of the 21st century [1,16,17]. For this application, these scintillation crystals of large volumes are produced at China (Shanghai Institue of Ceramics) and Russian Federation (Bogoroditsk). The properties of PWO crystals have been studied since 1994, since these crystals has been considered among the most promising candidates for the future electromagnetic calorimeters (EMC), together with CeF_3 and PbF_2, a Cherenkov light radiator. CeF_3 provides a high yield and PbF_2 has fast timing properties with a high stopping power. PWO crystals present a whole set of physical and chemical parameters which are attractive particularly for large EM calorimeters in the extreme conditions that are foreseen particularly at LHC. Despite a moderate mechanical hardness, it is easy to machine and polish. PWO is grown from a mixture of PbO and WO_3 oxides by Czochralski and Bridgeman method of crystal growth at a rather moderate temperature. Moreover, it has also been shown that light emission mechanism of PWO samples grown at various conditions alows to control the growth technology in order to optimise desirable characteristics [1]

Radiation hardness of lead tungstate crystals have been investigated very intensively in last ten years [1,17-19,38] . Light yield dependence on irradiation dose seem to be the most decisive parameter for the practical choice of this scintillation crystal particularly in high-energy physics detector applications, such LHC experiment at CERN. It has been found [17] that in some PWO crystals the light yield decreases sharply at doses below 100 Gy only. Therefore, the main goal of the radiation hardness investigations was to compare irradiation measurements performed with various large volume PWO samples from different manufacturers [19] . The transmission spectra before and after absorption of 10^4 and 10^5 Gy irradiations for large volume PWO samples from Bogoroditsk (Russian Federation), Turnov

(Czech Republic) and Kharkov (Ukraine) are illustrated in Fig.4.10 . The comparison of transmission spectra for a 3x3x6 cm^3 CeF_3 crystal for the same accumulated doses are displayed in Fig.4.11. The comparison of transmission of all three samples for doses 10^4 and 10^5 Gy at the peak emission wavelenght 480 nm are listed in Table 4.2 , the absolute degradation of transmission of all samples are compared in Table 4.3. The absolute degradation for La^{2+} - doped sample #1 was found to be about 12.3% and 14.2%, for accumulated doses 10^4 and 10^5 Gy, respectively. Radiation damage due to 10^5 Gy gamma-rays was found to be substantially better for La^{2+} - doped PWO crystal sample. It can be also seen that radiation damage of large volume crystals PWO and CeF_3 is comparable at 10^4 and 10^5 Gy accumulated gamma-ray doses, respectively. The recovery time has been estimated from measurements of transmisson during 2-20 days after irradiations at 10^5 Gy. A complete recovery of radiation damaged crystal samples was found between 10 and 15 days.

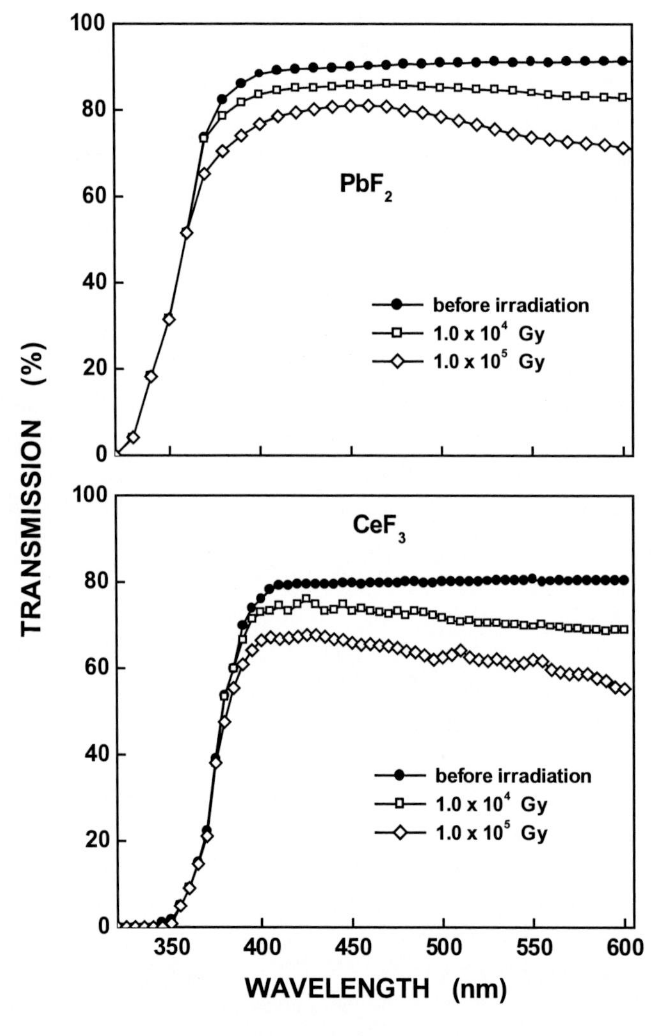

Figure 4.7. Comparison of transmission spectra before and after ^{60}Co gamma-ray irradiations for PbF_2 and CeF_3 crystals. The optical transmissions were measured longitudinally (through the crystals).

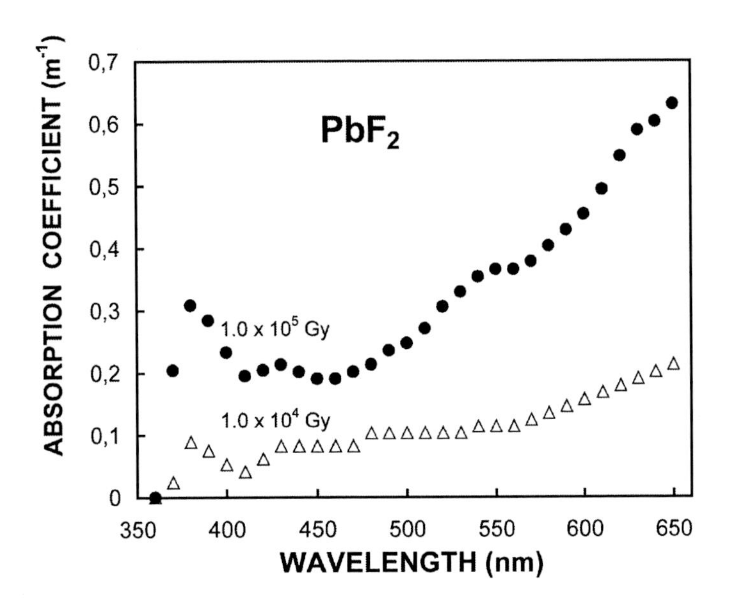

Figure 4.8. The induced absorption of PbF_2 at 1.0 x 10^4 Gy and 1.0 x 10^5 Gy, respectively.

Results on radiation sensitivity of large volume $CdWO_4$ crystals 2.0x2.0x19.5 cm^3 due to low energy gamma-rays [38] are displayed in Fig. 4.12. The corresponding variation of absorption coeficients Δk for $CdWO_4$ and $PbWO_4$ samples are compared in Fig.4.13 . The absolute degradation of $PbWO_4$ sample #1 (Lanthanum doped) and sample #2 (undoped), as well as $CdWO_4$ and CeF_3 samples is compared in Table 4.4 . It is evident that $CdWO_4$ scintillation crystal of large volume has a very good radiation resistance: for doses of 10^4 and 10^5 Gy gamma-rays, the optical transmission of the irradiated crystal decreases by less than 4.8 and 8.0%, respectively (see also Table 4.1). The measurement of transitions within 15 days showed a complete recovery of $CdWO_4$ after this period [38] .

Radiation hardness study of CsI (undoped) or CsI:Tl (thalium doped) large volume scintillation crystals is particularly motivated by the application of these crystals as detectors in EM calorimeters on coliders of luminosity up to 10^{34} $cm^{-2}s^{-1}$ [24] . It is estimated that radiation backround at these calorimeters corresponds to an absorbed dose of up to 100 rad (1.0 Gy) per year. Dependence of transmission light yield of large volume (2x2x30 cm^3) CsI:Tl sample crystals manufactured by SIC (Shanghai Institue of Ceramics), CRISMATEC and Kharkov (Institute of Single Crystals) on absorbed dose up to 10^4 Gy is displayed in Fig. 4.14 [39] . It is seen that radiation hardness of CsI:Tl scintillation crystals at about 1 Gy is satisfactory for this application. Nevertheless, it is also seen that radiation sensitivity of CsI:Tl sharply decreases up to 55% and 35% transmission at 10^3 and 10^4 Gy, respectively. Moreover, recovery time after 10^4 Gy irradiations was found to be large (few weeks). It should be also noted that CsI:Tl crystals manufactured by the same technology using the same raw material from the same manufacturer exhibit a large spread of the light output decrease by radiation.

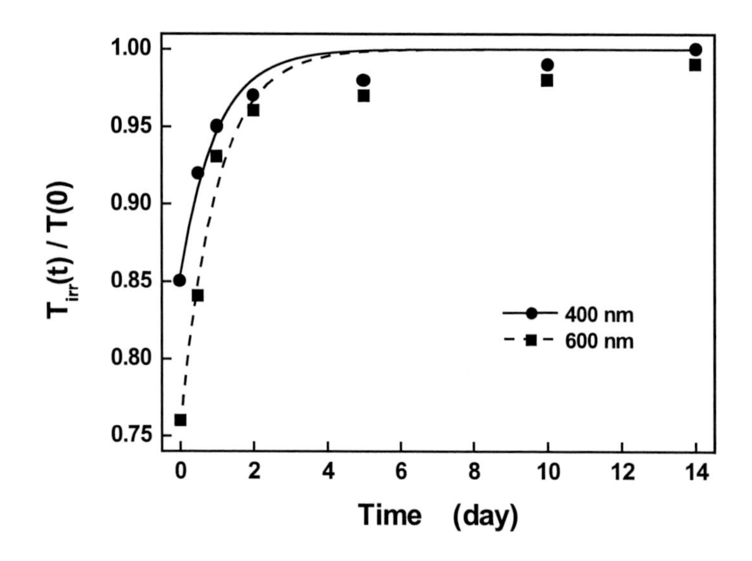

Figure 4.9. Recovery time for optical transmission of PbF_2 after ^{60}Co gamma-ray irradiation 1.0 x 10^5 Gy.

Table 4.2. Transmission of $PbWO_4$ crystals at peak emission (480 nm)

Dose [Gy]	Sample #1	Sample #2	Sample #3
0	77.5 %	72.0 %	65.0 %
10^4	68.0 %	67.0 %	56.0 %
10^5	66.5 %	56.0 %	51.5 %

Table 4.3. Degradation of transmissionfor of $PbWO_4$ crystals at peak emission (480 nm)

Dose [Gy]	Sample #1	Sample #2	Sample #3
10^4	12.3 %	7.0 %	13.8 %
10^5	14.2 %	22.2 %	20.8 %

Table 4.4. Degradation of transmission of fluoride and tungstate cry\stals at peak emission

Dose [Gy]	$PbWO_4$ #1 (480nm)	$PbWO_4$ #2 (480nm)	$CdWO_4$ (470nm)	CeF_3 (480nm)
10^4	15.0 %	7.5 %	4.8 %	2.8 %
10^5	27.5	12.5 %	8.0 %	10.0 %

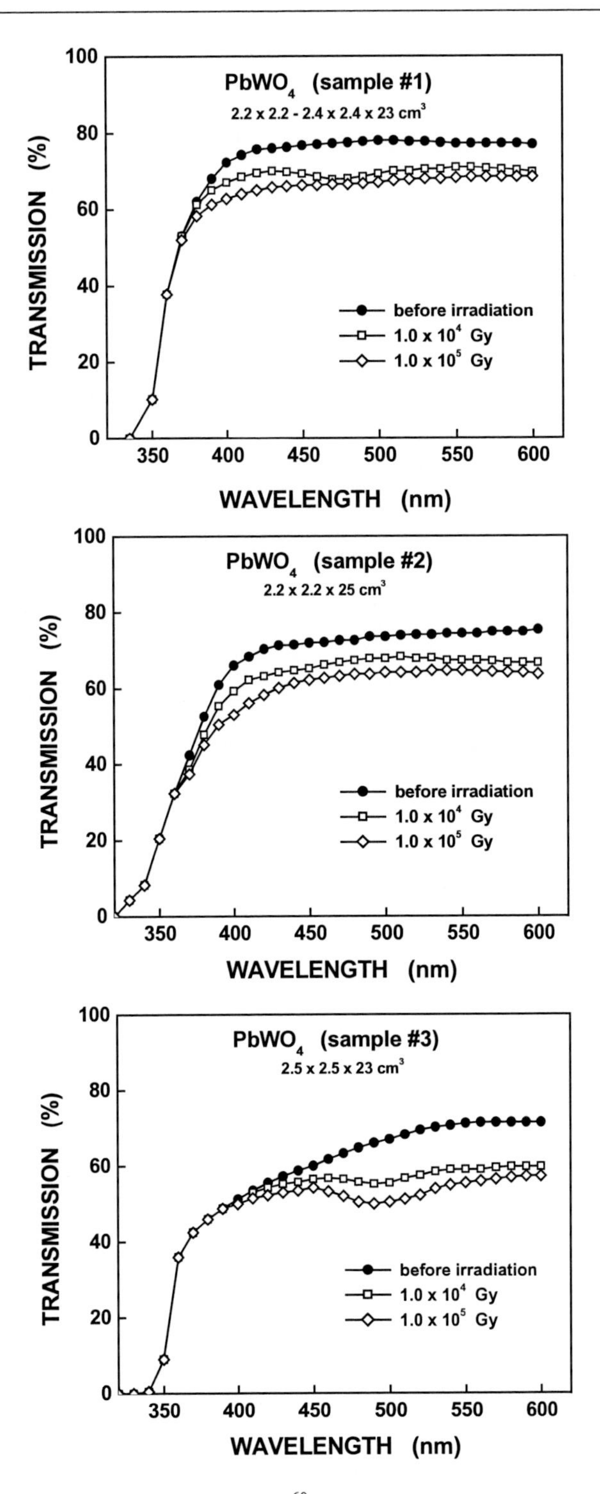

Figure 4.10. Transmission spectra before and after ^{60}Co gamma-ray irradiation for the different samples of $PbWO_4$:

Sample #1 – manufacturer Bogoroditsk, Russian Federation
Sample #2 – manufacturer Trunov, Czech Republic
Sample #3 – manufacturer Kharkov, Ukraine

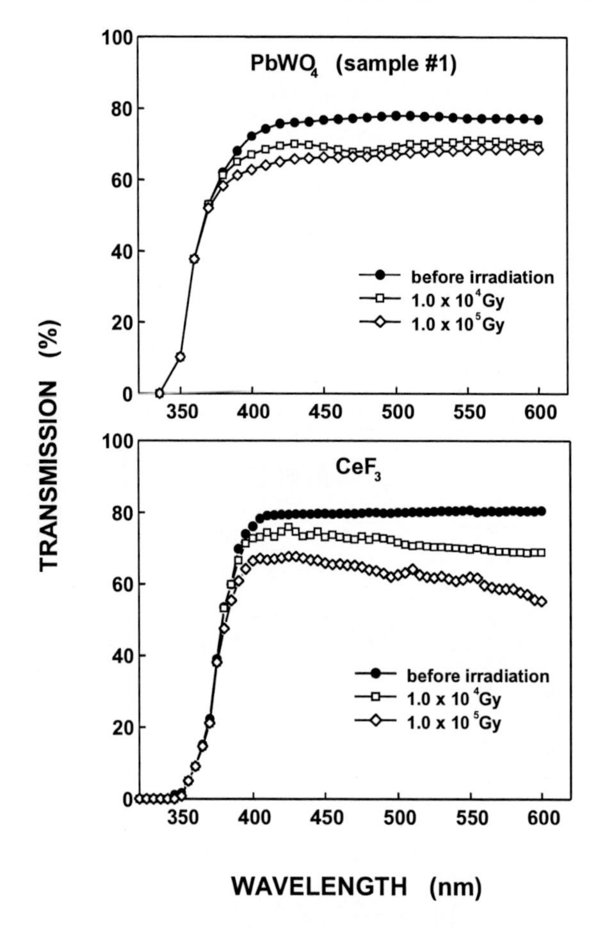

Figure 4.11. Comparison of transmission spectra before and after ^{60}Co gamma-ray irradiations for $PbWO_4$ and CeF_3 crystals. The optical transmissions were measured longitudinally (through the crystals).

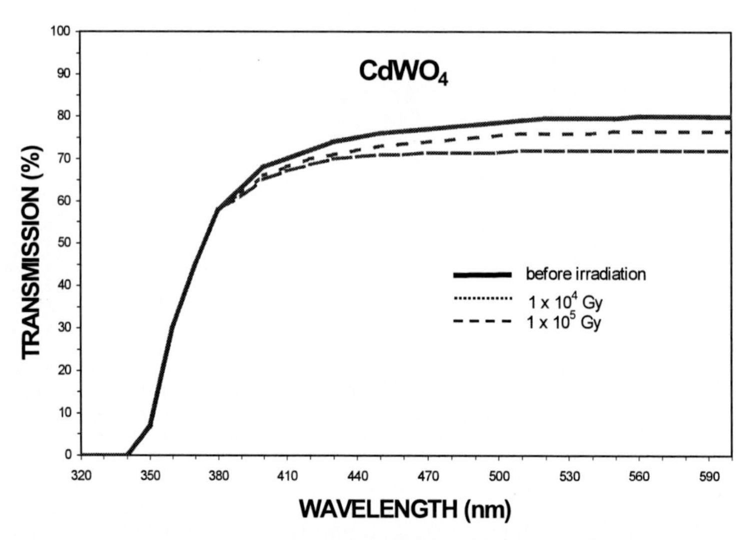

Figure 4.12. Transition spectra for $CdWO_4$ large volume (2 x 2 x 19.5 cm^3) crystal before and after ^{60}Co gamma-ray irradiation.

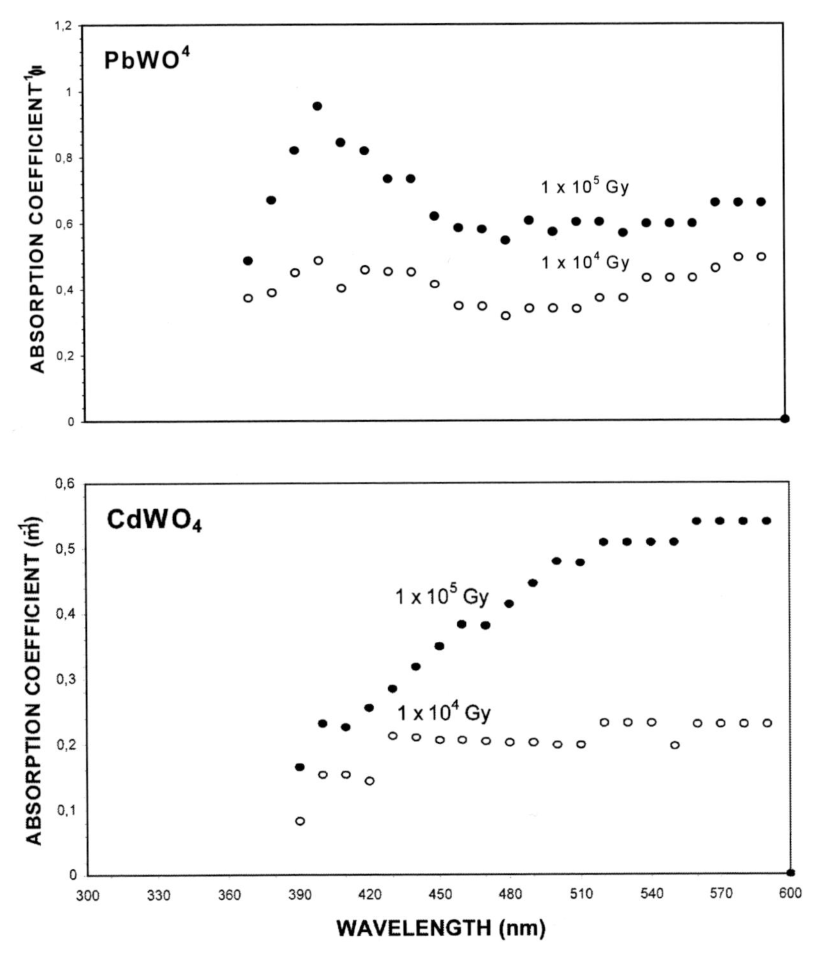

Figure 4.13. Comparsion of absorption coefficients on wavelenght for two large volume $PbWO_4$ and $CdWO_4$ crystals.

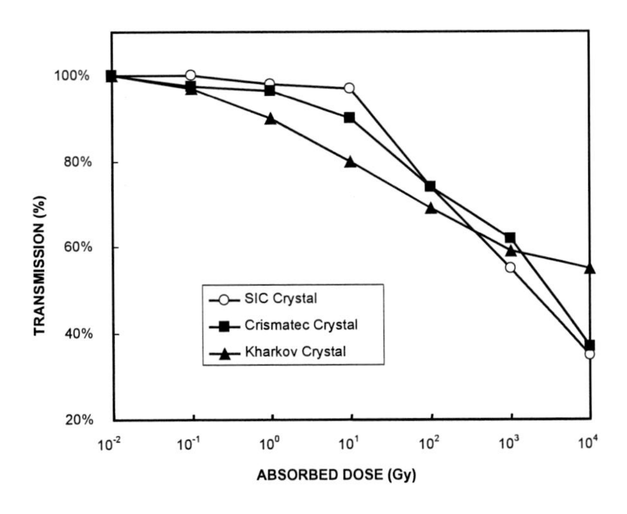

Figure 4.14. Transmission degradation in dependence of the absorbed gamma-ray dose for CsI:Tl crystals (manufactured by SIC, CRISMATEC and Kharkov) of a large volume 2 x 2 x 30 cm^3.

5. Radiation Sensitivity of Imaging Scintillation Crystals

The imaging with positron emission tompgraphy (PET) is feasible because of the unique characteristics of positron emiters which make it possible to detect accurately and non-invasively the in vivo concentration of radiopharmaceuticals. Positron emmiters are proton-rich isotopes (^{11}C, ^{13}Na, ^{15}O, etc.) that decay by escaping positrons from their nucleus. Positrons range a short distance during which they lost their energy by colliding with electrons in a human tissue. Reaching their resting state the positron combines with an electron and the two annihilate to produce two gamma-ray photons, each with energy of 511 keV. These two photons are liberated simultaneously in nearly opposite directions and can be detected by the pair of small volume detectors. Events are registrated in coincidence within a short time coincidence window (nanoseconds). Coincidence detection of positrons by an individual pair of detectors would be very inefficient. In order to optimize detection and to scan a volume each detector in a PET camera is made coincident with many other detectors simultaneously. Typically, small volume detectors are arranged around the object in a ring or multiple rings. Most of the PET cameras today have detectors arranged in a series of rings (or staggered rings) to image the whole human organ at one time. These systems use small volume scintillation crystals coupled to PMT or PD read-out systems and an appropriate "imaging" three dimensional visible electronics, as well. The total number of detectors, the number of rings, the size and geometry of the detectors, their configuration and the type of the scintillation crystal used in a detector as well as electronic circuits vary to according to the camera design which will have to affect high resolution, sensitivity and reliability of the measurements. In earlier stage of PET and CTS as well as gamma-camera systems, NaI:Tl (sodium iodide doped with thalium) crystals have been used. In 1970s, when BGO was invented and manufactured for aa acceptable price, it took place sodium iodide crystals as scintillators in these applications because of its promising chemical and physical characteristics (see Table 2.3, where the most important characteristics of imaging scintillation crystals are summarized).

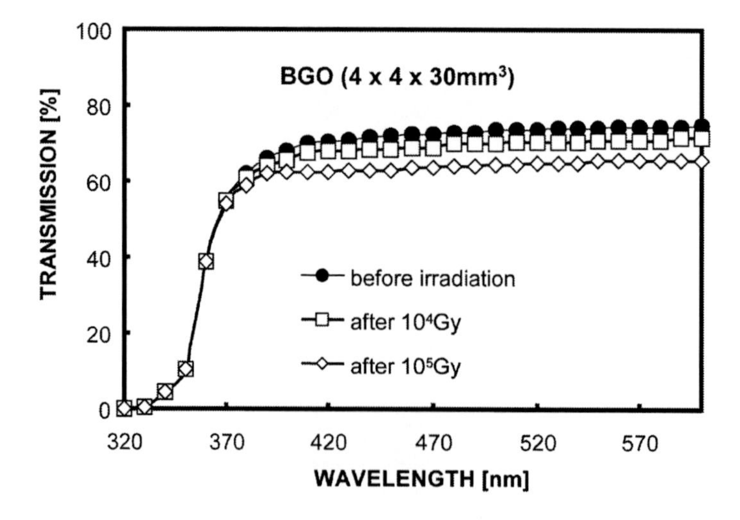

Figure 5.1. Transmission spectra before and after ^{60}Co gamma-ray irradiations. The optical transmissions were measured longitudinally.

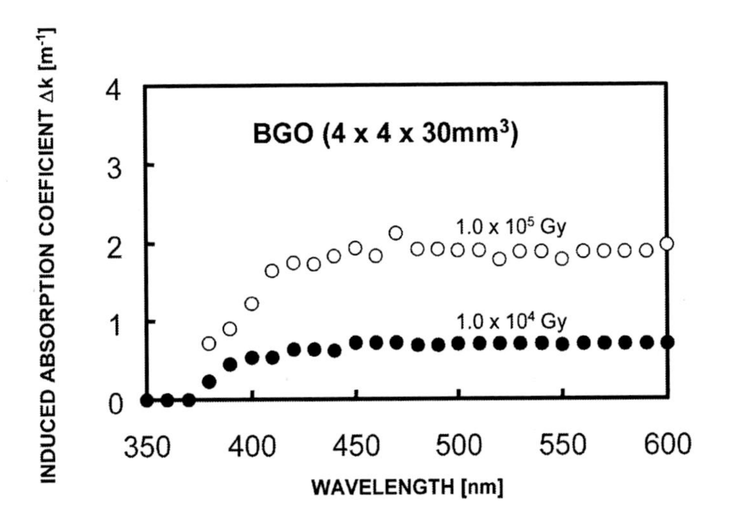

Figure 5.2. Induced absorption coeficients.

Radiation damage of small BGO scintillation crystals (4x4x30 mm^3) due to low energy gamma-rays have been studied by the measuring the change in transmittance of light [32]. The transmission spectra before and after absorption of 10^4 and 10^5 Gy accumulated gamma-ray doses for one of these samples are displayed in Fig.5.1, the appropriate induced absorption coefficients, Δk, are compared in Fig.5.2. The transmission at the peak emission per unit radiation length are compared in Table 5.1. As can be seen, the degradation of transmission for 10^4 Gy at the wavelength of the peak emission λ = 480 nm, was found to be

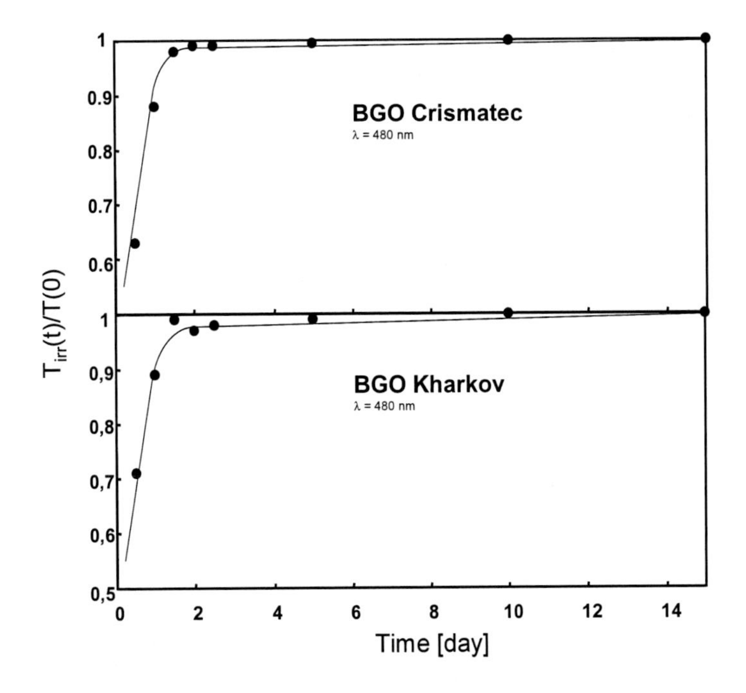

Figure 5.3. Recovery time for optical transmission of small (Crismatec) and large (Kharkov) BGO crystals after ^{60}Co gamma-ray irradiation 1.0 x 10^5 Gy

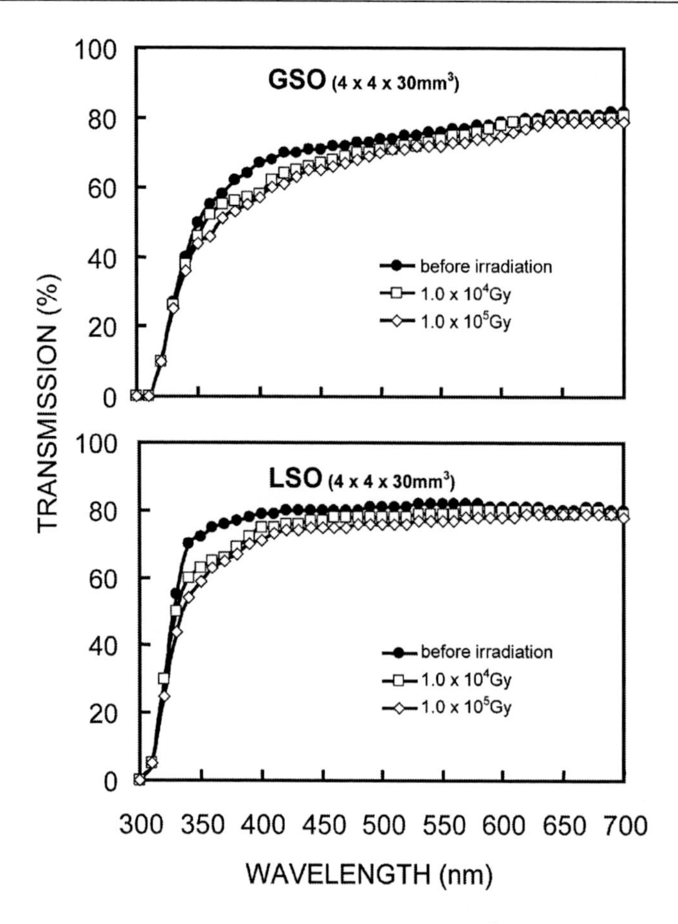

Figure 5.4. Comparison of transmission spectra before and after ^{60}Co gamma ray irradiation for GSO and LSO crystals. The optical transmission were measured longitudinally.

better than 3.4% per unit radiation length X_o = 1.1 cm .The degradation of transmission for 10^5 Gy at the same wavelength was found to be better than 7.5% per X_0. The recovery time estimated from data displayed in Fig.5.3 was found to be of about 2.5 day^{-1} . Radiation damage of small 30x4x4 mm^3 BGO crystals at 10^4 Gy accumulated low energy gamma-rays dose was found to be negligible. However, radiation damage of small BGO crystals can be significant at roughly about 10^5 Gy accumulated dose: absolute degradation of transmission per radiation unit was found to be lower than 7.5%. This value was found to be lower than the same value for large volume fluoride and tungstate crystals. Radiation damage of large Ø 30 x 30 mm^3 BGO crystals was found to be approximately the same as for small BGO crystals at 10^4 Gy dose. Radiation damage of the large volume BGO sample due to 10^5 Gy gamma-rays was found to be consistent with data published earlier for fluoride and tungstate large volume crystals. It can be concluded that small BGO crystals, used in nuclear medicine diagnostic PET and CTS systems, were found to have good radiation resistivity due to irradiation by low energy gamma-rays.

Scintillation crystals of GSO (cerium-doped gadolinium orthosilicate) and LSO (cerium-doped luthetium orthosilicate) are new scintillators considered as imaging detectors form nuclear medicine applications [2,3,40] . Both GSO and LSO are fast scinillation crystals characterized by very good scintillation, physical and chemical properties (see Table. 2.3).

They are not hygroscopic and can be grown and polished as crystals of small dimensions of 4 x 4 x 30 mm^3 , as required in PETs and SPECTs imaging techniques. Radiation sensitivity of GSO and LSO scintillation imaging crystals have been studied by measuring the change of transmittance of light before and after irradiations, results are illustrated in Fig. 5.4, appropriate radiation-induced absorption coefficients, Δk , are displayed in Fig.5.5. The transmission and relative degradation of transmission at the peak emission are compared in Table 5.2. As can be seen, the degradation of transmission of both new imaging scintillators GSO and LSO was found to be smaller than that repopred for BGO scintillation crystals of the same dimensions. The recovery time from the radiation damage due to 10^5 Gy gamma-ray dose has been determined from from measurements of transmission spectra during 0.5 – 10 days after irradiations. The results are summarized in Table 5.3, the recovery time was found to be very similar to small BGO samples of the same dimensions. It can be concluded that small GSO and LSO scintillation crystals, the most prospective candidates for applications as small-sized detectors in nuclear medicine imaging systems, have excellent radiation resistance to irradiations by low energy gamma-rays.

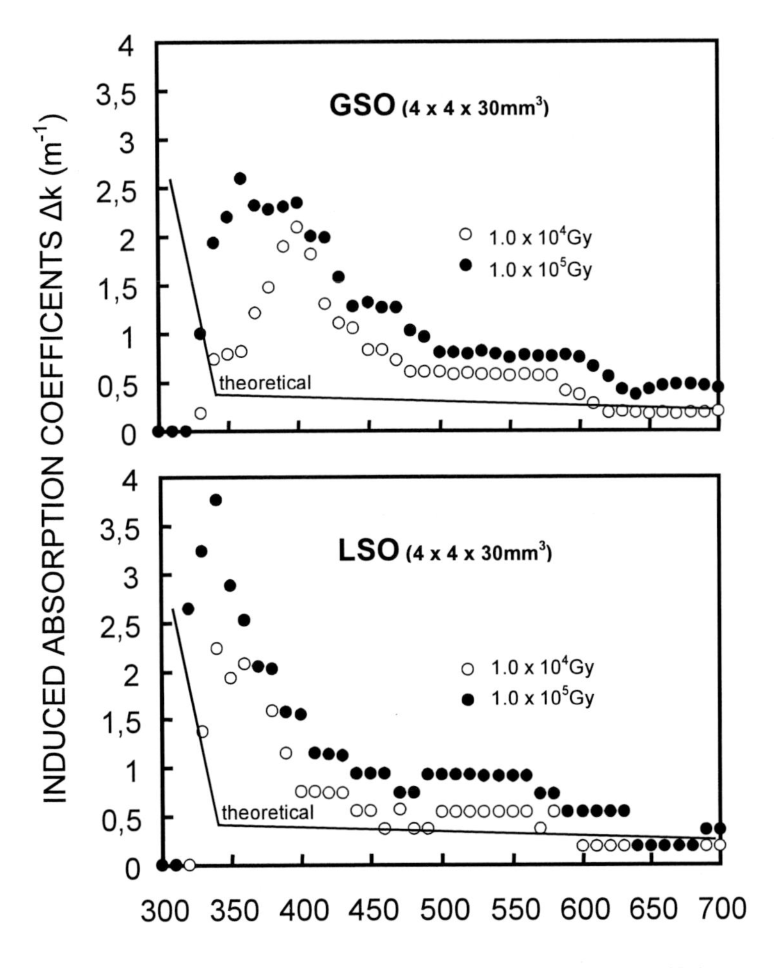

Figure 5.5. Comparison of the radiation induced absorption coefficients.

Table 5.1. Transmission and absolute degradation of BGO samples

Sample	Manufacturer	Size	Transmission [d]			Absolute degradation [e]	
		$[mm^3]$	0 Gy	10^4 Gy	10^5 Gy	10^4 Gy	10^5 Gy
#1	CRISMATEC [a]	4 x 4 x 30	74.0 %	71.2 %	67.8 %	3.4 %	7.6 %
#2	CRYTUR Ltd. [b]	4 x 4 x 30	70.0 %	67.1 %	63.8 %	3.7 %	7.9 %
#3	ISC Co. [c]	4 x 4 x 30	71.0 %	68.0 %	64.8 %	3.9 %	7.5 %
#4	CRISMATEC [a]	∅30 x 30	75.0 %	71.9 %	68.0 %	3.8 %	8.4 %

a) Saint – Gobain Crystals and Detectros, France
b) Turnov, Czech Republic
c) Institute for Single Crystals, Kharkov, Ukraine
d) Transmission at peak emission (λ = 480 nm)
e) Absolute degradation of transmission at peak emission (λ = 480 nm) per unit radiation length (X_0 = 1.1 cm)

Table 5.2. Transmission and absolute degradation of GSO, LSO and BGO crystals

Crystal	Manufacturer	Size	Transmission [b]			Absolute degradation [c]	
		$[mm^3]$	0 Gy	10^4 Gy	10^5 Gy	10^4 Gy	10^5 Gy
GSO	CRYTEX Corp.	4 x 4 x 30	70.0 %	65.2 %	63.4 %	3.2 %	5.2 %
LSO	CRYTEX Corp.	4 x 4 x 30	79.8 %	76.3 %	73.9 %	2.8 %	5.0 %
BGO	CRISMATEC [a]	4 x 4 x 30	71.0 %	68.0 %	64.8 %	3.4 %	7.5 %
BGO	CRISMATEC	∅30 x 30	75.0 %	71.9 %	68.0 %	3.8 %	8.4%

a) Saint – Gobain Crystals and Detectors, France
b) Transmission at peak emission
c) Absolute degradation of transmission at peak emission per unit radiation length

Table 5.3. Parameters of recovery time dependence $T_{irr}(t)/T(0) = 1 - e^{-\beta t}$ at peak wavelength emission

Crystal	Size	Peak wavelength, λ	Recovery parameter, β
GSO	4 x 4 x 30 mm^3	430 nm	(2.50 +/- 0.35) day^{-1}
LSO	4 x 4 x 30 mm^3	420 nm	(2.95 +/- 0.42) day^{-1}
BGO	4 x 4 x 30 mm^3	480 nm	(2.17 +/- 0.31) day^{-1}

6. Some Theoretical Remarks

Most of the papers devoted to a study of radiation hardness of scintillation crystals consider optical transparency deterioration as a main reason of light output loss after irradiations [41,42]. Other authors considered a mechanism of damage of an intrinsic scintillation efficiency of the material due to irradiations [43,44] . The detailed discussion of various physical models of radiation induced changes in the properties of scintillation crystals, as well as a relevant bibliography devoted to this subject can be found in [45] . In our work we do not pretend to explain theoretical studies concerning the mechanism of the light output degradation. Nevertheless, some remarks are of special importance:

The transmisson experiments have shown that the absorption spectrum keeps the same shape for different irradiation sources and accumulated doses, and during the recovering process. This indicates that probably one kind of colour centre in scintillation crystal is produced by irradiation. On the other hand, two recovery regimes proposed (fast and slow), as well as two thermoluminiscence peaks [5] , suggest two different modes of charge untrapping, one localized and the second through the valence band. The analysis of radiation-induced absorption coefficients of GSO and LSO crystals (see Fig.5.5, where theoretical model calculations are also drawn) proved that changes after irradiations were not caused by the scintillation mechanism. Optical transmission degradation of both crystals is caused particularly by creation of additional colour centres during gamma-ray irradiations. The complete recovery from the radiation damage due to gamma-rays is one of the most important features of GSO and LSO crystals, that makes these scintilators suitable for application in imagine techniques. The recovery from radiation damage has been treated [46] as a multistep process: a fast recovery with a time constant of an hour and a slower process with a time constant of a few days. This tendency has not been seen in our analysis: the recovery of GSO and LSO crystals as a function of time does not show two regimnes, one fast and one slow. Complete recovery of radiation damage of small GSO and LSO scintillation crystals irradiated by 10^5 Gy low energy gamma-rays was observed after few days only. The same evidence comes from the analysis of absorption coefficients of tungstates (see Fig.4.13), fluorides (see Fig.4.8) and BGO small-sized samples (see Fig.5.2), as well

It should be also noted that besides the loss of optical transparency, irradiations can produce both elastic and inelastic centres of photon scattering in the material. This effect can also cause the deterioration of the light collection efficiency. However, this effect is significant particulary at low irradiation doses below a few Gy. Taking into account greather irradiation doses, the secondary scattering effect could be negligible. Anyhow, this effect could be carefully studied for majority of scintillation crystals at various sources and different doses. In general, the inference that the main radiation stimulated effect is the loss of optical transparency of the crystal, seems to be well-grounded, at least for moderate doses of irradiations.

References

[1] Lecoq P.: The Challenge of new scintillator development for high-energy physics. In: Zhiwen Yin et al. (Eds.), Proceedings of the International Conference SCINT′97,

Shanghai, People´s Republic of China, September 22-26, 1997, CAS Shanghai Brand Press, p.22.
[2] Lu Z.M., Cheng-Sheng T.M.: Application of scintillation crystals in nuclear medicine. In: Zhiwen Yin et al. (Eds.), Proceedings of the International Conference SCVINT´97, Shanghai, People´s Republic of China, September 22-26, 1997, CSA Shanghai Brand Press, p.18.
[3] Lopes M.I., Chepel V.: *Radiation Phys. and Chemistry*, vol. 71/3-4 (2004) 683.
[4] L3 Collaboration: Adeva B. et al.: *Nucl.Instr. and Methods* **A 289** (1990) 35.
[5] Lecoq P., Li J.P., Rostaing B.: *Nucl. Instr. and Methods* **A300** (1991) 240.
[6] Laviron C., Lecoq P.: *Nucl. Instr and Methods* **227** (1984) 45 .
[7] Malakhov A.: Proc. XII Int. Conf. On Particles and Nuclei PANIC XII, Cambridge, 1990. Ed.: T.W.Donnelly, contributed paper, X-26.
[8] SPHERE Collaboration: Afanasiev S.V. et al.: Proc. Of the XI Int. Seminar on High Energy Physics Problems. *JINR Dubna*, September 7-12, 1992, p.49.
[9] Kozma P., Afanasiev S., Malakhov A., Schotanus P., Dorenbos P.: *Nucl.Instr. and Meth.* **A314** (1992) 26.
[10] Kozma P., Afanasiev S., Malakhov A., Povtoreiko A.:*Nucl.Instr. and Meth.* **A322** (1992) 302.
[11] Kozma P., Afanasiev S., Malakhov A., Povtoreiko A.: *Nucl.Instr. and Meth.* **A328** (1993) 599.
[12] Dally E.B, Hofstadter R.: *Rev. Sci. Instr.* **39** (1968) 658.
[13] Anderson D.F., Kobayashi M., Woody C.L., Yoshimura Y.: *Nucl.Instr. and Meth.* **A290** (1990) 385.
[14] Woody C.L. et al.: In: *IEEE Trans. Nucl. Sdci.* **NS-40** (1993) 543.
[15] Kozma P., Bajgar R., Kozma Jr. P.: *Nucl. Instr. and Meth.* **A484** (2002) 149.
[16] Lecoq P.: Large scale production of lead tungstate crystals in Russia. In: Baccaro S. et al. (Eds.) Proceedings of the International Workshop on Tungstate Crystals, Roma, Italy, October 12-14, 1998. La Sapienza Press, p.25.
[17] Annenkov A.N., Lecoq P.: Systematic study of the $PbWO_4$ crystal short term instability under irradiation. CMS Note 97-55, 1997, CERN.
[18] Kobayashi M. et al.: *Nucl.Instr. and Meth.* **A404** (1998) 149.
[19] Kozma P., Bajgar R., Kozma Jr. P.: **Rad. Phys. and Chem. 65** (2002) 127.
[20] Kessler R.S. et al.: *Nucl.Instr. and Meth.* **A368** (1996) 653.
[21] Miyabayashi K. (Belle Electromagnetic Group), Nucl.Instr. and Meth. A494 (2002) 298.
[22] Hitlin D.: *Nucl.Instr. and Meth.* **A494** (2002) 29.
[23] Yamauchi M.: *Nucl.Phys. B (Proc.Suppl.)* **111** (2002) 96.
[24] BELLE Collaboration: Abashian A. et al.: *Nucl.Instr. and Meth.* A479 (2002) 117.
[25] Rogers J.G.et al.: *IEE Trans. Nucl.Sci.* **NS-39** (1992) 1063 and NS-41 (1994) 1423.
[26] Moses W.W. et al.: In: van Eijk C.W.E. et al.(Eds.), Proceedings of the International Conference on Inorganic Scintillations and their Applications SCINT´95, Delft, The Netherlands, 1995, Delft University Press, 1996, p.25.
[27] van Eijk C.W.E.: New scintillators, new light sensors, new applications. In: Zhiwen Yin et al. (Eds.) Proceedings of the International Conference SCINT´97 , Shanghai, People´s Republic of China, September 22-26, 1997, CAS ASHanghai Brand Press, p.3
[28] Takagi K., Fukazawa T.: *Appl. Phys. Lett.*, **42** (1983) 43.
[29] Roscoe B.A. et al.: *IEEE Trans. Nucl. Sci.* **NS-39** (1992) 1412.

[30] Melcher C.L., Schweitzer J.S.: *Nucl. Instr. And Meth.* **A314** (1992) 212.
[31] Melcher C.L., Schweitzer J.S.: *IEEE Trans. Nucl. Sci.* **NS-39** (1992) 502.
[32] Kozma P., Kozma Jr. P.: *Nucl.Instr.and Meth.* **A501** (2003) 499.
[33] Kozma P., Kozma Jr. P.: *Rad. Phys. and Chem.* **71** (2004) 705.
[34] Schotanus P., Dorenbos P., Kozma P., Janovsky V.: Test of Large Volume BaF_2 Scintillators from JINR, TU Delft Report 90-1 (1990).
[35] Kozma P., Yanovsky V.V.: *Czech.J.Phys.* B40 (1990) 393.
[36] Kozma P., Janovsky V.V.: *Czech J.Phys.* **A40** (1990) 231.
[37] Kozma P.: *Czech J.Phys.* **B41** (1991) 530.
[38] Kozma P., Bajgar R., Kozma Jr.P.: *Rad.Phys.and Chem.* **59** (2000) 377.
[39] Kozma P., Kozma Jr.P.: *Nucl.Instr. and Meth.*, be appeared.
[40] Kozma P., Kozma Jr.P.: *Nucl.Instr. and Meth.* **A539** (2005) 132.
[41] Zhu R.Y. et al.: *Nucl.Insdtr. and Meth.* **A413** (1998) 297.
[42] Chowdhury M.A.H. et al.: *Nucl.Instr. and Meth.* **A254** (1999) 147.
[43] Globus M.E. , Grinyov B.V.: *Functional Materials* **3**, No.2 (1996) 231.
[44] Gektiv A.V.: Proc. of the 5th Int. Conference on Inorganic Scintillators and their Applications, Moscow, 1999, p.79.
[45] Globus M.E., Grinyov B.V.: Inorganic Scintillators, Ed. Acta Publishing Co., Kharkov, Ukraine, 2000, p.146.
[46] Fedorov A.A. et al.: *Rad.Meas.* **26** (1996) 215.

In: Radiation Physics Research Progress
Editor: Aidan N. Camilleri, pp. 355-384

ISBN: 978-1-60021-988-7

Chapter 10

CHEMICAL DOSIMETERS DEVELOPMENT AT IPEN FOR THE RADIATION PROCESSES QUALITY CONTROL

Ana Maria Sisti Galante [*] ***and Letícia Lucente Campos***
Instituto de Pesquisas Energéticas e Nucleares, IPEN – CNEN/SP

Abstract

Different chemical compounds have been studied aiming to optimize dosimetric systems in radiation processes. The absorbed dose measurement that induces beneficial changes in the irradiated material characteristics is the most efficient method to monitoring the radiation process quality. The chemical dosimetry is a dosimetric method very useful in high doses measurements. Chemical changes radiation induced in liquid, solid or gaseous can be measured and quantified by means of espectrophotomety, termoluminescence or electronic paramagnetic resonance techniques and then the absorbed dose can be determined. A variety of chemical dosimeters is available however, same time these materials are not adequate for monitoring all types of products and irradiation conditions. The necessity of developing high quality dosimeters whose cost is viable to the irradiation installations to assure the quality of their processes, have stimulated the researchers to study new compounds that can be used in the irradiations routine control. When the used dosimeter and the studied species have approximately the same density and atomic composition the dosimetry is facilitated; the chemical dosimeters offer these conditions. Aiming to improve the dosimetry processes used in Brazil the High Doses Laboratory - HDL of the Instituto de Pesquisas Energéticas e Nucleares – IPEN-CEN/SP developed and studied four different chemical dosimeters: bromocresol green solutions, potassium nitrate (KNO_3) pellets pure and mixed with manganese dioxide (MnO_2), Fricke gel dosimeter (FXG) and dyed polimethylmethacrylate radiochromic films. The optical absorption (OA) measures were chosen to study the dosimetric properties and evaluate the advantages and disadvantages of each dosimeter. The main dosimetric properties studied were: incident radiation energy and angle dependence, ambient conditions influence and dose rate response; stability before and after irradiation; reproducibility and response accuracy. The preparation method, the cost of raw materials and use easiness were evaluated. The obtained results shown that the addition of MnO_2 in the KNO_3 pellets extend the dose range of KNO_3 dosimeter from between 1 and 150 kGy to 150

[*] E-mail address: sgalante@ipen.br;

and 600 kGy. PMMA films produced using dye Macrolex® red 5B can be applied to measure dose range between 5 and 100 kGy, bromocresol green solution doses from 1 to 15 kGy and Fricke gel dosimeter (FXG) presents useful dose a range between 5 and 50 Gy. These four dosimeters are of easy obtaining, low cost and simple use and can be applied to ^{60}Co monitoring in industrial irradiation processes routine dosimetry.

Introduction

The XIX century marks the beginning of the atomic physics development and two discoveries contributed so that this occurred: in 1895, the German physicist Wilhelm Konrad Röentgen discovers a radiation of unknown nature calling it of radiation X or X-rays. One year later, the French physicist Antonie Henry Becquerel intensifies its studies with the uranium mineral and discovers radioactive emissions that can penetrate in the matter. The Polish physicist Marie Sklodowska Curie intrigued with Becquerel's discovery fractionated the uranium mineral and isolated two new elements, the radium and the polonium. The knowledge of the radioactive decay was not work of only one person, but it is contribution resulted of many studious [1].

Soon after the discoveries of X-rays and radioactivity it was observed that radiation could cause changes in matter and damage to health. The radiation chemical effect depends on the matter composition and the amount of deposited energy by the radiation.

Radiation processing is a rapidly growing industry involving the improvement of several materials by ionizing radiation: electron beam; X-rays; gamma radiation [2, 3].

Several dosimetric materials such as alanine [4,5], LiF [6-8], radiochromic films [9,10] and organic dyed polymer[11] are examples of solid dosimeters; ceric-cerous solution, ferrous - ceric solution, ferrous sulphate solution (Fricke dosimeter) [12,13] are liquid dosimeters and nitrogen oxide and carbon dioxide [14-16] are examples of gaseous dosimeters, among others, and different techniques thermoluminescence [17], spectrophotometry [18,19], electronic paramagnetic resonance [20,21], magnetic resonance image [22-24] has been studied and applied to optimize the quality control used in radiation processing.

The High Doses Laboratory - HDL of the Instituto de Pesquisas Energéticas e Nucleares – IPEN-CNEN/SP develops materials to be applied in high doses dosimetry. The purpose of this research is to obtain materials of simple preparation method, low cost of raw materials and use easiness that can improve the quality control of radiation process. In the last years different materials such as potassium nitrate (KNO_3) pellets pure and mixed with manganese dioxide (MnO_2), bromocresol green solutions, Fricke gel dosimeter (FXG) and dyed polimethylmethacrylate radiochromic films were developed and characterized to be applied as dosimetric material. Results on gamma radiation decomposition of mixtures that contain KNO_3 and compounds such as $Ba(NO_3)_2$, KBr and MnO_2, a new radiation-sensitive indicator in aqueous solution form, the development and characterization of a new PMMA detector produced in Brazil and a Fricke gel dosimeter performance are reported. The studied parameters were: absorption spectrum and signal stability of irradiated and non-irradiated detectors; batch reproducibility; effect of the environmental conditions, dose-response useful range and dose rate dependence response [25-28].

Radiation Dosimetry

The dosimeters are classified as primary, reference, transfer and routine according to their intrinsic accuracy and application. Primary standards enable an absolute measurement of absorbed dose and are held by national standards laboratories; there are two types – ionizing chambers and calorimeters. A reference dosimeter is defined as a dosimeter of high metrological quality that can be used as a reference standard to calibrate other dosimeters; examples of dosimeters include chemical dosimeters, such as the Fricke, ceric-cerous, dichromate solutions and alanine dosimeters. A transfer dosimeter is used for transferring dose information from an accredited or national laboratory to an irradiation facility in order to establish traceability to that standards laboratory; example alanine, Fricke and dichromate solutions. A routine dosimeter is a dosimeter whose performance is often not as good as that of a reference dosimeter, but whose cost and ease of use make it suitable for day-to-day measurements in a radiation processing facility; examples of commonly used routine dosimeters are systems based on polymethylmethacrylate (PMMA), both dyed and un-dyed, cellulose tri-acetate and thin radiochromic films. The ideal dosimeter must possess adequate reproducibility and stability before and after irradiation, independence response on dose rate, on radiation incidence angle and incident radiation energy, environmental conditions stability and simple preparation and analysis [29].

Chemical Dosimetry

In chemical dosimetry the radiation dose is determined from quantitative chemical changes in the irradiated medium that can be solid, gaseous and liquid [30-33].

Solid systems: The solid state materials consist normally of organic or inorganic crystalline compounds (amino acids [4, 5], lithium fluoride [6-8]), amorphous or almost crystalline (ceramics [6, 34], glasses [35-37], plastics dyed [11] and semiconductors [38, 39]). In routine activities it is more interesting to use the solid dosimeters, because they present adequate dimensions and mechanical resistance to the hard works in industrial radiation processing. The measurement techniques used are spectrophotometry, thermoluminescence, electron spin resonance and lyoluminescence.

Dosimetric systems based on radiolytic decomposition of inorganic nitrates are used in quality control programmes of radiation processing. The final products formed are NO_2^- and O_2 that can be related with the dose. The radiation induced effect of added inorganic compounds on nitrates pellets has been studied. When a sample is exposed to ionising radiation the energy is absorbed by all the components present in the system such as cation, anions and the added substances. The enhancement/retardation in the decomposition of nitrates by various additives has been explained in terms of their electron donor/acceptor properties [40-48].

Polymethylmethacrylate (PMMA) dosimetry systems are commonly applied in industrial radiation processing [49]. The PMMA detector provides a means of directly estimating the absorbed dose. Under the influence of ionizing radiation chemical reactions take place in the material, creating and/or enhancing absorption bands in the visible and/or ultraviolet regions of the spectrum. The absorbed dose is determined by using an experimental calibration curve obtained by irradiating of a set of detectors with known absorbed doses aiming the quality control of the radiation process and presenting excellent results. Dyed PMMA detectors produced at the High Doses Laboratory of IPEN using yellow, green, blue and red coloring compounds were previously characterized based on ^{60}Co gamma response. Their responses were compared with red 4034 Perspex and gammachrome YR detectors commercially available using the optical absorption (OA) technique [27, 28].

Gaseous systems: The gaseous systems have been applied for the quality control of irradiations of gases. Radiolysis of a gas may generate ions, excited atoms and molecules and free radicals. Many experimental techniques have been employed to study gas-phase reaction mechanisms, including permanent product analysis, measurements of ion currents, mass spectrometry, pulse radiolysis, the effects of scavengers, and isotopic labeling. The frequent types of reactions include direct formation of electronically excited molecules [14-16].

Liquids systems: When an ionizing radiation, i.e., gamma radiation or high-energy electrons, passes through a liquid medium, water, organic solvent or a combined mixture of water and organic solvent, ionizations and excitations occur along the trajectory. Aqueous solutions are the most used dosimeters to evaluate the absorbed dose. The dosimetric system is based on the reactions of solutes with the species formed in the radiolysis of water, as example can be cited – ferrous sulphate dosimeter (Fricke dosimeter); ferrous-cupric sulphate dosimeter; ceric sulphate (or ceric-cerous sulphate dosimeter; dichromate dosimeter; ethanol-chlorobenzene dosimeter and radiochromic dye solutions [33].

The radiation-chemical reactions with the water, produces free radicals and molecular products. Hydrogen atoms (H), hydroxyl radicals (OH) and hydrated electrons (e^-_{aq}) constitute the free radicals group; hydrogen (H_2) and hydrogen peroxide (H_2O_2) constitute the molecular products group [32].

The chemical species formed in an aqueous solution are the following:

$$H_2O \rightarrow H_2O^+ + e^-_{aq} \quad (1)$$
$$H_2O^+ \rightarrow H^+_{(aq)} + OH^\bullet \quad (2)$$
$$e^-_{aq} + H_2O \rightarrow H_2O^- \quad (3)$$
$$H_2O^- + H_2O \rightarrow H^\bullet + OH^- \quad (4)$$
$$H^\bullet + H^\bullet \rightarrow H_2 \quad (5)$$
$$OH^\bullet + OH^\bullet \rightarrow H_2O_2 \quad (6)$$
$$H^\bullet + OH^\bullet \rightarrow H_2O \quad (7)$$
$$H^\bullet + H_2O \rightarrow H_2 + OH^\bullet \quad (8)$$
$$H_2O_2 + OH^\bullet \rightarrow H_2O + HO_2^\bullet \quad (9)$$
$$H^\bullet + O_2H^\bullet \rightarrow HO_2^\bullet \quad (10)$$
$$HO_2^\bullet + H^\bullet \rightarrow H_2O_2 \quad (11)$$
$$R^\bullet + O_2 \rightarrow RO_2^\bullet \quad (12)$$

The Fricke dosimeter systems provides a reliable mean of absorbed doses measurement of in water, based on a process of oxidation of ferrous ions to ferric ions in acidic aqueous solutions by ionizing radiation [29].

When the gamma radiation interacts with Fe^{2+} ions transforms them into Fe^{3+} ions. The involved reactions are:

(1)	$Fe^{2+} + OH^{\bullet} \rightarrow Fe^{3+} + OH^{-}$
(2)	$H^{\bullet} + O_2 \rightarrow HO_2^{\bullet}$
(3)	$Fe^{2+} + HO_2^{\bullet} \rightarrow Fe^{3+} + HO_2^{-}$
(4)	$HO_2^{-} + H^{+} \rightarrow H_2O_2$
(5)	$Fe^{+2} + H_2O_2 \rightarrow Fe^{3+} + OH^{-} + OH^{\bullet}$

Each hydrogen atom forms a hydroperoxy radical, HO_2, and each of these radicals oxidizes three Fe^{2+} ions; one by reaction (3) and two by reactions (4), (5) and (1). Each hydroxyl radical oxidizes one Fe^{2+} ion, and each hydrogen peroxide molecule oxidizes two Fe^{2+} ions All reactions, except (5) are fast [32]. The organic impurities presence in the solution cause significant variation in the dosimeter response, for this reason it is essential to use very pure compounds.

Conventional Fricke dosimeter modified with gelatinous agent and the technique of image generation for Magnetic Resonance (MRI) has been used in dosimetric practices. The change of longitudinal proton relaxation time, T_1, due to the oxidation of ferrous ions (Fe^{2+}) to ferric ions (Fe^{3+}) in Fricke gel dosimeter after irradiation can be used for the dose distributions 3D determination [50-57].

An encapsulated liquid can be used in the dose control in irradiation process. If the reaction products radiation induced are stable and the changes in the color can be direct measured, with consequent variation in the optical absorption, and it is associated with the absorbed dose, this solution can be used us a dosimetric system. Dyed liquid systems have an advantage of being commercially available, cheap and easily handled, prepared and measured. The bromocresol green solution, an indicator of reaction, was prepared and characterized as a radiochromic dosimeter.

Methodology

Dosimeters Preparation

Potassium Nitrate

Pellets of KNO3 pure with different mass were cold pressed and the dosimetric properties were analyzed. It is known that same additives can increase the useful dose range of KNO_3 dosimeters. For this reason addition of compounds such potassium bromide (KBr), barium nitrate ($Ba(NO_3)_2$) and manganese dioxide (MnO_2) were studied. The *potassium nitrate pure* in powder form was dried in an oven at 80°C, samples were prepared by mixing required weights of each compound and grinding in agate mortar to obtain uniform grain size of 80 mesh. The homogeneous mixture was cold pressed in pellet form with diameter of 6 mm, using a Fred Frey press (model FC5; n° 2715) with applied load of 5 tons. The composition of the different mixtures is presented in the Table 1(a), (b), (c) and (d). Before irradiation the pellets were set in a badge, Figure 1, that consists of 2 Lucite® plates of 3 mm thick, that guarantee the electronic equilibrium condition to ^{60}Co gamma radiation, and 1 Lucite® plate of 1 mm thick with three holes, where the pellets are placed.

Table 1. Pellets composition of mixtures different of KNO_3 and additives.

(a)		
Pellet Mass (mg)	**-**	**KNO_3 %**
50	-	100
75	-	100
100	-	100
(b)		
Pellet mass (mg)	**MnO_2 %**	**KNO_3 %**
55.5	10	90
62.5	20	80
71.4	30	70
50.0	40	60
50.0	50	50
50.0	60	40
(c)		
Pellet mass (mg)	**KBr %**	**KNO_3 %**
50	40	60
50	50	50
50	60	40
(d)		
Pellet mass (mg)	**$Ba(NO_3)_2$ %**	**KNO_3 %**
50	40	60
50	50	50
50	60	40

(a) KNO_3 pure
(b)$KNO_3 + MnO_2$
(c) KNO_3 + KBr
(d) $KNO_3 + Ba(NO_3)_2$

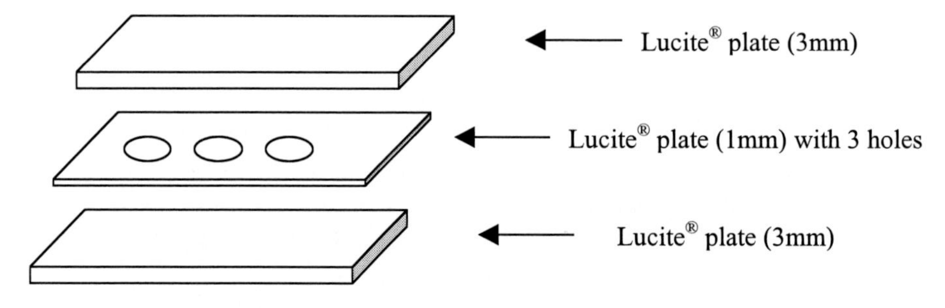

Figure 1. Schematic representation of the KNO_3 badge.

Bromocresol green solutions - VBC

The VBC is used in chemical analyses as acid - basic indicator and presents color change from yellow (pH 3,8) to blue (pH 5.4). Solutions with different VBC concentrations, 0.01%, 0.0075% and 0.005% in mass were prepared and evaluated. The indicator was dissolved in 5% ethyl alcohol and then added in right pureness water in the desired concentration. After dissolution the solutions were conditioned in amber glass bottles and stored in dark environment and room temperature.

Polimethylmethacrylate

A private firm using the dyes yellow, green, blue and red from Bayer produced PMMA sheets (120x60 cm^2 and thickness between 2 to 2.5 mm) especially to be used in this work. The PMMA sheets were cut into strips (10 mm width and 30 mm length) and sealed in special aluminum / paper / polyethylene pouches. The commercial Red 4034 Perspex and Gammachrome YR dosimeters were used for comparison with the new PMMA dosimeters.

Fricke gel dosimeter (FXG)

The Fricke gel (FXG) solutions were prepared using 50 mM sulphuric acid, 1 mM ferrous ammonium sulphate, 1 mM sodium chloride, 0.1 mM xylenol orange and 5% of porcine gelatine 270 Bloom. Sealed acrylic cuvettes of 10 mm optical path, 45 mm height, 1 mm wall thickness filled with Fricke gel solutions were used to calibrate the spectrophotometric and MRI signal intensity. A phantom in the breast shape was produced with FXG solution. The samples were always stored in dark environment at 5°C during 12 h to complete solidification.

Irradiation Facilities

Most of the irradiations were performed in electronic equilibrium conditions and room temperature and humidity using a Gammacell 220 source (38.9 x 10^{13} Bq) and a JOB-188 Dynamitron Inc. electron accelerator installed at the Radiation Technology Center CTR / IPEN-CNEN/SP.

Nitrate Potassium

Irradiations were performed using a ^{60}Co Gammacell source in the dose range between 1 and 150 kGy. The dose rate was approximately 7.48 kGy /h. The dose rate can be changed depending on the distance source-dosimeter.

Bromocresol green solutions

The solutions were irradiated in air at electronic equilibrium conditions with ^{60}Co Gammacell radiation and doses between 50 Gy and 15 kGy.

Polymethylmethacrylate

The PMMA dosimeters were irradiated with ^{60}Co Gammacell and electron from accelerator in air at electronic equilibrium. The absorbed dose rate corresponding to 4.9 kGy/h, was determined by Fricke dosimeter in the Gammacell source.

Fricke gel dosimeter

The gamma irradiations were performed in air at electronic equilibrium conditions using a Gammacell source with doses between 5 and 50 Gy and dose rate of 3.28 kGy/h.

Dose Evaluation

Each related point is the average of 3 measurements and the error bars the standard deviation of the mean (1σ).

Nitrate Potassium

For optical measurements potassium nitrate pellets were dissolved in 50 mL of high purity water in a 100 mL volumetric flask. After dissolution 20 mL of a coloring solution, composed by solutions A and B in the proportion of 5:1 was added. The solution A was prepared dissolving 2 g of sulfanilamide p.a. in 1 L of 30% of glacial acetic acid p.a. solution and solution B was prepared dissolving 1 g of N – 1 - Naphtylethylene - diamine dihydrochloride p.a. in 1 L of 30% of glacial acetic acid p.a. solution. Ten minutes after dissolution the absorbance was measured at 546 nm wavelength [58,59]. The absorbance of the solutions of irradiated and non-irradiated pellets of KNO_3 was measured against air, at the wavelength range from 500 to 600 nm using a Shimadzu UV1601PC (potassium nitrate pure) and UV-2101PC (mixtures) spectrophotometer.

Bromocresol green solutions

The solutions were evaluated immediately after irradiation using a Shimadzu UV1601PC spectrophotometer.

Polymethylmethacrylate

The absorbance changes were measured immediately after irradiation with a Shimadzu UV-2101PC spectrophotometer at the wavelength of maximum absorbance or maximum transmittance, depending on the dosimeter absorption spectrum at the wavelength range from 190 to 900 nm.

Fricke gel dosimeter

The Fricke gel solutions were measured immediately after irradiation using Shimadzu UV2101PC spectrophotometer at the wavelength from 190 to 900 nm. The MRI were obtained using a Philips Gyroscan S15/ACS (1.5 T) tomography and quadrature body coil using the IR sequence with following parameters: TR=367 ms, TI=300 ms and TE=18 ms.

Results

Potassium Nitrate

The dosimeter composed of pure potassium nitrate presents useful dose range between 5 and 150 kGy. The calibration curve of the dosimeter irradiated with gamma radiations is shown in Figure 2.

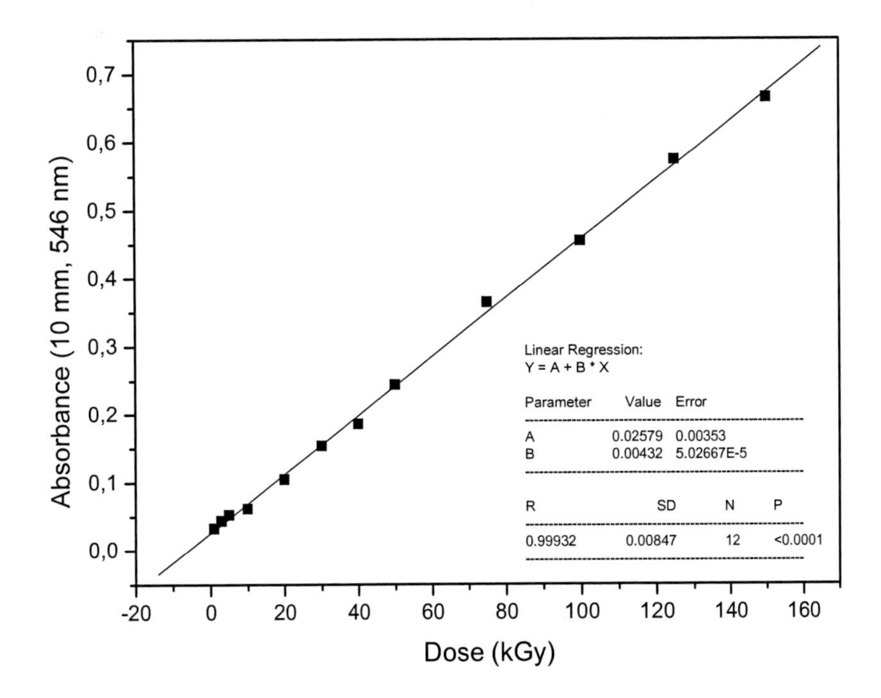

Figure 2. Potassium nitrate dosimeter calibration curve. Irradiation - 60Co gamma radiation and measurement - 546nm.

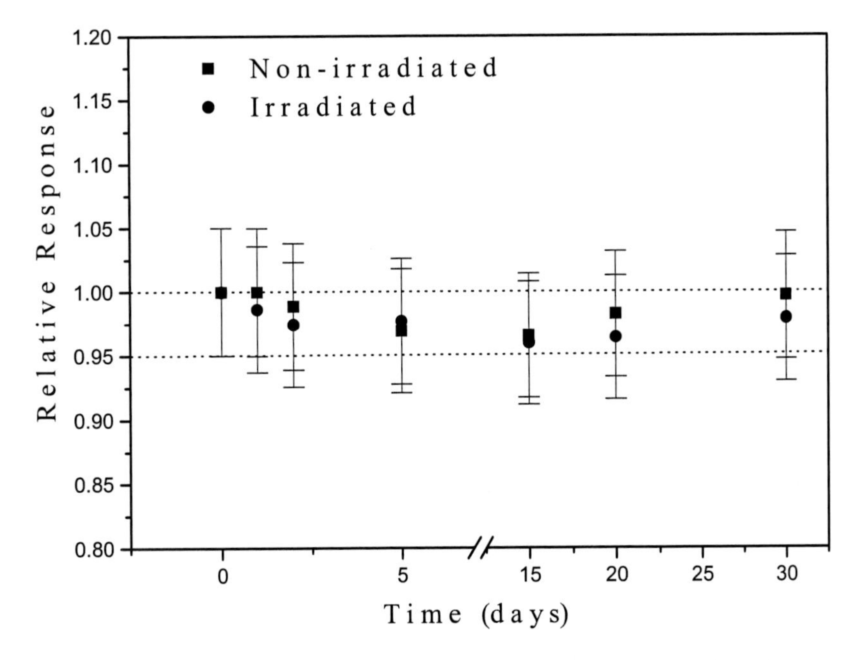

Figure 3. Relative response obtained with solutions (40% MnO_2 / 60% KNO_3) evaluated immediately after preparation and using pellets irradiated with 200 kGy and stored up to 30 days.

The maximum color intensity of the solution is obtained 10 minutes after preparation. The solutions are stable during a period of 1 h that permits the dose evaluation with high quality guarantee. The pellets irradiated and non-irradiated can be stored by long periods without any change in their characteristics.The results obtained with solutions evaluated immediately after preparation, using pellets irradiated and stored up to 30 days are shown in the Figure 3. The results obtained with solutions prepared at same time and evaluated periodically during 1 month of storage are shown in the Figure 4.

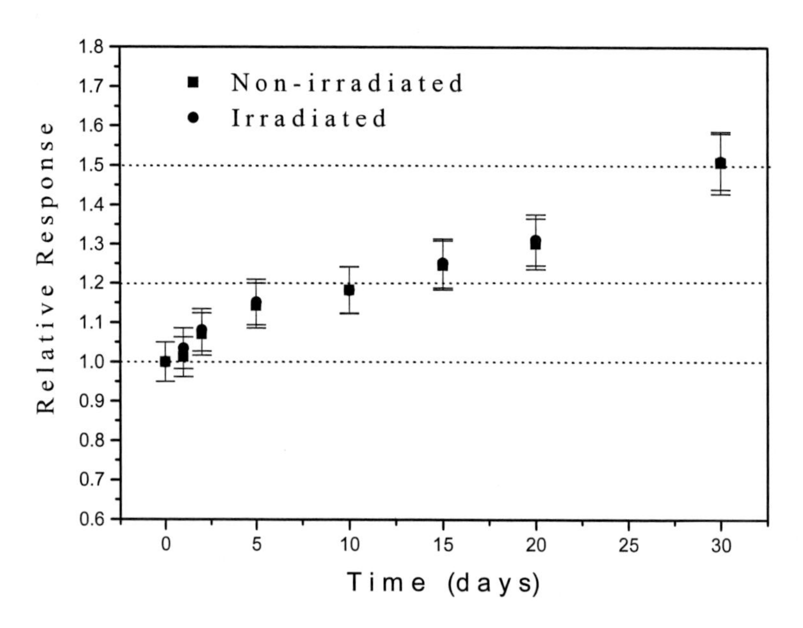

Figure 4. Relative response obtained with solutions prepared using pellets (40% MnO_2 / 60% KNO_3) irradiated with 200kGy, prepared and evaluated after up to 30 days of storage.

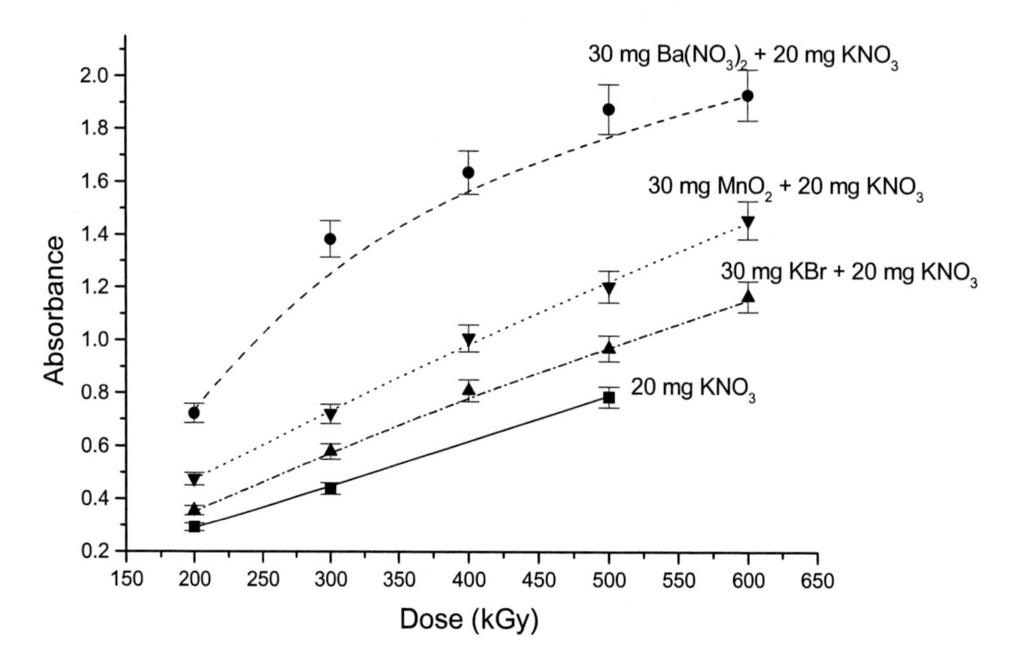

Figure 5. Dose-response (λ=540 nm) for mixtures of 30 mg $Ba(NO_3)_2$, KBr and MnO_2 and 20 mg KNO_3 irradiated with ^{60}Co gamma radiation.

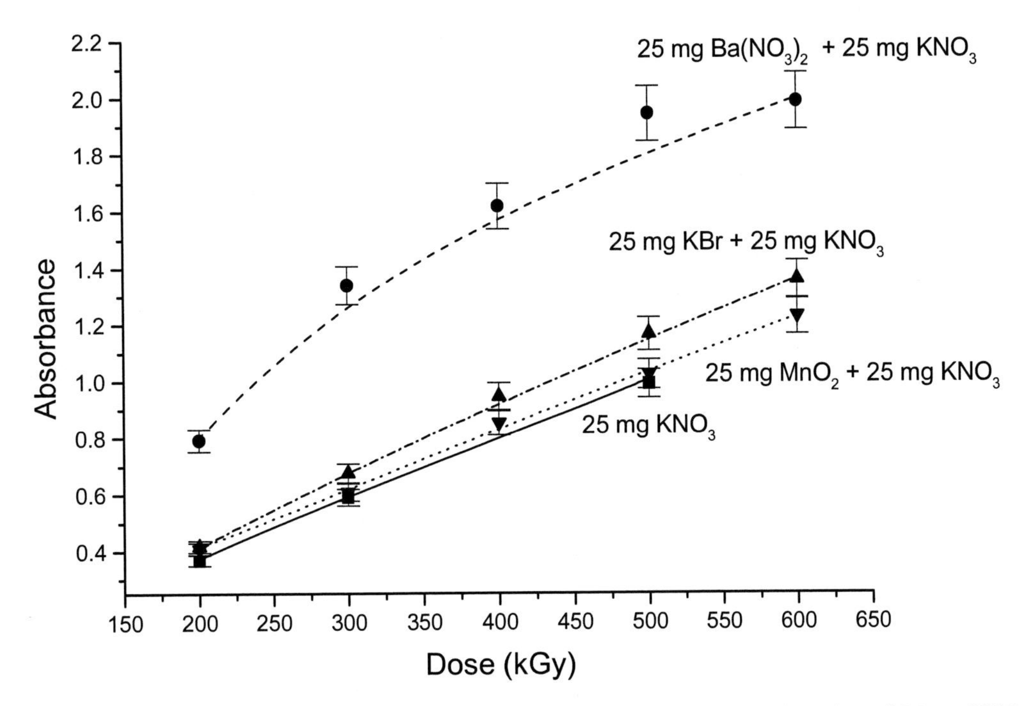

Figure 6. Dose-response (λ=540 nm) for mixtures of 25 mg $Ba(NO_3)_2$, KBr and MnO_2 and 25 mg KNO_3 irradiated with ^{60}Co gamma radiation.

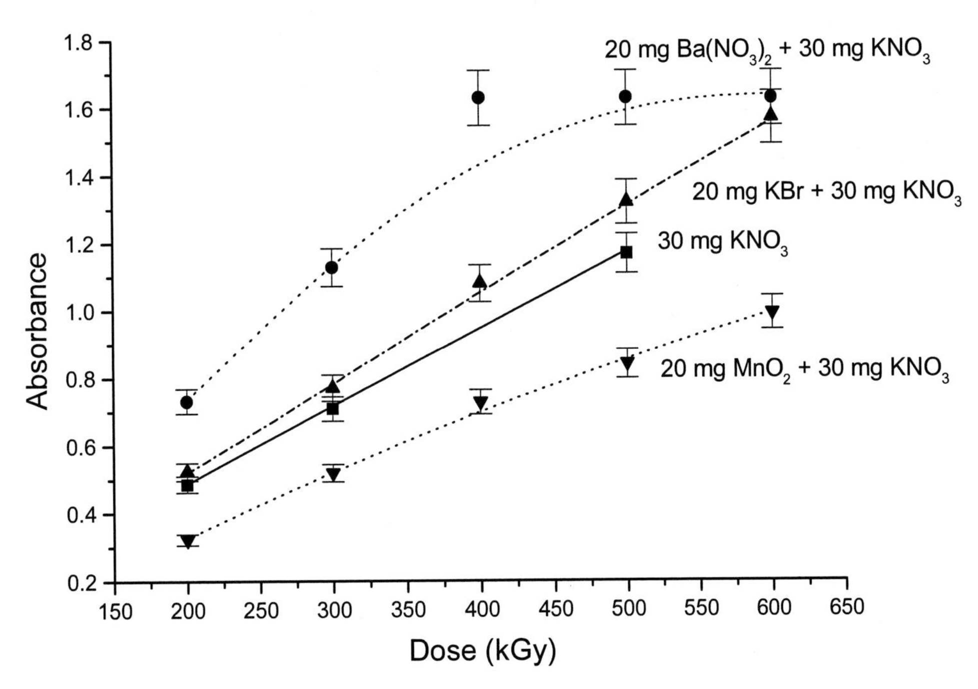

Figure 7. Dose-response (λ=540 nm) for mixtures of 20 mg $Ba(NO_3)_2$, KBr and MnO_2 and 30 mg KNO_3 irradiated with ^{60}Co gamma radiation.

The absorbance response of KNO_3 pure pellets and pellets of different compounds were evaluated for absorbed doses from 200 kGy to 600 kGy. The results are shown in Figures 5, 6 and 7 for the mixtures $Ba(NO_3)_2/KNO_3$, KBr/KNO_3 and MnO_2/KNO_3 in different concentrations.

The method reproducibility was studied preparing different batches of pellets and evaluating the optical response for different doses. The reproducibility was found to be better than 98% (1σ) from batch to batch.

The mixture of 20 mg MnO_2 / 30 mg KNO_3 presents better results in terms of decomposition rate. The decomposition rate of nitrate ions into nitrite ions is reduced, extending the useful range. In the case the dose range was extended up to 600 kGy, without signal saturation.

No dose rate dependence response in the range from 1.70 to 5.65 kGy/h was observed. The fluctuation observed was better than 95%.

Polymethylmethacrylate

The absorption spectrum obtained from non-irradiated PMMA dosimeters are shown in Figure 8 (a), (b), (c) and (d).

The batch reproducibility was evaluated and was found to be better than 98% from batch to batch, represented by a sheet produced in the dimensions 60 x 120 cm^2.

The effects of room temperature, humidity and light on the dosimeters response were studied.

Different humidity conditions between 0 and 100% were obtained by means of saturated salt solutions that are very useful in producing known relative humidity. Non-irradiated samples were exposed, during several hours, to atmospheres with relative humidity (R.H.) of 0%, 75.5%, 93.0% and 100.0%. The samples that contains the green dye were the ones that presented smaller variation in the optical response, the absorbance intensity decrease was less than 2.5% however, significant variation was not observed in the other samples, the largest variation found was ± 5.0% for the red H and red G dyed samples, for the standard dosimeter was found ± 5.0% of variation for Red Perspex and an absorbance value decrease of 10.0% for Gammachrome YR. To determine the effect of R.H. on irradiated samples, the samples were exposed to different R.H. conditions during 24 hours and, after the exposition were irradiated in the Gammacell source with dose of 20 kGy, in the same R.H conditions. The obtained responses are shown in the Figure 9.

The non-irradiated and irradiated samples were submitted to thermal treatments of 4, 40, 80 and 100°C during one hour. After temperature stabilization the absorbance of the non-treated sample (samples maintained at room temperature) and treated samples were measured. In the Figure 10 and Figure 11 are shown the relative responses obtained for non-irradiated and irradiated (20 kGy) PMMA samples, respectively. The presented values of each treated dye sample were normalized to the respective response of non treated sample.

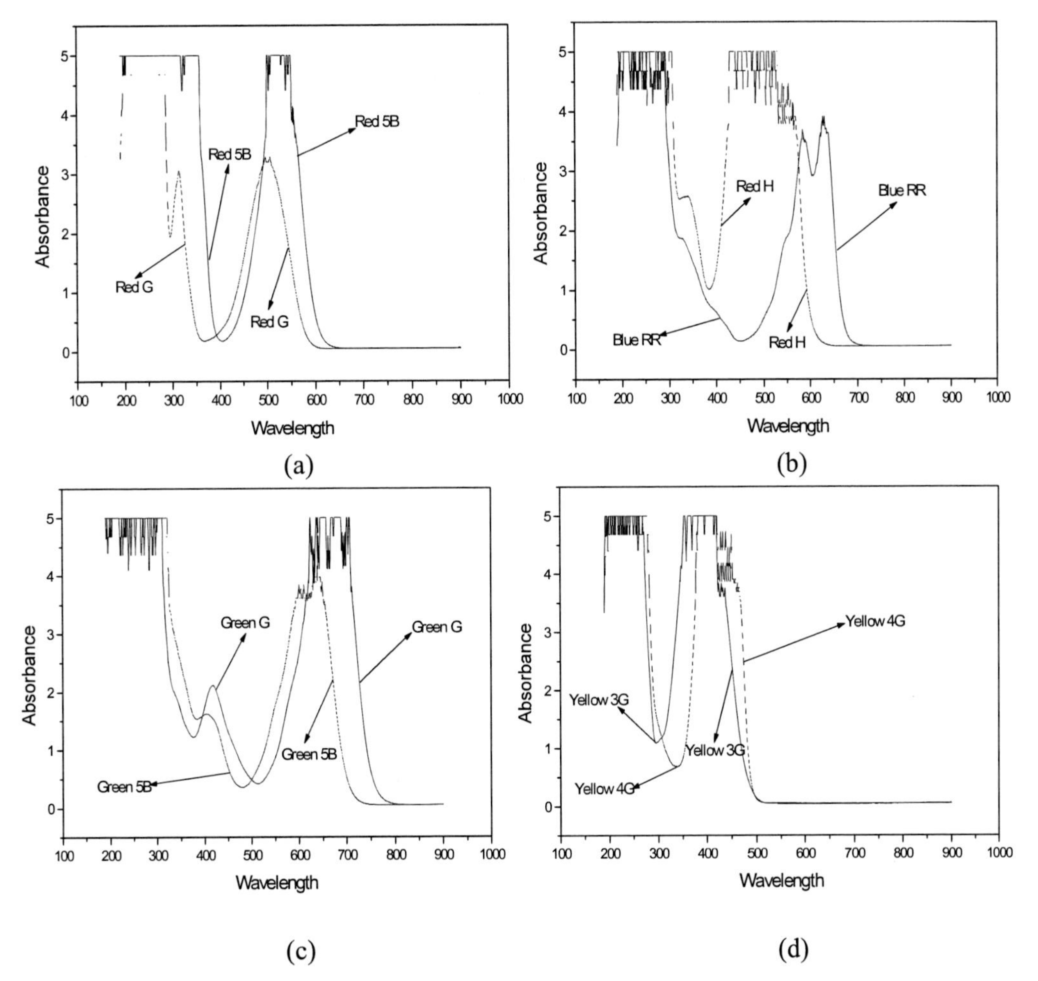

Figure 8. Absorption spectrum of the non-irradiated dyed PMMA dosimeters.
(a) Red 5B and Red G. (b) Red H and Blue RR. (c) Green 5B and Green G PMMA. (d) Yellow 3G and Yellow 4G PMMA.

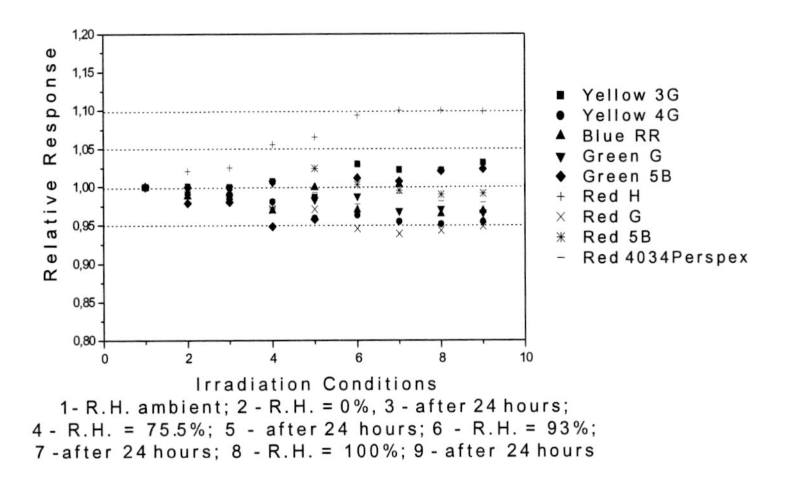

Figure 9. Relative response of the irradiated PMMA samples - dose 20 kGy of ^{60}Co gamma radiation as a function of relative humidity

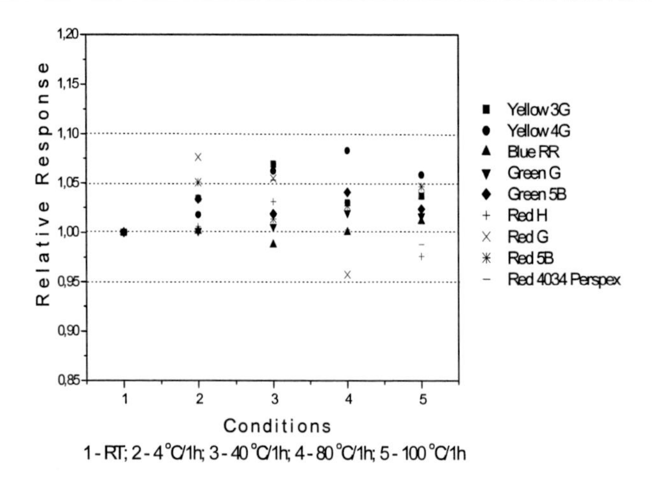

Figure 10. Optical response of non-irradiated PMMA samples maintained at room temperature (RT) and submitted to thermal treatments (TT) between 4 and 100°C /1h.

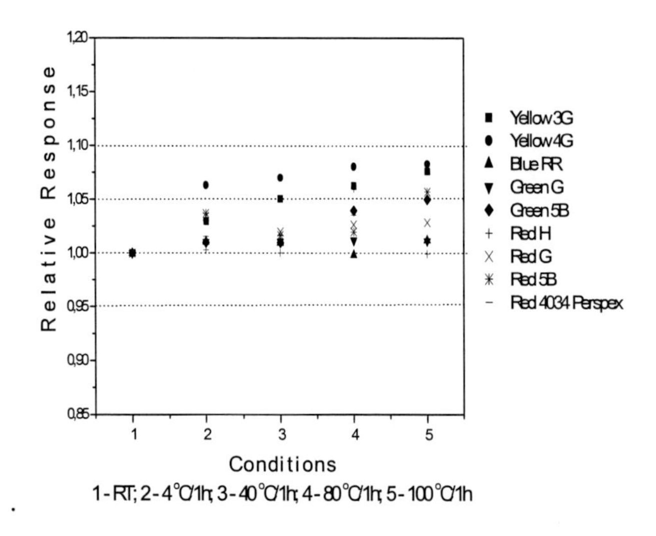

Figure 11. Optical response of irradiated PMMA samples - dose 20kGy maintained at room temperature (RT) and submitted to thermal treatments (TT) between 4 and 100°C /1h.

In the Figure 12 and Figure 13 are shown the relative responses obtained for non-irradiated samples exposed to ambient light, is observed an increase in the response of 5% in two months. Some materials are only photosensitive after irradiation, this way, it is necessary to determine how that factor can influence the performance of the dosimeter in order to avoid or to minimize this influence in the response. The PMMA samples were irradiated using the Gammacell source with doses of 10, 30 and 40 kGy and exposed to the laboratory ambient light. The samples presented media variation of ± 10% in the response, in that way the effect of the incident light in the irradiated samples is significant and appropriate careful has to be taken, packing the dosimeter for instance or applying correction factors. To avoid the application of corrections factors the samples has be maintained in a dark ambient and the measuring realized soon after irradiation or in the maximum between 1 or 2 days after irradiation.

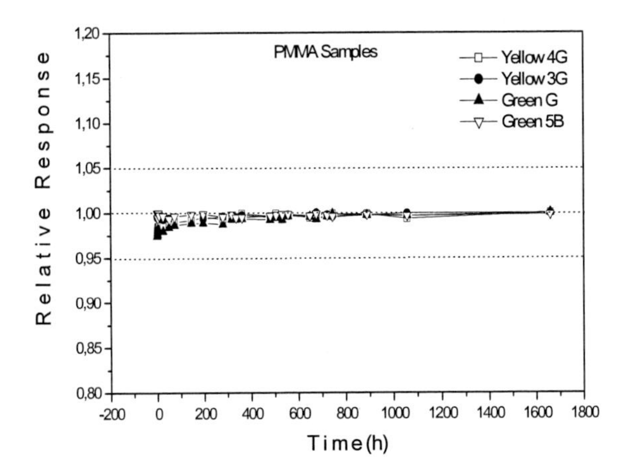

Figure 12. Relative response of non-irradiated PMMA samples – dyed with yellow 4G, yellow 3G, green G and green 5B – exposed to ambient light.

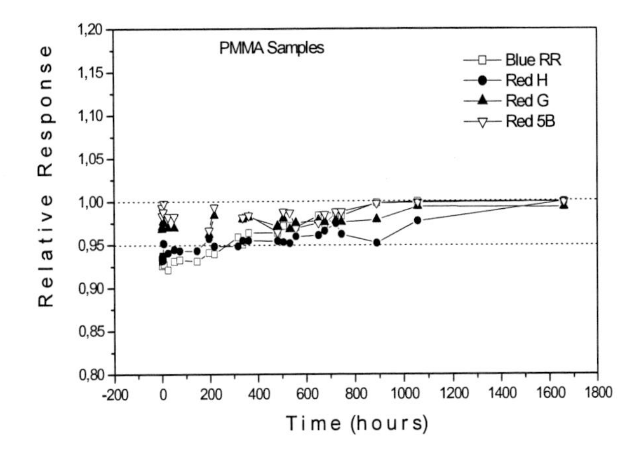

Figure 13. Relative response of non-irradiated PMMA samples – dyed with blue RR, red H, red G and red 5B – exposed to ambient light.

The dosimeters did not present significant variation on the absorbance response when exposed to the normal conditions of light and temperature, therefore, can be manipulated without special care. Although the dosimeters have not presented significant variation on the absorbance response for the different humidity conditions it is desirable that the dosimeters are maintained sealed in pouches.

Figure 14 (a), (b), (c), (d), (e), (f), (g) and (h) show dose-response curves to ^{60}Co gamma irradiation with doses between 0.5 and 100 kGy of the PMMA dosimeters produced with dyes yellow 3G, yellow 4G, blue RR, green G, green 5B, red H, red G and red 5B measured at 300 nm, 353 nm, 450 nm, 528 nm, 405 nm, 620 nm, 397 nm and 415 nm, respectively.

The wavelengths of yellow 3G, yellow 4G, blue RR, green G and red 5B PMMA dosimeters were taken from transmittance spectrum and green 5B, red H and red G PMMA dosimeters were taken from absorbance spectrum. The selected wavelengths presented higher sensitivity as a function of the dose and better stability as a function of environmental conditions.

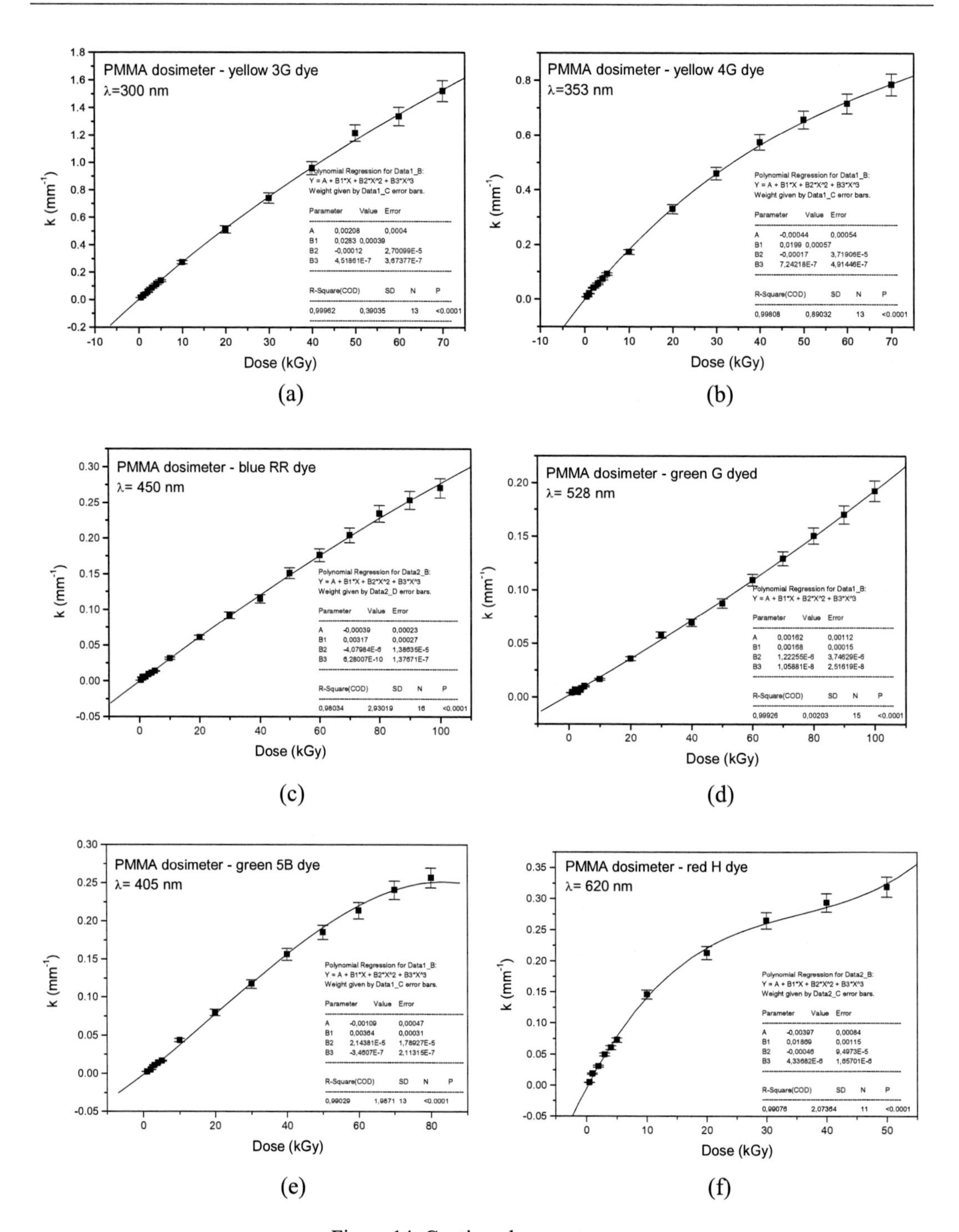

Figure 14. Continued on next page.

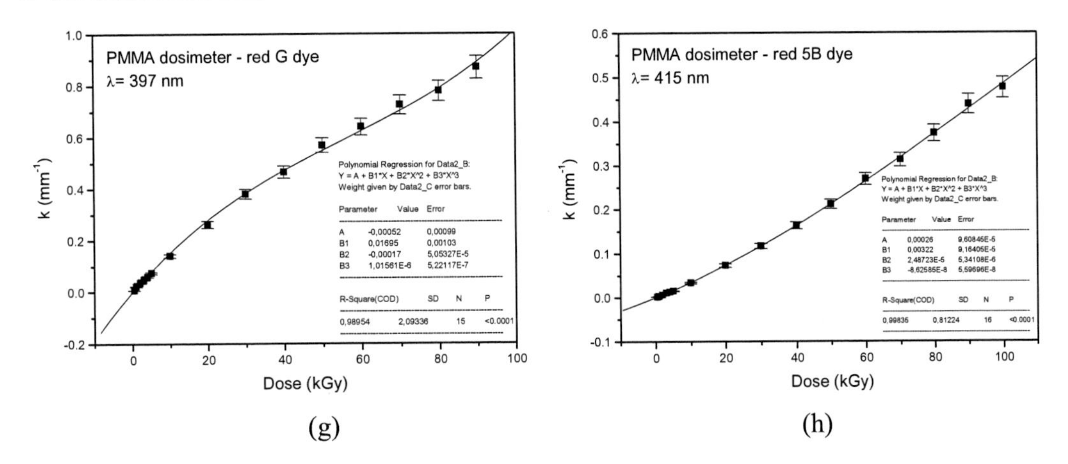

Figure 14. Dose-response curves for PMMA dosimeters irradiated with ^{60}Co gamma radiation.
(a) Yellow 3G, λ = 300 nm. (b) Yellow 4G, λ = 353 nm. (c) Blue RR, λ = 450 nm. (d) Green G, λ = 528 nm. (e) Green 5B, λ = 405 nm. (f) Red H, λ = 620 nm. (g) Red G, λ = 397 nm. (h) Red 5B, λ = 415nm.

Using lead attenuators, the samples were irradiated with 20 kGy of ^{60}Co gamma radiation with dose rates 4.90 kGy/h (without attenuator); 2.45 kGy/h (50% attenuation); 1.47 kGy/h (70% attenuation) and with 0.49 kGy/h (90% attenuation). The relative response of the dosimeters was determined in terms of the absorbance value by thickness unit normalized to the lower dose-rate. The samples containing the yellow 4G, green G, green 5B, red H and red G dye presented behavior similar to the comparative standard Red 4034 Perspex (variation between 5-10%). For the samples containing the yellow 3G, blue RR and red 5B dye the found variation is of approximately 20%, Figures 15 and 16, respectively.

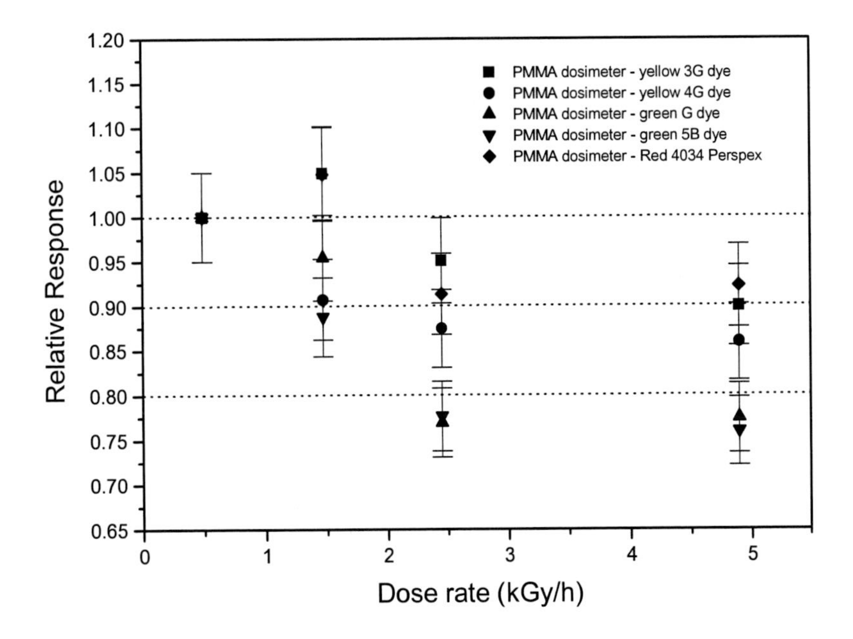

Figure 15. Absorbance dose-rate response of dyed PMMA dosimeters yellow 3G, yellow 4G, green G and green 5B: dose 20 kGy.

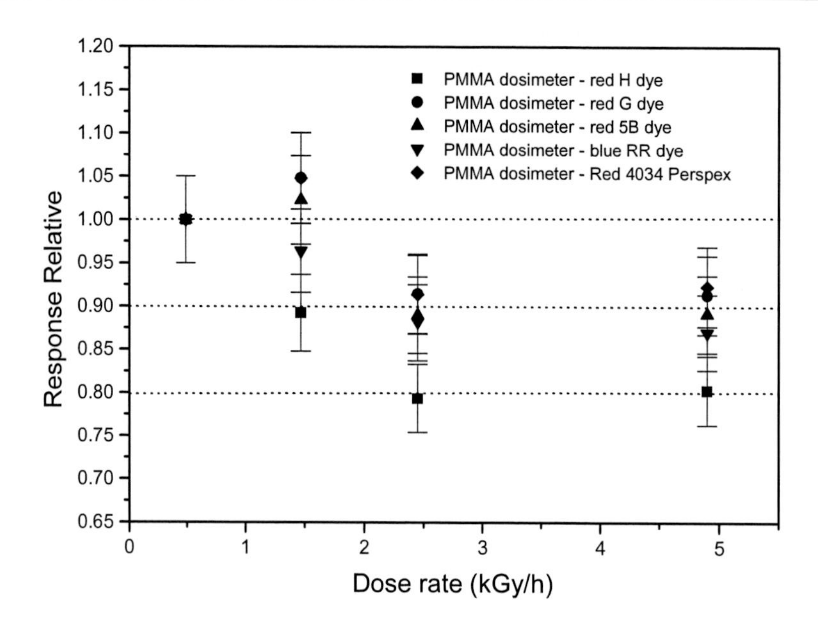

Figure 16. Absorbance dose-rate response of dyed PMMA dosimeters red H, red G, red 5B and blue RR: dose 20 kGy.

The dose response curves of the detectors produced with yellow 3G, yellow 4G, blue RR, green G, green 5B, red H, red G and red 5B, measured at 300 nm, 353 nm, 450 nm, 528 nm, 405 nm, 620 nm, 397 nm and 415 nm, respectively, irradiated in the electron accelerator (energy 1.25 MeV and current 0.6 mA) at doses between 1.2 and 110 kGy are presented in Figure 17 (a), (b), (c), (d), (e), (f), (g) and (h), respectively.

To analyze the electron dose-rate dependence response the samples were irradiated with dose rates 2.66 kGy/s, 22.61 kGy/s and 45.22 kGy/s with electron energy of 1.25 MeV obtained varying the current between 0.6 e 10.2 mA. The results obtained are shown in Figure 18 (a) and (b). The relative response of the detectors was determined in terms of the value of the absorbance by thickness unit normalized to the lower dose-rate.

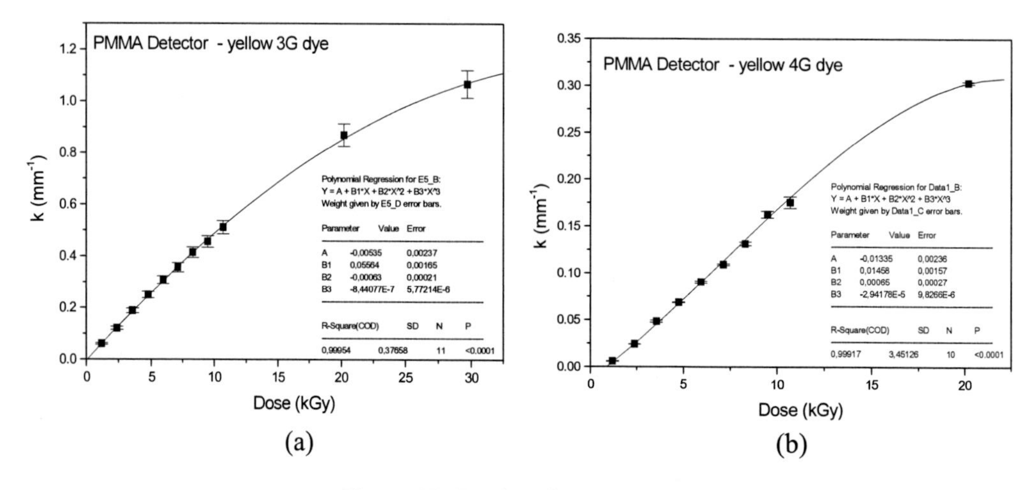

Figure 17. Continued on next page.

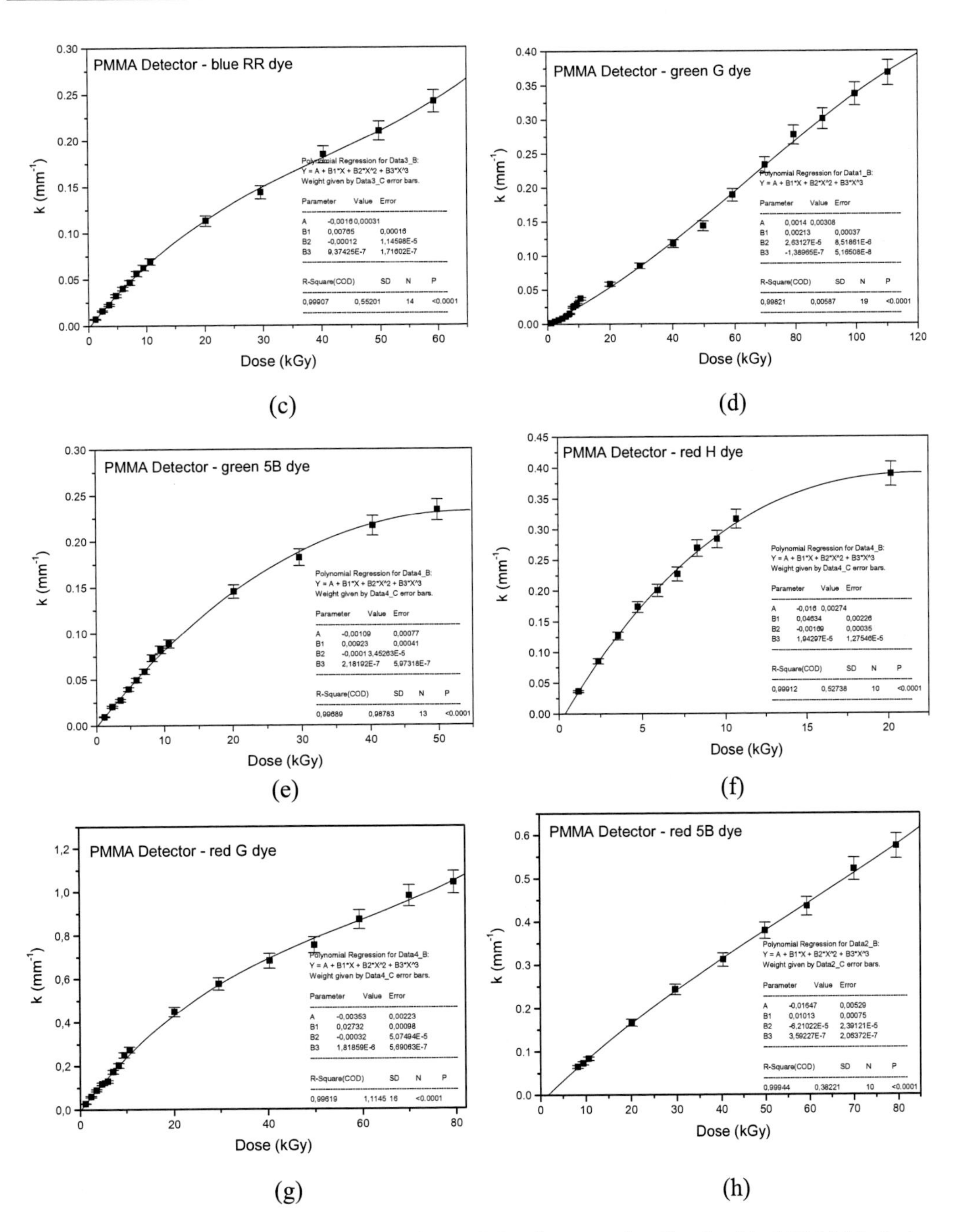

Figure 17 (cont.). Dose-response curves for PMMA detectors irradiated with 1.25 MeV electrons (JOB188).
(a) yellow 3G PMMA detector, at 300 nm. (b) yellow 4G PMMA detector, at 353 nm. (c) blue RR PMMA detector, at 450 nm. (d) green G PMMA detector, at 528 nm. (e) green 5B PMMA detector, at 405 nm. (f) red H PMMA detector, at 620 nm. (g) red G PMMA detector, at 397 nm. (h) red 5B PMMA detector, at 415nm.

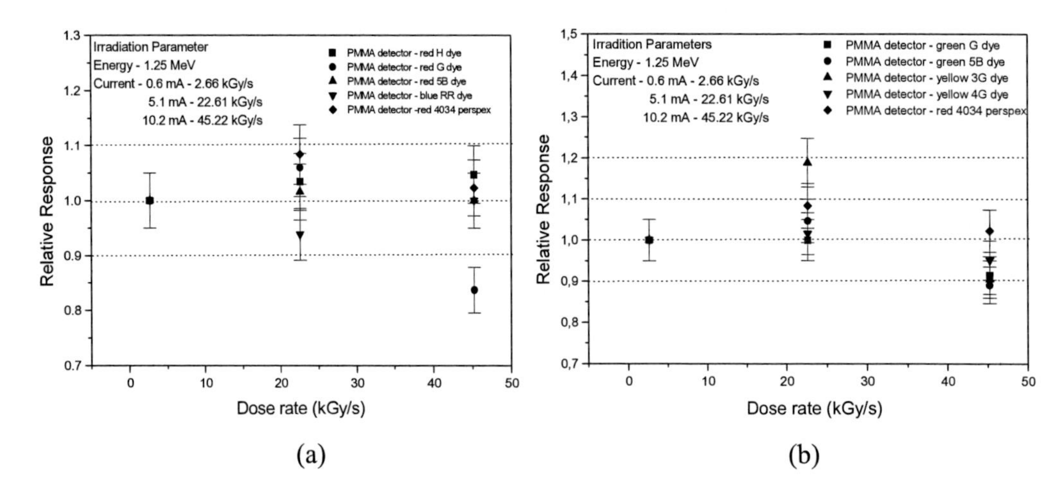

Figure 18. Electron dose-rate dependence response. (a)PMMA detectors – Dye red H (λ=620 nm); dye red G (λ=397 nm); dye red 5B (λ=415 nm) and dye blue RR (λ=450 nm); red 4034 Perspex. (b) PMMA detectors – Dye yellow 3G (λ=300 nm); dye yellow 4G (λ=353 nm); dye green G (λ=528 nm) and dye green 5B (λ= 405 nm); red 4034 Perspex.

The electron energy dependence response was determined using samples 2.4 mm thick. The samples were irradiated with electron energies of 0.8, 1.25 and 1.5 and current of 0.6 mA. The results obtained are show in Figure 19 (a) and (b). For measurements using detector thickness of 1.2, 2.5, 3.1, 4.3 and 5.6 mm and electron energies of 0.8, 1.25, 1.5 MeV no energy dependence was observed when the thickness was maintained inside of the electron range. The relative response of the detectors was determined in terms of absorbance value by thickness unit normalized to the lower energy.

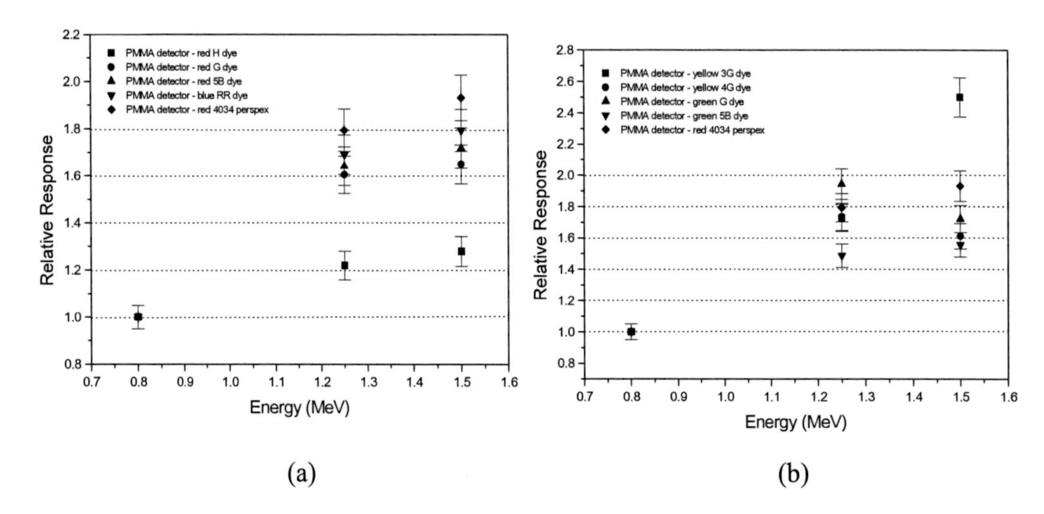

Figure 19. Electron energy dependence response. (a)PMMA detectors – Dye red H (λ=620 nm); dye red G (λ=397 nm); dye red 5B (λ=415 nm), dye blue RR (λ=450 nm) and red 4034 Perspex (λ=640 nm). (b) PMMA detectors – Dye yellow 3G (λ=300 nm); dye yellow 4G (λ=353 nm); dye green G (λ=528 nm), dye green 5B (λ= 405 nm) and red 4034 Perspex (λ=640 nm).

Bromocresol green

The absorption spectra of non-irradiated solutions prepared with different bromocresol green concentrations between 0.005 and 0.01% are shown in Figure 20, two absorption bands with maximums at 450 and 620 nm can be observed.

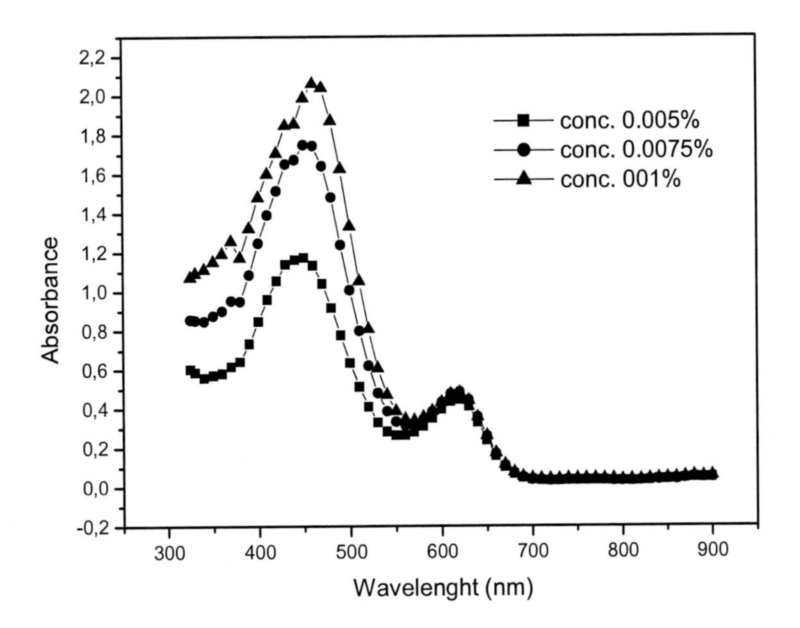

Figure 20. Bromocresol green aqueous solution absorption spectra: non-irradiated samples.

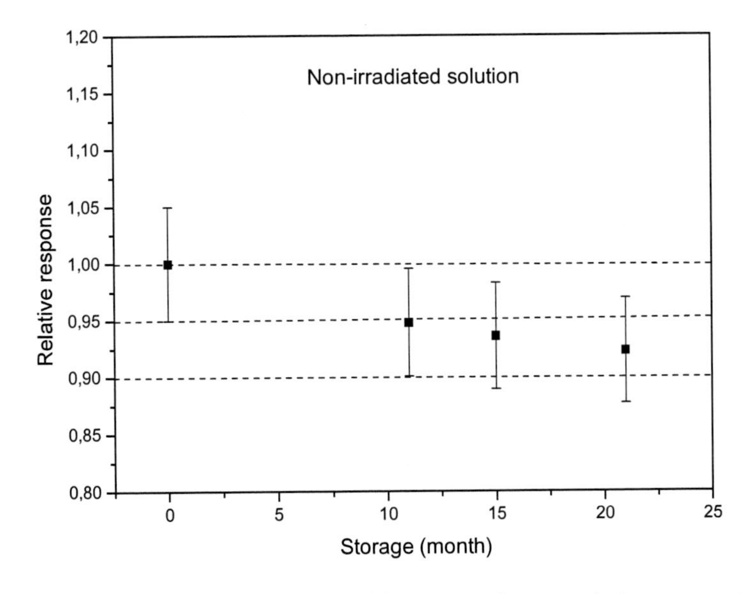

Figure 21. Response stability of non-irradiated bromocresol green solution at room temperature.

Non-irradiated aqueous solutions of bromocresol green were stored during 21 months under room temperature and evaluated in different times after preparation. No significant variations were observed as shown in the Figure 21. These results indicate that it is not necessary fresh solutions for each experiment, the solutions are stable for longer storage time. The post-irradiation stability was observed during a period of about 48 hours, no absorbance change larger than 3% was observed, Figure 22.

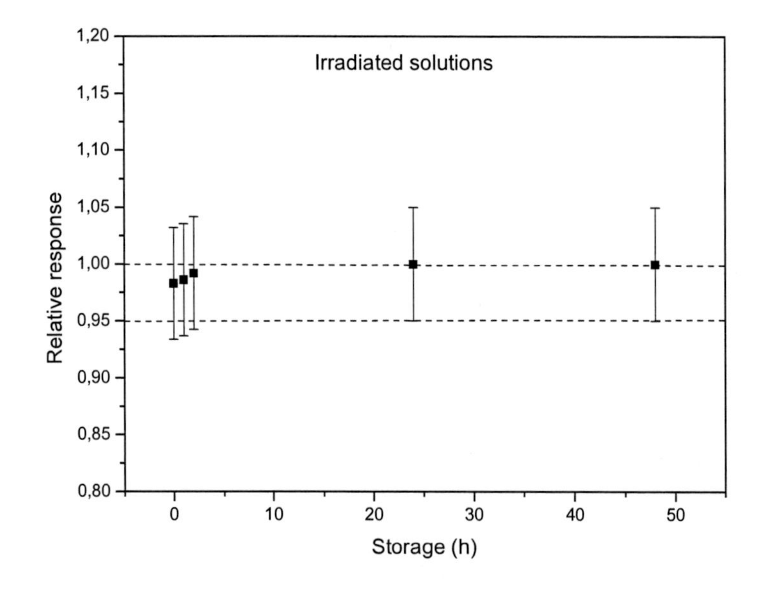

Figure 22. Post-irradiation response of bromocresol green solutions irradiated with gamma radiation with 8 kGy and storage during 48 hours.

The absorbance values of the 0.075% bromocresol green solutions was plotted as a function of absorbed dose between 50 Gy and 15 kGy at 450nm. A grade extinction of optical absorption intensity is shown in Figure 23 and a linear response in the studied range is observed.

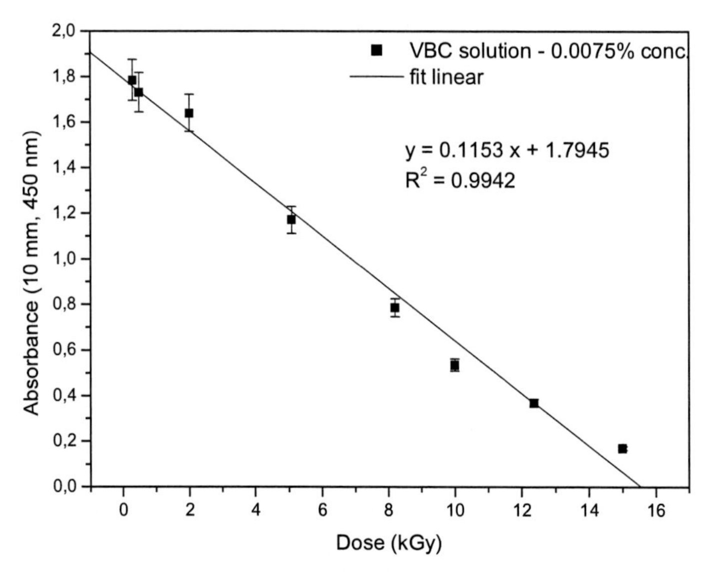

Figure 23. Dose response curve of 0.0075% bromocresol green solution.

Fricke gel (FXG)

Spectrophotometric measurements in the wavelength range between 190 and 900 nm were performed for non-irradiated Fricke gel samples. In the Figure 24 is shown the spectrum obtained in the range 350-650 nm. The non-irradiated (0 Gy) gel sample presents two absorption bands at 440 and 585 nm originated by Fe^{2+} and Fe^{3+} presence respectively in the sample.

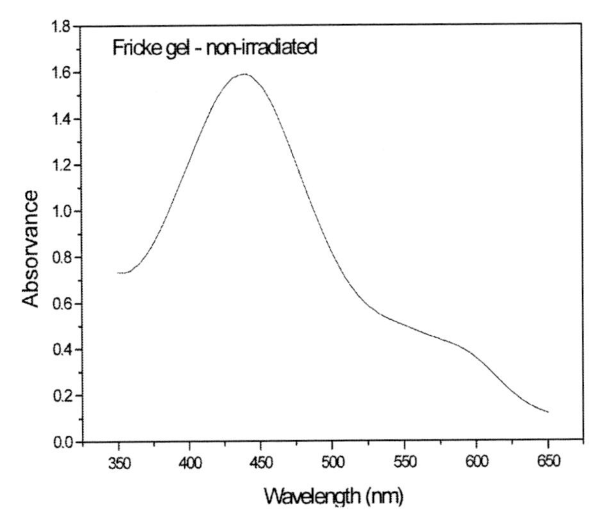

Figure 24. Non-irradiated Fricke gel spectrum: wavelength range 350 - 650 nm.

Absorption spectrum of irradiated FXG solutions with doses between 5 and 40 Gy were recorded in the range of 400 and 625 nm and presents the same two bands; in this case the 585 nm band increases with the dose (Fe^{3+} increasing concentration) and the 440 nm band (Fe^{2+} decreasing concentration) tends to disappear. In the Figure 25 is shown the spectra of Fricke gel solutions as a function of the ^{60}Co radiation dose.

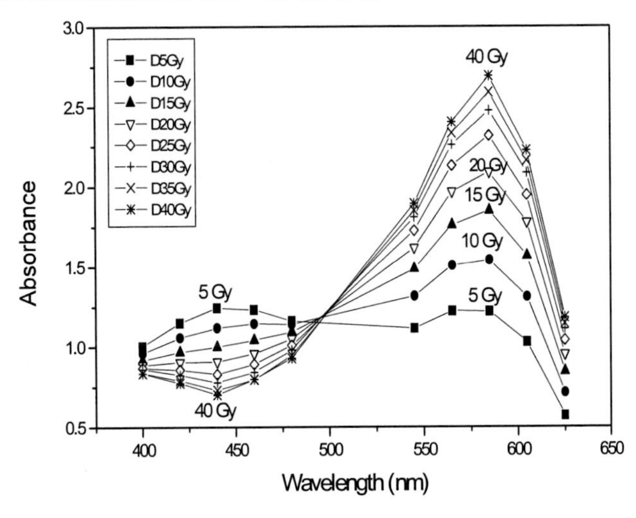

Figure 25. Fricke gel spectrum. Gamma doses between 5 – 40 Gy, wavelength range between 400 and 625 nm.

In the Figure 26 is shown the dose-response curve of Fricke gel samples irradiated with doses between 5 – 40 Gy of ^{60}Co gamma radiation. The 0 Gy absorbance value of 585 nm band (Fe^{3+}) was subtracted. The obtained results are close with that presented by Bero [50-51].

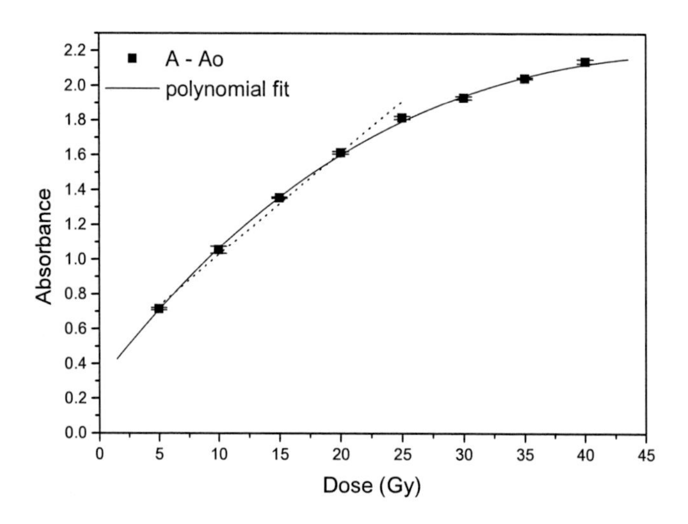

Figure 26. Dose-response curve of Fricke gel dosimeter irradiated with ^{60}Co gamma radiation.

Brest simulators were prepared using the gel solution and preliminary measures using spectrophotometry technique were performed introducing the cuvettes containing the gel solution in the simulator. Figure 27 (a) and (b) shows the breast phantom and the dosimeter positioning, respectively.

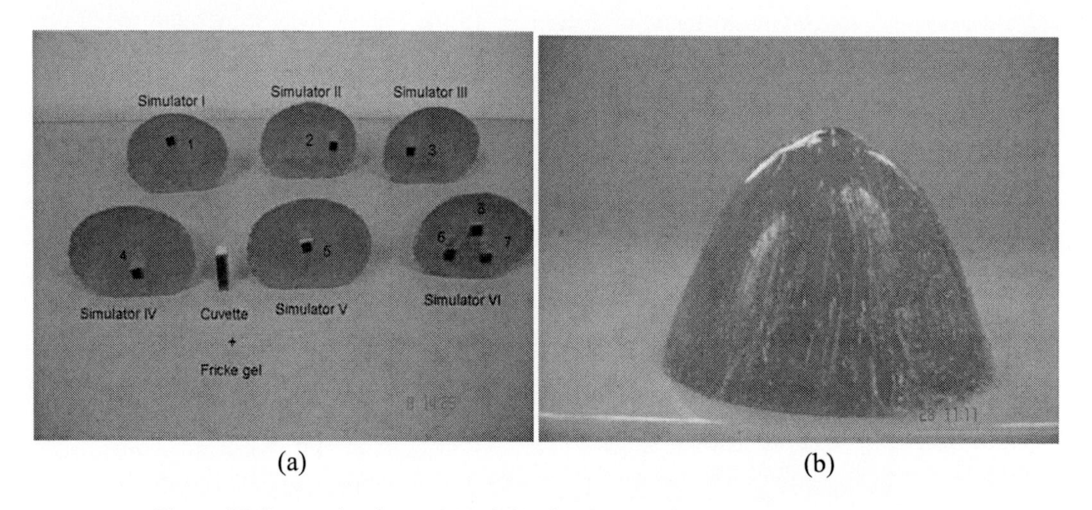

Figure 27. Breast simulators 340ml (a) simulator and (b) dosimeter positioning.

The simulator was irradiated with 15 Gy at point 2. The obtained values obtained was 15.2 Gy at point 2 and 12.21 Gy at point 1 This result shows that the dosimeter presents good sensitivity and can be used to dose distribution determination.

Spectrophotometric analysis has proved to be very reliable. Moreover, Nuclear Magnetic Resonance (NMR) analysis gives the possibility of spatial determination of paramagnetic species, because of influence of the ferric ions on the spin relaxation times of the hydrogen nuclei of the solution.

Changes in proton relaxation time related with absorbed dose due ferric ion presence were determinate using samples irradiated with 20 Gy and measures performed using a Gyroscan S 15/ASC (1.5 T) equipment. A linear relation was found between the relaxation rate of the solution and the absorbed dose. The correlation coefficient was 0.99 shown in the Figure 28.

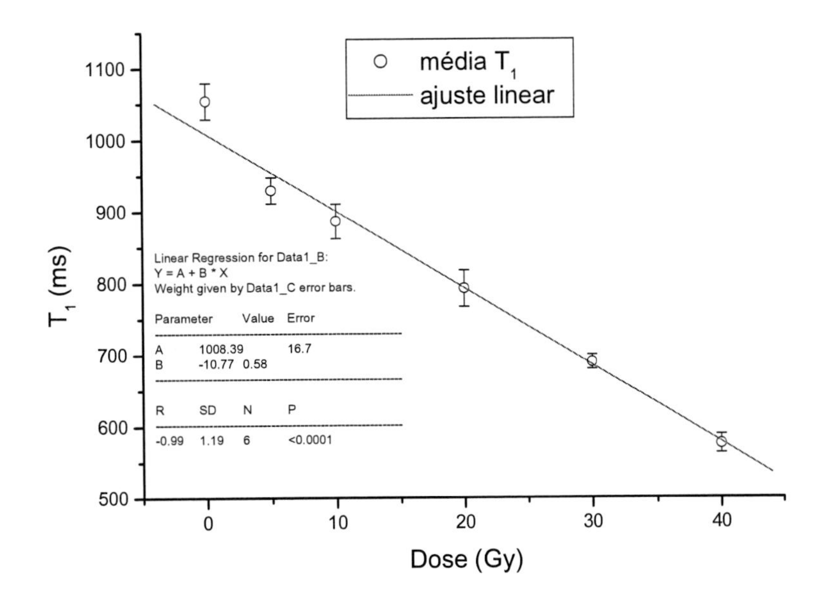

Figure 28. Proton relaxation time as a function of absorbed dose for FXG solution.

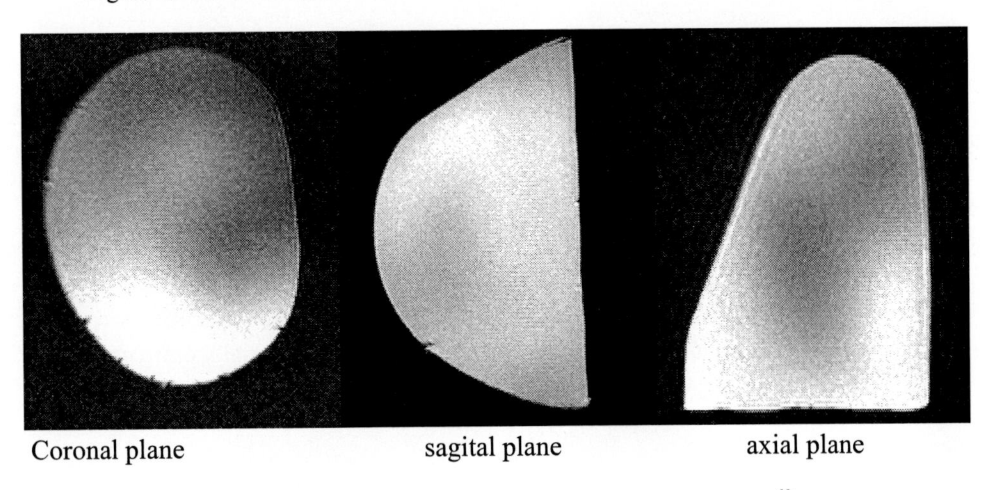

Coronal plane sagital plane axial plane

Figure 29. Magnetic resonance image to FXG breast simulator irradiated with ^{60}Co gamma radiation.

Preliminary magnetic resonance images were performed using a breast simulator prepared with Fricke gel solution, Figure 29. Changes in the image intensity are observed as a result of the dose distribution on the simulator. The brightness area corresponds with the higher absorbed dose. Dose calibrations are in gray scale and color scale are in progress aiming to construct the isodose curves.

Conclusion

The potassium nitrate can be used in pellet form for irradiation and processed for optical measures in the solution form. Environmental conditions do not affect the response, but the pellets should be maintained sealed and handled in humidity atmosphere below 60%. The dosimetric properties of this compound are appropriated for high doses applications. The system presents a reproducible response in the dose range between 1 and 150 kGy. The addition of MnO_2 in the pellet composition increase dose range for 600 kGy.

The use of dyed PMMA detectors for electron beam and gamma dosimetry is cheap and the absorbance analysis is very simple. The detectors are easy to prepare, manipulate and analyze. The useful dose range of the eight dyed PMMA detectors developed is large and interesting for dose evaluation in radiation processing and their performance is according to the commercially available Red 4034 Perspex detector. The yellow 4G, green G and red 5B detectors were shown to be the most promising, considering their response characteristics and dose range, although the others detectors can also be used for different dose ranges. All types of dyed PMMA detectors demonstrated high sensitivity and reproducibility. They do not present energy dependence when thicknesses equal to or smaller than the range of the electrons with different energies. These data also show that the dyed detectors present promising characteristics and can be useful for electron dosimetry.

Aqueous solution of bromocresol green can be used for a dose interval between 0.3 and 15 kGy. The solutions are stable and can be used after preparation for various months.

The FXG dosimeter is easy to prepare and cheap. The gelatine is tissue equivalent and can be prepared in a variety of the shapes and volumes. The effect of the natural oxidation needs to be known and separated of the contribution radiation induced. The best dosimeter storage condition is to maintain the solution at dark environment and low temperature (5°C). The spectrophotometry technique is cheaper and faster than MRI technique and provides good results. The gel dosimeter can be also used to determination of absorbed dose distribution in radiotherapy treatments. The improvement of this technique using MRI will provide the obtaining of the isodose curves in the irradiated organ simulators.

All developed materials are available for high dose evaluation and can be used according to the characteristics of the installation, dose level required and material atomic number.

References

[1] Martins, J. B. (2007). História da energia nuclear. Apostilas CNEN. Available from: http://www.cnen.gov.br.

[2] McLaughlin, W. L.; Desrosiers, M. C. Dosimetry systems for radiation processing *Radiat Phys Chem*. 1995, 46, 4-6, 1163-1174.

[3] Mehta, K.; Fuochi, P. G.; Kovacs A.; Lavalle M.; Hargittai P. Dose distribution in electron-irradiated PMMA: effect of dose and geometry. *Radiat Phys Chem.* 1999, 55, 773-779.

[4] Regula D. F.; Deffener, U. Dosimetry by ESR spectroscopy of alanine. *Int J Appl Radiat Isot.* 1982, 33, 1101-1114.

[5] Galante, O. L.; Rodrigues Jr., O.; Campos, L. L. Development of a dosimeter for high doses assessment base don alanine/ EPR. Available from: http://www.irpa.net/irpa10/cdrom/00531.pdf.

[6] Campos, L. L. Termoluminescência de materiais e sua aplicação em dosimetria da radiação. Cerâmica. 1998, 44, 290.

[7] Breen, S. L.; Battista, J. J. Feasibility of reading LiF thermoluminescent dosimeters by electron spin resonance. *Phys Med Biol.* 1999, 44, 8, 2063-9.

[8] Carrilo, R. E.; Pearson, D. W.; Deluca Jr, P. M.; MacKay, J. F.; Lagally, M. G. Response of thermoluminescent lithium fluoride (TLD-100) to photon beams of 275, 400, 500, 600, 730, 900, 1200, 1500, and 2550 eV. *Phys. Med. Biol.* 1994, 39, 1875-1894.

[9] DiMauro, T. Radiochromic Film Dosimetry. Radiation Physics. 2000.

[10] American Society for Testing and Materials. Standard practice for use of a radiochromic film dosimetry system. 2004, ISO/ASTM51275-04, 12.02.

[11] Harwell Dosimeters References . Available from: http://www.harwell-dosimeters.co.uk/references.php

[12] American Society for Testing and Materials. Standard Practice for Use of a Ceric-Cerous Sulfate Dosimetry System. 2002, ISO/ASTM51205-02, 12.02.

[13] American Society for Testing and Materials. Standard Practice for Using the Fricke Reference-Standard Dosimetry System. 2004, E1026-04e1, 12.02.

[14] Harteck, P.; Dondes, S. Nitrous oxide dosimeter for high levels of betas, gammas and thermal neutrons. *Nucleonics.* 1956, 13, 3, 66-72.

[15] Hearne, J. A.; Hummel, R. W. Nitrous as a dosimeter for ionizing radiations. *Radiation Research.* 1961, 15, 254-267.

[16] Harteck, P.; Dondes, S. Radiation chemistry of gases. *Rensselaer Polytechnic Inst.* 1969, 321, 19.

[17] Daros, K. A. C.; Campos, L. L.; Medeiros, R. B. TL response study of the $CaSO_4$:Dy pellets with graphite for dosimetry in beta radiation and low-energy photons fields, *Applied Radiation and Isotopes.* 2001, 54, 6, 957-960.

[18] Optical properties of materials. Available from: http://www.npl.co.uk/optical_radiation/primary_standards/spectrophotometry.html

[19] Spectrophotometry. Available from: http://physics.nist.gov/Divisions/Div844/facilities/specphoto/specphoto.html.

[20] Encyclopedia Britannica online. Electron paramagnetic resonance (EPR). Available from: http://www.britannica.com/eb/article-9032325/electron-paramagnetic-resonance

[21] Bergstrand, E. S.; Shortt, K. R.; Ross, C.K.; Hole, E. O. An investigation of the photon energy dependence of the EPR alanine dosimetry systems. *Phys Med Biol.* 2003, 48, 1753-1771.

[22] Andrew, E. R. Nuclear magnetic resonance, Boston, Cambridge, University Press, 1969.

[23] Advanced undergraduated laboratory. Nuclear Magnetic Resonance. Available from: http://www.physics.utoronto.ca/~phy326/nmr/nmr.pdf.

[24] Stark, D. D.; Bradley, W. G. Magnetic resonance imaging, ed. 2, St. Louis, *Mosby-Tyear Book*, 1992.
[25] Galante, A. M. S.; Rzyski, B. M.; Villavicencio, A. L. C. H.; Campos L. L. The response of potassium nitrate for high-dose radiation dosimetry. *Radiat Phys Chem.* 2002, 63, 719-722
[26] Galante, A. M. S.; Villavicencio, A. L. C. H.; Campos, L. L. Dosimetric properties of KNO_3 pellets mixed with sensitizing compounds. *Radiat Phys Chem.* 2004, 71 (1), 387-389.
[27] Galante, A. M. S.; Villavicencio, A. L. C. H.; Campos, L. L. Preliminary investigations of several new dyed PMMA. *Radiat Phys Chem.* 2004, 71 (1), 393-396.
[28] Galante, A. M. S.; Campos, L. L. Electron dose radiation response of dyed PMMA detectors at IPEN. *Radiat Prot Dos.* 2006, 120, 1-4, 113-116.
[29] Technical Reports Series no. 409. Dosimetry for food irradiation. Printed by the IAEA, Vienna, 2002.
[30] Holm, N. W.; Berry, R. J. Manual on radiation dosimetry. Marcel Dekker, New York, 1970.
[31] Attix, F. H. Introduction to radiological physical and radiation dosimetry. John Wiley & Sons, London, 1986.
[32] Fricke, H.; Hart, E. J. Chemical dosimetry. In: Attix, F. H.; Roesch, W. C. Radiation dosimetry. New York, *Academic*, **3**, 167, 1968.
[33] McLaughlin, W. L.; Boyd, A. W.; McDonald, J. C.; Miller, A. Dosimetry for radiation processing. Taylor & Francis, New York, 1989.
[34] Constantinescu, B. Gamma and proton induced degradation in ceramics material – a Proposal. Available from: http://www.iaea.org/programmes/ripc/physics/fec2000/pdf/ftp1_11.pdf
[35] Teixeira M.I.; Ferraz G.M.; Caldas L.V. EPR dosimetry using commercial glasses for high gamma doses. *Appl Radiat Isot.* 2005, 62, 2, 365-70.
[36] Caldas, L. V. E.; Teixeira, M. I. Comercial glass for high doses using different dosimetric techniques. *Radiat Protec Dosim.* 2002, 101, 1-4, 149-152.
[37] Hsu, S. M.; Yeh, S. H.; Lin, M. S. and Chen, W. L. Comparison on characteristics of radiophotoluminescent glass dosemeters and thermoluminescent dosemeters. *Radiat Prote Dosim.* 2006, 119, 1-4, 327-331.
[38] Semiconductor detector. Available from: http://www.radiofreeithaca.net/search/Semiconductor_detector
[39] Rasolonjatovo, A. H. D.; Shiomi, T.; Nakamura, T. Development of gamma-ray monitor using CdZnTe semiconductor detector. Available from: http://www.irpa.net/irpa10/cdrom/00110.pdf
[40] Patil, S. F.; Chiplunkar, N. R. Influence of oxides on the radiolysis of $RbNO_3$ e $CsNO_3$. *Radiat Phys Chem.* 1991, 37, 2, 241-244.
[41] Joshi, N. G.; Garg, A. N.; Natarajan, V.; Sastry, M. Effect of oxide additives on radiolytic decomposition of zirconium and thorium nitrates. *Radiation Measurements.* 1996, 26, 1, 131-137.
[42] Pogge, H. B.; Jones F. T. The effect of temperature and additives in the radiolysis of potassium nitrate. *J Phys Chem.* 1970, 74, 8, 1700-1705.
[43] Parwate, D. V.; Garg A. N. Effect of outer cations and water of crystallization on the radiolytic decomposition of nitrates. *J Radioanal Nucl Chem Lett.* 1984, 85, 4, 203-212.

[44] Patil, S. F.; Bedekar, A. G. Radiation decomposition of pure and barium doped potassium nitrate and effect of oxides thereon. *Radiochimica Acta*. 1985, 38, 165-168.
[45] Kulkarni, S. P.; Garg, A. N. Effect of additives with common cation on the radiolysis of ammonium, sodium and potassium nitrates in admixtures. *Radiat Phys Chem*. 1988, 32, 4, 609-614.
[46] Batra, R. J.; Garg, A. N. Gamma radiolytic decomposition of solid binary mixtures of potassium nitrate with halide. *J Radioanal Nucl Chem Art*. 1989, 129, 1, 155-162.
[47] Joshi, N. G.; Dhoble, S. J.; Moharil, S. V.; Garg, A. N. Effect of particle size on gamma ray induced decomposition in KI-KNO_3 crystals. *Radiat Phys Chem*. 1994, 44, 3, 317-322.
[48] Agrawal, N.; Garg, A. N. Effects of oxide additives on radiolytic decomposition of potassium nitrate. In: TSRP-98 Trombay. Symposium on Radiation and Photochemistry. Mumbai, India, 1999.
[49] American Society for Testing and Materials. 1996. Standard practice for use of a polymethylmethacrylate dosimetry system, (ASTM E 1276 – 96), Annual Book, 12.02.
[50] Bero, M. A.; Gilboy W. B. and Glover P. M. Radiochromic gel dosemeter for three-dimensional dosimetry. *Radiat Phys Chem*. 2001, 61, 433-435.
[51] Bero M. A. and Kharita M. H. Effects of ambient temperature on the FXG radiochromic gels used for 3-D dosimetry. *Journal of Physics*: Conference Series 3, 2004, 213-216.
[52] Bero M. A.; Gilboy W. B. High-resolution optical tomography for 3-D radiation dosimetry with radiochromic gels. *Journal of Physics*: Conference Series 3, 2004, 261-264.
[53] Gore J. C.; Kang Y. S.; Shulz, R. J. Measurement of radiation dose distributions by nuclear magnetic resonance (NMR) imaging. *Phys Med Biol*. 1984, 29, 10, 1189-1197.
[54] Healy, B.; Brindha S.; Zahmatkesh M.; Baldock, C. Characterization of the ferrous-xylenol orange-gelatin (FXG) gel dosimeter. *Journal of Physics*, Conference Series 3, 2004, 142-145.
[55] Jayachandran C. A. Calculated effective atomic number and kerma values for tissue-equivalent and dosimetry materials". *Phys Med Biol*. 1971, 16, 4, 617-623.
[56] Olsson L. E.; Appleby A.; Sommer, J. A new Dosimeter Based on Ferrous Sulphate Solution and Agarose Gel. *Appl Radiat* Isot. 1991, 42, 11, 1081-1086.
[57] Schulz R. J.; DeGuzman A. F.; Nguyen D. B.; Gore J. C. Dose-Response Curves for Fricke-Infused Agarose Gels as Obtained by Nuclear Magnetic Resonance. *Phys Med Biol*. 1990, 35, 12, 1611-1622.
[58] Dorda, E. M.; Muñoz, S. S. Potassium Nitrate/ Nitrite Dosimeter for high-dose. In: High-dose dosimetry, Proceeding of an International Symposium. 1984, IAEA SM 272/1, Vienna.
[59] Torres C., R. R. Medición de altas dosis de radiación gamma por el sistema nitrate-nitrito de potasio. Callao, Peru. 1993, Tese, Universidade del Callao.

In: Radiation Physics Research Progress
Editor: Aidan N. Camilleri, pp. 385-405
ISBN: 978-1-60021-988-7

Chapter 11

PREPARATION OF FUNCTIONAL POLYMERS VIA γ-RADIATION

Zhengpu Zhang
Key Lab. of Functional Polymer materials, Ministry of Education,
Institute of Polymer Chemistry, Nankai University, Tianjin 300071, P.R.China

Abstract

In this chapter PTFE acidic cationic exchange fiber have been prepared via the radiation grafting of acrylic acid, maleic acid or the sulfonation of the grafted polystyrene via radiation. It shown the special properties for the radiation grafted PTFE fiber for example, the acidity and superacidity, the pH titration curves and the excellent adsorption, desorption and separation properties for some metal ions.

The NIPA grafted chitosan also have been prepared via irradiation. The grafted products shown good thermo and pH activity, it has the potential use in the drug releasing and the other areas.

Introduction

It becomes more general method for the preparation of functional polymers from an inert polymer for example polyethylene (PE), polypropylene (PP) and polytetrafluoroethylene (PTEF) et al., as more and more γ-radiation station have been established in the world. Recently in my lab some special functional polymers have been prepared via γ-radiation. They included PTFE fiber grafted acrylic acid and maleic acid, grafted polystyrene and its sulfonated product, and chitosan grafted N-isopropylacrylamide. Our research proved these novel materials have special and interesting chemical and physical properties. In this chapter we'll introduce these special functional polymer materials.

1. γ-Radiation Induced PTFE Fiber Grafted Acrylic Acid.[1-6]

Because of very good chemical stability, thermal stability either at high or low temperature and electric insulation et al., PTFE have been widely be used in many areas, such as chemical and petrol industry, textile, electrical engineering, medicinal apparatus et al. Generally, this material is very difficulty for the chemical modification by the normal chemical reactions. γ-Radiation is one of the effective methods for him. In the previous literatures it could be found the chemical modification of the PTFE membrane via γ-radiation [7-9], but the chemical modification of the PTFE fiber via γ-radiation is seldom be found.

1.1. The Preparation of Acrylic Acid Grafted PTFE Fiber Via γ-Radiation

For the preparation of acrylic acid grafted PTFE fiber, the simultaneous ^{60}Co γ-radiation induced acrylic acid grafting was carried out in a 500 ml container in which there were 100 ml acrylic acid, 10 g PTFE fiber (φ= 20μm) and 300 ml deioned water containing 2.15 % Mohr's salt. The grafting solution was bubbled with nitrogen for 20 min. to create an inert atmosphere. The container was irradiated at room temperature using γ-rays from a ^{60}Co source at dose rate 0.5 KGy/h for 48 h. The grafted fiber was washed with deioned water, 1 M NaOH, deioned water, 1 M HCl, deioned water for the removing of residual monomer, homopolymer and metal ions adhered to the PTFE fiber surface after irradiation. Finally, the grafted fiber was washed with acetone and dried in a vacuum oven at 50□ for 48 h and weighed before characterization. The relation of the concentration of the Mohr's salt on the grafting percentage was listed in Tab.1

Table 1. The relation of the concentration of the Mohr's salt on the grafting percentage

$C_{Mohr's}$ (%)	<0.5	1.24	2.15	2.48	2.78
G (%)	0	28.7	37.3	35.0	33.6

Q (radiation dosage) = 24 kGy

The grafting percentage G % is calculated from the following equation:

$$G\% = (Wg—Wo)/Wo\times 100\% \quad (1)$$

Wg and Wo are the weight of the PTFE fiber after and before the irradiation respectively.

It could be found from Tab.1, for the achieving a good grafting percentage, the concentration of the Mohr's salt is very important, when the concentration is lower than 1.24%, it couldn't inhibit the self polymerization of monomer and couldn't got higher grafting percentage, on the other side, the more Mohr's salt may capture the free radicals, so the higher concentration of the Mohr's salt also couldn't got higher grafting percentage.

The radiation dosage is another effective factor for the grafting percentage. The related data were listed in Tab.2. It could be found from Tab.2 the grafting percentage increased rapidly as the increase of the radiation dosage from zero to 16 KGy, but after this dosage the grafting percentage was increased slowly even if the irradiation dosage still increased.

Table 2. The effect of the radiation dose (Q) to the grafting percentage

Q (KGy)	8	16	24	48
G (%)	16.7	32.3	37.3	37.7

$C_{Mohr's} = 2.15\%$

The monomer concentration is also the effective factor for the grafting percentage. Tab.3 listed the related data. It could be found, the grafting percentage was increased as the monomer concentration increased. When the monomer concentration is more than 15%, the increase of the grafting percentage is not very obvious.

Table 3. The effect of the monomer concentration (Cm) to the grafting percentage

Cm (%)	3.33	8	15	25
G (%)	9.9	24.4	33.4	37.3

Q = 24KGy□ $C_{Mohr's} = 2.15\%$

1.2. The Characterization of Acrylic Acid Grafted PTFE Fiber

The FT-IR spectrum, contact angle, ion exchange capacity and pH titration curve of the grafted PTFE fiber was characterized.

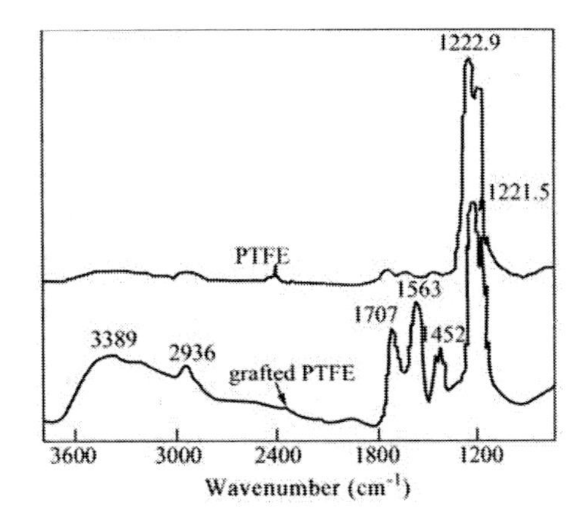

Figure 1. The FT-IR spectrum of the PTFE fiber and grafted PTFE fiber

The FT-IR spectrum of the PTFE fiber and grafted PTFE fiber was conducted in FTS 135 FT-IR, (Bio-Rad Co. Ltd), by microspectroscopy and transmission technology shown in Fig.1. It could be found from Fig.1 that both the strong adsorption band at 3389 cm^{-1} and 1707 cm^{-1} are the stretching vibration of OH and C=O groups respectively. It is greatly difference from the original PTFE IR spectrum.

The determination of the contact angle was conducted in DSI-10 drop shape analysis (German , KRUSS Co. Ltd). For the sample preparation, a PTFE membrane together with PTFE fiber was simultaneously irradiated. The contact angle of the PTFE fiber should be the same as PTFE membrane. The contact angle of some polymer materials with water were listed in Tab.4, except for the polymers, the stearic acid also listed as a ref. It could be found, the contact angle of the grafted PTFE membrane is lower than the initial PTFE membrane.

Table 4. The contact angle of some materials with water

Sample	grafted PFTE	PFTE	PP	PE	Stearic acid
contact angle	69.11°	102.8°	108°	103°	80°

The ion exchange capacity of the grafted PTFE fiber was determined by direct acid-base titration, it is difference from the commercial weakly acidic ion exchange resin (e.g. D-152, manufactured by the chemical plant of Nankai University, Tianjin, China) because of quite strong acidity of the COOH group grafted in PTFE fiber comes from the electro negativity of the F atom. A certain H type grafted PTFE fiber in deioned water was titrated by standard NaOH solution directly, methyl orange was used as indicator, the content of the COOH group in the grafted PTFE fiber could be calculated based on the volume of the standard NaOH solution consumed in the titration. In our experiment, determined ion exchange capacity of the grafted PTFE fiber is 3.07mmol/g.

(The normal method for the determination of the weakly acidic ion exchange resin is the indirect titration, i.e. quantitative H type weakly acidic ion exchange resin was neutralized with excessive standard NaOH solution, the residue of the NaOH was titrated by standard HCl solution, the exchange capacity of the weakly acidic ion exchange resin should be the differences of both the standard NaOH and HCl.) [10]

The pH titration curve of both the acrylic acid grafted PTFE fiber and the D-152 weakly acidic ion exchange resin was shown in Fig.2. There is a sudden jump when the pH value was changed from 6 to 10 for the grafted PTFE fiber. It is difference from the D-152 titration curve, even if they have the same functional group.

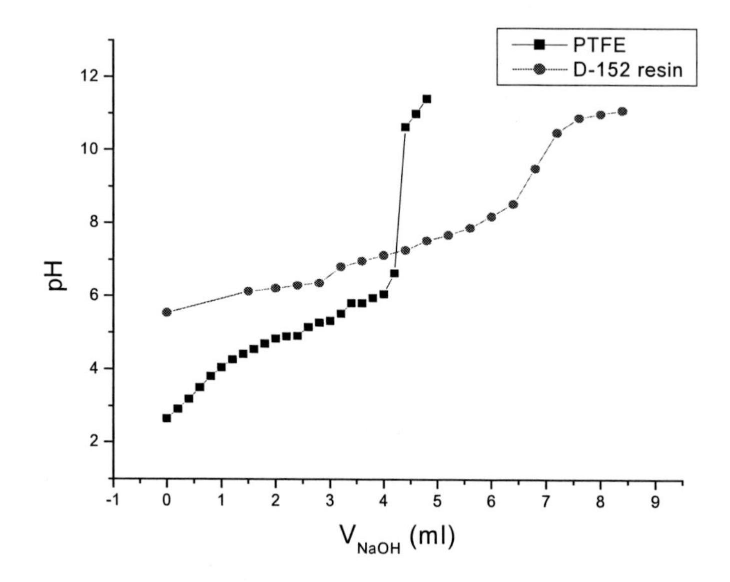

Figure 2. pH titration curve of both grafted PTFE fiber and D-152 resin

The pk° value of some weak acid calculated based on the Enderson-Hasselbach equation modified by V.S.Soldatov [11] was listed in Tab.5. It could be found the acidity of the grafted PTFE fiber is about 10 times of the acrylic acid.

Table 5. The pk° value of some weak acid

Weak acid	Grafted PTFE	Acrylic acid	HF	H_2CO_3	HAC	H_3PO_4
pk°	3.37	4.26	3.17	6.37	4.75	2.16 (pK_1)

1.3. The adsorption and desorption properties of acrylic acid grafted PTFE fiber for metal ions

1.3.1. The adsorption and desorption behavior of the acrylic acid grafted PTFE fiber for Cu^{2+}

The static adsorption curve of the acrylic acid grafted PTFE fiber for Cu^{2+} is shown in Fig. 3, the ordinate in Fig.3 is the residue Cu^{2+} concentration after adsorption, it could be found from Fig.3 that the adsorption rate is very quickly, it only need a couple of minutes for reaching the adsorption balance. The optimum pH for the adsorption of Cu^{2+} is from 3 to 5. The precipitate of $Cu(OH)_2$ is found, when the pH is over 5, it will greatly influence the adsorption of the grafted fiber. On the other hand, when the pH is less than 2, the acidity of the solution is quite higher, it is the disadvantage for the adsorption of Cu^{2+}.

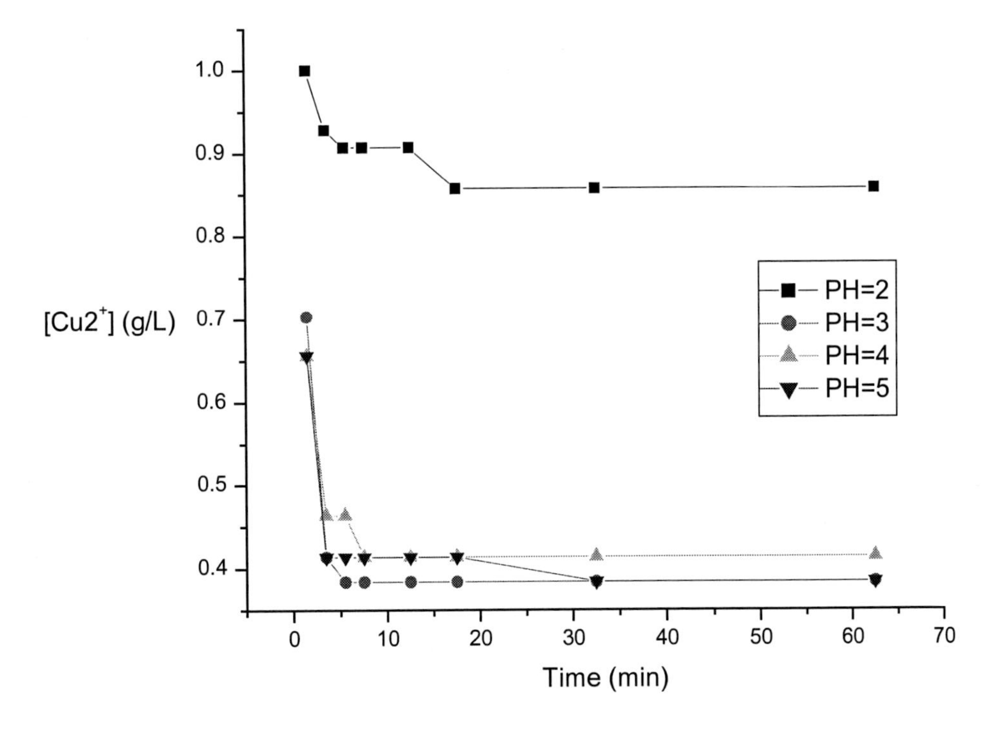

Figure 3. effect of pH on static adsorption on Cu^{2+}

The dynamic adsorption experiment was carried out in a glass column □φ=1.3 cm□, acrylic acid grafted PTFE fiber was cut into pieces and packed into the column. A pH=5,1mg /ml $CuSO_4$ solution was passed through the column by the SV=7.5 BV/h. The dynamic adsorption curve of the acrylic acid grafted PTFE fiber for Cu^{2+} is shown in Fig. 4, curve A is the instant adsorption curve and curve B is the cumulative adsorption curve. It could be found the cumulative quantity of the Cu^{2+} adsorbed onto the grafted PTFE fiber almost is the proportion with the volume passed through the column at the initial 80 ml. The dynamic adsorption comes to balance, when the cumulative volume is 160 ml. The saturated dynamic adsorption of the grafted fiber for Cu(II) is 107.48 mg/g.

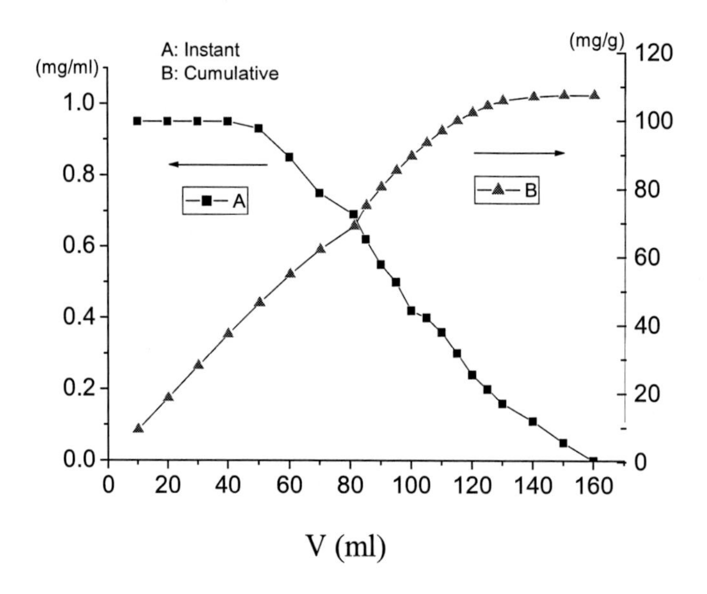

Figure 4. The dynamic adsorption of the acrylic acid grafted PTFE fiber for Cu^{2+}

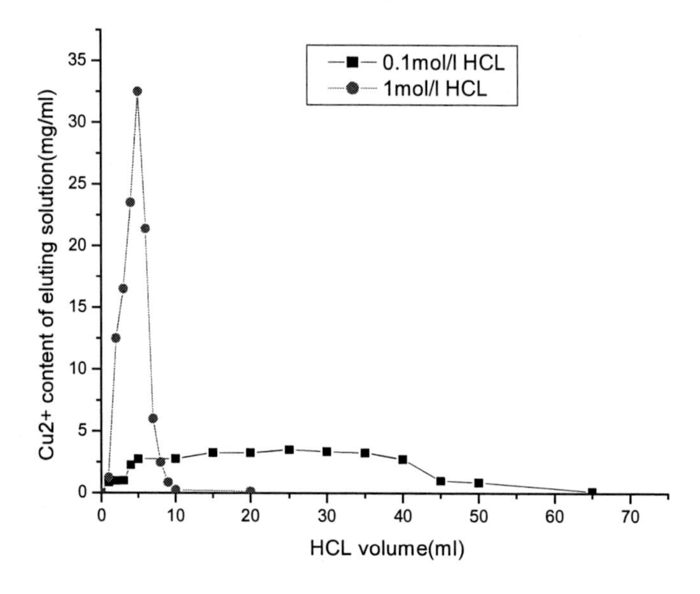

Fig. 5 The desorption of Cu^{2+} from the grafted PTFE fiber eluted by different concentrated HCL The desorption condition: SV=15BV/h, □φ=1.3 cm□

Desorption is another very important property for an adsorbent. The grafted PTFE fiber has excellent desorption property for Cu^{2+}. Limited desorption agent, very short time, could make the Cu^{2+} desorption completely. Fig. 5 gives the desorption curve. Both the eluted ratio of Cu^{2+} is 100 % and 95.87 %, when 1mol/L and 0.1mol/L HCL was used as eluent respectively.

1.3.2. The Competitive Adsorption of 9 Kinds of Metal Ions by the Grafted PTFE Fiber.

The influence of pH on the adsorption of Ni(II), Ca(II), Pb(II), Mg(II), Cr(III), Co(II), Ag(I), Cu(II), La(III) 9 kinds of metal ions by the grafted PTFE fiber was given in Fig.6, it could be found from Fig.6, that the optimum pH value for the adsorption of the most of the metal ions is at 5, except for Cr(III). So, the pH = 5 was chosen for the dynamic competitive adsorption.

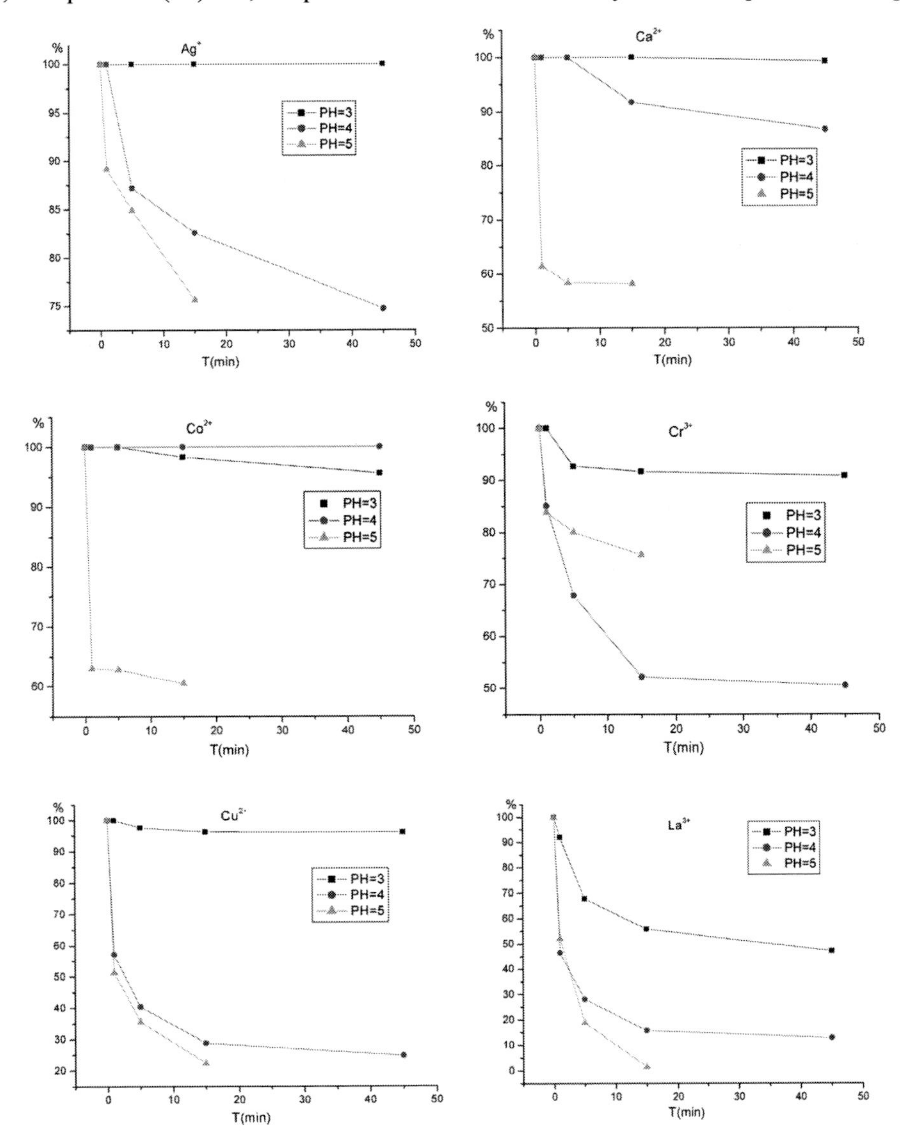

Figure 6. Continued on next page.

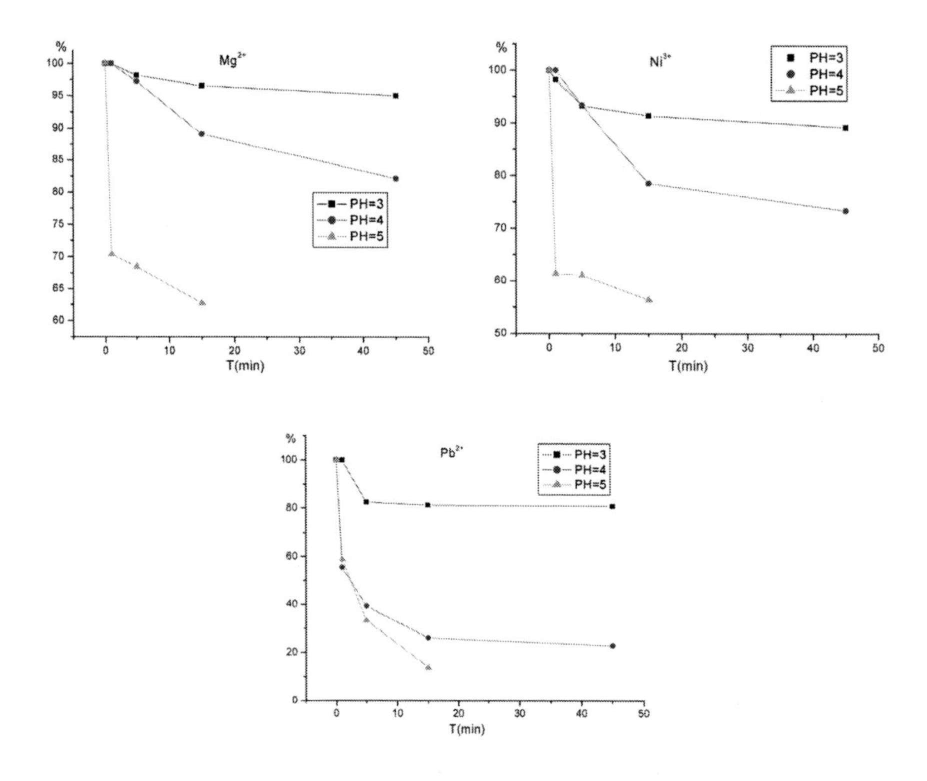

Fig.6 The static adsorption of 9 kinds of metal ions by grafted PTFE fiber at different pH (The ordinate of the Fig.6 is the percentage of the residue metal ions respectively.)

The dynamic adsorption experiment was carried out in a φ=1.3cm glass column packed 1.04 g of the grafted fiber. A pH = 5 NaAc/HAc buffer containing 9 kinds of nitrate including Ni(II), Ca(II), Pb(II), Mg(II), Cr(III), Co(II), Ag(I), Cu(II), La(III) (the concentration of all the 9 kinds of metal ions are the same as 40 mg/ml) were passed through the column. The residue of the 9 kinds of metal ions was determined by ICP (ICP-9000, N+M, TGA Co. USA). The quality adsorbed onto the grafted fiber were calculated and listed in Tab.6. It could be found from Tab.6, the adsorption for Cu(II), Pb(II) and La(III) is better than the others, the quality adsorbed onto the grafted fiber of the three ions is about 90% of the total adsorbed ions.

Table 6. The dynamic competitive adsorption of 9 kinds of metal ions by grafted PTFE fiber

V(ml)	Q_{Mg}	Q_{Ca}	Q_{Cr}	Q_{Co}	Q_{Ni}	Q_{Cu}	Q_{Ag}	Q_{La}	Q_{Pb}
20	38.34	39.63	12.52	39.75	39.14	39.69	38.30	39.91	38.23
40	14.98	37.70	1.22	39.37	38.60	39.68	31.10	39.84	38.17
60	0.68	23.45	0	36.20	35.05	39.55	6.20	39.77	38.11
80	0	13.90	0	21.00	24.28	39.50	3.60	39.75	38.10
100	0	9.50	0	11.80	15.58	39.11	0.40	39.75	37.58
120	0	3.36	0	4.80	5.18	38.95	0	39.74	37.06
140	0	0	0	0	0	37.58	0	39.74	36.78
160	0	0	0	0	0	37.55	0	39.73	36.74
180	0	0	0	0	0	37.48	0	39.73	36.46

Table 6. Continued.

V(ml)	Q_{Mg}	Q_{Ca}	Q_{Cr}	Q_{Co}	Q_{Ni}	Q_{Cu}	Q_{Ag}	Q_{La}	Q_{Pb}
200	0	0	0	0	0	37.47	0	39.72	36.16
220	0	0	0	0	0	37.31	0	39.72	35.91
240	0	0	0	0	0	37.24	0	39.71	35.65
250	0	0	0	0	0	37.09	0	39.71	35.40
270	0	0	0	0	0	36.93	0	39.71	35.37
290	0	0	0	0	0	36.74	0	39.69	35.31
310	0	0	0	0	0	36.64	0	39.68	35.30
340	0	0	0	0	0	36.13	0	39.68	35.15
370	0	0	0	0	0	34.63	0	39.67	35.03
400	0	0	0	0	0	33.12	0	39.66	34.88
430	0	0	0	0	0	30.14	0	39.65	35.17
460	0	0	0	0	0	28.94	0	39.64	34.35
480	0	0	0	0	0	28.16	0	39.64	33.38
500	0	0	0	0	0	27.58	0	39.64	33.24
520	0	0	0	0	0	25.83	0	39.63	32.87
540	0	0	0	0	0	21.42	0	39.62	32.78
560	0	0	0	0	0	21.19	0	39.61	32.09
580	0	0	0	0	0	19.9	0	39.58	30.53
600	0	0	0	0	0	17.26	0	39.56	29.18
630	0	0	0	0	0	16.40	0	39.50	29.00
660	0	0	0	0	0	15.26	0	39.47	27.02
690	0	0	0	0	0	11.84	0	39.45	24.05
720	0	0	0	0	0	10.02	0	39.29	20.74
Ratio(%)	1.3	3.08	0.33	3.69	3.81	25.96	1.92	34.5	29.12

Q: The quality of the metal ions adsorbed onto the grafted fiber. (The weight of the grafted fiber is 1.04 g; pH = 5; SV (Speed Velocity) = 7.5 BV/h)

1.3.3. The dynamic saturating adsorption and desorption of the grafted PTFE fiber for Pb^{2+}, La^{3+}

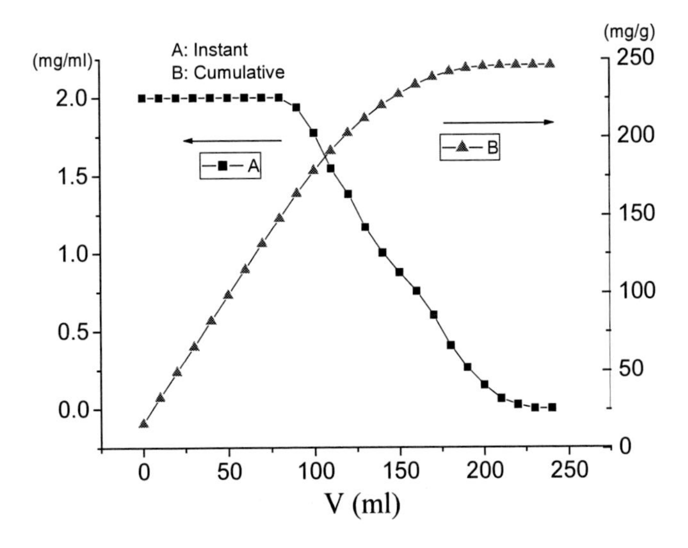

Figure 7. The dynamic adsorption of Pb^{2+} by grafted PTFE fiber; The adsorption condition: $[Pb^{2+}] = 2$ mg/ml; pH = 5; SV(Speed Velocity) = 7.5 BV/h

The experimental condition of the dynamic saturating adsorption and desorption of the grafted PTFE fiber for Pb^{2+}, La^{3+} were the same as Cu^{2+} in 1.3.1. The instant and cumulative adsorption curves for Pb(II) and La(III) were shown in Fig. 7 and 9 respectively. It could be found in Fig. 7, that at the first 100 ml almost all the Pb(II) could be adsorbed onto the fiber, the saturated dynamic adsorption for Pb(II) is 246.41mg/g. About 98.1% Pb(II) could be eluted by 7 ml 1 mol/L HNO_3. (Fig. 8)

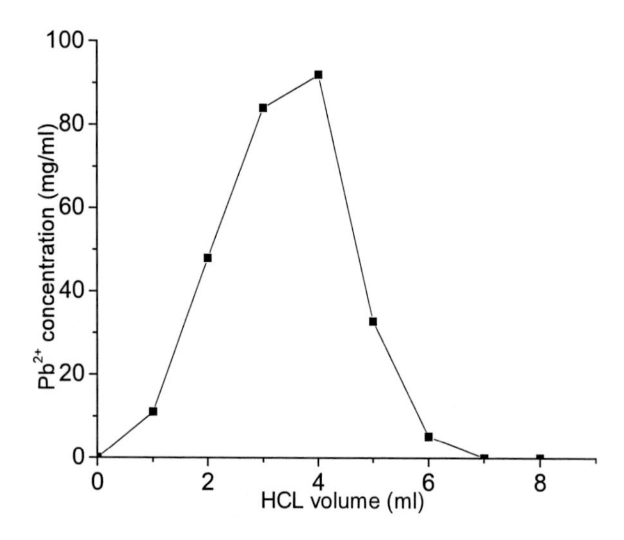

Figure 8. The desorption of Pb^{2+} from the grafted PTFE fiber by 1 mol/L HNO_3 elutions. The desorption condition: SV = 15BV/h□eluted ratio (%) = 98.1%

It also could be found from Fig.9 and Fig.10 that the quality of the dynamic adsorption for La(III) is 150.20mg/g, 99.5% of adsorbed La(III) could be eluted by 8 ml 1mol/L HNO_3.

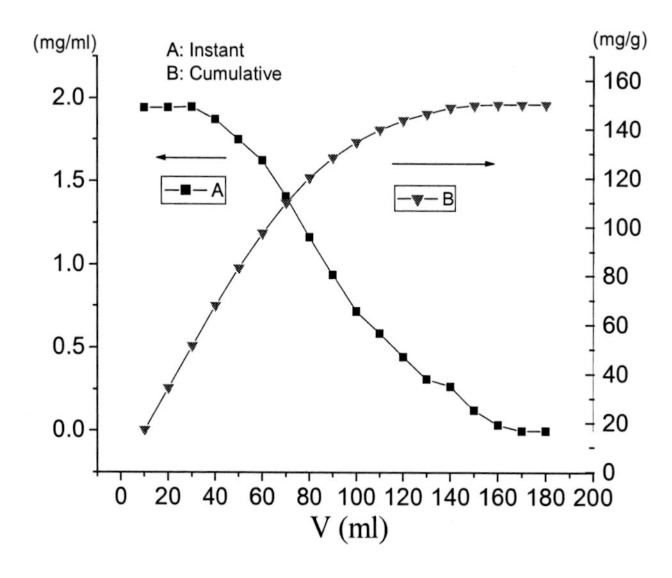

Figure 9. The dynamic adsorption of La^{3+} by grafted PTFE fiber; The adsorption condition: $[La^{3+}]_0$=2 mg/ml; pH=5;SV(Speed Velocity)=7.5 BV/h

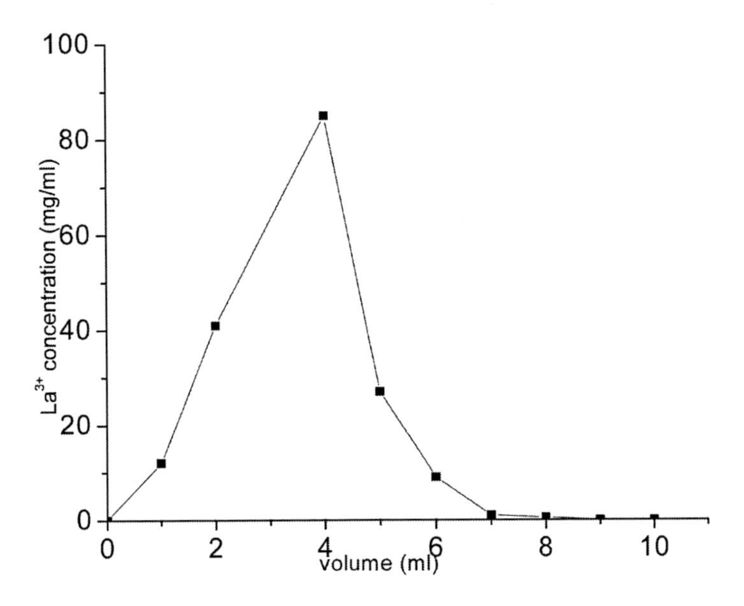

Fig.10 The desorption of La^{3+} from the grafted PTFE fiber by 1mol/L HNO_3 elutions; The desorption condition: SV=15BV/h; eluted ratio%=99.5%

2. γ-Radiation Induced PTFE Fiber Grafted Maleic Acid[12]

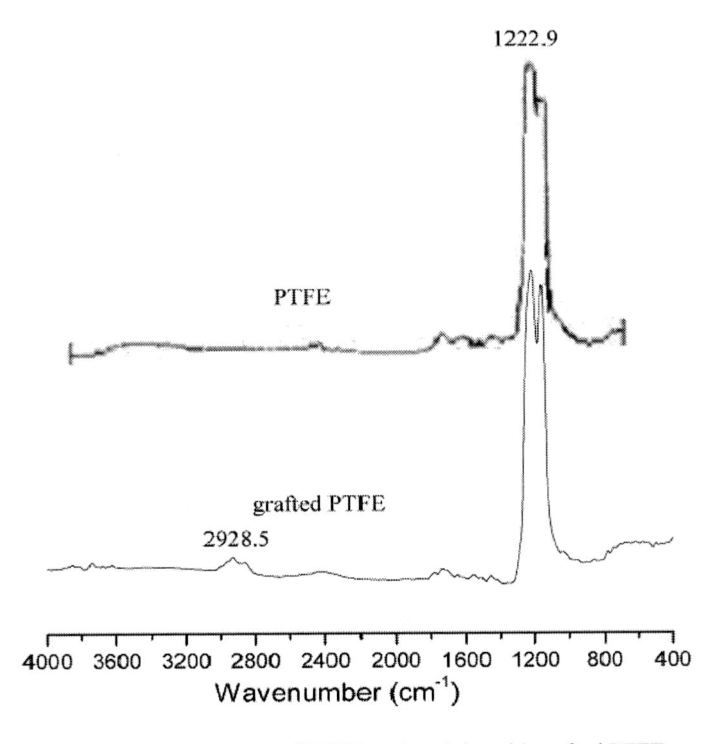

Figure 11. IR spectra of PTFE and maleic acid grafted PTFE

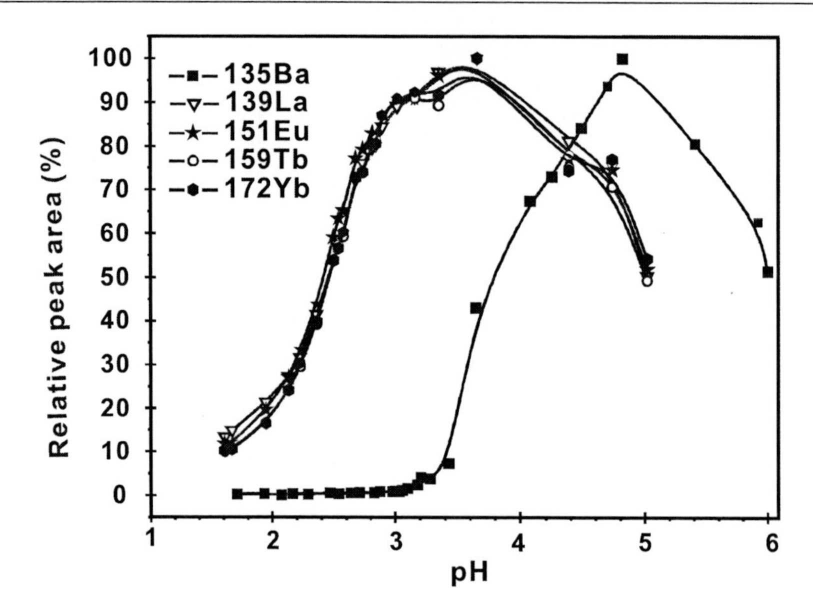

Figure 12. Effect of the pH of sample solution on the on-line solid phase extraction of trace REEs and Ba(II)

The γ-Radiation induced maleic acid grafting to PTFE fiber was carried out in acetone, AIBN was used as co-initiator, the irradiation dose is 40 kGy□0.8 kGy × 50 h□, the grafting percentage is 6.3%. The microscopic FT-IR spectrum of the PTFE fiber grafted maleic acid was shown in Fig.11, compared with the two curves in the Fig11□it could be found the peak at 2928 cm^{-1} is the stretching vibration of OH groups of aliphatic bi-carboxylic acid, but it was very difficulty to found the adsorption peak of C=O group, the reason is on the one side there are some interfere peaks happen to be at the same wave number as C=O group, on the another side the grafting efficiency is not very high and only limited sample could be reflected in the microscopic FT-IR spectrum experiments. The exchange capacity of the grafted maleic acid PTFE fiber is 0.61mmol/g by titration.

The PTFE grafted maleic acid fiber was used for the separation of Ba(II) with rare earth ions in the microanalysis of rare earth elements, for the removing of the interference of Ba(II). It could be found in Fig.12, Ba(II) could be separated with La(III), Eu(IV), Tb, Yb completely, when the pH is round 3.

3. Sulphonated PTFE-g-styrene Fiber via γ-Radiation [13,14]

The PTFE-g-styrene (PTFE-g-ST) fibers were prepared by radiation-induced grafting of styrene onto PTFE fibers using a simultaneous irradiation technique. A 100 ml glass flask containing PTFE fiber of known weight was immersed into the styrene monomer with concentrations in the range of 10-70 % (wt) and diluted with dichloromethane. The grafting solution was bubbled with nitrogen for 20 min. to create an inert atmosphere, then the flask was sealed. The flask was irradiated at ambient temperature using γ-rays from a ^{60}Co source to a total dose of 30Gy at dose rate 0.375 kGy/h for 80 h. The grafted fibers was removed, washed thoroughly with toluene and soaked therein overnight in order to remove the residual monomer and homopolymers adhered to the fiber surface. The grafted fibers were washed with acetone and dried in a vacuum oven at 60-70□ for 24 hours and weighed.

The PTFE-g-ST fiber was sulfonated by a mixture composed of 40% chlorosulfonic acid and 1,1,2,2-tetrachloroethane (v/v). The sulfonation reaction was carried out at 90□ for 4 hs under nitrogen. The sulfonated fiber was washed several times with 1,1,2,2–tetrachloroethane and dichloromethane to remove the excess chlorosulfonic acid after the sulfonation. The sulfonated fiber was hydrolyzed with 0.5 N KOH and regenerated with 1M HCL for 2 h. The fiber was washed with deioned water to remove the acid residul completely.

Fig.13 shown the TG analysis of PTFE, PTFE-g-ST and PTFE-g-ST-SO_3 fiber respectively, it could be found form Fig.13, the PTFE fiber was decomposed at ca. 500□, the styrene in PTFE -g - ST fiber was decomposed at ca. 400□, the sulfonic group in styrene ring was decomposed at ca. 250□

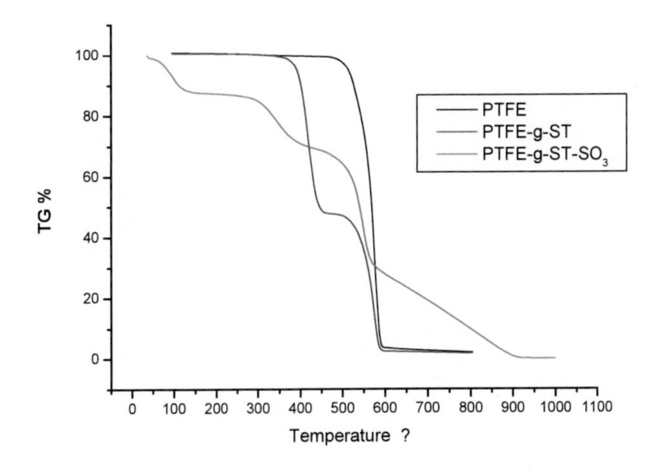

Figure 13. TG analysis of the 3 kinds of fibers (For the curve of the PTFE-g-ST-SO_3 fiber, the initial weight loss is the water contained in the fiber, when the temperature risen to ca. 100□)

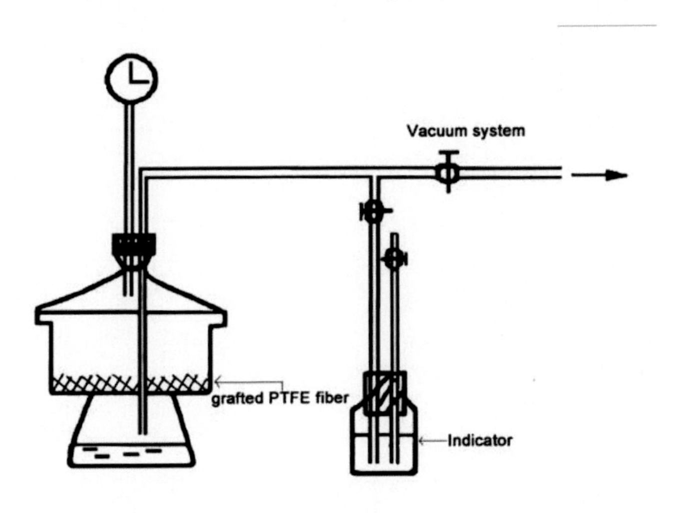

Figure 14. The apparatus for the determination of the acidity by Hammett indicators.

In order to investigate the influence of strong electronegativity of fluorine atom in PTFE matrix on the sulfonic acid group in the benzene ring of the grafted styrene, Hammett superacid indicators were used to estimate the superacidity of PTFE-g-ST-SO_3 fibers with different degree of grafting.

Hammett indicator is a sensitive, efficient and valuable method for the determination of the acidity of solid superacids.[15] The experiment was carried out in a vacuum drier shown as Fig 14. The indicator was store in a small bottle and drawing to the vacuum drier when the system at vacuum. The grafted PTFE fiber was set above the indicator. The system was put into 60□ hot water at the vacuum in order to make the indicator vapor charged to all the vacuum drier and contacted with the grafted PTFE fiber, the colour of the grafted PTFE fiber will be changed, when the acidity of the grafted PTFE fiber is strong enough.

Table 7 gives the relationship between the grafting degrees of the PTFE fiber and the Hammett acidity constants measured by superacid indicators, Hammett acidity indicators used in our's experiments included m-nitrotoluene (pKa = -11.99), p-nitrochlorobenzene (pKa = -12.70) and n-notrochlorobenzene (pKa = -13.16). The results showed that the acidity of the grafted fiber is decreased with the increase of the grafting degree of styrene onto the PTFE fiber, and the Hammett acidity constant is lower than –12.70, when the grafting degree is 21.7 %. In the other words, at this condition, the acidity of PTFE-co-St-SO_3H fiber is stronger than 100 % H_2SO_4. The experimental result indicated that the strong electronegativity of fluorine atom in PTFE matrix could induce the sulfonic acid group result in the superacidity.

Table 7. The relationship of the acidity of grafted PTFE fiber and the grafting percentage of polystyrene characterized by Hammett Indicator*

G% \ Indi.	**m-NT pka = -11.99**	**p-CNB pka = -12.70**	**m-CNB pka = -13.16**
21.7 %	+	+	—
23.0 %	+	—	—
28.2 %	—	—	—
45.6 %	—	—	—

- m-NT (m-nitrotoluene); p-CNB (p-chloronitrobenzene); m-CNB (m-chloronitrobenzene); "+"positive, "-"negative , (100% H_2SO_4 Hammett acidity function H_0 = -11.94)

There are great advantages for the polymeric superacids as catalysts, for example, the mild reaction conditions, less of the reaction byproducts, easy of the separation for the reaction products et al. When the novel PTFE-co-St-SO_3H fiber have been used as the polymer superacids catalyst for the esterification of HAc and 1-butanol, it could be found that both the reaction rate and the chemical conversion are better than D-72 (a commercial macroporous strong acidic ion exchange resin, made from the chemical plant of Nankai University). Fig. 15 showed the results.

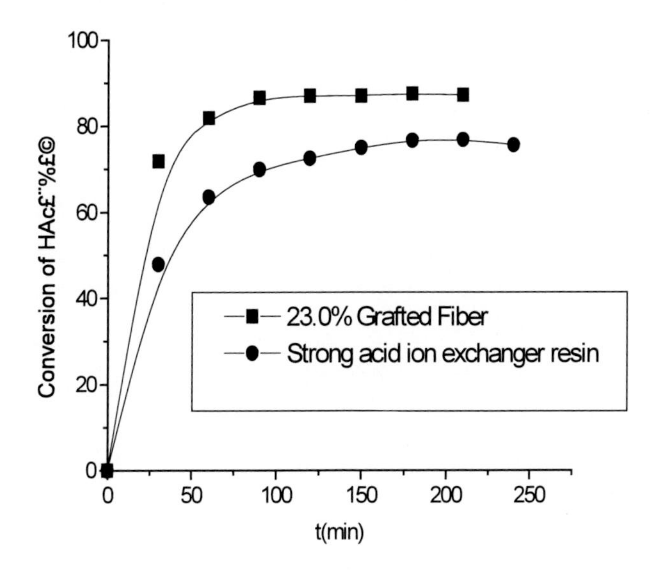

Figure 15. The comparison of the catalytic activity for the esterification of HAc and 1-butanol.

4. γ-Radiation Induced Chitosan Grafted N-isopropylacrylamide[16,17]

Hydrogels are crosslinked three-dimensional hydrophilic polymer networks. Poly-N-isopropyl- acrylamide (PNIPA) is the most widely studied thermosensitive hyrdogels. PNIPA in aqueous medium has its lower critical solution temperature (LCST) at 32□. The volume phase transition is reversible, which led to some potential applications such as temperature sensors, controlled drug releasing devices, aritificial muscles, and enzyme-activity controller.

Chitosan (CS) is a natural cationic polymer obtained from N-deacetylation of chitin [(1-4)-2-acetamido-2-deoxy-d-glucose], which is the second most abundant natural polymer on earth after cellulose. This polysaccharide is considered to be nontoxic, biodegradable and biocompatible. It has been used as an anticoagulant, a wound-healing accelerator, and drug delivery materials. CS exhibits a pH-sensitive behavior as a weak polymer base due to the large quantities of amino groups on its chain.

Graft copolymerization of vinyl monomers onto CS can introduce desirable properties and enlarge the field of the potential applications by choosing various types of side chains. In recent years, a number of initiator systems have been developed to initiate grafting copolymerization.

In our initial experiment the chemical grafting initiated by ammonium cerium nitrate have been carried out, but only limited NIPA monomer have been grafted onto chitosan successfully, even if the grafting reaction have been carried out at very serious conditions. The highest grafting percentage (PG) reached is 28 %, the thermo sensitive properties of the grafted CS is not very obviously.

For the preparation of NIPA highly grafted chitosan sample, the ^{60}Co induced co-radiation grafting copolymerization have been carried out. A series of CS-g-NIPA hydrogels were prepared in the following procedures: pure CS (0.5 g) dissolved in 5% aqueous acetic acid (25 ml) in a glass reaction bottle, the monomer was added to the CS solution. Mohr's salt (ammonium ferrous sulphate) was added to the mixture to minimize homopolymerization during irradiation. The solution was deoxygenated by purging with nitrogen for 30 mins. The sealed reaction bottles were irradiation at a dose from 3 KGy to 10 KGy. After irradiation, the product was extracted with methanol in a Soxhlet extractor for 48 hrs, in order to remove the unreacted monomer, homopolymer and other impurities. The hydrogel was dried at 40□ in a vacuum oven overnight. The grafting percentage (G %) could be calculated by the equation (1). The experimental results shown the NIPA monomer grafted onto the chitosan very successfully. Very high grafting percentage could be achieved, for example 400%, even if 600%. The highly grafted CS shown very special physical and chemical properties.

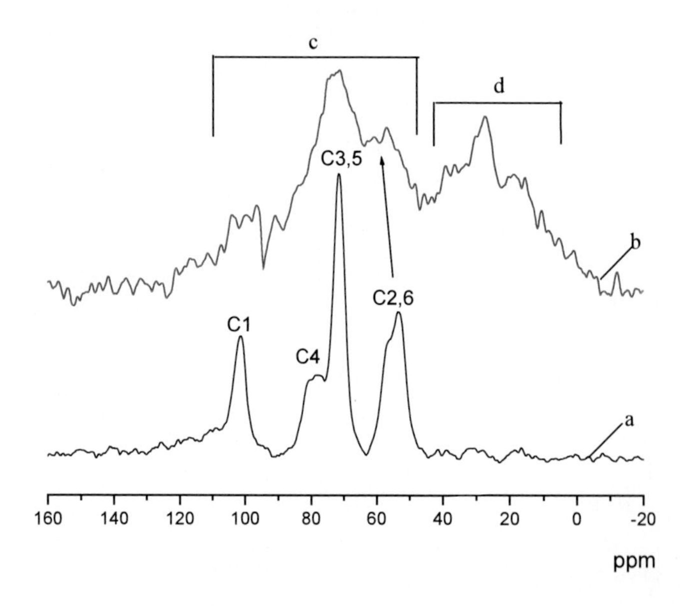

Figure 16. ^{13}C-NMR spectra of (a) CS and (b) CS-g-NIPA.

The ^{13}C CP/MAS NMR spectra of CS and CS-g-NIPA were recorded on a Varian UNITY plus 400 NMR spectrometer in solid state at a frequency of 100.6 MHz for ^{13}C nuclei and shown in Fig. 16. The peak assignment and ^{13}C chemical shifts of chitosan were 53.4 (C2), 56.8(C6), 71.4 (C3, 5), 79.2 (C4), and 101.6 (C1) ppm respectively, which is close to the previous NMR experimental result of chitosan[18]. Except for the signals of chitosan denoted as ‘c’ in Figure 16, several new peaks denoted as ‘d’ in Figure 16 were found at range from 10 – 45 ppm in grafted chitosan, which should be assigned to aliphatic signals of grafted PNIPA, this further confirms the grafting of PNIPA onto chitosan. On the other hand, it is known that chitosan molecule contains two reactive groups at C2 and C6 positions[19], compared with original chitosan there are signals in grafted chitosan down-field shift of C2 and C6, this implies that the grafting reaction may have taken placed on C2 and C6 positions. In addition, we also noticed that all peaks in grafted chitosan were broadened as the result of the paramagnetic effect of Fe^{3+} in ammonium ferrous sulphate on the ^{13}C NMR signals.

The dynamic thermogravimetric analysis of the CS-g-NIPA was carried out with a NETZSCH TG 209 (Germany). The experiments were performed at a heating rate of 10□/min under nitrogen from room temperature to 600□. The results are shown in Fig. 17.

The thermogram of chitosan exhibits two distinct stages. One is in the range of 40-120□ due to water loss and decomposition of polymers of low molecular weights, the other in the range of 220-480□ is ascribed to a complex process including dehydration of the sugar rings, depolymerization and decomposition of the acetylated and deacetylated units of the polymer. The differential thermogravimetric curve of the grafted chitosan shows three degradation steps. For the grafted chitosan, the second degradation stage starts at about 210□ which is lower than that of chitosan. In the third stage of degradation in the range of 370-480□, the grafted chitosan degrades less slowly than that of chitosan. The appearance of this stage indicates the structure of chitosan chains has been changed due to the grafting of NIPA chains.

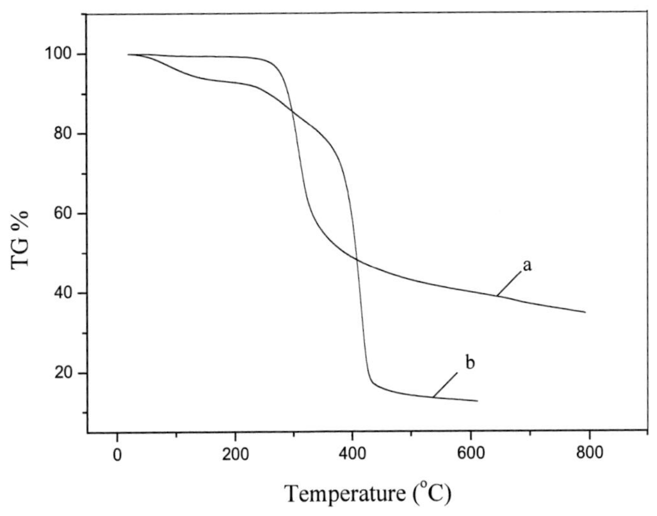

Figure 17. TGA curves of (a) CS and (b) CS-g-NIPA.

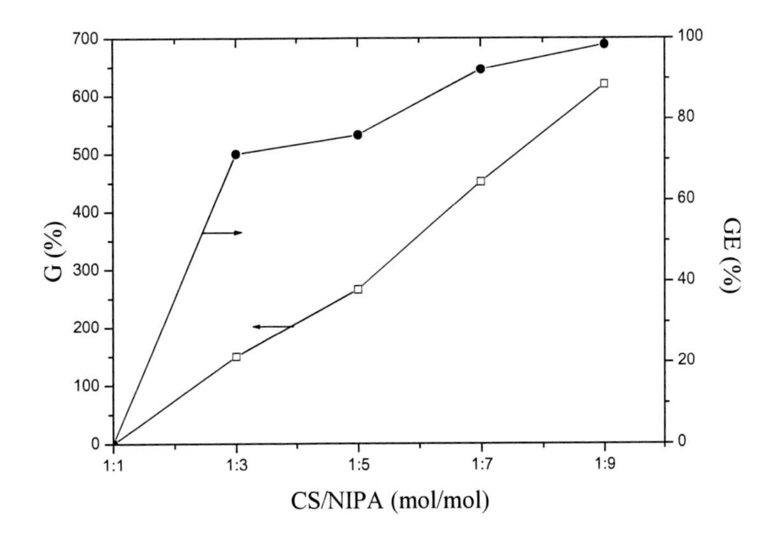

Figure 18. The effect of monomer concentration on the degree of grafting and grafting efficiency.

Fig.18 gives the variation in grafting percentage and grafting efficiency with monomer concentration. Grafting was carried out at room temperature at a total dose of 10 KGy. The results showed that there is an increase in the grafting percentage and grafting efficiency with increasing monomer concentration. During the grafting process, the monomers continuously diffused into the polymer. The ability of chitosan macroradicals to capture NIPA depends on the availability of NIPA molecules in their vicinity. Therefore, the increase in monomer concentration leads to an increase in grafting percentage and grafting efficiency. (grafting efficiency (GE %) were defined as $GE\%=\square(W_g-W_{o)}/W_m\square\times 100$, Where W_g, W_o and W_m are the weights of grafted copolymer, chitosan and monomer, respectively.)

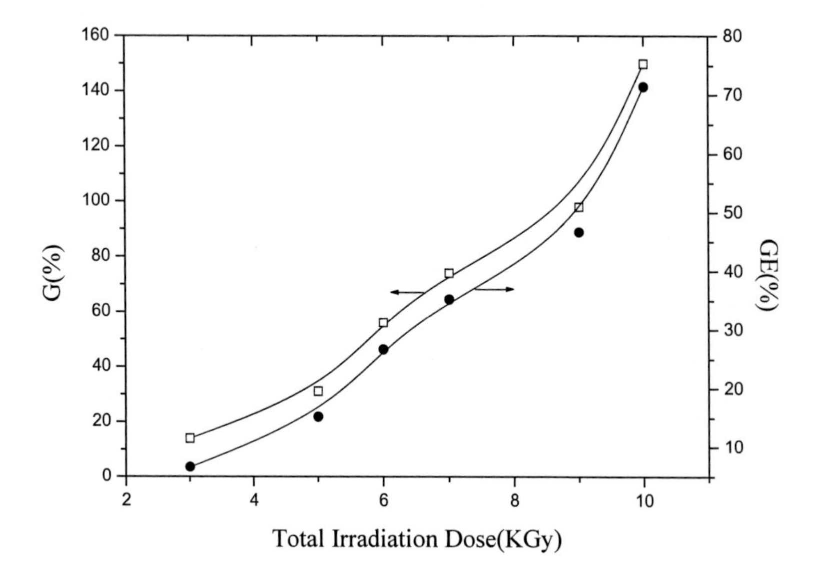

Figure 19. The effect of total irradiation dose on the degree of grafting and grafting efficiency.

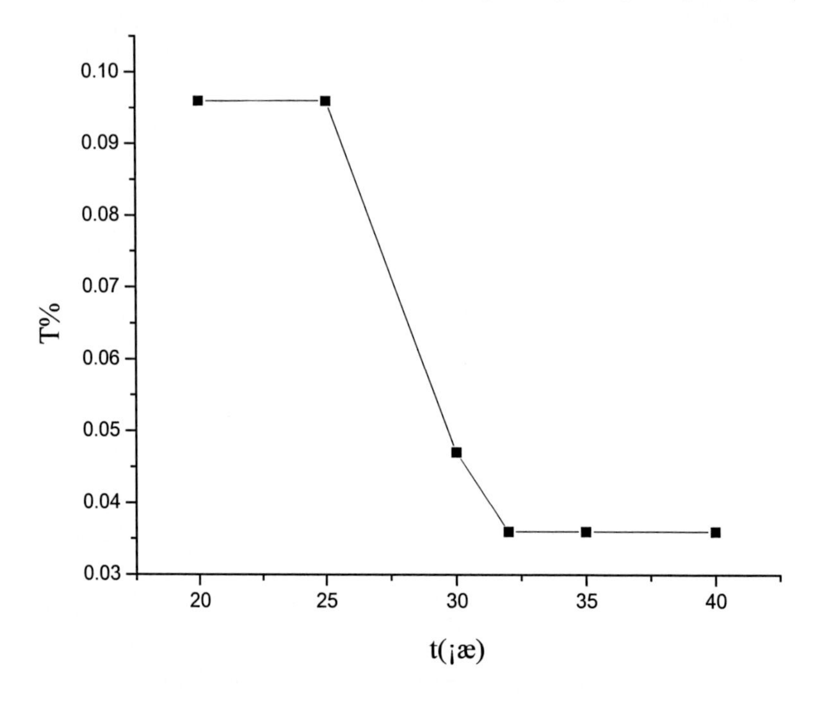

Figure 20. Light transmission at different temperatures (λ= 500 nm).

The effect of total irradiation dose on grafting percentage and grafting efficiency is presented in Fig 19. The graph exhibited an increase in grafting percentage and grafting efficiency with the increase in the dose range from 3 to 10 KGy. The increase of grafting percentage and grafting efficiency may be due to the increasing concentration of free radicals formed in the polymer substrate.

The thermosensitivity of the hydrogels is shown in Fig 20, measured by the transparency of the hydrogels under different temperature. There is a sharp decrease in the transparency when the temperature is about 28□, it is the LCST of the grafting hydrogels (620% grafting percentage).

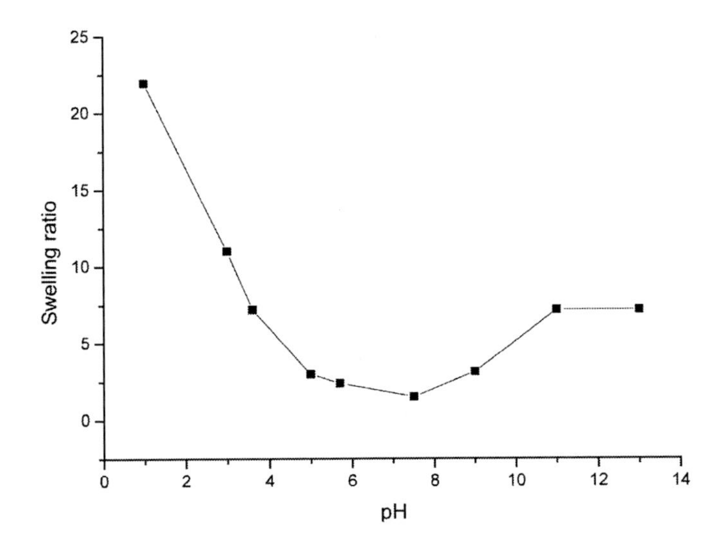

Figure 21. Effect of pH on swelling ratios.

The swelling ratios for hydrogels with grafting percentage in different pH solutions shown in Fig 21 indicate that the swelling ratios decrease with the increase of pH value of the buffer solutions. This is because PNIPA gels do not have pH sensitivity in the buffer solution but chitosan has.

Acknowledgement

I'm grateful my students Dr. Zhipeng Wang, Dr. Junfu Wei, Dr. Hong Cai, Mr. Wenbo Pei and my colleague Professor Xiuping Yan and my friend Senior Engineer (Professor) Yangeng Zhang.

References

[1] Zhang Zhengpu, Wang Zhipeng, Wu Qiang, Zhang Baogui, Zhang Yangeng, (2004), The application of polytetrafluoroethylene fiber grafted acrylic acid as a cation exchanger for the removing of Cu^{2+}, *Acta Polymerica Sinica*, *(***1***)*, 84-87

[2] Zhaohui Wang, Zhengpu Zhang, Zhipeng Wang, Liwen Liu, Xiuping Yan, (2004), Acrylic acid grafted polytetrafluoroethylene fiber as new packing for flow injection on-line microcolumn preconcentration coupled with flame atomic absorption spectrometry for determination of lead and cadmium in environmental and biological samples. *Analytica Chimica Acta*, **514**, 151-7

[3] Junfu Wei, Zhipeng Wang, Jing Zhang, Yueying Wu, Zhengpu Zhang, Chunhua Xiong, (2005), The preparation and the application of grafted polytetrafluoroethylene fiber as a cation exchanger for adsorption of heavy metals, *Reactive & Functional Polymers*, **65**, 127-134

[4] Zheng Pu ZHANG, Zhi Peng WANG, Bao Gui ZHANG, Yan Geng ZHANG, (2003), The Application of Polytetrafluoroethylene (PTFE) Fiber Grafted Acrylic Acid as a Cation Exchanger for Removing Cu^{2+} , *Chinese Chemical Letter*, **14** (6), 609

[5] Wang ZH, Wang ZP, Zhang ZP, Yan XP, (2005), Determination of trace copper and nickel in environmental and biological samples by flow on-line microcolumn preconcentration flame AAS using acrylic acid-grafted polytetrafluoroethylene fiber for column packing, *Atomic Spectroscopy*, **26**(1), 34-39

[6] Cai-Yan Lu, Xiu-Ping Yan, Zheng-Pu Zhang, Zhi-Peng Wang, and Li-Wen Liu, (2004), Flow injection on-line sorption preconcentration coupled with hydride generation atomic fluorescence spectrometry using a polytetrafluoroethylene fiber-packed microcolumn for determination of Se(IV) in natural water, *J. of Analytical Atomic Spectrometry*, **19**, 277-281.

[7] T Yamaki, M Asano, Y Maekawa, Y Morita, T Suwa, J Chen, N Tsubokawa, K Kobayashi, H Kubota, M Yoshida (2003), Radiation grafting of styrene into crosslinked PTEE films and subsequent sulfonation for fuel cell applications, *Radiation Physics and Chemistry* , **67**, 403–407

[8] M M Nasef (2002), Structural investigation of polystyrene grafted and sulfonated poly (tetrafluoroethylene) membranes, *European Polymer J. 38*, 87-95

[9] M M Nasef, E A Hegazy (2004), Preparation and applications of ion exchange membranes by radiation-induced graft copolymerization of polar monomers onto non-polar films, *Prog. Polym. Sci.*, **29**, 499-561,

[10] Determination method for exchange capacity of cation exchange resins, National standard of P.R.China, GB 8144-87

[11] V. S. Soldatov, (2000), A simple method for the determination of the acidity parameters of ion exchangers, *React. & Funct. Polym.* **46**(1), 55-58

[12] Wang ZH, Yan XP, Wang ZP, Zhang ZP, Liu LW, (2006), Flow Injection On-Line Solid Phase Extraction Coupled with Inductively Coupled Plasma Mass Spectrometry for Determination of (Ultra)Trace Rare Earth Elements in Environmental Materials Using Maleic Acid Grafted Polytetrafluoroethylene Fibers as Sorbent, *Journal of The American Society for Mass Spectrometry,* **17** (9): 1258-1264

[13] Wei Junfu, Li Hui, Wang Zhipeng, Zhang Zhengpu, Zhang Yangen, (2006), Preparation of strong acidic polytetrafluoroethylene cation exchange fibers, *Ion Exchange and Adsorption,* **22**(1),16-24

[14] Wei Junfu, Li Hui, Zhang Zhengpu, Hu Rongxia, A method for the preparation of a strong acidic ion exchange fiber. *Chinese Pat.* ZL 2004 1 0072017.4

[15] a. Hashimoto K., Masuda T., Motoyama H., Yakushiji H., Ono M., (1986) Method for measuring acid strength distribution on solid acid catalysts by use of chemisorption

isotherms of Hammett indicators. Industrial & Engineering Chemistry Product Research and Development , **25**(2), 243-50.
b. Ghosh A. K., Curthoys G., (1983), Characterization of zeolite acidity by n-butylamine titration. A comparative study using Hammett indicators and calorimetry. Journal of the Chemical Society, Faraday Transactions 1: Physical Chemistry in Condensed Phases, **79**(1), 147-53.

[16] Hong Cai, Zhengpu Zhang, Pingchuan Sun, Yangen Zhang, Binglin He, (2005), Preparation of hydrogels based on poly-N-isopropylacrylamide grafted chitosan via γ-radiation and their properties, *Acta Polymerica Sinica*, **(5),**709-713

[17] Hong Cai, Zhengpu Zhang, Pingchuan Sun, Binglin He, Xiaoxia Zhu (2005) Synthesis and characterization of thermo- and pH-sensitive hydrogels based on Chitosan-grafted N-isopropylacrylamide via γ-radiation, *Radiation Physics and Chemistry*, **74**, 26-30

[18] Yoksan, R., Akashi, M., Biramontri, S., Chirachanchai, S., (2001) Hydrophobic Chain Conjugation at Hydroxyl Group onto γ-Ray Irradiated Chitosan. *Biomacromolecules* **2**, 1038.

[19] Liu, P.F., Zhai ,M.L., Wu, J.L., (2001), Study on radiation-induced grafting of styrene onto chitin and chitosan. *Radiat. Phys.Chem.* **61**, 149

In: Radiation Physics Research Progress
Editor: Aidan N. Camilleri, pp. 407-423

ISBN: 978-1-60021-988-7

Chapter 12

CAN WE ACCELERATE NEUTRONS? EXPERIMENTAL AND MONTE CARLO STUDY OF THERMAL TO 14MEV NEUTRON CONVERSION IN TRIGA REACTOR BY MEANS OF SECONDARY NUCLEAR REACTIONS

T. El Bardouni[1], and E. Chakir[2]

[1] ERSN/LMR, Dept of Physics, Faculty of Sciences,
POB 2121 Tetuan 93000 – Morocco
[2] LRM/EPTN, Physical department, Faculty of sciences Kenitra (Morocco)

Abstract

This work is devoted to the theoretical and experimental study of the feasibility of fast 14MeV neutron production from thermal neutrons by using an in-core cylindrical converter. Prediction of the characteristics of this alternate 14MeV neutron source was performed by means of analytical calculations and by our Monte Carlo based code. The neutron flux density and energy dependent spectrum have been measured, at the irradiation position, by foil activation detectors in conjunction with the iterative unfolding method adopted in the code SANDII. Measurements show that the efficiency of the hollow cylindrical ^{6}LiD converter, under study, is $1.3 10^{-4}$ when considering neutrons of energy above 12MeV and $1.5 10^{-4}$ while the energy range is extended to 10MeV. The validity of our Monte Carlo based code was proved by comparison with the available data in the literature. In fact, our code reproduces correctly the efficiency and the neutron energy spectrum due to an ^{6}LiD converter, in plane geometry, designed to be used in thermal column of reactor. Good agreement was obtained between our Monte Carlo calculation and the converter efficiency we have measured. However, slight discrepancy was observed in spectra. An application of the in-core converter to averaged cross sections of some (n,γ) captures and threshold reactions was performed.

Key-words: thermal-, 14MeV- neutrons, neutron- spectra, LiD- converter, efficiency, foil activation detectors, Monte Carlo calculation, averaged cross section.

Introduction

In a reactor, the neutron field determines the power production and the generation of radioactive nuclides. It has also an important contribution to radiation damage and biological effects. To estimate the magnitude of this contribution, knowledge of neutron field strength and the spectrum shape is fundamental. The aim of the present work is to characterise the fast neutron environment produced by use of an in-core ^{6}LiD converter in conjunction with thermal neutrons from the TRIGA MARK II reactor of Atominstitut of Vienna. Both measurement and Monte Carlo calculation of the yield are carried and compared. Energy dependent neutron flux measurement can be made with activation detector technique which is based on the production of radioactive nuclides through nuclear reactions induced by neutrons in suitable materials. This method is completed by high resolution γ-ray spectrometry. Measured neutron flux is used to determine some averaged reaction cross sections.

Theoretical Analysis

14MeV neutrons are usually generated by means of the fusion process T(d,n)^{4}He. Generally, this reaction is realised by accelerator devices (Csikai *et al.* 1987). Some authors (Frigerio *et al.* 1971, Perry *et al.* 1985, Kobayashi *et al.* 1988) proposed an alternative flat source of 14MeV neutrons which can simulate adequately the neutrons escaping from the fusion plasma in thermonuclear reactor. This alternative source is based on the design of a neutron converter, by using compounds of enriched lithium and deuterium, in conjunction with thermal neutrons of the reactor.

Data

During exposure of the converter to a thermal neutron flux, the following processes take place successively inside the lattice. The production of tritons in the primary reaction

$$^6\mathrm{Li} + \mathrm{n_{th}} \rightarrow {}^4\mathrm{He} + \mathrm{t} \qquad Q = 4.79\mathrm{MeV}$$

and the fusion reaction (secondary reaction)

$$\mathrm{D} + \mathrm{t} \rightarrow {}^4\mathrm{He} + \mathrm{n_{14MeV}} \qquad Q = 17.58\mathrm{MeV}$$

that competes with slowing down of triton and generates 14MeV neutrons.

The absorption of thermal neutrons on ^{6}Li in the primary reaction produces monokinetic triton which is sufficiently energetic (2.73MeV) to induce fusion reaction during slowing down through the ^{6}LiD layer. The first reaction, due to its large cross section (936b) (IAEA, 1987) at thermal energy, represents the main source of tritons. The energy dependent cross section of the second reaction, represented in Fig.1a, is given in the (IAEA,1987) reference as a function of deuteron energy and has been converted to a function of triton energy.

Fast neutrons, could also be induced by absorption of tritons on ^{6}Li and ^{7}Li according to the following reactions whose contribution will be taken into account.

$$^6\text{Li} + \text{t} \rightarrow {}^8\text{Be} + \text{n} \qquad Q = 16.02 \text{ MeV}$$

$$^7\text{Li} + \text{t} \rightarrow {}^9\text{Be} + \text{n} \qquad Q = 10.44 \text{ MeV}$$

The energy dependent cross sections of these two reactions, given by Lone (1980), are represented in Fig.1b and 1c.

Energy loss of tritons through the LiD sample is calculated by means of TRIM code published by Andersen and Ziegler (1977).

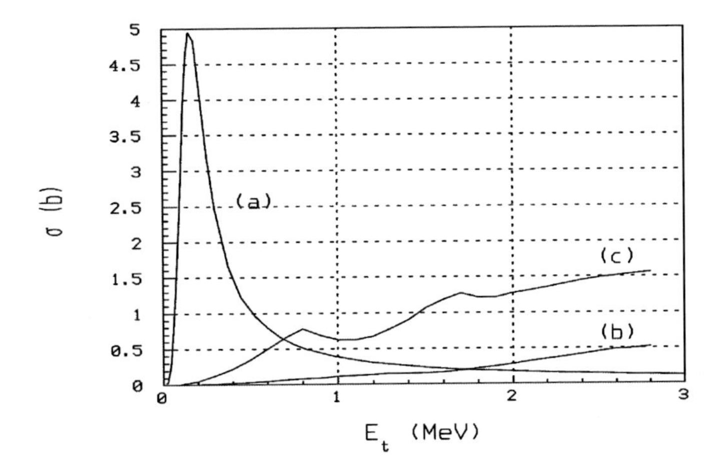

Figure 1. Energy dependent cross sections for (t,n) reactions on D (a), ^{6}Li (b) and ^{7}Li (c).

Yield and spectrum calculation

If we define

$$Y_1 = \frac{n_1\sigma_1}{\Sigma_t^1}\left(1 - e^{-\Sigma_t^1 L_0}\right)(\Phi_0 S) \tag{1}$$

as the number of tritons produced by the primary reaction $^6\text{Li}(n_{th},t)$ and

$$Y_2^i = \int_0^{E_t^{max}} \frac{n_2^i \sigma_2^i(E_t)}{\rho S(E_t)} dE_t \tag{2}$$

as the thick target yield produced by (t,n) reaction on nuclide i, the fast neutron yield can be calculated, for one thermal neutron absorbed, by the following equation

$$Y = \frac{n_1\sigma_1}{\Sigma_t^1}\left(1 - e^{-\Sigma_t^1 L_0}\right)\int_0^{E_t^{max}} \sum_{i=1}^{3} \frac{n_2^i}{\rho} \frac{\sigma_2^i(E_t)}{S(E_t)} dE_t \tag{3}$$

where n_1 and n_2 denote the atom densities of targets in the primary and secondary reactions respectively; Σ_t^1 and σ_1 are the total and the primary reaction cross section at thermal energy respectively; $\sigma_2^i(E_t)$ denotes the energy dependent cross section of (t,n) reaction on the ith isotope; L_0 is the converter thickness; ($\Phi_0 S$) is the total number of thermal neutrons arriving on the area S of the converter; $E_t^{max} = 2.73$MeV is the maximum energy of tritons; ρ is the converter density and $S(E_t)$ is the stopping power of the material through which tritons are slowing down.

The neutron spectrum, normalized to unity, can be derived from equation (3) and written for the particular direction as

$$N(E_n,\theta)dE_n d\Omega = \frac{n_1\sigma_1}{\Sigma_t^1}\left(1 - e^{-\Sigma_t^1 L_0}\right)\sum_{i=1}^{3} \frac{n_2^i}{\rho S(E_t)} \frac{d\sigma_2^i(E_t)}{d\Omega}\left(\frac{dE_t}{dE_n}\right)_i dE_n d\Omega \tag{4}$$

Calculation of the neutron yield can be performed by numerical integration of equation (3). In our previous works (El Bardouni *et al.* 1995, 1996) we presented a theoretical study about the feasibility of conversion of thermal to 14MeV neutrons with an enriched ^{6}LiD converter. Our analytical results, obtained by using the known cross sections of (t,n) reactions and the stopping power of the material for tritons, have been compared to some experiments (Lone *et al.* 1980, Kobayashi *et al.* 1988) achieved under thermal neutron beam from the thermal column of a reactor. The analytical calculation slightly overestimates the fast neutron yield. This can be explained by the fact that, in such calculations, neither the leakage nor the multiple scattering of different particles arriving on the converter or created inside it could be taken into account.

Knowing that such a complicated problem can not be tackled easily by simple analytical methods, we decided to apply more detailed calculations based on the Monte Carlo technique. This allowed us to simulate the physical events occurring within the converter when irradiated in thermal neutron field. We established a computing program to Characterise the fast neutron field so generated.

In their experiments, Lone *et al.* (1980) and Kobayashi *et al.* (1988) used a simple square converter and exposed it to a pure thermal flux. These conditions made it easier to perform both measurement and calculation. A more complicated situation was encountered when Napier *et al.* (1976) designed an in-core converter containing a mixture of D_2O and ^{6}LiD, and measured abnormally high yield. Unfortunately, Wysocki *et al.* (1978) found an error in these measurements which was later confirmed by Eckhoff *et al.* (1978).

As discussed in our earlier papers, square shape converter has proved satisfactory for use in the conversion of thermal to 14MeV neutrons and the encouraging results in testing the validity of our code were shown. In the present work, we aim at determining the characteristics of fast neutrons produced in reactor core by means of an enriched ^{6}LiD cylindrical converter. Since this converter is movable, we performed measurements with and without it (El Bardouni *et al.* 1997, El Bardouni 1997), at the same irradiation position, to prove its efficiency in extending the use of reactor neutrons to produce high threshold reactions. The obtained results were used to test the validity of our Monte Carlo calculations under such conditions.

Monte Carlo Calulation

In order to simulate the characteristics of the ^{6}LiD converter, the physical events summarized in Fig.2 are considered. The particle history is determined by sampling the problem variables by means of random numbers uniformly distributed between 0 and 1. These variables are constructed according to physics laws. Among the variables of our study, there are co-ordinates of the starting point, path length of particles, their direction, energy of interaction, etc...

In our detailed simulation, processes other than neutron emission that contribute to the decrease in the energy and the number of tritons are considered as well as absorption and scattering of fast neutrons on the sample components.

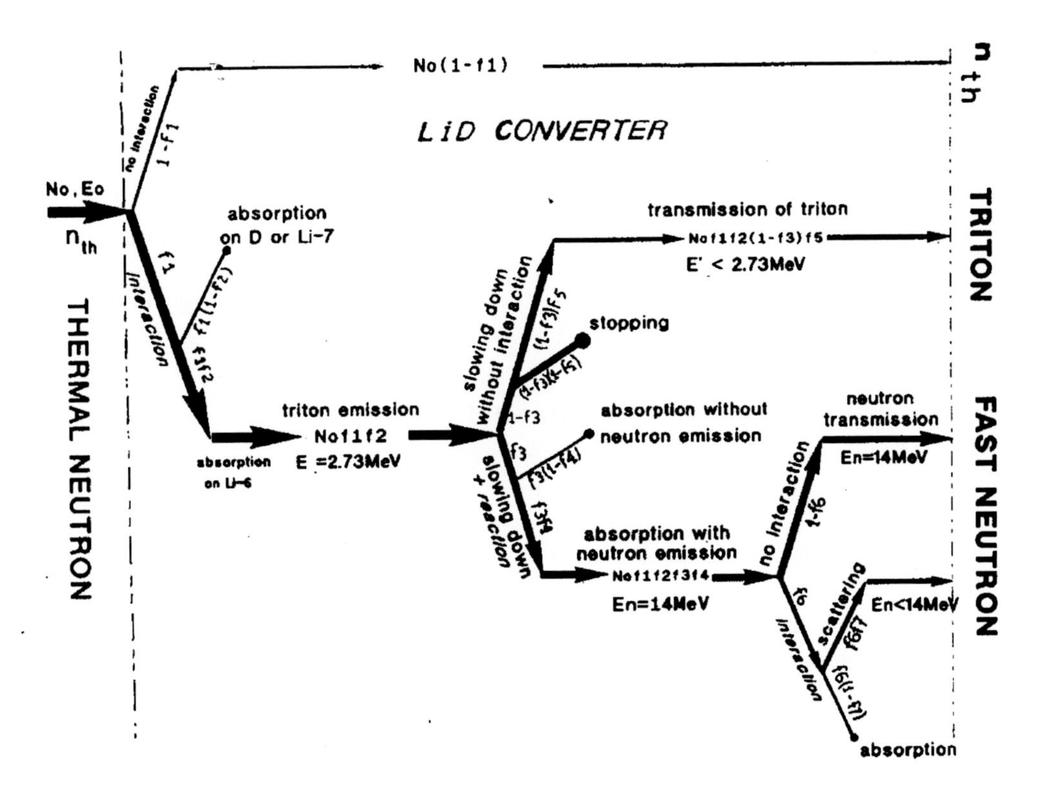

Figure 2. Flow diagram of the main processes occuring in the ^{6}LiD converter irradiated by thermal neutrons.

Experimental Procedure

Conventional neutron activation analysis is widely used in different fields and particularly in trace element determination. In general, this analysis can be achieved by the comparative method and rarely by the absolute one. In the first method, the use of appropriate standards renders a large number of correcting factors related to irradiation and activity measurements unnecessary. Unfortunately, in some cases, the choice of standards can be limited and many problems must be resolved by turning to the absolute method. This method is fruitful especially if some data are known with good accuracy. Among these data, we can outline the spatial flux density and the energy dependent spectrum of the neutron field.

Experimental Arrangement

In the Atominstitut TRIGA MARK II Reactor, there are five irradiation facilities. The detailed description of these systems is given by Salahi et al (1988) and Grass et al (1994). In our experimental study, we used the Vertical Fast Irradiation and Measurement System (FIMS) that leads to the irradiation position in the ring F of the core. In this system the transfer time, from irradiation to measurement position, is 300ms when using nitrogen as expelling-gas and 120ms while replacing it by helium (Grass 1996). The neutron environment, at this irradiation facility, has a large component of fast flux which can be enhanced by pneumatically inserting a removable ^{6}LiD converter. The heat produced in the converter, at 250kW reactor power, is removed by cooling. The production of ^{6}LiD-converter was achieved according to the method described by Salahi *et al.* (1988) which takes into account the recommendations of Zillner *et al.* (1981). According to Zillner the optimum size of the hollow cylindrical converter is 150 mm in height and 1 mm in thickness.

Materials to be irradiated are wrapped in polyethylene tubes of 1mm thickness, 10 mm outer diameter and about 30 mm length. Polyethylene tubes minimise the activity induced in the container material and avoid sample contamination while passing through highly activated metallic transfer pipes. During an experiment, the sample container passes through the counting chamber before irradiation and stops in front the detector after irradiation. A slowing-down and stopping device, installed in this facility, prevents sample destruction. After measurement, the sample is transferred to a shielded sample storage container.

In the activity measurements, we used the high-rate high-resolution γ-spectrometry system. It includes an n-type high purity Germanium detector of 14% relative efficiency and an improved pre-loaded amplifier (Westphal 1990, Westphal 1993) combined with the Westphal's Loss Free Counting (LFC) system. The LFC system enables real time correction of pileup and dead time counting losses. It is based on the virtual pulse generator method (Westphal 1987). A PC based ACCUSPEC 8000 channels analyser was used to acquire and store gamma spectra. In the case where long lived nuclei are investigated, a 100ccm Ge-Li detector is used for long measurements of moderate activities. The efficiency of the two detectors had been calibrated with Amersham standard sources at the counting positions. For each sample, the γ self-absorption correction factor was determined by means of a Monte Carlo program.

Neutron Flux and Spectrum Measurement

Two kinds of activation measurements have been performed, one for the measurement of thermal and fast neutron fluxes in the ring F of the reactor core without the converter and another for the determination of fast flux and spectrum shape by means of the converter inserted in the same irradiation position. In both measurements, high purity dosimetry materials that were available either in the form of thin foils, wires or powder were irradiated.

Table 1. Nuclear reaction data corresponding to materials irradiated

a- without converter.

Material	Reaction	Nuclide	$T_{1/2}$ [a)]	Eγ (keV) [a)]	Iγ (%) [a)]
Al-Au (0.112%)	$^{197}Au(n,\gamma)^{198}Au$	^{198}Au	2.69d	412	95.6
Cu	$^{63}Cu(n,\gamma)^{64}Cu$	^{64}Cu	12.7h	1346	0.6
Al-Au (0.112%)	$^{27}Al(n,p)^{27}Mg$	^{27}Mg	9.46min	1014	28
Zr	$^{90}Zr(n,2n)^{89}Zr$	^{89}Zr	3.27d	909	99.4
Nb	$^{93}Nb(n,2n)^{92m}Nb$	^{92m}Nb	10.13d	935	99.2

b- with converter.

Material	Reaction	Nuclide	$T_{1/2}$ [a]	Eγ (keV) [a]	Iγ (%) [a]
Al-Au (0.112%)	$^{197}Au(n,\gamma)^{198}Au$	^{198}Au	2.69d	412	95.6
	$^{27}Al(n,\gamma)^{28}Al$	^{28}Al	2.24min	1779	100
Cu	$^{63}Cu(n,\gamma)^{64}Cu$	^{64}Cu	12.7h	1346	0.6
NaF	$^{23}Na(n,\gamma)^{24}Na$	^{24}Na	15.02h	1369	100
				2754	99.9
CaF_2	$^{19}F(n,p)^{19}O$	^{19}O	26.91sec	1357	50.4
NaF	$^{23}Na(n,p)^{23}Ne$	^{23}Ne	37.24sec	440	33
NaCl	$^{23}Na(n,\alpha)^{20}F$	^{20}F	11.16sec	1634	100
	$^{24}Mg(n,p)^{24}Na$	^{24}Na	15.02h	1369	100
Mg				2754	99.9
	$^{25}Mg(n,p)^{25}Na$	^{25}Na	59.1s	585	13.8
	$^{27}Al(n,p)^{27}Mg$	^{27}Mg	9.46min	1014	28
Al-Au (0.112%)	$^{27}Al(n,\alpha)^{24}Na$	^{24}Na	15.02h	1369	100
				2754	99.9
	$^{47}Ti(n,p)^{47}Sc$	^{47}Sc	3.42d	159	68
Ti	$^{48}Ti(n,p)^{48}Sc$	^{48}Sc	1.83d	984	100
				1038	97.5
				1312	100
Zr	$^{90}Zr(n,2n)^{89}Zr$	^{89}Zr	3.27d	909	99.4
Nb	$^{93}Nb(n,2n)^{92m}Nb$	^{92m}Nb	10.13d	935	99.2
KIO_3	$^{127}I(n,2n)^{126}I$	^{126}I	12.93d	388	35
				666	33.9

a) These data are taken from "Table of Isotopes (Leederer et al. 1978)"

The selection of a best set of activation detectors depends on the neutron energy of interest. Slow neutrons are usually detected by (n,γ) reactions. For medium and high energy neutron detection, threshold activation foils are commonly used. They have different responses, at different energies, that are proportional to the excitation function of the induced reactions. The difference in the response of activation foils is the basis for the unfolding methods (McELroy et al. 1967) in the determination of spectrum shape from activation integrals. In order to cover the whole energy range of the spectrum, a large number of nuclear reactions induced in different materials, irradiated under converter, have been studied. Table 1 summarises the materials used in our spectral shape evaluation and the corresponding nuclear reactions.

Activation Technique

Consider a target containing N_0 isotopes X. During the irradiation, the number N of isotopes Y produced per unit of time is proportional to the particle flux ϕ ($cm^{-2}\ s^{-1}$), the number N_0 and the cross section σ of the particular nuclear reaction X(n,b)Y. Isotope Y decays with a decay constant λ. If the target is irradiated during a time t_i and t_w is the waiting time, elapsed between the end of irradiation and the beginning of activation measurement, the activity of the produced nuclide of interest Y will be given by the activation formula

$$A = N_0\sigma\phi\left(1 - e^{-\lambda t_i}\right)e^{-\lambda t_w} \tag{5}$$

If the sample activity is measured, by means of a gamma spectrometer, between the times t_1 and t_2 ($t_c= t_2-t_1$), its average value $\overline{A}$, during the counting time t_c, is given by

$$\overline{A} = \frac{N_0\sigma\phi}{t_c} I_\gamma \varepsilon\left(1 - e^{-\lambda t_i}\right)\int_{t_1}^{t_2} e^{-\lambda t}dt \tag{6}$$

where $I\gamma$ is the number of photons per disintegration and ε the detector efficiency. This equation can be written as

$$\overline{A} = \frac{N_0\sigma\phi}{\lambda t_c} I_\gamma \varepsilon\left(1 - e^{-\lambda t_i}\right)e^{-\lambda t_w}\left(1 - e^{-\lambda t_c}\right) \tag{7}$$

The saturation activity, defined as the value obtained at the end of a very long irradiation ($t_i \rightarrow \infty$), has the expression

$$A_{sat} = N_0\sigma\phi = \frac{\overline{A}\lambda t_c e^{\lambda t_w}}{I_\gamma \varepsilon\left(1 - e^{-\lambda t_i}\right)\left(1 - e^{-\lambda t_c}\right)} \tag{8}$$

The quantity

$$R = \sigma\phi = \frac{\overline{A}\lambda t_c e^{\lambda t_w}}{N_0 I_\gamma \varepsilon\left(1-e^{-\lambda t_i}\right)\left(1-e^{-\lambda t_c}\right)} \tag{9}$$

defined as the saturation activity for one target nucleus, is called the infinitely dilute saturation activity or the reaction rate.

If the cross section σ of the considered reaction is known the monokinetic particle flux can easily be determined from the measurement of the reaction rate. In the case of flux density that changes with particle energy, the reaction rate is defined by the activation integral equation

$$R = \int_{E_{\min}}^{E_{\max}} \sigma(E)\varphi(E)dE \tag{10}$$

where $\sigma(E)$ is the energy depending cross section, $\varphi(E)$ the flux density and E_{min}, E_{max} the energy range of spectrum. The averaged cross section $\overline{\sigma}$ of the observed process can be defined as

$$\overline{\sigma} = \frac{\int_{E_{\min}}^{E_{\max}} \sigma(E)\varphi(E)dE}{\int_{E_{\min}}^{E_{\max}} \varphi(E)dE} \tag{11}$$

Thermal and epithermal flux profiles are commonly determined by (n,γ) reaction rate measurement. These reactions have very known cross section σ_{th}, at thermal energy, and resonance integral I (Mughabghab et al. 1973, IAEA 1987). Following the convention of Hogdahl (1962), the reaction rate can be written as

$$R = \sigma_{th}\phi_{th} + I\phi_{ep} \tag{12}$$

and by adequate choice of two (n,γ) reactions, the thermal flux ϕ_{th} and the epithermal flux ϕ_{ep} can be evaluated. Fast spectrum determination is achieved by use of several activation reactions with different energy dependence of cross sections. A set of linear integral equations is obtained for the flux $\varphi(E)$.

Unfolding Method

A large number of unfolding programs can be found in the literature. SANDII (McELroy 1967) is one of the best known programs. It is one of the most often referred to iterative code which is designed to provide a best fit of neutron spectrum for a given input set of infinitely dilute foil activities. In the code, the neutron energy range between 10^{-10} and 18 MeV is represented in 620 intervals and the solution spectrum is given for 621 points. The main problem is to solve a system of m linear activity equations for 621 unknowns, where m (m << 621) is the number of foils used. The procedure consists of selecting an initial approximated spectral form, based on all the available physical information, which is modified, by iterative method, to a form acceptable as appropriate solution.

Analysis and Results

Calculation Results

From Fig.1 we can see that the T-D reaction is the reaction of primary interest. As we considered an enriched ^{6}LiD converter, the production of neutrons from T-^{6}Li reaction is large enough to render its inclusion in the calculations necessary. In this case, the contribution of the T-^{7}Li reaction becomes less significant. As the energy dependent cross section of the fusion reaction T-D peaks at low energies of tritons and as this reaction dominates the 14MeV neutron production, the emission of the majority of these neutrons is isotropic in the centre of the mass system. So we assumed that more detailed calculation was not of high interest. Our assumption is corroborated by the isotropy of triton emission from the primary reaction, this means that the integral emission of neutrons will occur with an isotropic angular distribution. Since the primary reaction, which is the main source of tritons, has a large cross section, more than 99% of the thermal neutrons are absorbed by the converter. Calculation shows that the amount of tritons escaping from the sample is about 13%.

In order to study the convergence of the required results, we show in Fig.3 the evolution of the fast neutrons yield as a function of the number of histories. The calculations were performed for the conditions proposed by Kobayashi (1988). No significant variations are seen after about 15.10^6 histories.

Prior to Monte Carlo simulation, we estimated, by analytical integration, the yield to be 1.910^{-4} fast neutrons per absorbed triton. This value shows a good agreement with the value 2.0810^{-4} obtained by Perry et al. (1985). In Table 2, the results of our code are compared to some values reported in the literature.

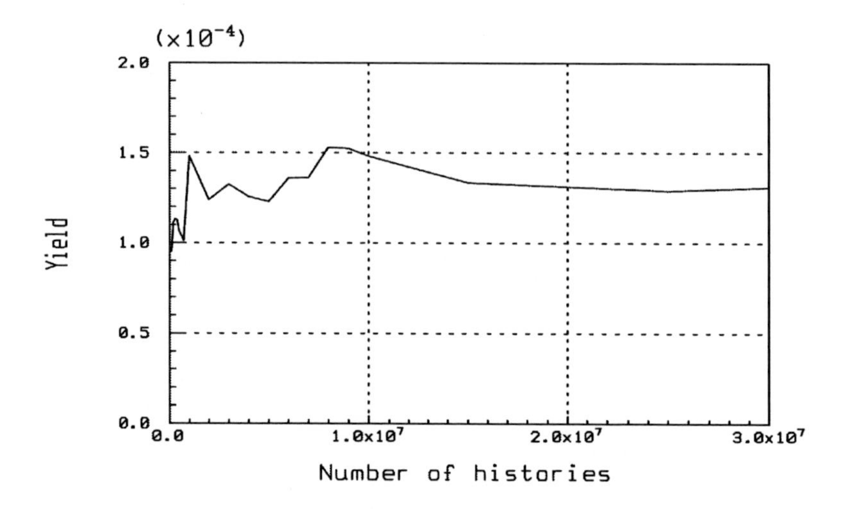

Figure 3. Study of convergence of the results of our code.

Table 2. Yield of fast neutrons produced by an enriched ^{6}LiD converter.

Authors (date)	% ^{6}Li	Yield	Notes	Our code
Lone *et al.* (1980)		$2.83\ 10^{-4}$	calculation	
		$1.56\ 10^{-4}$	measurement	
	95.5			$1.31\ 10^{-4}$
Kobayashi *et al.* (1988)		$1.67\ 10^{-4}$	measurement / $^{27}Al(n,\alpha)$	
Perry *et al.* (1985)	99	$2.08\ 10^{-4}$	Calculation	$1.64\ 10^{-4}$

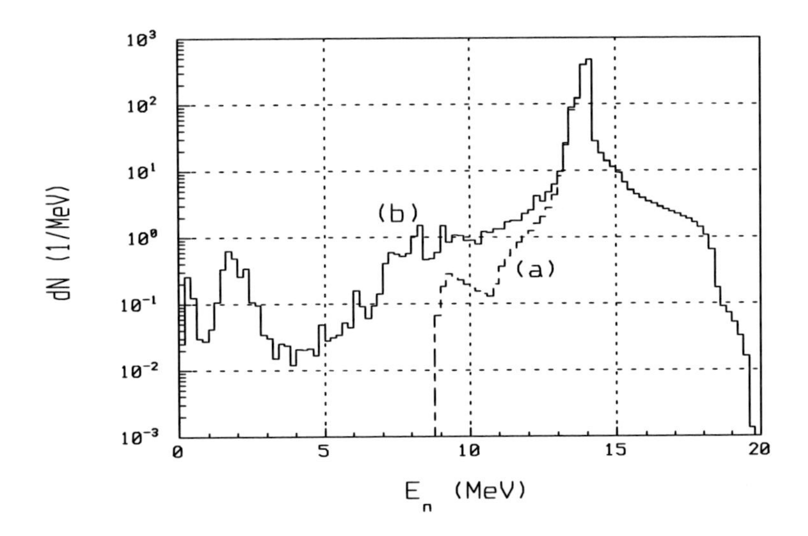

Figure 4. Fast neutron energy spectrum simulated by our code. (a) direct spectrum, (b) total spectrum.

Application of our code to the conditions proposed by Perry et al. (1985) predicts the production of 1.6410^{-4} fast neutron per thermal neutron absorbed in the converter. This value coincides well with our analytical calculation yield and that obtained by Perry et al. (1985) provided the two values are corrected to the triton leakage of 13%. According to our code, the values 1.5610^{-4} and 1.6710^{-4} measured by Lone (1980) and Kobayashi et al. (1988) respectively, are slightly underestimated by about 16% and 20%, respectively.

Fig.4 shows the energy dependent spectrum of fast neutrons, as simulated by our Monte Carlo computing program, under the conditions proposed by Kobayashi (1988). The peak, at high energies, is characterized by a mean energy and full-width at half-maximum of about 14MeV and 1MeV, respectively.

Experimental Results

Before rendring final results, some preliminary measurements were made to test the reproducibility and the validity of the procedure. In order to determine the reproducibility of our method, the same sample was irradiated and analysed two or three times under the same conditions. We found that the determined reaction rate was always the same within our experimental errors. For each nuclide, the activity measurement was repeated at different cooling times to compute the radioisotope half life from the experimental corresponding decay curve. Whenever the nuclide had more than one γ-ray, the calculation was made independently for each line in order to estimate the contribution from any possible interference. Correction for γ-ray attenuation in our samples was achieved, by means of a Monte Carlo code, with greater accuracy because the geometry was well defined. In all cases this attenuation was estimated to 2 to 7%. According to the method proposed by Beckurts K.H. and Wirtz K. (1964) calculations show that no self shielding correction is significant except in the case of cooper at thermal energy for which the correction factor was about 0.96.

Our samples were irradiated at a reactor power of 250kW in the ring F of the TRIGA reactor. At this irradiation position thermal and fast neutron fluxes were measured by means of the five reactions in Table 1a induced when corresponding samples were irradiated without converter. The measured reaction rates were analysed by SAND II code to deduce the integral values of the neutron flux shown in Table 4. As not enough activation foils were used to cover correctly the neutron energy range, a smooth solution spectrum could not be obtained especially in the resonance range. Input guess spectrum to SAND II was that measured by Weber *et al.* (1986). Measured and calculated reaction rates are compared in Table 3.

The second set of samples in Table 1b were irradiated with the ^{6}LiD converter inserted in ring F of reactor at the same power. To determine fluxes and neutron spectrum shape at this irradiation position, the code SANDII was run for the 15 reactions listed in table 1b. The spectrum n° 29, in the reference spectrum library (McElroy 1967), which contains a fusion peak at the high energy end and matched to thermal spectrum at low energy end, was used as an initial approximated guess spectral shape. In addition to these data, nuclear cross sections are needed as input parameters in the spectral adjustment. For (n,γ) reactions, cross sections have been taken from the CSTAPE package accompanying SAND II except for $^{27}Al(n,\gamma)^{28}Al$ where values from reference (McLane *et al.* 1988) were used. Threshold reaction cross sections applied to this analysis were those from the evaluation published by IAEA (1987).

Table 3. Measured reaction rates analysis as an output of SANDII code (without converter)

Foil-reaction	Measured Saturation activity (Bq/nucleus)	Calculated saturation activity (Bq/ nucleus)	Ratio of measured to calculated activities	Deviation of measured from calculated activity (%)
$^{197}Au(n,\gamma)^{198}Au$	$5.45\ 10^{-10}$(± 5.4%)	$5.487\ 10^{-10}$	0.993	-0.72
$^{63}Cu(n,\gamma)^{64}Cu$	$1.10\ 10^{-11}$(±5.1%)	$1.095\ 10^{-11}$	1.007	0.70
$^{27}Al(n,p)^{27}Mg$	$3.78\ 10^{-15}$(±11.%)	$3.764\ 10^{-15}$	1.003	0.34
$^{90}Zr(n,2n)^{89}Zr$	$1.54\ 10^{-16}$(±5.8%)	$1.535\ 10^{-16}$	1.001	0.05
$^{93}Nb(n,2n)^{92m}Nb$	$3.35\ 10^{-16}$(±5.7%)	$3.365\ 10^{-16}$	0.996	-0,37
Standard deviation activities				0.56%

The results of the analysis of our experimental activities obtained with the converter are summarised in table 4 where both measured and calculated reaction rates are given for each reaction. These results have been reached after 7 iterations that were sufficient to find one solution spectrum which is smooth enough and can reproduce the measured activities within the experimental errors and a standard deviation less than 5%. This solution is shown in figs 5 and 6. Table 5 summarises the evaluation of neutron flux densities within the thermal and fast neutron energy range.

Table 4. Analysis of the reaction rates as an output of SAND II code (with converter)

Foil-reaction	Measured Saturation activity (DPS/nucleus)	Calculated saturation activity (DPS/ nucleus)	Ratio of measured to calculated activities	Deviation of measured from calculated activity (%)
$^{23}Na(n,\gamma)^{24}Na$	$6.40\ 10^{-14}$(±5.2%)	$6.36\ 10^{-14}$	1.005	0.52
$^{197}Au(n,\gamma)^{198}Au$	$9.48\ 10^{-11}$(±6.2%)	$9.49\ 10^{-11}$	0.999	-0.08
$^{63}Cu(n,\gamma)^{64}Cu$	$6.46\ 10^{-13}$(±5.2%)	$6.51\ 10^{-13}$	0.993	-0.75
$^{27}Al(n,\gamma)^{28}Al$	$3.17\ 10^{-14}$(±5.1%)	$3.16\ 10^{-14}$	1.003	0.30
$^{19}F(n,p)^{19}O$	$9.16\ 10^{-16}$(±6.4%)	$8.94\ 10^{-16}$	1.025	2.48
$^{23}Na(n,p)^{23}Ne$	$1.18\ 10^{-15}$(±5.2%)	$1.07\ 10^{-15}$	1.102	10.21
$^{23}Na(n,\alpha)^{20}F$	$6.03\ 10^{-16}$(±6.1%)	$6.22\ 10^{-16}$	0.969	-3.07
$^{24}Mg(n,p)^{24}Na$	$1.18\ 10^{-15}$(±5.5%)	$1.19\ 10^{-15}$	0.990	-1.04
$^{25}Mg(n,p)^{25}Na$	$1.13\ 10^{-15}$(±5.9%)	$1.03\ 10^{-15}$	1.100	10.00
$^{27}Al(n,p)^{27}Mg$	$2.49\ 10^{-15}$(±5.2%)	$2.65\ 10^{-15}$	0.942	-5.84
$^{27}Al(n,\alpha)^{24}Na$	$5.72\ 10^{-16}$(±7.2%)	$6.08\ 10^{-16}$	0.940	-6.04
$^{47}Ti(n,p)^{47}Sc$	$1.01\ 10^{-14}$(±5.5%)	$1.06\ 10^{-14}$	0.951	-4.92
$^{48}Ti(n,p)^{48}Sc$	$2.26\ 10^{-16}$(±6.8%)	$2.30\ 10^{-16}$	0.984	-1.63
$^{90}Zr(n,2n)^{89}Zr$	$2.82\ 10^{-16}$(±6.2%)	$2.92\ 10^{-16}$	0.966	-3.41
$^{93}Nb(n,2n)^{92m}Nb$	$5.15\ 10^{-16}$(±5.1%)	$5.14\ 10^{-16}$	1.003	0.29
$^{127}I(n,2n)^{126}I$	$1.57\ 10^{-15}$(±6.7%)	$1.52\ 10^{-15}$	1.030	2.97
Standard deviation activities				4.76%

Table 5. Experimental results of the neutron flux densities in ring F of the TRIGA reactor at a power of 250kW.

Energy range	Integral flux ($cm^{-2}s^{-1}$)		Ratio of integral flux with and without converter
	with converter	without converter	
Thermal E < 0.55 eV	$1.46\ 10^{11}$ (7.1-8%)	$2.56\ 10^{12}$ (7-7.4%)	0.057
Fast E > 10 MeV	$1.36\ 10^{9}$(9.7-13.7%)	$9.84\ 10^{8}$(12.5-21.3%)	1.382
E > 12 MeV	$6.55\ 10^{8}$ (7.8-12%)	$3.19\ 10^{8}$ (14-18.2%)	2.053

Values in brackets are relative uncertainty ranges on flux density calculated from error limits of experimental reaction rates and cross sections.

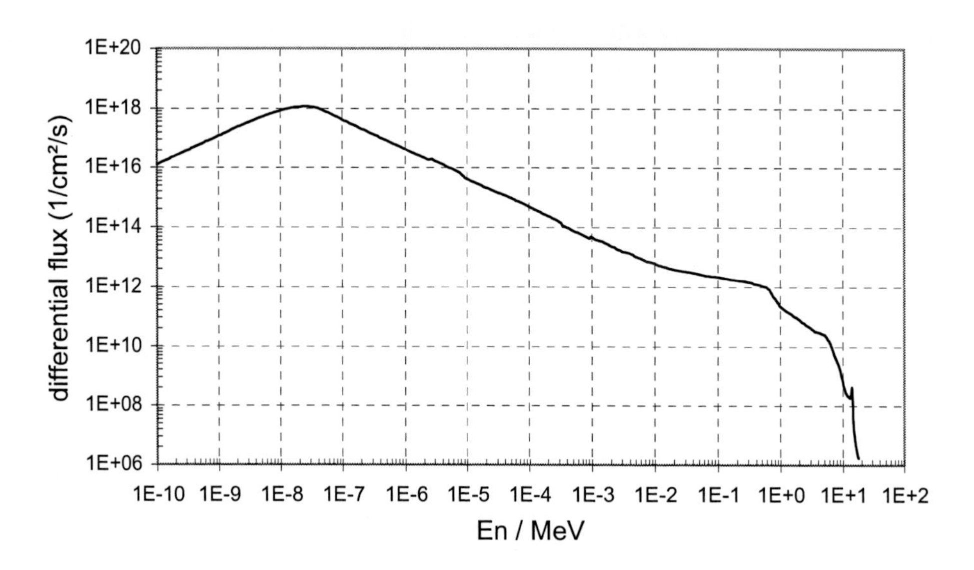

Figure 5. Neutron energy dependent flux versus energy group at the irradiation position inside the ^{6}LiD in-core converter in the TRIGA reactor at a power of 250kW.

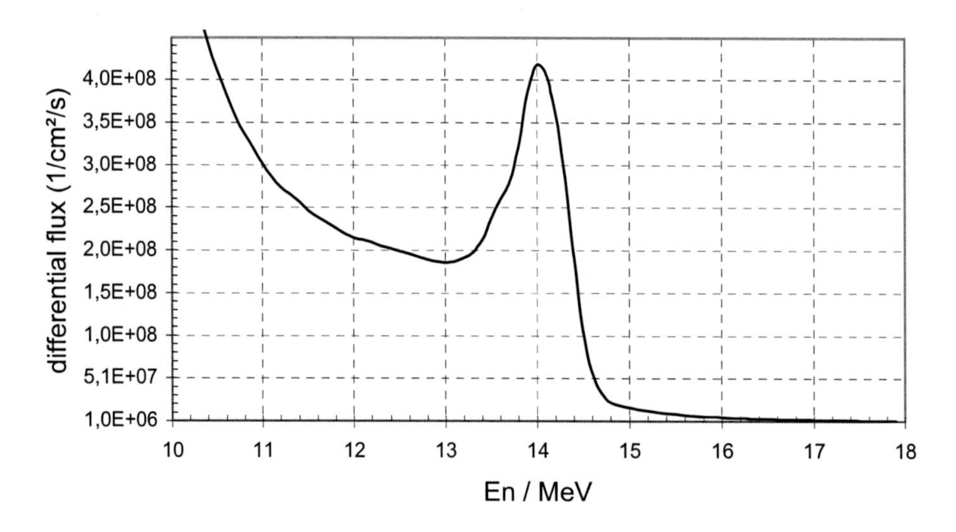

Figure 6. Neutron energy dependent flux around 14MeV at the irradiation position inside the ^{6}LiD in-core converter.

From Table 4, we observe that for our in-core converter the conversion factor, which is defined as ratio of fast flux (3.36 10^{8} $cm^{-2}s^{-1}$, $E_n > 12$MeV) due to ^{6}LiD to thermal flux (2.56 10^{12} $cm^{-2}s^{-1}$) due to reactor, is estimated to 1.31 10^{-4}. This result is close to the values 1.56 10^{-4} and 1.67 10^{-4} measured in the thermal column by Lone *et al.* (1980) and Kimura *et al.* (1990) respectively for a converter in plane geometry. On the other hand, our Monte Carlo code (El Bardouni *et al.* 1996) has been adapted to describe cylindrical converter. Calculations made for the pure thermal neutron field led to a conversion factor of 1.4 10^{-4}, which agrees well with measured value, and 9.4% as statistical error in 10^{7} histories. The mean neutron energy under the peak and its FWHM, according to Monte Carlo calculation, are 14.0MeV and 1.2MeV, respectively. These values are in good agreement with the average fusion peak energy 13.9MeV and its spread of 1MeV calculated from the measured spectrum of fig. 6. On the fusion peak, a slight asymmetry is seen at low energies which is due to neutron scattering and contribution of fast neutrons in the reactor.

Error Analysis

The measured reaction rate uncertainties are due to errors in efficiency calibration, errors in branching ratio and half-life and statistical errors in peak area determination. The errors in branching ratio and half-life, found in the literature, are usually small enough that we could neglect their contribution. In all our measurements, peak area estimation involves typically 1 to 7% as statistical errors. Combined with the uncertainty in efficiency calibration that was estimated at 4 to 5%, these errors lead to relative uncertainties in experimental saturated activities that are limited to 5 to 8%. The largest error source in the uncertainty in flux determination should be due to insufficiently known cross sections and could amount to 5 to 20% (IAEA 1987) in most cases. Combination of these uncertainties during iterative process is not performed by SAND II algorithm to estimate errors in calculated spectrum. The problem can be avoided by use of an approximate iterative technique to obtain an appropriate solution to the system of integral equations. This technique minimises the effects of experimental error propagation. SAND II iterative procedure was designed to accomplish this kind of minimisation. Independently of the SAND II iterative process, relative uncertainty intervals of measured integral flux, reported in Table 5, were calculated from limits of both experimental and cross section error ranges. These values should exceed those predicted by the authors of SAND II.

Conclusion

By the present work, we show that the ^{6}LiD converter efficiently to produce fast 14MeV neutrons from a thermal field. Under our conditions, since the conversion factor is not very high, the fast flux (E_n > 12MeV) is only enhanced by a factor of 2. In spite of this, the converter can be useful to induce, through high threshold reactions, activities with better accuracy than without it because its insertion in the irradiation position provides also adequate shielding against thermal neutrons. The converter decreases the thermal flux by a factor of about 18 rendring it possible to attain satisfactory and readable γ-ray spectra by irradiating longer than without converter.

The flux density has been measured in the thermal and fast neutron energy ranges with and without converter in ring F of the TRIGA reactor. In the case of irradiation with converter, we have used a large number of activation detectors to gain also information on the shape of neutron spectrum. The measured activities were unfolded by the iterative SAND II code and one solution for differential flux was obtained. Profiles of differential flux in thermal and fast regions are obtained with acceptable accuracy, but in the resonance region difficulties were encountered. The code cannot produce real structure in solution spectrum. It produces artificial structure that is very approximated particularly if large percentage of activity can be induced by one single major resonance. The measured neutron spectrum is of great interest in prediction of activity levels that could be induced in samples and thus for the optimisation of experiment conditions. It is also useful for the measurement of reaction cross sections, for absolute activation analysis and for calculation of material damage induced by 14MeV neutrons in 4π irradiation geometry.

Acknowledgments

The authors would like to express their thanks to Professors F. Grass and H. Böck of Atominstitut of Austrian University for their kind support and help during this work.

References

[1] Beckurts K.H. and Wirtz K., Neutron Physics, Springer Verlag ed. NewYork, (1964).

[2] Csikai, *HandBook of Fast Neutron Generators*, CRC Press ed, (1987).

[3] Eckhoff N.D. and Merklin J.F., On the utility of an in-core fast neutron generator. Response, *Nuclear Instruments and Methods,* **156**, 607, (1978).

[4] El Bardouni T., Mouadili A., Aït Haddou A., Monte Carlo calculation of 14MeV neutrons characteristics produced in reactor thermal column by an enriched ^{6}LiD compound, *Appl. Radiat. Isot,* **46** No 6/7, 505-506, (1995).

[5] El Bardouni T., Mouadili A., Aït Haddou A., Böck H., Grass F., Monte Carlo calculation of the conversion of thermal to 14MeV neutrons by enriched ^{6}LiD converter, *J. Trace and Microprobe Techniques*, **14**(1), 37-43, (1996).

[6] T. El Bardouni, F. Grass, S.S. Ismail, A. Mouadili, Characterisation of the fusion neutron field produced by an in-core 6LiD-converter, *J. Trace and Microprobe Techniques*, **15**(2), 145-155, (1997).

[7] T. El Bardouni, Etude théorique et expérimentale de la possibilité de production des neutrons de 14MeV dans un réacteur de recherche à partir des réactions secondaires, *Thèse de Doctorat d'Etat Es-Sciences Physique, Université Abdelmalek Essaadi, Tétouan – Maroc,* (1997).

[8] Frigerio N.A., *ANL-7870, Argone National Laboratory Report 10,* (1971).

[9] Grass F., Dorner J. Gwozdz R. and Naumauv A.P., Short-time activation analysis with and without the ^{6}LiD converter, *J. Trace and Microprobe Techniques,* **14**(1), 293-300, (1996).

[10] Grass F., Holzner H., Ritschel A., Ismail S.S., Dorner J., Short-time activation analysis with and without a ^{6}LiD converter, *J. Radioanalytical and Nuclear Chemistry Articles,* **179**, no 1, 13-25, (1994).

[11] Hogdahl O.T., Neutron absorption in pile neutron activation analysis. MM PP -226 –1, (1962).

[12] IAEA, *Handbook on Nuclear Activation Data,* TRS no 273, Vienna, (1987).

[13] Kimura I. and Kobayashi K., Calibrated fission and fusion neutron fields at Kyoto University Reactor, *Nuclear Science and Engineering*, **106**, 332-344, (1990).

[14] Kobayashi K. and Kimura, I., Application of ^{6}LiD thermal-14MeV neutron converter to the measurement of activation cross sections, *Nuclear Data for Science and Technology*, 261-265, JAERI, (1988).

[15] Leederer C.M. and Shirley V.S., *Table of Isotopes,* 7th ed., John Willey and Sons Inc New York, (1978).

[16] Lone M.A., Santry D.C. and Inglis W.M., MeV neutron production from thermal neutron capture in Li and B compound, *Nuclear Instruments and Methods,* **174**, 521-529, (1980).

[17] McElroy W.N. Berg S., A Computer Automated Iterative Method for Neutron Flux Spectra Determination by Foil Activation, Report AFWL - TR 67-41, **Vol. 1-4,** (1967).

[18] McLane V., Dunford C.L. and Rose P.F., Neutron Cross Section, **Vol 2** NNDC-BNL Academic Press. New York, (1988).

[19] Mughabghab S.F., Garber D.I., Neutron cross sections, **Vol 1** Resonance parameters, BNL 325, (1973).

[20] Napier B.A. et al, Design of an in-core fast neutron generator, *Nuclear Instruments and Methods,* **138**, 463-465, (1976).

[21] Perry, R.T. and Parish, T.A., A 14MeV neutron source, *Fusion Technology,* **8**, 1454-1459, (1985).

[22] Salahi A., Grass F, Bensh F., Kasa T., Seidel E., Roth S., A helium-driven fast transfer system with a removable ^{6}LiD-converter, *J. Trace and Microprobe Techniques*, **6**(2); 229-245, (1988).

[23] Weber H.W., Böck H. and Greenwood L.R., Neutron flux density distribution in the central irradiation thimble of the TRIGA MARK II reactor in Vienna, Proceeding TRIGA Users Conference. College Station USA April 1986.

[24] Westphal G.P., Kasa T., Roth W. (1987), Trends in instrumentation for activation analysis of short-lived nuclides, *J. Radioanalytical and Nuclear Chemistry Articles* **110** no 1, 9-31.

[25] Westphal G.P., High -count -rate gamma spectroscopy, TANSAO 62, 198-199, (1990).

[26] Westphal G.P., Gamma spectrometry shatter the 100kc/s throughput barrier, *J. Radioanalytical and Nuclear Chemistry Articles,* **168**, 513-517, (1993).

[27] Wysocki C.M. and Driffin H.C., On the utility of an in-core fast neutron generator, *Nuclear Instruments and Methods,* **156**, 605-606, (1978).

[28] J.F. Ziegler and J.P. Biersack, TRIM The Transport of Ions in Matter, Instruction Manual, Version 95.4, March 1995.

[29] Zillner K. Bensch F., Neutron-Physical aspects of an in-core generator for 14 MeV neutrons, *J. Radioanalytical Chemistry*, **61**, No. 1-2, 191-194, (1981).

INDEX

A

D

E

H

I

J

K

L

M

N

O

P

Q

R

T

U

V

W

X

Y

Z